Nonlinear Solid Mechanics

A Continuum Approach for Engineering

Nonlinear Solid Mechanics

A Continuum Approach for Engineering

Gerhard A. Holzapfel

Graz University of Technology, Austria

JOHN WILEY & SONS, LTD

Chichester · Weinheim · New York · Brisbane · Singapore · Toronto

Copyright © 2000 John Wiley & Sons Ltd,
The Atrium, Southern Gate, Chichester,
West Sussex PO19 8SQ, England

Telephone   (+44) 1243 779777

Email (for orders and customer service enquiries): cs-books@wiley.co.uk
Visit our Home Page on www.wileyeurope.com or www.wiley.com

Reprinted with corrections 2001, July 2006
Reprinted March 2004, February and September 2005, December 2007, April and December 2008,
April 2010

Other Wiley Editorial Offices

John Wiley & Sons Inc., 111 River Street, Hoboken, NJ 07030, USA

Jossey-Bass, 989 Market Street, San Francisco, CA 94103-1741, USA

Wiley-VCH Verlag GmbH, Boschstr. 12, D-69469 Weinheim, Germany

John Wiley & Sons Australia Ltd, 33 Park Road, Milton, Queensland 4064, Australia

John Wiley & Sons (Asia) Pte Ltd, 2 Clementi Loop #02-01, Jin Xing Distripark, Singapore 129809

John Wiley & Sons Canada Ltd, 22 Worcester Road, Etobicoke, Ontario, Canada M9W 1L1

Library of Congress Cataloging-in-Publication Data

Holzapfel, Gerhard A.
    Nonlinear solid mechanics : a continuum approach for engineering/ Gerhard A. Holzapfel
    p. cm
    Includes bibliographical references and index.
    ISBN 0-471-82304-X—ISBN 0-471-82319-8
    I. Continuum mechanics. I. Title

QA808.2.H655 2000
531--dc21   00-027315

British Library Cataloguing in Publication Data

A catalogue record for this book is available from the British Library

ISBN: 978-0471-82304-9 (H/B)   978-0-471-82319-3 (P/B)

Produced from camera-ready copy supplied by the author

Contents

Preface

My desire in writing this textbook was to show the fascination and beauty of nonlinear solid mechanics and thermodynamics from an engineering computational point of view. My primary goal was not only to offer a modern introductory textbook using the continuum approach to be read with interest, enjoyment and curiosity, but also to offer a reference book that incorporates some of the recent developments in the field. I wanted to stimulate and invite the reader to study this exciting science and take him on a pleasant journey in the wonderful world of nonlinear mechanics, which serves as a solid basis for a surprisingly large variety of problems arising in practical engineering.

Linear theories of solid mechanics are highly developed and are in a satisfactory state of completion. Most processes in nature, however, are highly nonlinear. The approach taken has the aim of providing insight in the basic concepts of solid mechanics with particular reference to the nonlinear regime. Once familiar with the main ideas the reader will be able to specialize in different aspects of the subject matter. I felt the need for a self-contained textbook intended primarily for beginners who want to understand the correspondence between nonlinear continuum mechanics, nonlinear constitutive models and variational principles as essential prerequisites for finite element formulations.

Of course, no single book can cover all aspects of the broad field of solid mechanics, so that many topics are not discussed here at all. The selection of the material for inclusion is influenced strongly by current curricula, trends in the literature and the author's particular interests in engineering and science. Here, a particular selection and style was chosen in order to highlight some of the more inspiring topics in solid mechanics. I hope that my choice, which is of course subjective, will be found to be acceptable.

My ultimate intention was to present an introduction to the subject matter in a didactically sound manner and as clearly as possible. I hope that the text provides enough insights for understanding of the terminology used in scientific state-of-the art papers and to find the 'right and straightforward path' in the scientific world through the effective use of figures, which are very important learning tools. They are designed in order to attract attention and to be instructive and helpful to the reader. Necessary mathematics and physics are explained in the text. The approach used in each of the eight chapters will enable the reader to work through the chapters in order of appearance, each topic being presented in a logical sequence and based on the preceding topics.

A proper understanding of the subject matter requires knowledge of tensor algebra and tensor calculus. For most of the derivations throughout the text I have used symbolic notation with those clear bold-faced symbols which give the subject matter a distinguished beauty. However, for higher-order tensors and for final results in most of the derivations I have used index notation, which provides the reader with more insight. Terminology is printed in bold-face where it appears for the first time while the notation used in the text is defined at the appropriate point.

For those who have not been exposed to the necessary mathematics I have included a chapter on tensor algebra and tensor calculus. It includes the essential ideas of linearization in the form of the concept of the directional derivative. Chapter 1 summarizes elementary properties which are needed for the vector and tensor manipulations performed in all subsequent chapters and which are necessary to many problems that arise frequently in engineering and physics.

It is the prime consideration of Chapter 2 to use tensor analysis for the description of the motion and finite deformation of continua. The continuum approach is introduced along with the notion 'Lagrangian' (material) and 'Eulerian' (spatial) descriptions. In a systematic way the most important kinematic tensors are provided and their physical significance explained. The push-forward and pull-back operations for material and spatial quantities and the concept of the Lie time derivative are introduced. The concept of stress is the main topic of Chapter 3. Cauchy's stress theorem is introduced, along with the Cauchy and first Piola-Kirchhoff traction vectors, and the essential stress tensors are defined and their interrelationships discussed. In Chapter 4 attention is focused on the discussion of the balance principles. Both statics and dynamics are treated. Based upon continuum thermodynamics the entropy inequality principle is provided and the general structure of all principles is summarized as the master balance (inequality) principle. Chapter 5 deals with important aspects of objectivity, which plays a crucial part in nonlinear continuum mechanics. A discussion of change of observer and superimposed rigid-body motions is followed by a development of objective (stress) rates and invariance of elastic material response.

Chapters 6 and 7 form the central part of the book and provide insight in the construction of nonlinear constitutive equations for the description of the mechanical and thermomechanical behavior of solids. These two chapters show the essential richness of the field. They are written for those who want to gain experience in handling material models and deriving stress relations and the associated elasticity tensors that are fundamental for finite element methods. Several examples and exercises are aimed at enabling the reader to think in terms of constitutive models and to formulate more complex material models. All of the types of constitutive equations presented are accessible for use within finite element procedures.

The bulk of Chapter 6 is concerned with finite elasticity and finite viscoelasticity. It includes a discussion of isotropic, incompressible and compressible hyperelastic materials and provides constitutive models for transversely isotropic and composite materials which are suitable for a large number of applications in practical engineering. An approach to inelastic materials with internal variables is given along with instructive examples of hyperelastic materials that involve relaxation and/or creep effects and isotropic damage mechanisms at finite strains. The main purpose of Chapter 7 is to provide an introduction to the thermodynamics of materials. This chapter is devoted not only to the foundation of continuum thermodynamics but also to selected topics of statistical thermodynamics. It starts with a statistical approach by summarizing important physical aspects of the thermoelastic behavior of molecular networks (for example, amorphous solid polymers), based almost entirely on an entropy concept, and continues with a systematic phenomenological approach including finite thermoelasticity and finite thermoviscoelasticity. The stress-strain-temperature response of so-called entropic elastic materials is discussed in more detail and based on a representative example which is concerned with the adiabatic stretching of a rubber band. Typical thermomechanical coupling effects are studied.

Chapter 8 is designed to cover the essential features of the most important variational principles that are very useful in formulating approximation techniques such as the finite element method. Although finite elements are not treated in this text, it is hoped that this chapter will be attractive to those who approach the subject from the computational side. It shows the relationship between the strong and weak forms of initial boundary-value problems, presents the classical principle of virtual work in both spatial and material descriptions and its linearized form. Two- and three-field variational principles are also discussed. The present text ends where conventionally a book on the finite element method would begin.

There are numerous worked examples adjacent to the relevant text. These have the goal of clarifying and supplementing the subject matter. In many cases they are straightforward, but provide an essential part of the text. The symbol ∎ is used to denote the end of an exercise or a proof. The end-of chapter exercises are for homework. The (almost) 200 exercises provided are designed to supplement the text and to consolidate concepts discussed in the text. Most of them serve the purpose of stimulating the reader to further study and to reinforce and develop practical skills in nonlinear continuum and solid mechanics, towards the direction of computational mechanics. In many cases the solutions of selected exercises are given and frequently used later in further developments. Therefore, it should be instructive for the reader to work through a reasonable number of exercises.

Numerous references to supplementary material are suggested and discussed briefly throughout the book. However, for a book of this kind it is not possible to give a comprehensive bibliography of the field. Some of the references listed serve as a starting point for more advanced studies.

The material in this book is based on a sequence of courses that I have taught at the University of Technology in Graz and Vienna. The mechanics and thermodynamics of solids are relevant to all branches of engineering, to applied mechanics, mathematics, physics and material science, and it is a central field in biomechanics. This book is primarily addressed to graduate students, researchers and practitioners, although it has also proven to be of interest to advanced undergraduates. Although I have tried to provide a textbook that is self-contained and appropriate for self-instruction, it is desirable that the reader has a reasonable background in elementary mechanics and thermodynamics.

I feel that Chapters 2-5 and parts of the remaining chapters are well suited for a complete course on nonlinear continuum mechanics lasting two semesters or three quarters. A one semester or one quarter course in the nonlinear mechanics of solids would focus on Chapter 6, while a one quarter course in the thermodynamics of solids could be based on Chapter 7. Chapter 1 is the core for a course that provides the student with the necessary background in vector and tensor analysis. Chapter 8 is certainly not designed to train specialists in variational principles, but to form a basic one quarter course at the graduate level.

I believe that the present textbook, in providing many applications to engineering science, is not too advanced mathematically. Of course, some of the results presented may be derived with the help of more advanced mathematics using theorems and proofs. I hope that this book will help pure engineers to teach nonlinear continuum mechanics and solid mechanics.

Naturally, as the author, I take full responsibility for not doing it better. Comments and criticisms will be welcome and greatly appreciated. I have learnt that the spirit of modern continuum mechanics and the underlying mathematics are as important to the design of powerful finite element models as are insights in the theoretical foundation of constitutive models and variational principles. A successful transfer of that combination to the reader would indicate that my objective has been achieved.

Graz, Austria, August 1999   *Gerhard A. Holzapfel*

Preface to the Second Printing

The focus of the revision for this second printing is the insertion of additional equations on pages 50, 51, 74, and an additional exercise on page 76. This has caused changes in the numbers of some equations in parts of the book as compared with the first printing. I have also done some minor re-wording and have added a few more references.

Graz, Austria, October 2001   *Gerhard A. Holzapfel*

Acknowledgements

When I was a postdoctoral student at Stanford University I worked with the late *Juan C. Simo*, Professor of Mechanical Engineering, to whom I owe my deepest thanks. He stimulated, influenced and focused my study and writing in recent years; his friendship, versatility and dedication to scientific excellence provided a unique learning experience for me.

I am particularly indebted to *Ray W. Ogden*, Professor of Mathematics at the University of Glasgow, who spent a lot of time in reading the entire manuscript and rectifying certain ineptnesses. His outstanding expertise in the field made working with him an inspiring pleasure. His detailed scientific criticism and suggestions for improvements of the text were of immeasurable help.

Many others have contributed to the book. Here I mention my collaborators, whose gentle encouragement and support during the course of the preparation of the manuscript I gratefully acknowledge. In particular, the inspiring and detailed comments of *Dr. Christian A.J. Schulze-Bauer*, from a background in physics and medicine, were extremely helpful. He let me filter this text through his sharp mind. Also, *Christian T. Gasser*, whose background is in mechanical engineering, has suggested a number of valuable improvements to the substance of the text. His profound remarks in class prevented me from getting away with anything. Also, *Elisabeth Pernkopf*, a mathematician, deserves special thanks for her generous assistance. She gave much helpful advice on the preparation of this text and offered many suggestions for improving it. Special thanks are due to *Michael Stadler* for his productive discussions and to *Mario Ch. Palli* for his patience in preparing the figures. I am grateful to each of these individuals without whose contributions the book would not have taken this shape.

I want to thank the Department of Civil Engineering, Graz University of Technology, for providing an environment in which this project could be completed. My gratitude goes to *Gernot Beer*, Professor of Civil Engineering, for his outspoken support. I also wish to acknowledge the Austrian Science Foundation, which has influenced my scientific agenda through the financial support of several grants over the past eight years.

The enjoyment I experienced in writing this textbook would not have been the same without the moral support of numerous friends. My thanks belong to all of them who tolerated my absence when I disappeared for many evenings and weekends in order to bring the ideas of this fascinating field to you, the reader.

1 Introduction to Vectors and Tensors

The aim of this chapter is to present the fundamental rules and standard results of tensor algebra and tensor calculus permanently used in nonlinear continuum mechanics. Some readers may prefer to pass directly to Chapter 2 leaving the present part for reference as needed.

Many of the statements are given without proofs. For a more detailed exposition see the standard texts by HALMOS [1958], TRUESDELL and NOLL [2004] or the textbooks by GURTIN [1981a], SIMMONDS [1994], DANIELSON [1997], OGDEN [1997] and CHADWICK [1999], among many other references on vectors and tensors. To recall the elements of linear algebra see, for example, the book by STRANG [1988a].

In this text we use lowercase Greek letters for *scalars*, lowercase bold-face Latin letters for *vectors*, uppercase bold-face Latin letters for *second-order tensors*, uppercase bold-face calligraphic letters for *third-order tensors* and uppercase blackboard Latin letters for *fourth-order tensors*; for example,

$$\left.\begin{array}{ll} \alpha, \beta, \gamma, \ldots \ \text{(scalars)} & \mathbf{a}, \mathbf{b}, \mathbf{c}, \ldots \ \text{(vectors)} \\ \mathbf{A}, \mathbf{B}, \mathbf{C}, \ldots \ \text{(2nd-order tensors)} & \boldsymbol{\mathcal{A}}, \boldsymbol{\mathcal{B}}, \boldsymbol{\mathcal{C}}, \ldots \ \text{(3rd-order tensors)} \\ \mathbb{A}, \mathbb{B}, \mathbb{C}, \ldots \ \text{(4th-order tensors)} \end{array}\right\} \quad (1.1)$$

For equations such as

$$\mathbf{u} = \alpha\mathbf{v} = \beta\mathbf{a} = \gamma\mathbf{b} \ , \tag{1.2}$$

we agree that $(1.2)_2$ refers to $\mathbf{u} = \beta\mathbf{a}$.

Often the derivations of formulas need relations introduced previously. If this is the case we refer to these relations in a particular order, reflecting the consecutive steps necessary for deriving the formula in question.

1.1 Algebra of Vectors

A physical quantity, completely described by a single *real* number, such as *temperature*, *density* or *mass*, is called a **scalar** designated by $\alpha, \beta, \gamma, \ldots$ A **vector** designated by $\mathbf{u}$, $\mathbf{v}$, $\mathbf{w}$, ... (or in other texts frequently designated by $\underset{\sim}{u}$, $\underset{\sim}{v}$, $\underset{\sim}{w}$, ..., or $\vec{u}$, $\vec{v}$, $\vec{w}$, ...), is a directed line element in space. It is a model for physical quantities having both direction and length, for example, *force, velocity* or *acceleration*. Two vectors that have the same direction and length are said to be **equal**.

The **sum** of vectors yields a new vector, based on the **parallelogram law** of addition. The following properties

$$\mathbf{u} + \mathbf{v} = \mathbf{v} + \mathbf{u} \ , \tag{1.3}$$

$$(\mathbf{u} + \mathbf{v}) + \mathbf{w} = \mathbf{u} + (\mathbf{v} + \mathbf{w}) \ , \tag{1.4}$$

$$\mathbf{u} + \mathbf{o} = \mathbf{u} \ , \tag{1.5}$$

$$\mathbf{u} + (-\mathbf{u}) = \mathbf{o} \tag{1.6}$$

hold, where $\mathbf{o}$ denotes the unique **zero vector** with *unspecified* direction and *zero* length.

Let $\mathbf{u}$ be a vector and α be a real number (a scalar). Then the **scalar multiplication** $\alpha\mathbf{u}$ produces a new vector with the same direction as $\mathbf{u}$ if $\alpha > 0$ or with the opposite direction to $\mathbf{u}$ if $\alpha < 0$.

Further properties are:

$$(\alpha\beta)\mathbf{u} = \alpha(\beta\mathbf{u}) \ , \tag{1.7}$$

$$(\alpha + \beta)\mathbf{u} = \alpha\mathbf{u} + \beta\mathbf{u} \ , \tag{1.8}$$

$$\alpha(\mathbf{u} + \mathbf{v}) = \alpha\mathbf{u} + \alpha\mathbf{v} \ . \tag{1.9}$$

Dot product.   The **dot** (or **scalar** or **inner**) **product** of $\mathbf{u}$ and $\mathbf{v}$, denoted by $\mathbf{u} \cdot \mathbf{v}$ (or $\langle \mathbf{u}, \mathbf{v} \rangle$), is

$$\mathbf{u} \cdot \mathbf{v} = |\mathbf{u}||\mathbf{v}|\cos\theta(\mathbf{u}, \mathbf{v}) \ , \qquad 0 \le \theta(\mathbf{u}, \mathbf{v}) \le \pi \ , \tag{1.10}$$

where $\theta(\mathbf{u}, \mathbf{v})$ is the **angle** between two nonzero vectors $\mathbf{u}$ and $\mathbf{v}$ when their origins coincide. This product gives a scalar quantity with the properties

$$\mathbf{u} \cdot \mathbf{v} = \mathbf{v} \cdot \mathbf{u} \ , \tag{1.11}$$

$$\mathbf{u} \cdot \mathbf{o} = 0 \ , \tag{1.12}$$

$$\mathbf{u} \cdot (\alpha\mathbf{v} + \beta\mathbf{w}) = \alpha(\mathbf{u} \cdot \mathbf{v}) + \beta(\mathbf{u} \cdot \mathbf{w}) \ , \tag{1.13}$$

$$\mathbf{u} \cdot \mathbf{u} > 0 \quad \Leftrightarrow \quad \mathbf{u} \ne \mathbf{o} \qquad \text{and} \qquad \mathbf{u} \cdot \mathbf{u} = 0 \quad \Leftrightarrow \quad \mathbf{u} = \mathbf{o} \ . \tag{1.14}$$

The quantity $|\mathbf{u}|$ (or $\|\mathbf{u}\|$) is called the **length** (or **norm** or **magnitude**) of a vector $\mathbf{u}$, which is a non-negative real number. It is defined by the square root of $\mathbf{u} \cdot \mathbf{u}$, i.e.

$$|\mathbf{u}| = (\mathbf{u} \cdot \mathbf{u})^{1/2} \ge 0 \ , \qquad \mathbf{u}^2 = \mathbf{u} \cdot \mathbf{u} \ . \tag{1.15}$$

A vector **e** is called a **unit vector** if $|\mathbf{e}| = 1$. A nonzero vector **u** is said to be **orthogonal** (or **perpendicular**) to a nonzero vector **v** if

$$\mathbf{u} \cdot \mathbf{v} = 0 \qquad \text{with} \qquad \theta(\mathbf{u}, \mathbf{v}) = \pi/2 \; . \qquad (1.16)$$

Thus, using (1.10) we find the **projection** of a vector **u** along the direction of a vector **e** with unit length, i.e.

$$\mathbf{u} \cdot \mathbf{e} = |\mathbf{u}| \cos\theta(\mathbf{u}, \mathbf{e}) \; . \qquad (1.17)$$

For a geometrical interpretation of eq. (1.17) see Figure 1.1.

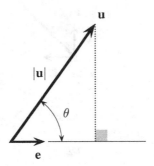

Figure 1.1 Projection of **u** along a unit vector **e**.

Index notation.   So far algebra has been presented in **symbolic** (or **direct** or **absolute**) **notation** exclusively employing bold-face letters. It represents a very convenient and concise tool to manipulate most of the relations used in continuum mechanics. It will be the preferred representation in this text. However, particularly in computational mechanics, it is essential to refer vector (and tensor) quantities to a *basis*. Additionally, to gain more insight in some quantities and to carry out mathematical operations among (higher-order) tensors more readily (see next section) it is often helpful to refer to components.

In order to present **coordinate** (or **component**) expressions relative to a **right-handed** (or **dextral**) and **orthonormal system** we introduce a fixed set of three **basis vectors** $\mathbf{e}_1, \mathbf{e}_2, \mathbf{e}_3$ (sometimes introduced as $\mathbf{i}, \mathbf{j}, \mathbf{k}$), called a **(Cartesian) basis**, with properties

$$\mathbf{e}_1 \cdot \mathbf{e}_2 = \mathbf{e}_1 \cdot \mathbf{e}_3 = \mathbf{e}_2 \cdot \mathbf{e}_3 = 0 \; , \qquad \mathbf{e}_1 \cdot \mathbf{e}_1 = \mathbf{e}_2 \cdot \mathbf{e}_2 = \mathbf{e}_3 \cdot \mathbf{e}_3 = 1 \; . \qquad (1.18)$$

These vectors of unit length which are mutually orthogonal form a so-called orthonormal system.

Then any vector **u** in the **three-dimensional Euclidean space** is represented **uniquely** by a linear combination of the basis vectors $\mathbf{e}_1, \mathbf{e}_2, \mathbf{e}_3$, i.e.

$$\mathbf{u} = u_1\mathbf{e}_1 + u_2\mathbf{e}_2 + u_3\mathbf{e}_3 \quad , \tag{1.19}$$

where the three real numbers u_1, u_2, u_3 are the uniquely determined **Cartesian** (or **rectangular**) **components** of vector **u** along the given directions $\mathbf{e}_1, \mathbf{e}_2, \mathbf{e}_3$, respectively (see Figure 1.2). The components of $\mathbf{e}_1, \mathbf{e}_2, \mathbf{e}_3$ are $(1, 0, 0), (0, 1, 0), (0, 0, 1)$, respectively.

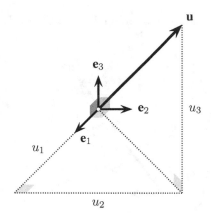

Figure 1.2 Vector **u** with its Cartesian components u_1, u_2, u_3.

Using **index** (or **subscript** or **suffix**) **notation**, relation (1.19) can be written as $\mathbf{u} = \sum_{i=1}^{3} u_i\mathbf{e}_i$ or, in an abbreviated form by leaving out the summation symbol, simply as

$$\mathbf{u} = u_i\mathbf{e}_i \quad , \qquad (\text{sum over } i = 1, 2, 3) \quad , \tag{1.20}$$

where we have adopt the **summation convention**, invented by *Einstein*. The summation convention says that whenever an index is repeated (only once) in the same term, then, a summation over the *range* of this index is implied unless otherwise indicated. We consider only the *three-dimensional Euclidean space*, which we characterize by means of Latin indices $i, j, k \ldots$ running over $1, 2, 3$. We denote the basis vectors by $\{\mathbf{e}_i\}_{i \in \{1,2,3\}}$ collectively. Subsequently, in this text, the braces $\{\bullet\}$ will stand for a fixed set of basis vectors and the symbol $\bullet$ for any tensor element.

The index i that is summed over is said to be a **dummy** (or **summation**) **index**, since a replacement by any other symbol does not affect the value of the sum. An index that is not summed over in a given term is called a **free** (or **live**) **index**. Note that in the same equation an index is either dummy or free. Thus, relations (1.18) can be written in a more convenient form as

$$\mathbf{e}_i \cdot \mathbf{e}_j = \delta_{ij} \equiv \begin{cases} 1 \ , & \text{if} \quad i = j \ , \\ 0 \ , & \text{if} \quad i \neq j \ , \end{cases} \tag{1.21}$$

which defines the **Kronecker delta** δ_{ij}. The useful properties

$$\delta_{ii} = 3 \ , \qquad \delta_{ij}u_i = u_j \ , \qquad \delta_{ij}\delta_{jk} = \delta_{ik} \tag{1.22}$$

hold. Note that δ_{ij} also serves as a **replacement operator**; for example, the index on u_i becomes an j when the components u_i are multiplied by δ_{ij}.

The projection of a vector $\mathbf{u} = u_i\mathbf{e}_i$ onto the basis vectors $\mathbf{e}_j$ yields the j-th component of $\mathbf{u}$. Thus, from eq. (1.20) and properties (1.13), (1.21) and (1.22)$_2$ we have

$$\mathbf{u} \cdot \mathbf{e}_j = (u_i\mathbf{e}_i) \cdot \mathbf{e}_j = u_i\delta_{ij} = u_j \ . \tag{1.23}$$

Taking the basis $\{\mathbf{e}_i\}$ and eqs. (1.13), (1.20), (1.21) and (1.22)$_2$, the component expression for the dot product (1.10) gives

$$\mathbf{u} \cdot \mathbf{v} = u_i\mathbf{e}_i \cdot v_j\mathbf{e}_j = u_iv_j\mathbf{e}_i \cdot \mathbf{e}_j = u_iv_j\delta_{ij} = u_iv_i$$
$$= u_1v_1 + u_2v_2 + u_3v_3 \ , \tag{1.24}$$

which is commonly used as the definition of the dot product. Thus, we may derive the dot product of $\mathbf{u}$ and $\mathbf{v}$ without knowledge of the angle between $\mathbf{u}$ and $\mathbf{v}$.

In an analogous manner, the component expression for the square of the length of $\mathbf{u}$, i.e. (1.15), is

$$|\mathbf{u}|^2 = \mathbf{u} \cdot \mathbf{u} = u_i\mathbf{e}_i \cdot u_j\mathbf{e}_j = u_iu_j\delta_{ij} = u_iu_i$$
$$= u_1^2 + u_2^2 + u_3^2 \ . \tag{1.25}$$

Note that in eqs. (1.24) and (1.25) *one* index is repeated, indicating summation over $1, 2, 3$. In symbolic notation this is indicated by *one* dot.

Cross product. The **cross** (or **vector**) **product** of $\mathbf{u}$ and $\mathbf{v}$, denoted by $\mathbf{u} \times \mathbf{v}$ (in the literature also $\mathbf{u} \wedge \mathbf{v}$), produces a new vector. The cross product is *not* commutative. It is defined as

$$\mathbf{u} \times \mathbf{v} = -(\mathbf{v} \times \mathbf{u}) \ , \tag{1.26}$$

$$\mathbf{u} \times \mathbf{v} = \mathbf{o} \qquad \Leftrightarrow \qquad \mathbf{u} \text{ and } \mathbf{v} \text{ are linearly dependent} \ , \tag{1.27}$$

$$(\alpha\mathbf{u}) \times \mathbf{v} = \mathbf{u} \times (\alpha\mathbf{v}) = \alpha(\mathbf{u} \times \mathbf{v}) \ , \tag{1.28}$$

$$\mathbf{u} \cdot (\mathbf{v} \times \mathbf{w}) = \mathbf{v} \cdot (\mathbf{w} \times \mathbf{u}) = \mathbf{w} \cdot (\mathbf{u} \times \mathbf{v}) \ , \tag{1.29}$$

$$\mathbf{u} \times (\mathbf{v} + \mathbf{w}) = (\mathbf{u} \times \mathbf{v}) + (\mathbf{u} \times \mathbf{w}) = \mathbf{u} \times \mathbf{v} + \mathbf{u} \times \mathbf{w} \ . \tag{1.30}$$

If relation (1.27) holds with **u** and **v** assumed to be nonzero vectors, we say that vector **u** is **parallel** to vector **v**. From eq. (1.29) we learn that

$$\mathbf{u} \cdot (\mathbf{u} \times \mathbf{v}) = 0 \ . \tag{1.31}$$

The magnitude of the cross product is defined to be

$$|\mathbf{u} \times \mathbf{v}| = |\mathbf{u}||\mathbf{v}|\sin\theta(\mathbf{u},\mathbf{v}) \ , \qquad 0 \le \theta(\mathbf{u},\mathbf{v}) \le \pi \ . \tag{1.32}$$

It characterizes the area of a parallelogram spanned by the vectors **u** and **v** (see Figure 1.3). The right-handed cross product of **u** and **v**, i.e. the vector **u** × **v**, is perpendicular to the plane spanned by **u** and **v** (θ is the angle between **u** and **v**).

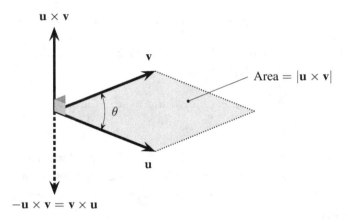

Figure 1.3 Cross product of vectors **u** and **v**.

In order to express the cross product in terms of components we introduce the **permutation** (or **alternating** or **Levi-Civita** ε-) **symbol** ε_{ijk}, which is defined as

$$\varepsilon_{ijk} = \begin{cases} 1 \ , & \text{for **even** permutations of } (i,j,k) \text{ (i.e. } 123, 231, 312) \ , \\ -1 \ , & \text{for **odd** permutations of } (i,j,k) \text{ (i.e. } 132, 213, 321) \ , \\ 0 \ , & \text{if there is a **repeated** index } , \end{cases} \tag{1.33}$$

with the properties $\varepsilon_{ijk} = \varepsilon_{jki} = \varepsilon_{kij}$, $\varepsilon_{ijk} = -\varepsilon_{ikj}$ and $\varepsilon_{ijk} = -\varepsilon_{jik}$, respectively.

Consider the right-handed and orthonormal basis $\{\mathbf{e}_i\}$, then

$$\left.\begin{array}{lll} \mathbf{e}_1 \times \mathbf{e}_2 = \mathbf{e}_3 \ , & \mathbf{e}_2 \times \mathbf{e}_1 = -\mathbf{e}_3 \ , & \mathbf{e}_1 \times \mathbf{e}_1 = \mathbf{e}_2 \times \mathbf{e}_2 \\ \mathbf{e}_2 \times \mathbf{e}_3 = \mathbf{e}_1 \ , & \mathbf{e}_3 \times \mathbf{e}_2 = -\mathbf{e}_1 \ , & \qquad = \mathbf{e}_3 \times \mathbf{e}_3 = \mathbf{0} \ , \\ \mathbf{e}_3 \times \mathbf{e}_1 = \mathbf{e}_2 \ , & \mathbf{e}_1 \times \mathbf{e}_3 = -\mathbf{e}_2 \ , & \end{array}\right\} \tag{1.34}$$

or in a more convenient short-hand notation, with (1.33),

$$\mathbf{e}_i \times \mathbf{e}_j = \varepsilon_{ijk}\mathbf{e}_k \quad . \tag{1.35}$$

With relations (1.21), (1.34), (1.35) it is easy to verify that ε_{ijk} may be expressed as the **determinant** of a matrix

$$\varepsilon_{ijk} = \det \begin{bmatrix} \delta_{i1} & \delta_{i2} & \delta_{i3} \\ \delta_{j1} & \delta_{j2} & \delta_{j3} \\ \delta_{k1} & \delta_{k2} & \delta_{k3} \end{bmatrix} \quad , \tag{1.36}$$

where we have introduced the square brackets [●] for a matrix. The product of the permutation symbols $\varepsilon_{ijk}\,\varepsilon_{pqr}$ is related to the Kronecker delta by the relation

$$\varepsilon_{ijk}\,\varepsilon_{pqr} = \det \begin{bmatrix} \delta_{ip} & \delta_{iq} & \delta_{ir} \\ \delta_{jp} & \delta_{jq} & \delta_{jr} \\ \delta_{kp} & \delta_{kq} & \delta_{kr} \end{bmatrix} \quad . \tag{1.37}$$

With $(1.22)_1$ and $(1.22)_3$ we deduce from (1.37) the important relations

$$\varepsilon_{ijk}\,\varepsilon_{pqk} = \delta_{ip}\delta_{jq} - \delta_{iq}\delta_{jp} \quad , \qquad \varepsilon_{ijk}\,\varepsilon_{pjk} = 2\delta_{pi} \quad , \qquad \varepsilon_{ijk}\,\varepsilon_{ijk} = 6 \quad . \tag{1.38}$$

EXAMPLE 1.1   Obtain the coordinate expression for the cross product $\mathbf{w} = \mathbf{u} \times \mathbf{v}$.

Solution.   Taking advantage of eqs. (1.20) and (1.35) we find that

$$\mathbf{w} = \mathbf{u} \times \mathbf{v} = u_i\mathbf{e}_i \times v_j\mathbf{e}_j = u_iv_j(\mathbf{e}_i \times \mathbf{e}_j) = \varepsilon_{ijk}u_iv_j\mathbf{e}_k = w_k\mathbf{e}_k \quad , \tag{1.39}$$

with the three components

$$w_1 = u_2v_3 - u_3v_2 \quad , \tag{1.40}$$

$$w_2 = u_3v_1 - u_1v_3 \quad , \tag{1.41}$$

$$w_3 = u_1v_2 - u_2v_1 \quad . \tag{1.42}$$

Consequently, $(\mathbf{u} \times \mathbf{v}) \cdot \mathbf{e}_k$ equals $\varepsilon_{ijk}u_iv_j$.   ■

Now, using expressions (1.36), $(1.39)_4$, the vector product $\mathbf{u} \times \mathbf{v}$ relative to $\{\mathbf{e}_k\}$ may be written as

$$\mathbf{u} \times \mathbf{v} = \det \begin{bmatrix} \mathbf{e}_1 & \mathbf{e}_2 & \mathbf{e}_3 \\ u_1 & u_2 & u_3 \\ v_1 & v_2 & v_3 \end{bmatrix} = \varepsilon_{ijk}u_iv_j\mathbf{e}_k \quad . \tag{1.43}$$

The **triple scalar** (or **box**) **product** $(\mathbf{u} \times \mathbf{v}) \cdot \mathbf{w}$ represents the volume V of a parallelepiped 'spanned' by $\mathbf{u}, \mathbf{v}, \mathbf{w}$ forming a *right-handed* triad (see Figure 1.4).

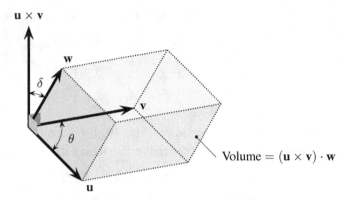

Figure 1.4 Triple scalar product.

By recalling definitions (1.29), (1.10) and (1.32), we have

$$V = (\mathbf{u} \times \mathbf{v}) \cdot \mathbf{w} = (\mathbf{v} \times \mathbf{w}) \cdot \mathbf{u} = (\mathbf{w} \times \mathbf{u}) \cdot \mathbf{v}$$
$$= |\mathbf{u} \times \mathbf{v}||\mathbf{w}|\cos\delta$$
$$= \underbrace{|\mathbf{u}||\mathbf{v}|\sin\theta}_{\text{base area}}\, \underbrace{|\mathbf{w}|\cos\delta}_{\text{height}} \quad . \tag{1.44}$$

Using index notation, then, from eqs. (1.20) and (1.35) we find with (1.21) the volume V to be

$$V = (\mathbf{u} \times \mathbf{v}) \cdot \mathbf{w} = \varepsilon_{ijk} u_i v_j w_k$$
$$= (u_2 v_3 - u_3 v_2)w_1 + (u_3 v_1 - u_1 v_3)w_2 + (u_1 v_2 - u_2 v_1)w_3 \quad . \tag{1.45}$$

Hence, the triple scalar product $(1.45)_3$ can be written in the convenient determinant form

$$(\mathbf{u} \times \mathbf{v}) \cdot \mathbf{w} = \det \begin{bmatrix} u_1 & v_1 & w_1 \\ u_2 & v_2 & w_2 \\ u_3 & v_3 & w_3 \end{bmatrix} \quad . \tag{1.46}$$

Note that the vectors $\mathbf{u}, \mathbf{v}, \mathbf{w}$ are linearly dependent *if and only if* their triple scalar product vanishes (the parallelepiped has no volume).

The product $\mathbf{u} \times (\mathbf{v} \times \mathbf{w})$ is called the **triple vector product** and may be verified with $(1.39)_4$, $(1.38)_1$, $(1.22)_2$ and representations (1.20) and $(1.24)_4$

$$\mathbf{u} \times (\mathbf{v} \times \mathbf{w}) = \varepsilon_{ijk} u_i (\varepsilon_{mnj} v_m w_n) \mathbf{e}_k = \varepsilon_{kij} \varepsilon_{mnj} u_i v_m w_n \mathbf{e}_k$$
$$= (\delta_{km}\delta_{in} - \delta_{kn}\delta_{im}) u_i v_m w_n \mathbf{e}_k$$
$$= u_n v_k w_n \mathbf{e}_k - u_m v_m w_k \mathbf{e}_k$$
$$= (\mathbf{u} \cdot \mathbf{w})\mathbf{v} - (\mathbf{u} \cdot \mathbf{v})\mathbf{w} \quad , \tag{1.47}$$

which is the so-called *'back-cab'* rule well-known from vector algebra. Similarly,

$$(\mathbf{u} \times \mathbf{v}) \times \mathbf{w} = (\mathbf{u} \cdot \mathbf{w})\mathbf{v} - (\mathbf{v} \cdot \mathbf{w})\mathbf{u} \quad . \tag{1.48}$$

The triple vector product is, in general, *not* associative, i.e. $(\mathbf{u} \times \mathbf{v}) \times \mathbf{w} \neq \mathbf{u} \times (\mathbf{v} \times \mathbf{w})$.

EXERCISES

1. Use the properties (1.5), (1.6), (1.8) and (1.9) to show that

$$\alpha\mathbf{o} = \mathbf{o} \quad , \qquad 0\mathbf{u} = \mathbf{o} \quad , \qquad (-\alpha)\mathbf{u} = \alpha(-\mathbf{u}) \quad .$$

2. By means of (1.30) and (1.28) derive the property

$$(\alpha\mathbf{u} + \beta\mathbf{v}) \times \mathbf{w} = \alpha(\mathbf{u} \times \mathbf{w}) + \beta(\mathbf{v} \times \mathbf{w}) \quad .$$

3. Prove the triple vector product (1.48) and show that the vector $(\mathbf{u} \times \mathbf{v}) \times \mathbf{w}$ lies in the plane spanned by the vectors $\mathbf{u}$ and $\mathbf{v}$.

1.2 Algebra of Tensors

A **second-order tensor A**, denoted by **A, B, C** . . . (or in the literature sometimes written as A, B, C . . .), may be thought of as a **linear operator** that acts on a vector **u** generating a vector **v**. Thus, we may write

$$\mathbf{v} = \mathbf{A}\mathbf{u} \quad , \tag{1.49}$$

which defines a **linear transformation** that assigns a vector **v** to each vector **u**. Since **A** is linear we have

$$\mathbf{A}(\alpha\mathbf{u} + \mathbf{v}) = \alpha\mathbf{A}\mathbf{u} + \mathbf{A}\mathbf{v} \tag{1.50}$$

for all vectors **u**, **v** and all scalars α. Since most tensors used in this text are of order two, we shall often omit the adjective 'second-order'.

If **A** and **B** are two (second-order) tensors, we can define the **sum A + B**, the **difference A − B** and the **scalar multiplication** α**A** by the rules

$$(\mathbf{A} \pm \mathbf{B})\mathbf{u} = \mathbf{Au} \pm \mathbf{Bu} \quad, \tag{1.51}$$

$$(\alpha\mathbf{A})\mathbf{u} = \alpha(\mathbf{Au}) \quad, \tag{1.52}$$

where **u** denotes an arbitrary vector. The important **second-order unit** (or **identity**) **tensor I** and the **second-order zero tensor O** are defined, respectively, by the relations **Iu = uI = u** and **Ou = uO = o** for all vectors **u**. Note that tensor **O** maps every **u** to the zero vector **o**.

Tensor product.   The **tensor** (or **direct** or **matrix**) **product** or the **dyad** of the vectors **u** and **v** is denoted by **u** ⊗ **v** (some authors use the notation **uv**). It is a second-order tensor which linearly *transforms* a vector **w** into a vector with the direction of **u** following the rule

$$(\mathbf{u} \otimes \mathbf{v})\mathbf{w} = \mathbf{u}(\mathbf{v} \cdot \mathbf{w}) = (\mathbf{v} \cdot \mathbf{w})\mathbf{u} \quad. \tag{1.53}$$

The dyad, not to be confused with the *dot* or *cross* product, has the linearity property

$$(\mathbf{u} \otimes \mathbf{v})(\alpha\mathbf{w} + \mathbf{x}) = \alpha(\mathbf{u} \otimes \mathbf{v})\mathbf{w} + (\mathbf{u} \otimes \mathbf{v})\mathbf{x} \quad. \tag{1.54}$$

In addition, note the relations

$$(\alpha\mathbf{u} + \beta\mathbf{v}) \otimes \mathbf{w} = \alpha(\mathbf{u} \otimes \mathbf{w}) + \beta(\mathbf{v} \otimes \mathbf{w}) \quad, \tag{1.55}$$

$$\mathbf{u}(\mathbf{v} \otimes \mathbf{w}) = (\mathbf{u} \cdot \mathbf{v})\mathbf{w} = \mathbf{w}(\mathbf{u} \cdot \mathbf{v}) \quad, \tag{1.56}$$

$$(\mathbf{u} \otimes \mathbf{v})(\mathbf{w} \otimes \mathbf{x}) = (\mathbf{v} \cdot \mathbf{w})\mathbf{u} \otimes \mathbf{x} = \mathbf{u} \otimes \mathbf{x}(\mathbf{v} \cdot \mathbf{w}) \quad, \tag{1.57}$$

$$\mathbf{A}(\mathbf{u} \otimes \mathbf{v}) = (\mathbf{Au}) \otimes \mathbf{v} \quad. \tag{1.58}$$

Generally, the dyad is *not* commutative, i.e. **u** ⊗ **v** ≠ **v** ⊗ **u** and (**u** ⊗ **v**)(**w** ⊗ **x**) ≠ (**w** ⊗ **x**)(**u** ⊗ **v**).

A **dyadic** is a linear combination of dyads with scalar coefficients, for example, $\alpha(\mathbf{u} \otimes \mathbf{v}) + \beta(\mathbf{w} \otimes \mathbf{x})$. Note further that no tensor may be expressed as a single tensor product, in general, **A** = **u** ⊗ **v** + **w** ⊗ **x** ≠ **y** ⊗ **z**.

Any second-order tensor may be expressed as a *dyadic*. As an example, the second-order tensor **A** may be represented by a linear combination of dyads formed by the (Cartesian) basis $\{\mathbf{e}_i\}$, i.e.

$$\mathbf{A} = A_{ij}\mathbf{e}_i \otimes \mathbf{e}_j \quad. \tag{1.59}$$

We call **A**, which is resolved along basis vectors that are orthonormal, a **Cartesian tensor** of order two (more general tensors will be considered in Section 1.6). The nine

Cartesian components of $\mathbf{A}$ with respect to $\{\mathbf{e}_i\}$, represented by A_{ij}, form the entries of the matrix $[\mathbf{A}]$, i.e.

$$[\mathbf{A}] = \begin{bmatrix} A_{11} & A_{12} & A_{13} \\ A_{21} & A_{22} & A_{23} \\ A_{31} & A_{32} & A_{33} \end{bmatrix} \quad , \tag{1.60}$$

where by analogy with $(1.22)_2$

$$A_{ij}\delta_{jk} = A_{ik} \tag{1.61}$$

holds. Relation (1.60) is known as the **matrix notation** of tensor $\mathbf{A}$.

EXAMPLE 1.2   Let $\mathbf{A}$ be a *Cartesian tensor* of order two. Show that the projection of $\mathbf{A}$ onto the orthonormal basis vectors $\mathbf{e}_i$ is according to

$$A_{ij} = \mathbf{e}_i \cdot \mathbf{A}\mathbf{e}_j \quad , \tag{1.62}$$

where A_{ij} are the nine Cartesian components of tensor $\mathbf{A}$.

Solution.   Using representation (1.59) and properties $(1.53)_2$, (1.21) and (1.61), we find that

$$\begin{aligned} \mathbf{A}\mathbf{e}_j &= A_{lk}(\mathbf{e}_l \otimes \mathbf{e}_k)\mathbf{e}_j = A_{lk}(\mathbf{e}_k \cdot \mathbf{e}_j)\mathbf{e}_l \\ &= A_{lk}\delta_{kj}\mathbf{e}_l = A_{lj}\mathbf{e}_l \quad . \end{aligned} \tag{1.63}$$

On taking the dot product of $(1.63)_4$ with $\mathbf{e}_i$, the nine Cartesian components A_{ij} are completely determined, namely

$$\begin{aligned} \mathbf{e}_i \cdot \mathbf{A}\mathbf{e}_j &= \mathbf{e}_i \cdot A_{lj}\mathbf{e}_l = A_{lj}(\mathbf{e}_i \cdot \mathbf{e}_l) \\ &= A_{lj}\delta_{il} = A_{ij} = (\mathbf{A})_{ij} \quad . \end{aligned} \tag{1.64}$$

In (1.64) we have used the notation $(\mathbf{A})_{ij}$ for characterizing the components of $\mathbf{A}$.   ∎

If the relation $\mathbf{v} \cdot \mathbf{A}\mathbf{v} \geq 0$ holds for all vectors $\mathbf{v} \neq \mathbf{o}$, then $\mathbf{A}$ is said to be a **positive semi-definite** tensor. If the stronger condition $\mathbf{v} \cdot \mathbf{A}\mathbf{v} > 0$ holds for all nonzero vectors $\mathbf{v}$, then $\mathbf{A}$ is said to be a **positive definite** tensor. Tensor $\mathbf{A}$ is called **negative semi-definite** if $\mathbf{v} \cdot \mathbf{A}\mathbf{v} \leq 0$ and **negative definite** if $\mathbf{v} \cdot \mathbf{A}\mathbf{v} < 0$ for all vectors $\mathbf{v} \neq \mathbf{o}$, respectively.

The Cartesian components of the unit tensor $\mathbf{I}$ form the Kronecker delta symbol introduced in eq. (1.21). Thus,

$$\mathbf{I} = \delta_{ij}\mathbf{e}_i \otimes \mathbf{e}_j = \mathbf{e}_j \otimes \mathbf{e}_j \quad . \tag{1.65}$$

We derive further the components of $\mathbf{u} \otimes \mathbf{v}$ along an orthonormal basis $\{\mathbf{e}_i\}$. Using representation (1.62) and properties (1.53) and $(1.23)_3$, we find that

$$
\begin{aligned}
(\mathbf{u} \otimes \mathbf{v})_{ij} &= \mathbf{e}_i \cdot (\mathbf{u} \otimes \mathbf{v})\mathbf{e}_j \\
&= \mathbf{e}_i \cdot \mathbf{u}(\mathbf{v} \cdot \mathbf{e}_j) = (\mathbf{e}_i \cdot \mathbf{u})v_j \\
&= u_i v_j \quad,
\end{aligned}
\tag{1.66}
$$

where the coefficients $u_i v_j$ define the nine Cartesian components of $\mathbf{u} \otimes \mathbf{v}$. Writing eq. $(1.66)_4$ in the convenient matrix notation we have

$$
(\mathbf{u} \otimes \mathbf{v})_{ij} =
\begin{bmatrix} u_1 \\ u_2 \\ u_3 \end{bmatrix}
[\, v_1 \; v_2 \; v_3 \,] =
\begin{bmatrix}
u_1 v_1 & u_1 v_2 & u_1 v_3 \\
u_2 v_1 & u_2 v_2 & u_2 v_3 \\
u_3 v_1 & u_3 v_2 & u_3 v_3
\end{bmatrix} \quad.
\tag{1.67}
$$

Here, the 3×1 **column matrix** $[u_i]$ and the 1×3 **row matrix** $[v_j]$ represent the vectors $\mathbf{u}$ and $\mathbf{v}$, while the 3×3 **square matrix** represents the second-order tensor $\mathbf{u} \otimes \mathbf{v}$. A scalar would be represented by a 1×1 matrix.

EXAMPLE 1.3   Show that the linear transformation (1.49) that maps $\mathbf{v}$ to $\mathbf{u}$ may be written as

$$
v_i = A_{ij} u_j \quad,
\tag{1.68}
$$

where index j is a dummy index.

Solution.   Using expressions (1.20), (1.59) and rules (1.53), (1.21) and (1.61), we find from (1.49) that

$$
\begin{aligned}
v_i \mathbf{e}_i &= A_{ij} u_k (\mathbf{e}_i \otimes \mathbf{e}_j)\mathbf{e}_k = A_{ij} u_k (\mathbf{e}_j \cdot \mathbf{e}_k)\mathbf{e}_i \\
&= A_{ij} u_k \delta_{jk} \mathbf{e}_i = A_{ik} u_k \mathbf{e}_i \quad.
\end{aligned}
\tag{1.69}
$$

A replacement of the dummy index k in eq. $(1.69)_4$ by j gives the desired result (1.68). ∎

Dot product.   The **dot product** of two second-order tensors $\mathbf{A}$ and $\mathbf{B}$, denoted by $\mathbf{AB}$, is again a second-order tensor. It follows the requirement

$$
(\mathbf{AB})\mathbf{u} = \mathbf{A}(\mathbf{Bu})
\tag{1.70}
$$

for all vectors $\mathbf{u}$.

The summation, multiplication by scalars and dot products of tensors are governed mainly by properties known from ordinary arithmetic, for example,

$$\mathbf{A} + \mathbf{B} = \mathbf{B} + \mathbf{A} \ , \tag{1.71}$$

$$\mathbf{A} + \mathbf{O} = \mathbf{A} \ , \tag{1.72}$$

$$\mathbf{A} + (-\mathbf{A}) = \mathbf{O} \ , \tag{1.73}$$

$$\mathbf{A} + (\mathbf{B} + \mathbf{C}) = (\mathbf{A} + \mathbf{B}) + \mathbf{C} \ , \tag{1.74}$$

$$(\alpha \mathbf{A}) = \alpha(\mathbf{A}) = \alpha \mathbf{A} \ , \tag{1.75}$$

$$(\mathbf{AB})\mathbf{C} = \mathbf{A}(\mathbf{BC}) = \mathbf{ABC} \ , \tag{1.76}$$

$$\mathbf{A}^2 = \mathbf{AA} \ , \tag{1.77}$$

$$(\mathbf{A} + \mathbf{B})\mathbf{C} = \mathbf{AC} + \mathbf{BC} \ . \tag{1.78}$$

Note that, in general, the dot product of second-order tensors is *not* commutative, i.e. $\mathbf{AB} \neq \mathbf{BA}$ and also $\mathbf{Au} \neq \mathbf{uA}$. Moreover, relations $\mathbf{AB} = \mathbf{O}$ and $\mathbf{Au} = \mathbf{o}$ do not, in general, imply that $\mathbf{A}$, $\mathbf{B}$ or $\mathbf{u}$ are zero.

The components of the dot product $\mathbf{AB}$ along an orthonormal basis $\{\mathbf{e}_i\}$, as introduced in (1.18), read, by means of representation (1.62) and relations (1.70), (1.63)$_4$,

$$\begin{aligned} (\mathbf{AB})_{ij} &= \mathbf{e}_i \cdot (\mathbf{AB})\mathbf{e}_j = \mathbf{e}_i \cdot \mathbf{A}(\mathbf{Be}_j) \\ &= \mathbf{e}_i \cdot \mathbf{A}(B_{kj}\mathbf{e}_k) = (\mathbf{e}_i \cdot \mathbf{Ae}_k)B_{kj} \\ &= A_{ik}B_{kj} \end{aligned} \tag{1.79}$$

or equivalently

$$(\mathbf{AB})_{ij} = A_{ik}B_{kj} = A_{i1}B_{1j} + A_{i2}B_{2j} + A_{i3}B_{3j} \ . \tag{1.80}$$

For convenience, we adopt the convention that a repetition of only one index between a tensor and a vector (see, for example, eq. (1.68) with the dummy index j) or between two tensors (see, for example, eq. (1.80)$_1$ with the dummy index k) will *not* be indicated by a dot when symbolic notation is applied. Specifically that means for $\mathbf{v} = \mathbf{Au}$ ($v_i = A_{ij}u_j$) and $\mathbf{A} = \mathbf{BC}$ ($A_{ij} = B_{ik}C_{kj}$) we do not write $\mathbf{v} = \mathbf{A} \cdot \mathbf{u}$ and $\mathbf{A} = \mathbf{B} \cdot \mathbf{C}$, respectively.

Transpose of a tensor. The unique **transpose** of a second-order tensor $\mathbf{A}$ denoted by $\mathbf{A}^{\mathrm{T}}$ is governed by the identity

$$\mathbf{v} \cdot \mathbf{A}^{\mathrm{T}}\mathbf{u} = \mathbf{u} \cdot \mathbf{Av} = \mathbf{Av} \cdot \mathbf{u} \tag{1.81}$$

for all vectors $\mathbf{u}$ and $\mathbf{v}$. The useful properties

$$(\mathbf{A}^{\mathrm{T}})^{\mathrm{T}} = \mathbf{A} \ , \tag{1.82}$$

$$(\alpha\mathbf{A} + \beta\mathbf{B})^{\mathrm{T}} = \alpha\mathbf{A}^{\mathrm{T}} + \beta\mathbf{B}^{\mathrm{T}} \quad, \tag{1.83}$$

$$(\mathbf{AB})^{\mathrm{T}} = \mathbf{B}^{\mathrm{T}}\mathbf{A}^{\mathrm{T}} \quad, \tag{1.84}$$

$$(\mathbf{u} \otimes \mathbf{v})^{\mathrm{T}} = \mathbf{v} \otimes \mathbf{u} \tag{1.85}$$

hold. Hence, from identity (1.81) we obtain $\mathbf{e}_i \cdot \mathbf{A}^{\mathrm{T}}\mathbf{e}_j = \mathbf{e}_j \cdot \mathbf{A}\mathbf{e}_i$, which gives, in regard to (1.62), the important index relation $(\mathbf{A}^{\mathrm{T}})_{ij} = A_{ji}$.

Trace and contraction.   The **trace** of a tensor $\mathbf{A}$ is a scalar denoted by $\mathrm{tr}\mathbf{A}$. Taking, for example, the dyad $\mathbf{u} \otimes \mathbf{v}$ and summing up the diagonal terms of the matrix form of that second-order tensor, we get the dot product $\mathbf{u} \cdot \mathbf{v} = u_i v_i$, which is called the **trace** of $\mathbf{u} \otimes \mathbf{v}$. We write

$$\mathrm{tr}(\mathbf{u} \otimes \mathbf{v}) = \mathbf{u} \cdot \mathbf{v} = u_i v_i \tag{1.86}$$

for all vectors $\mathbf{u}$ and $\mathbf{v}$. Thus, with representation (1.59) and eqs. $(1.86)_1$, (1.21), (1.61) the trace of a tensor $\mathbf{A}$ with respect to the orthonormal basis $\{\mathbf{e}_i\}$ is given by

$$\begin{aligned}
\mathrm{tr}\mathbf{A} &= \mathrm{tr}(A_{ij}\mathbf{e}_i \otimes \mathbf{e}_j) = A_{ij}\mathrm{tr}(\mathbf{e}_i \otimes \mathbf{e}_j) \\
&= A_{ij}(\mathbf{e}_i \cdot \mathbf{e}_j) = A_{ij}\delta_{ji} \\
&= A_{ii} \quad,
\end{aligned} \tag{1.87}$$

or equivalently

$$\mathrm{tr}\mathbf{A} = A_{ii} = A_{11} + A_{22} + A_{33} \quad. \tag{1.88}$$

We have the properties of the trace:

$$\mathrm{tr}\mathbf{A}^{\mathrm{T}} = \mathrm{tr}\mathbf{A} \quad, \tag{1.89}$$

$$\mathrm{tr}(\mathbf{AB}) = \mathrm{tr}(\mathbf{BA}) \quad, \tag{1.90}$$

$$\mathrm{tr}(\mathbf{A} + \mathbf{B}) = \mathrm{tr}\mathbf{A} + \mathrm{tr}\mathbf{B} \quad, \tag{1.91}$$

$$\mathrm{tr}(\alpha\mathbf{A}) = \alpha\,\mathrm{tr}\mathbf{A} \quad. \tag{1.92}$$

In index notation a **contraction** means to identify two indices and to sum over them as dummy indices. In symbolic notation a contraction is characterized by a dot. A double contraction of two tensors $\mathbf{A}$ and $\mathbf{B}$, characterized by two dots, yields a scalar. It is defined in terms of the trace by

$$\begin{aligned}
\mathbf{A} : \mathbf{B} &= \mathrm{tr}(\mathbf{A}^{\mathrm{T}}\mathbf{B}) = \mathrm{tr}(\mathbf{B}^{\mathrm{T}}\mathbf{A}) \\
&= \mathrm{tr}(\mathbf{AB}^{\mathrm{T}}) = \mathrm{tr}(\mathbf{BA}^{\mathrm{T}}) \\
&= \mathbf{B} : \mathbf{A} \qquad \text{or} \qquad A_{ij}B_{ij} = B_{ij}A_{ij} \quad.
\end{aligned} \tag{1.93}$$

The useful properties of double contractions

$$\mathbf{I} : \mathbf{A} = \mathrm{tr}\mathbf{A} = \mathbf{A} : \mathbf{I} \ , \tag{1.94}$$

$$\mathbf{A} : (\mathbf{BC}) = (\mathbf{B}^{\mathrm{T}}\mathbf{A}) : \mathbf{C} = (\mathbf{AC}^{\mathrm{T}}) : \mathbf{B} \ , \tag{1.95}$$

$$\mathbf{A} : (\mathbf{u} \otimes \mathbf{v}) = \mathbf{u} \cdot \mathbf{Av} = (\mathbf{u} \otimes \mathbf{v}) : \mathbf{A} \ , \tag{1.96}$$

$$(\mathbf{u} \otimes \mathbf{v}) : (\mathbf{w} \otimes \mathbf{x}) = (\mathbf{u} \cdot \mathbf{w})(\mathbf{v} \cdot \mathbf{x}) \ , \tag{1.97}$$

$$(\mathbf{e}_i \otimes \mathbf{e}_j) : (\mathbf{e}_k \otimes \mathbf{e}_l) = (\mathbf{e}_i \cdot \mathbf{e}_k)(\mathbf{e}_j \cdot \mathbf{e}_l) = \delta_{ik}\delta_{jl} \tag{1.98}$$

hold. Note that if we have the relation $\mathbf{A} : \mathbf{B} = \mathbf{C} : \mathbf{B}$, in general, we cannot conclude that $\mathbf{A}$ equals $\mathbf{C}$.

The **norm** of a tensor $\mathbf{A}$ is denoted by $|\mathbf{A}|$ (or $\|\mathbf{A}\|$). It is a non-negative real number and is defined by the square root of $\mathbf{A} : \mathbf{A}$, i.e.

$$|\mathbf{A}| = (\mathbf{A} : \mathbf{A})^{1/2} = (A_{ij}A_{ij})^{1/2} \geq 0 \ . \tag{1.99}$$

Determinant and inverse of a tensor. The **determinant** of a tensor $\mathbf{A}$ yields a scalar. It is defined by the determinant of the matrix $[\mathbf{A}]$ of components of the tensor, i.e.

$$\det\mathbf{A} = \det[\mathbf{A}] = \det \begin{bmatrix} A_{11} & A_{12} & A_{13} \\ A_{21} & A_{22} & A_{23} \\ A_{31} & A_{32} & A_{33} \end{bmatrix} \ , \tag{1.100}$$

with properties

$$\det(\mathbf{AB}) = \det\mathbf{A}\det\mathbf{B} \ , \tag{1.101}$$

$$\det\mathbf{A}^{\mathrm{T}} = \det\mathbf{A} \ . \tag{1.102}$$

A tensor $\mathbf{A}$ is said to be **singular** *if and only if* $\det\mathbf{A} = 0$. We assume that $\mathbf{A}$ is a **nonsingular** tensor, i.e. $\det\mathbf{A} \neq 0$. Then there exists a unique **inverse** $\mathbf{A}^{-1}$ of $\mathbf{A}$ satisfying the reciprocal relation

$$\mathbf{AA}^{-1} = \mathbf{I} = \mathbf{A}^{-1}\mathbf{A} \ . \tag{1.103}$$

If tensors $\mathbf{A}$ and $\mathbf{B}$ are **invertible**, then the properties

$$(\mathbf{AB})^{-1} = \mathbf{B}^{-1}\mathbf{A}^{-1} \ , \tag{1.104}$$

$$(\mathbf{A}^{-1})^{-1} = \mathbf{A} \ , \tag{1.105}$$

$$(\alpha\mathbf{A})^{-1} = 1/\alpha\,\mathbf{A}^{-1} \ , \tag{1.106}$$

$$(\mathbf{A}^{-1})^{\mathrm{T}} = (\mathbf{A}^{\mathrm{T}})^{-1} \ , \tag{1.107}$$

$$\mathbf{A}^{-2} = \mathbf{A}^{-1}\mathbf{A}^{-1} \ , \tag{1.108}$$

$$\det(\mathbf{A}^{-1}) = (\det\mathbf{A})^{-1} \tag{1.109}$$

hold. Subsequently, in this text, we use the abbreviation

$$(\mathbf{A}^{-1})^{\mathrm{T}} = \mathbf{A}^{-\mathrm{T}} \tag{1.110}$$

for notational convenience.

Orthogonal tensor.   An **orthogonal tensor Q** is a linear transformation satisfying the condition

$$\mathbf{Q}\mathbf{u} \cdot \mathbf{Q}\mathbf{v} = \mathbf{u} \cdot \mathbf{v} \tag{1.111}$$

for all vectors **u** and **v** (see Figure 1.5). As can be seen, the dot product $\mathbf{u} \cdot \mathbf{v}$ is invariant during that transformation, which means that both the *angle* θ between **u** and **v** and the *lengths* of the vectors, $|\mathbf{u}|$, $|\mathbf{v}|$, are preserved.

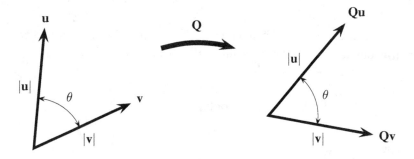

Figure 1.5 Orthogonal tensor.

Hence, an orthogonal tensor has the property $\mathbf{Q}^{\mathrm{T}}\mathbf{Q} = \mathbf{Q}\mathbf{Q}^{\mathrm{T}} = \mathbf{I}$, which means that $\mathbf{Q}^{-1} = \mathbf{Q}^{\mathrm{T}}$. Another important property is that $\det(\mathbf{Q}^{\mathrm{T}}\mathbf{Q}) = (\det\mathbf{Q})^2$ with $\det\mathbf{Q} = \pm 1$. If $\det\mathbf{Q} = +1\,(-1)$, $\mathbf{Q}$ is said to be **proper** (**improper**) orthogonal corresponding to a **rotation** (**reflection**), respectively.

Symmetric and skew tensors.   Any tensor **A** can always be uniquely decomposed into a **symmetric** tensor, here denoted by **S**, and a **skew** (or **antisymmetric**) tensor, here denoted by **W**. Hence, $\mathbf{A} = \mathbf{S} + \mathbf{W}$, while

$$\mathbf{S} = \frac{1}{2}(\mathbf{A} + \mathbf{A}^{\mathrm{T}}) \ , \qquad \mathbf{W} = \frac{1}{2}(\mathbf{A} - \mathbf{A}^{\mathrm{T}}) \ . \tag{1.112}$$

In this text, we also use the notation sym**A** for **S** and skew**A** for **W**. Tensors **S** and **W** are governed by properties such as

$$\mathbf{S} = \mathbf{S}^{\mathrm{T}} \quad \text{or} \quad S_{ij} = S_{ji} \ , \qquad \mathbf{W} = -\mathbf{W}^{\mathrm{T}} \quad \text{or} \quad W_{ij} = -W_{ji} \ , \tag{1.113}$$

which in matrix notation reads

$$
[\mathbf{S}] = \begin{bmatrix} S_{11} & S_{12} & S_{13} \\ S_{12} & S_{22} & S_{23} \\ S_{13} & S_{23} & S_{33} \end{bmatrix} \quad , \qquad [\mathbf{W}] = \begin{bmatrix} 0 & W_{12} & W_{13} \\ -W_{12} & 0 & W_{23} \\ -W_{13} & -W_{23} & 0 \end{bmatrix} \quad . \quad (1.114)
$$

In accord with the notation introduced, the useful properties

$$
\mathbf{S} : \mathbf{B} = \mathbf{S} : \mathbf{B}^{\mathrm{T}} = \mathbf{S} : \frac{1}{2}(\mathbf{B} + \mathbf{B}^{\mathrm{T}}) \quad , \tag{1.115}
$$

$$
\mathbf{W} : \mathbf{B} = -\mathbf{W} : \mathbf{B}^{\mathrm{T}} = \mathbf{W} : \frac{1}{2}(\mathbf{B} - \mathbf{B}^{\mathrm{T}}) \quad , \tag{1.116}
$$

$$
\mathbf{S} : \mathbf{W} = 0 \tag{1.117}
$$

hold, where $\mathbf{B}$ denotes any second-order tensor.

A skew tensor with property $\mathbf{W} = -\mathbf{W}^{\mathrm{T}}$ has zero diagonal elements and only three independent scalar quantities as seen from eq. $(1.114)_2$. Hence, every skew tensor $\mathbf{W}$ behaves **like a vector** with *three* components. Indeed, the relation holds:

$$
\mathbf{W}\mathbf{u} = \boldsymbol{\omega} \times \mathbf{u} \quad , \tag{1.118}
$$

where $\mathbf{u}$ is any vector and $\boldsymbol{\omega}$ characterizes the **axial** (or **dual**) **vector** of skew tensor $\mathbf{W}$, with property $|\boldsymbol{\omega}| = (1/\sqrt{2})|\mathbf{W}|$ (the proof is omitted).

The components of $\mathbf{W}$ follow from (1.62), with the help of (1.118), (1.20), (1.35), (1.21) and the properties of the permutation symbol

$$
\begin{aligned}
W_{ij} &= \mathbf{e}_i \cdot \mathbf{W}\mathbf{e}_j = \mathbf{e}_i \cdot (\boldsymbol{\omega} \times \mathbf{e}_j) = \mathbf{e}_i \cdot (\omega_k \mathbf{e}_k \times \mathbf{e}_j) \\
&= \mathbf{e}_i \cdot (\omega_k \varepsilon_{kjl} \mathbf{e}_l) = \omega_k \varepsilon_{kjl} \delta_{il} \\
&= \varepsilon_{kji} \omega_k = -\varepsilon_{ijk} \omega_k \quad .
\end{aligned} \tag{1.119}
$$

Therefore, with definition (1.33) we have

$$
W_{12} = -\varepsilon_{12k}\omega_k = -\omega_3 \quad , \tag{1.120}
$$

$$
W_{13} = -\varepsilon_{13k}\omega_k = \omega_2 \quad , \tag{1.121}
$$

$$
W_{23} = -\varepsilon_{23k}\omega_k = -\omega_1 \quad , \tag{1.122}
$$

where the components W_{12}, W_{13}, W_{23} form the entries of the matrix $[\mathbf{W}]$ as characterized in $(1.114)_2$.

The inversion of $(1.119)_7$ follows with relations $(1.38)_2$ and $(1.22)_2$ after multiplication with the permutation symbol ε_{ijp}

$$
\varepsilon_{ijp} W_{ij} = - \underbrace{\varepsilon_{ijp} \varepsilon_{ijk}}_{2\delta_{pk}} \omega_k = -2\omega_p \quad , \tag{1.123}
$$

which, after a change of the free index, gives finally

$$\omega_k = -\frac{1}{2}\varepsilon_{ijk}W_{ij} \qquad \text{and} \qquad \omega = -\frac{1}{2}\varepsilon_{ijk}W_{ij}\mathbf{e}_k \quad . \tag{1.124}$$

Projection, spherical and deviatoric tensors.   Consider any vector $\mathbf{u}$ and a unit vector $\mathbf{e}$. With reference to Figure 1.6, we write $\mathbf{u} = \mathbf{u}_{\parallel} + \mathbf{u}_{\perp}$, with $\mathbf{u}_{\parallel}$ and $\mathbf{u}_{\perp}$ characterizing the projection of $\mathbf{u}$ onto the line spanned by $\mathbf{e}$ and onto the plane normal to $\mathbf{e}$, respectively.

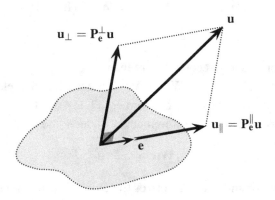

Figure 1.6 Projection tensor.

With (1.53) we deduce that

$$\mathbf{u}_{\parallel} = (\mathbf{u} \cdot \mathbf{e})\mathbf{e} = \underbrace{(\mathbf{e} \otimes \mathbf{e})}_{\mathbf{P}_{\mathbf{e}}^{\parallel}}\mathbf{u} = \mathbf{P}_{\mathbf{e}}^{\parallel}\mathbf{u} \quad , \tag{1.125}$$

$$\mathbf{u}_{\perp} = \mathbf{u} - \mathbf{u}_{\parallel} = \mathbf{u} - (\mathbf{e} \otimes \mathbf{e})\mathbf{u} = \underbrace{(\mathbf{I} - \mathbf{e} \otimes \mathbf{e})}_{\mathbf{P}_{\mathbf{e}}^{\perp}}\mathbf{u} = \mathbf{P}_{\mathbf{e}}^{\perp}\mathbf{u} \quad , \tag{1.126}$$

where $\mathbf{P}_{\mathbf{e}}^{\parallel}$ and $\mathbf{P}_{\mathbf{e}}^{\perp}$ are **projection tensors** of order two. The projection tensor $\mathbf{P}_{\mathbf{e}}^{\parallel}$ applied to any vector $\mathbf{u}$ maps $\mathbf{u}$ into the direction of $\mathbf{e}$, while $\mathbf{P}_{\mathbf{e}}^{\perp}$ applied to $\mathbf{u}$ gives the projection of $\mathbf{u}$ onto the plane normal to $\mathbf{e}$ (see Figure 1.6).

A tensor $\mathbf{P}$ is a **projection** if $\mathbf{P}$ is **symmetric** and $\mathbf{P}^n = \mathbf{P}$ (n is a positive integer), with the properties

$$\mathbf{P}_{\mathbf{e}}^{\perp} + \mathbf{P}_{\mathbf{e}}^{\parallel} = \mathbf{I} \quad , \tag{1.127}$$

$$\mathbf{P}_{\mathbf{e}}^{\parallel}\,\mathbf{P}_{\mathbf{e}}^{\parallel} = \mathbf{P}_{\mathbf{e}}^{\parallel} \quad , \tag{1.128}$$

$$\mathbf{P}_{\mathbf{e}}^{\perp}\,\mathbf{P}_{\mathbf{e}}^{\perp} = \mathbf{P}_{\mathbf{e}}^{\perp} \quad , \tag{1.129}$$

$$\mathbf{P}_{\mathbf{e}}^{\parallel}\,\mathbf{P}_{\mathbf{e}}^{\perp} = \mathbf{O} \quad . \tag{1.130}$$

Every tensor $\mathbf{A}$ can be decomposed into its so-called *spherical part* and its *deviatoric part*, i.e.

$$\mathbf{A} = \alpha\mathbf{I} + \mathrm{dev}\mathbf{A} \qquad \text{or} \qquad A_{ij} = \alpha\delta_{ij} + \mathrm{dev}\,A_{ij} \tag{1.131}$$

$$\alpha = \frac{1}{3}\mathrm{tr}\mathbf{A} = \frac{1}{3}(\mathbf{I}:\mathbf{A}) = \frac{1}{3}A_{ii} \ . \tag{1.132}$$

Any tensor of the form $\alpha\mathbf{I}$, with α denoting a scalar, is known as a **spherical tensor**, while $\mathrm{dev}\mathbf{A}$ is known as a **deviator** of $\mathbf{A}$, or a **deviatoric tensor**. The **deviatoric operator** $\mathrm{dev}(\bullet)$ is denoted by the short-hand notation, i.e. $\mathrm{dev}(\bullet) = (\bullet) - (1/3)\mathrm{tr}(\bullet)\mathbf{I}$, or $\mathrm{dev}(\bullet)_{ij} = (\bullet)_{ij} - (1/3)(\bullet)_{kk}\delta_{ij}$.

Computing the trace of (1.131) we deduce with (1.92), (1.87)$_5$, (1.22)$_1$ and (1.132)$_3$ that $A_{ii} = 3(A_{ii}/3) + \mathrm{dev}\,A_{ii}$. This relation yields the important property

$$\mathrm{tr}(\mathrm{dev}\mathbf{A}) = 0 \ , \tag{1.133}$$

which means that the trace of the deviator of $\mathbf{A}$ is always zero.

EXERCISES

1. Let $\mathbf{u}$ be any vector and $\mathbf{e}$ any unit vector. Use eqs. (1.125), (1.126) and the triple vector product (1.47) to show that $\mathbf{u}$ can be resolved into components parallel, i.e. $\mathbf{u}_{\parallel}$, and perpendicular, i.e. $\mathbf{u}_{\perp}$, to unit vector $\mathbf{e}$, according to

$$\mathbf{u} = \mathbf{u}_{\parallel} + \mathbf{u}_{\perp} = (\mathbf{u}\cdot\mathbf{e})\mathbf{e} + \mathbf{e}\times(\mathbf{u}\times\mathbf{e}) \ .$$

2. Prove eqs. (1.57) and (1.58) and show that

$$(\mathbf{u}\otimes\mathbf{v})\mathbf{A} = \mathbf{u}\otimes(\mathbf{A}^{\mathrm{T}}\mathbf{v}) \ .$$

3. Let $\mathbf{A}$ be a tensor with matrix $[\mathbf{A}]$ and $\mathbf{A}^{\mathrm{T}}$ the transpose of $\mathbf{A}$ with matrix $[\mathbf{A}]^{\mathrm{T}}$. Show that $[\mathbf{A}^{\mathrm{T}}] = [\mathbf{A}]^{\mathrm{T}}$.

4. Show by means of representation (1.59) and properties (1.98) and (1.61) that the double contraction of the two tensors $\mathbf{A}$ and $\mathbf{B}$ yields the component form $A_{ij}B_{ij}$.

5. Starting from eq. (1.100), verify that $\det\mathbf{A} = \varepsilon_{ijk}A_{i1}A_{j2}A_{k3}$ and show property (1.101).

6. Consider eq. (1.93)$_1$ and relations (1.103), (1.88) in order to obtain the property

$$\mathbf{A}^{-\mathrm{T}}:\mathbf{A} = 3 \ . \tag{1.134}$$

7. Given that $\mathbf{S}$ is a symmetric tensor and $\mathbf{W}$ is antisymmetric, prove property (1.117).

8. Two tensors $\mathbf{A}$ and $\mathbf{B}$ are given in the form of their matrix representations

$$[\mathbf{A}] = \begin{bmatrix} \cos\theta & \sin\theta & 0 \\ -\sin\theta & \cos\theta & 0 \\ 0 & 0 & 1 \end{bmatrix} \quad , \qquad [\mathbf{B}] = \begin{bmatrix} -1 & 0 & 0 \\ 0 & 1 & 0 \\ 0 & 0 & 1 \end{bmatrix} \quad .$$

Verify that these matrices are orthogonal. Show that $[\mathbf{A}]$ describes a rotation and $[\mathbf{B}]$ a reflection. Give a geometrical interpretation.

9. Find the axial vector of the skew tensor $\mathbf{W} = 1/2(\mathbf{u} \otimes \mathbf{v} - \mathbf{v} \otimes \mathbf{u})$.

10. Let $\mathbf{A}$ be a tensor whose matrix is

$$[\mathbf{A}] = \begin{bmatrix} 1 & 1 & 1 \\ 1 & 1 & 1 \\ 1 & 1 & 1 \end{bmatrix} \quad .$$

Find the spherical and the deviatoric parts of $\mathbf{A}$.

1.3   Higher-order Tensors

In the following we discuss higher-order tensors. Any tensor of **order** (or **rank**) n may be expressed in the form

$$A_{i_1 i_2 \dots i_n} \mathbf{e}_{i_1} \otimes \mathbf{e}_{i_2} \otimes \cdots \otimes \mathbf{e}_{i_n} \quad . \tag{1.135}$$

As (1.135) shows, a tensor of order n has 3^n components $A_{i_1 i_2 \dots i_n}$, provided with n indices $i_1, i_2, \dots, i_n$. In particular, a **tensor of order zero** has $3^0 = 1$ component with 0 (no) index, which simply is a scalar α (i.e. a single real number). A **tensor of order one** has $3^1 = 3$ components characterizing any vector $\mathbf{v}$, i.e. v_i as indicated in (1.23). A **tensor of order two,** for example, $\mathbf{A}$, has $3^2 = 9$ components, i.e. A_{ij} as indicated in (1.60).

Tensor of order three.   A **tensor of order three** we distinguish by the notation $\mathcal{A}$, $\mathcal{B}, \mathcal{C}, \dots$ According to (1.135), $\mathcal{A}$ may be expressed as

$$\mathcal{A} = \mathcal{A}_{ijk}\mathbf{e}_i \otimes \mathbf{e}_j \otimes \mathbf{e}_k \quad , \tag{1.136}$$

where $\mathcal{A}_{ijk}$ characterizes the $3^3 = 27$ components of $\mathcal{A}$.

A particular example for a third-order tensor is the so-called **triadic product** of three vectors $\mathbf{u}, \mathbf{v}, \mathbf{w}$, denoted by $\mathbf{u} \otimes \mathbf{v} \otimes \mathbf{w}$, with the properties

$$(\mathbf{u} \otimes \mathbf{v}) \otimes \mathbf{w} = \mathbf{u} \otimes \mathbf{v} \otimes \mathbf{w} \quad , \tag{1.137}$$

$$(\mathbf{u} \otimes \mathbf{v} \otimes \mathbf{w})\mathbf{x} = (\mathbf{w} \cdot \mathbf{x})\mathbf{u} \otimes \mathbf{v} \quad , \tag{1.138}$$

$$(\mathbf{u} \otimes \mathbf{v} \otimes \mathbf{w}) : (\mathbf{x} \otimes \mathbf{y}) = (\mathbf{v} \cdot \mathbf{x})(\mathbf{w} \cdot \mathbf{y})\mathbf{u} \quad , \tag{1.139}$$

$$(\mathbf{u} \otimes \mathbf{v} \otimes \mathbf{w}) : \mathbf{I} = (\mathbf{v} \cdot \mathbf{w})\mathbf{u} \quad . \tag{1.140}$$

Hence, by analogy with the procedure which led to eq. (1.62) we may determine the components of $\mathcal{A}$, i.e.

$$\mathcal{A}_{ijk} = (\mathbf{e}_i \otimes \mathbf{e}_j) : \mathcal{A}\mathbf{e}_k \quad . \tag{1.141}$$

The double contraction of a third-order tensor $\mathcal{A}$ with a second-order tensor $\mathbf{B}$ produces a vector. It is denoted by $\mathcal{A} : \mathbf{B}$ and may be found with representations (1.136), (1.59) and properties (1.139), (1.21) and (1.61) as

$$
\begin{aligned}
\mathcal{A} : \mathbf{B} &= \mathcal{A}_{ijk}B_{lm}(\mathbf{e}_i \otimes \mathbf{e}_j \otimes \mathbf{e}_k) : (\mathbf{e}_l \otimes \mathbf{e}_m) \\
&= \mathcal{A}_{ijk}B_{lm}(\mathbf{e}_j \cdot \mathbf{e}_l)(\mathbf{e}_k \cdot \mathbf{e}_m)\mathbf{e}_i = \mathcal{A}_{ijk}B_{lm}\delta_{jl}\delta_{km}\mathbf{e}_i \\
&= \mathcal{A}_{ijk}B_{jk}\mathbf{e}_i \quad .
\end{aligned}
\tag{1.142}
$$

EXAMPLE 1.4   Let $\mathcal{E}$ denote the third-order **permutation tensor**, which may be expressed as

$$\mathcal{E} = \varepsilon_{ijk}\mathbf{e}_i \otimes \mathbf{e}_j \otimes \mathbf{e}_k \quad . \tag{1.143}$$

Here, $\varepsilon_{ijk} = (\mathbf{e}_i \times \mathbf{e}_j) \cdot \mathbf{e}_k$ are the 3^3 components of $\mathcal{E}$ which we have introduced as the permutation symbol (recall eqs. (1.33) and (1.35)).

Show that the double contraction of $\mathcal{E}$ with the dyad of any vectors $\mathbf{u}$ and $\mathbf{v}$, i.e. $\mathbf{u} \otimes \mathbf{v}$, gives the cross product of $\mathbf{v}$ and $\mathbf{u}$.

Solution.   With expressions $(\mathcal{E})_{ijk} = \varepsilon_{ijk}$ and $(\mathbf{u} \otimes \mathbf{v})_{lm} = u_l v_m$ we may write

$$\mathcal{E} : (\mathbf{u} \otimes \mathbf{v}) = \varepsilon_{ijk}u_l v_m(\mathbf{e}_i \otimes \mathbf{e}_j \otimes \mathbf{e}_k) : (\mathbf{e}_l \otimes \mathbf{e}_m) \quad . \tag{1.144}$$

Using properties (1.139), (1.21) and (1.22)$_2$, we find that

$$\mathcal{E} : (\mathbf{u} \otimes \mathbf{v}) = \varepsilon_{ijk}u_l v_m \delta_{jl}\delta_{km}\mathbf{e}_i = \varepsilon_{ijk}u_j v_k\mathbf{e}_i \quad , \tag{1.145}$$

which is a vector with components $\varepsilon_{ijk}u_j v_k$. Finally, by means of the permutation property $\varepsilon_{ijk} = -\varepsilon_{kji}$ and expressions $(1.43)_2$, (1.26) we obtain

$$\mathcal{E} : (\mathbf{u} \otimes \mathbf{v}) = \mathbf{v} \times \mathbf{u} \ , \tag{1.146}$$

which is the desired result.　■

Tensor of order four.　　Any **tensor of order four**, denoted by $\mathbb{A}, \mathbb{B}, \mathbb{C}, \ldots$, has $3^4 = 81$ components, i.e. $A_{ijkl}, B_{ijkl}, C_{ijkl}, \ldots$, respectively. We may express any tensor $\mathbb{A}$ in terms of the three Cartesian basis vectors, we write

$$\mathbb{A} = A_{ijkl}\mathbf{e}_i \otimes \mathbf{e}_j \otimes \mathbf{e}_k \otimes \mathbf{e}_l \ , \tag{1.147}$$

in which A_{ijkl} are the components of $\mathbb{A}$.

One example for a fourth-order tensor is the product of the four vectors $\mathbf{u}, \mathbf{v}, \mathbf{w}, \mathbf{x}$, denoted by $\mathbf{u} \otimes \mathbf{v} \otimes \mathbf{w} \otimes \mathbf{x}$. We have the useful properties:

$$(\mathbf{u} \otimes \mathbf{v}) \otimes (\mathbf{w} \otimes \mathbf{x}) = \mathbf{u} \otimes \mathbf{v} \otimes \mathbf{w} \otimes \mathbf{x} \ , \tag{1.148}$$

$$(\mathbf{u} \otimes \mathbf{v} \otimes \mathbf{w} \otimes \mathbf{x}) : (\mathbf{y} \otimes \mathbf{z}) = (\mathbf{w} \cdot \mathbf{y})(\mathbf{x} \cdot \mathbf{z})(\mathbf{u} \otimes \mathbf{v}) \ . \tag{1.149}$$

With these properties we can show that the components A_{ijkl} of $\mathbb{A}$ can be expressed as

$$A_{ijkl} = (\mathbf{e}_i \otimes \mathbf{e}_j) : \mathbb{A} : (\mathbf{e}_k \otimes \mathbf{e}_l) \ . \tag{1.150}$$

Hence, the double contraction of $\mathbb{A}$ with a second-order tensor $\mathbf{B}$ is a second-order tensor, denoted by $\mathbb{A} : \mathbf{B}$. With representations (1.147), (1.59), and properties (1.149), (1.21) and (1.61) we find that

$$\begin{aligned}
\mathbb{A} : \mathbf{B} &= A_{ijkl}B_{mn}(\mathbf{e}_i \otimes \mathbf{e}_j \otimes \mathbf{e}_k \otimes \mathbf{e}_l) : (\mathbf{e}_m \otimes \mathbf{e}_n) \\
&= A_{ijkl}B_{mn}\delta_{km}\delta_{ln}\mathbf{e}_i \otimes \mathbf{e}_j \\
&= A_{ijkl}B_{kl}\mathbf{e}_i \otimes \mathbf{e}_j \ .
\end{aligned} \tag{1.151}$$

Following the definitions above and by analogy with (1.53), (1.56) and (1.57), note the important rules

$$(\mathbf{A} \otimes \mathbf{B}) : \mathbf{C} = \mathbf{A}(\mathbf{B} : \mathbf{C}) = (\mathbf{B} : \mathbf{C})\mathbf{A} \ , \tag{1.152}$$

$$\mathbf{A} : (\mathbf{B} \otimes \mathbf{C}) = (\mathbf{A} : \mathbf{B})\mathbf{C} = \mathbf{C}(\mathbf{A} : \mathbf{B}) \ , \tag{1.153}$$

$$(\mathbf{A} \otimes \mathbf{B}) : (\mathbf{C} \otimes \mathbf{D}) = (\mathbf{B} : \mathbf{C})(\mathbf{A} \otimes \mathbf{D}) = (\mathbf{A} \otimes \mathbf{D})(\mathbf{B} : \mathbf{C}) \ , \tag{1.154}$$

and the property

$$\mathbf{A}(\mathbf{B} \otimes \mathbf{C})\mathbf{D} = (\mathbf{AB}) \otimes (\mathbf{CD}) \ . \tag{1.155}$$

EXAMPLE 1.5 Show eq. (1.152) which is the analogue of eq. (1.53).

Solution. Express tensor $\mathbf{A}, \mathbf{B}, \mathbf{C}$ in terms of their basis vectors according to (1.59). With properties (1.148), (1.149), (1.21) and (1.61) we find that

$$
\begin{aligned}
(\mathbf{A} \otimes \mathbf{B}) : \mathbf{C} &= A_{ij}B_{kl}C_{mn}(\mathbf{e}_i \otimes \mathbf{e}_j) \otimes (\mathbf{e}_k \otimes \mathbf{e}_l) : (\mathbf{e}_m \otimes \mathbf{e}_n) \\
&= A_{ij}B_{kl}C_{mn}(\mathbf{e}_k \cdot \mathbf{e}_m)(\mathbf{e}_l \cdot \mathbf{e}_n)(\mathbf{e}_i \otimes \mathbf{e}_j) \\
&= A_{ij}B_{kl}C_{mn}\delta_{km}\delta_{ln}\mathbf{e}_i \otimes \mathbf{e}_j \\
&= A_{ij}B_{kl}C_{kl}\mathbf{e}_i \otimes \mathbf{e}_j \quad .
\end{aligned}
\tag{1.156}
$$

By using representation (1.59) and definition (1.93) we obtain $A_{ij}B_{kl}C_{kl}\mathbf{e}_i \otimes \mathbf{e}_j = \mathbf{A}(\mathbf{B} : \mathbf{C}) = (\mathbf{B} : \mathbf{C})\mathbf{A}$, which is the desired result. ∎

A further example for a fourth-order tensor is the tensor product $\mathbb{D} = \mathbf{A} \otimes \mathbf{B}$, while $\mathbf{A}$ and $\mathbf{B}$ are second-order tensors with components A_{ij} and B_{ij}, respectively. In index notation we may write $D_{ijkl} = A_{ij}B_{kl}$.

The unique **transpose** of a fourth-order tensor $\mathbb{A}$ denoted by $\mathbb{A}^{\mathrm{T}}$ is governed, by analogy with (1.81), by the identity

$$
\mathbf{B} : \mathbb{A}^{\mathrm{T}} : \mathbf{C} = \mathbf{C} : \mathbb{A} : \mathbf{B} = (\mathbb{A} : \mathbf{B}) : \mathbf{C}
\tag{1.157}
$$

for all second-order tensors $\mathbf{B}$ and $\mathbf{C}$. The properties of a fourth-order tensor

$$
(\mathbb{A}^{\mathrm{T}})^{\mathrm{T}} = \mathbb{A} \quad ,
\tag{1.158}
$$

$$
(\mathbf{A} \otimes \mathbf{B})^{\mathrm{T}} = \mathbf{B} \otimes \mathbf{A}
\tag{1.159}
$$

hold. From the identity (1.157) we deduce the index relation, namely $(\mathbb{A}^{\mathrm{T}})_{ijkl} = A_{klij}$.

We now define **fourth-order unit tensors** $\mathbb{I}$ and $\bar{\mathbb{I}}$ so that

$$
\mathbf{A} = \mathbb{I} : \mathbf{A} \quad , \qquad \mathbf{A}^{\mathrm{T}} = \bar{\mathbb{I}} : \mathbf{A} \quad ,
\tag{1.160}
$$

for any second-order tensor $\mathbf{A}$. These transformations assign to any $\mathbf{A}$ the tensor $\mathbf{A}$ itself, and its transpose, respectively. The fourth-order unit tensors may be represented by

$$
\mathbb{I} = \delta_{ik}\delta_{jl}\mathbf{e}_i \otimes \mathbf{e}_j \otimes \mathbf{e}_k \otimes \mathbf{e}_l = \mathbf{e}_i \otimes \mathbf{e}_j \otimes \mathbf{e}_i \otimes \mathbf{e}_j \quad ,
\tag{1.161}
$$

$$
\bar{\mathbb{I}} = \delta_{il}\delta_{jk}\mathbf{e}_i \otimes \mathbf{e}_j \otimes \mathbf{e}_k \otimes \mathbf{e}_l = \mathbf{e}_i \otimes \mathbf{e}_j \otimes \mathbf{e}_j \otimes \mathbf{e}_i \quad ,
\tag{1.162}
$$

where $(\mathbb{I})_{ijkl} = \delta_{ik}\delta_{jl}$ and $(\bar{\mathbb{I}})_{ijkl} = \delta_{il}\delta_{jk}$ define the Cartesian components of $\mathbb{I}$ and $\bar{\mathbb{I}}$, respectively. Note that $\bar{\mathbb{I}} \neq \mathbb{I}^{\mathrm{T}}$.

The deviatoric part of a second-order tensor $\mathbf{A}$ may be described by means of a fourth-order tensor. From relations (1.131), (1.132)$_2$ and transformations (1.160)$_1$, (1.152) we conclude that

$$\mathrm{dev}\mathbf{A} = \mathbf{A} - \frac{1}{3}(\mathbf{I}:\mathbf{A})\mathbf{I} = (\mathbb{I} - \frac{1}{3}\mathbf{I}\otimes\mathbf{I}):\mathbf{A} \quad, \tag{1.163}$$

or with the definition of a fourth-order **projection tensor** $\mathbb{P}$

$$\mathrm{dev}\mathbf{A} = \mathbb{P}:\mathbf{A} \quad, \qquad \mathbb{P} = \mathbb{I} - \frac{1}{3}\mathbf{I}\otimes\mathbf{I} \quad. \tag{1.164}$$

The components of $\mathrm{dev}\mathbf{A}$ and $\mathbf{A}$ are related through the expression $\mathrm{dev}A_{ij} = P_{ijkl}A_{kl}$, with $P_{ijkl} = \delta_{ik}\delta_{jl} - (1/3)\delta_{ij}\delta_{kl}$.

EXERCISE

1. Two fourth-order tensors $\mathbb{S}, \mathbb{W}$ are given in the form

$$\mathbb{S} = \frac{1}{2}(\mathbb{I} + \bar{\mathbb{I}}) \quad, \qquad \mathbb{W} = \frac{1}{2}(\mathbb{I} - \bar{\mathbb{I}})$$

(observe the similar structure to eq. (1.112)). Show that

$$\frac{1}{2}(\mathbf{A} + \mathbf{A}^{\mathrm{T}}) = \mathbb{S}:\mathbf{A} \quad, \qquad \frac{1}{2}(\mathbf{A} - \mathbf{A}^{\mathrm{T}}) = \mathbb{W}:\mathbf{A} \quad,$$

which assign to any second-order tensor $\mathbf{A}$ its symmetric and skew parts, respectively.

1.4   Eigenvalues, Eigenvectors of Tensors

The scalars λ_i characterize **eigenvalues** (or **principal values**) of a tensor $\mathbf{A}$ if there exist corresponding nonzero normalized **eigenvectors** $\hat{\mathbf{n}}_i$ (or **principal directions** or **principal axes**) of $\mathbf{A}$, so that

$$\mathbf{A}\hat{\mathbf{n}}_i = \lambda_i\hat{\mathbf{n}}_i \quad, \qquad (i = 1, 2, 3; \text{ no summation}) \quad. \tag{1.165}$$

To identify the eigenvectors of a tensor, we use subsequently a hat on the vector quantity concerned, for example, $\hat{\mathbf{n}}$.

Thus, a set of homogeneous algebraic equations for the unknown eigenvalues λ_i, $i = 1, 2, 3$, and the unknown eigenvectors $\hat{\mathbf{n}}_i$, $i = 1, 2, 3$, is

$$(\mathbf{A} - \lambda_i\mathbf{I})\hat{\mathbf{n}}_i = \mathbf{o} \quad, \qquad (i = 1, 2, 3; \text{ no summation}) \quad. \tag{1.166}$$

Eigenvalues characterize the physical nature of a tensor. They do not depend on coordinates. For a *positive definite symmetric* tensor $\mathbf{A}$, all eigenvalues λ_i are (real and) **positive** since, using (1.165), we have $\lambda_i = \hat{\mathbf{n}}_i \cdot \mathbf{A}\hat{\mathbf{n}}_i > 0$, $i = 1, 2, 3$. Moreover, the set of eigenvectors of a *symmetric* tensor $\mathbf{A}$ form a **mutually orthogonal (orthonormal) basis** $\{\hat{\mathbf{n}}_i\}$ (the proof of this statement is omitted).

Principal scalar invariants.   For the system (1.166) to have solutions $\hat{\mathbf{n}}_i \neq \mathbf{o}$ the determinant of the system must vanish. Thus,

$$\det(\mathbf{A} - \lambda_i \mathbf{I}) = 0 \quad , \tag{1.167}$$

where

$$\det(\mathbf{A} - \lambda_i \mathbf{I}) = -\lambda_i^3 + I_1 \lambda_i^2 - I_2 \lambda_i + I_3 \quad . \tag{1.168}$$

This requires that we solve a cubic equation in λ, usually written as

$$\lambda^3 - I_1 \lambda^2 + I_2 \lambda - I_3 = 0 \quad , \tag{1.169}$$

called the **characteristic polynomial** (or **equation**) for $\mathbf{A}$, the solutions of which are the eigenvalues λ_i, $i = 1, 2, 3$.

Here, $I_i(\mathbf{A})$, $i = 1, 2, 3$, are the so-called **principal scalar invariants** of $\mathbf{A}$. In terms of $\mathbf{A}$ and its principal values λ_i, $i = 1, 2, 3$, these are given by

$$I_1(\mathbf{A}) = A_{ii} = \text{tr}\mathbf{A} = \lambda_1 + \lambda_2 + \lambda_3 \quad , \tag{1.170}$$

$$I_2(\mathbf{A}) = \frac{1}{2}(A_{ii} A_{jj} - A_{ji} A_{ij}) = \frac{1}{2}\left[(\text{tr}\mathbf{A})^2 - \text{tr}(\mathbf{A}^2)\right] = \text{tr}\mathbf{A}^{-1}\det\mathbf{A}$$
$$= \lambda_1 \lambda_2 + \lambda_1 \lambda_3 + \lambda_2 \lambda_3 \quad , \tag{1.171}$$

$$I_3(\mathbf{A}) = \varepsilon_{ijk} A_{1i} A_{2j} A_{3k} = \det\mathbf{A} = \lambda_1 \lambda_2 \lambda_3 \quad . \tag{1.172}$$

Sometimes we shall use the alternative notation

$$I_1(\mathbf{A}) = I_{\mathbf{A}} \quad , \qquad I_2(\mathbf{A}) = II_{\mathbf{A}} \quad , \qquad I_3(\mathbf{A}) = III_{\mathbf{A}} \tag{1.173}$$

for the three principal invariants interchangeably.

A repeated application of tensor $\mathbf{A}$ to eq. (1.165) yields $\mathbf{A}^\alpha \hat{\mathbf{n}}_i = \lambda_i^\alpha \hat{\mathbf{n}}_i$, $i = 1, 2, 3$, for any positive integer α. Using this relation and (1.169) multiplied by $\hat{\mathbf{n}}_i$, we obtain the well-known **Cayley-Hamilton equation**

$$\mathbf{A}^3 - I_1 \mathbf{A}^2 + I_2 \mathbf{A} - I_3 \mathbf{I} = \mathbf{O} \quad . \tag{1.174}$$

It states that every (second-order) tensor $\mathbf{A}$ satisfies its own characteristic equation.

Spectral decomposition of a tensor.   Any symmetric tensor $\mathbf{A}$ may be represented by its eigenvalues λ_i, $i = 1, 2, 3$, and the corresponding eigenvectors of $\mathbf{A}$ forming an orthonormal basis $\{\hat{\mathbf{n}}_i\}$. Using the unit tensor, by analogy with $(1.65)_2$, i.e. $\mathbf{I} = \hat{\mathbf{n}}_i \otimes \hat{\mathbf{n}}_i$, and relations (1.58), (1.165) and (1.55) we obtain an expression which is known as the **spectral decomposition** (or **spectral representation**) of $\mathbf{A}$, i.e.

$$\mathbf{A} = \mathbf{AI} = (\mathbf{A}\hat{\mathbf{n}}_i) \otimes \hat{\mathbf{n}}_i = \sum_{i=1}^{3} \lambda_i \hat{\mathbf{n}}_i \otimes \hat{\mathbf{n}}_i \quad . \tag{1.175}$$

The components A_{ij} of tensor $\mathbf{A}$ relative to a *basis of principal directions* follow with (1.59) by replacing $\mathbf{e}_i$ with the three orthonormal basis vectors $\hat{\mathbf{n}}_i$. With eqs. (1.165) and (1.21) we obtain

$$A_{ij} = \hat{\mathbf{n}}_i \cdot \mathbf{A}\hat{\mathbf{n}}_j = \hat{\mathbf{n}}_i \cdot \lambda_j \hat{\mathbf{n}}_j = \lambda_j \delta_{ij} \quad , \qquad (j = 1, 2, 3; \text{ no summation}) \quad , \tag{1.176}$$

which produces a **diagonal matrix** $[\mathbf{A}]$ in the form

$$[\mathbf{A}] = \begin{bmatrix} \lambda_1 & 0 & 0 \\ 0 & \lambda_2 & 0 \\ 0 & 0 & \lambda_3 \end{bmatrix} \quad , \tag{1.177}$$

where the diagonal elements are the eigenvalues of $\mathbf{A}$. This result may be obtained directly from the spectral decomposition (1.175) of $\mathbf{A}$.

For a pair of equal roots, i.e. $\lambda_1 = \lambda_2 \neq \lambda_3$ ($\lambda_1 = \lambda_2 = \lambda$), with one unique eigenvector $\hat{\mathbf{n}}_3$ associated with λ_3, we deduce from (1.175) that

$$\mathbf{A} = \lambda_3 \underbrace{\hat{\mathbf{n}}_3 \otimes \hat{\mathbf{n}}_3}_{\mathbf{P}_{\hat{\mathbf{n}}_3}^{\|}} + \lambda \underbrace{(\mathbf{I} - \hat{\mathbf{n}}_3 \otimes \hat{\mathbf{n}}_3)}_{\mathbf{P}_{\hat{\mathbf{n}}_3}^{\perp}} = \lambda_3 \mathbf{P}_{\hat{\mathbf{n}}_3}^{\|} + \lambda \mathbf{P}_{\hat{\mathbf{n}}_3}^{\perp} \quad , \tag{1.178}$$

where $\mathbf{P}_{\hat{\mathbf{n}}_3}^{\|}$, $\mathbf{P}_{\hat{\mathbf{n}}_3}^{\perp}$ denote projection tensors introduced in (1.125) and (1.126). In (1.178) the eigenvector $\hat{\mathbf{n}}_3$ is perpendicular to a plane spanned by the mutually orthogonal eigenvectors $\hat{\mathbf{n}}_1$ and $\hat{\mathbf{n}}_2$.

The third case to distinguish, namely for equal roots ($\lambda_1 = \lambda_2 = \lambda_3 = \lambda$), is

$$\mathbf{A} = \lambda \mathbf{I} \quad , \tag{1.179}$$

where every direction is a principal direction and every set of mutually orthogonal basis denotes principal axes.

EXERCISES

1. Show that the quantities in eqs. (1.170)–(1.172), i.e. I_1, I_2, I_3, are indeed principal invariants of a tensor $\mathbf{A}$ under an orthogonal transformation of coordinates.

2. Let a tensor $\mathbf{A}$ be given by

$$\mathbf{A} = \alpha(\mathbf{I} - \mathbf{e}_1 \otimes \mathbf{e}_1) + \beta(\mathbf{e}_1 \otimes \mathbf{e}_2 + \mathbf{e}_2 \otimes \mathbf{e}_1) \quad ,$$

where α, β are scalars and $\mathbf{e}_1, \mathbf{e}_2$ are orthogonal unit vectors.

  (a) Show that the eigenvalues λ_i, $i = 1, 2, 3$, of $\mathbf{A}$ are

$$\lambda_1 = \alpha \quad , \qquad \lambda_2 = \frac{\alpha}{2} + \left(\frac{\alpha^2}{4} + \beta^2\right)^{1/2} , \qquad \lambda_3 = \frac{\alpha}{2} - \left(\frac{\alpha^2}{4} + \beta^2\right)^{1/2} .$$

  (b) Derive the associated normalized eigenvectors $\hat{\mathbf{n}}_i$, $i = 1, 2, 3$, which in matrix representation read

$$[\hat{\mathbf{n}}_1] = \begin{bmatrix} 0 \\ 0 \\ 1 \end{bmatrix} \quad , \qquad [\hat{\mathbf{n}}_2] = \frac{1}{\sqrt{1 + \frac{\lambda_2^2}{\beta^2}}} \begin{bmatrix} 1 \\ \lambda_2/\beta \\ 0 \end{bmatrix} \quad ,$$

$$[\hat{\mathbf{n}}_3] = \frac{1}{\sqrt{1 + \frac{\lambda_3^2}{\beta^2}}} \begin{bmatrix} 1 \\ \lambda_3/\beta \\ 0 \end{bmatrix} \quad .$$

3. Let a non-symmetric tensor $\mathbf{A}$ be given by the tensor product $\mathbf{e}_1 \otimes \mathbf{e}_2$, with the property $\mathbf{e}_1 \cdot \mathbf{e}_2 = 0$. Determine the eigenvalues and eigenvectors of tensor $\mathbf{A}$.

4. Let $\mathbf{A}$ be a given second-order tensor. Show that the invariants of its deviatoric part, according to (1.163), are

$$I_{\mathrm{devA}} = 0 \quad , \qquad II_{\mathrm{devA}} = -\frac{1}{2}\mathrm{tr}(\mathrm{devA})^2 \quad , \tag{1.180}$$

$$III_{\mathrm{devA}} = \det(\mathrm{devA}) = \frac{1}{3}\mathrm{tr}(\mathrm{devA})^3 \quad .$$

5. Use the relations (1.170)–(1.172), (1.180) and $(1.163)_1$ to show that II_{devA}, III_{devA} may be expressed in terms of $I_{\mathbf{A}}, II_{\mathbf{A}}$ and $III_{\mathbf{A}}$ as

$$II_{\mathrm{devA}} = II_{\mathbf{A}} - \frac{1}{3}I_{\mathbf{A}}^2 \quad ,$$

$$III_{\mathrm{devA}} = III_{\mathbf{A}} - \frac{1}{3}I_{\mathbf{A}}II_{\mathbf{A}} + \frac{2}{27}I_{\mathbf{A}}^3 \quad . \tag{1.181}$$

Hint: To derive eq. (1.181) apply the *Cayley-Hamilton* equation in the form of $(\mathrm{devA})^3 - I_{\mathrm{devA}}(\mathrm{devA})^2 + II_{\mathrm{devA}}\,\mathrm{devA} - III_{\mathrm{devA}}\mathbf{I} = \mathbf{O}$.

1.5   Transformation Laws for Basis Vectors and Components

Consider two sets of mutually orthogonal basis vectors which share a common origin. They correspond to a *'new'* and an *'old'* (original) Cartesian coordinate system which we assume to be right-handed characterized by the two sets $\{\tilde{\mathbf{e}}_i\}$ and $\{\mathbf{e}_i\}$ of basis vectors, respectively. Hence, the new coordinate system may be obtained from the original one by a **rotation** of the basis vectors $\mathbf{e}_i$ about their origin (see Figure 1.7). We find from (1.10) that the cosine of the angle between $\mathbf{e}_i$ and $\tilde{\mathbf{e}}_j$ is represented by $Q_{ij} = \cos\theta(\mathbf{e}_i, \tilde{\mathbf{e}}_j) = \mathbf{e}_i \cdot \tilde{\mathbf{e}}_j$. Note that the first index on Q_{ij} indicates the 'old' components whereas the second index holds for the 'new' components.

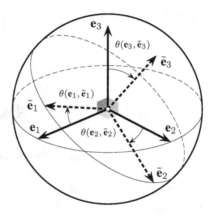

Figure 1.7 Rotation of mutually orthogonal basis vectors $\mathbf{e}_i$ into $\tilde{\mathbf{e}}_i$.

It is worthwhile to mention that vectors and tensors themselves remain *invariant* upon a change of basis – they are said to be independent of any coordinate system. However, their respective *components* do depend upon the coordinate system introduced, which is arbitrary. The components change their magnitudes by a *rotation* (and/or *reflection*) of the basis vectors, but are independent of any *translation* (for the characterization of a translation, see p. 62).

We now set up the **transformation laws** for various components of vectors and tensors under a change of basis. Following the considerations above we have

$$\tilde{\mathbf{e}}_i = \mathbf{Q}\mathbf{e}_i \qquad \text{and} \qquad \mathbf{e}_i = \mathbf{Q}^{\mathrm{T}}\tilde{\mathbf{e}}_i \ , \qquad i = 1, 2, 3 \ , \qquad (1.182)$$

where $\mathbf{Q}$ denotes an orthogonal tensor (see p. 16), with components Q_{ij} which are the same in either basis. The components describe the orientation of the two sets of basis vectors relative to each other. In particular, $\mathbf{Q}$ rotates the basis vectors $\mathbf{e}_i$ into $\tilde{\mathbf{e}}_i$, while $\mathbf{Q}^T$ rotates $\tilde{\mathbf{e}}_i$ back to $\mathbf{e}_i$. Using eqs. (1.59), (1.53)$_2$, (1.21) and rule (1.61), we find that

$$\mathbf{Q}\mathbf{e}_i = Q_{ji}\mathbf{e}_j \qquad \text{and} \qquad \mathbf{Q}^T\tilde{\mathbf{e}}_i = Q_{ij}\tilde{\mathbf{e}}_j \ , \qquad (1.183)$$

which is the analogue of eq. (1.63)$_4$.

By comparing the equations above we may extract the **orthogonality condition** of the cosines, characterized by $\mathbf{Q}^T\mathbf{Q} = \mathbf{Q}\mathbf{Q}^T = \mathbf{I}$. Equivalently, expressed in index or in matrix notation

$$Q_{ij}Q_{ik} = Q_{ji}Q_{ki} = \delta_{jk} \ , \qquad [\mathbf{Q}]^T[\mathbf{Q}] = [\mathbf{Q}][\mathbf{Q}]^T = [\mathbf{I}] \ , \qquad (1.184)$$

where $[\mathbf{Q}]$ contains the collection of the components Q_{ij}. It is an **orthogonal matrix** which is referred to as the **transformation matrix**. Note that $[\mathbf{Q}]^T = [\mathbf{Q}^T]$ (compare with Exercise 3 on p.19). In order to maintain the right-handedness of the basis vectors we have admitted only rotations of the basis vectors, consequently $\det[\mathbf{Q}] = +1$.

Vectorial transformation law.   We consider any vector $\mathbf{u}$ resolved along the two sets $\{\tilde{\mathbf{e}}_i\}$ and $\{\mathbf{e}_i\}$ of basis vectors, i.e.

$$\tilde{u}_i = \mathbf{u} \cdot \tilde{\mathbf{e}}_i \qquad \text{in} \qquad \{\tilde{\mathbf{e}}_i\} \ , \qquad (1.185)$$

$$u_i = \mathbf{u} \cdot \mathbf{e}_i \qquad \text{in} \qquad \{\mathbf{e}_i\} \ . \qquad (1.186)$$

We assume that the relation between the basis vectors $\tilde{\mathbf{e}}_i$ and $\mathbf{e}_j$ is known by the cosines Q_{ij}. Combining (1.185) with (1.182)$_1$ and using (1.183)$_1$ and (1.186), we obtain the **vectorial transformation law** for the Cartesian components of the vector $\mathbf{u}$, i.e.

$$\tilde{u}_i = \mathbf{u} \cdot \tilde{\mathbf{e}}_i = Q_{ji}(\mathbf{u} \cdot \mathbf{e}_j) = Q_{ji}u_j \qquad \text{or} \qquad [\tilde{\mathbf{u}}] = [\mathbf{Q}]^T[\mathbf{u}] \qquad (1.187)$$

(given here in both index and matrix notation). In an analogous manner, we find that

$$u_i = Q_{ij}\tilde{u}_j \qquad \text{or} \qquad [\mathbf{u}] = [\mathbf{Q}][\tilde{\mathbf{u}}] \ . \qquad (1.188)$$

These equations determine the relationship between the components of a vector associated with the (old) basis $\{\mathbf{e}_i\}$ and the components of the *same* vector associated with another (new) basis $\{\tilde{\mathbf{e}}_i\}$.

It is important to emphasize that the equations $[\tilde{\mathbf{u}}] = [\mathbf{Q}]^T[\mathbf{u}]$ and $[\mathbf{u}] = [\mathbf{Q}][\tilde{\mathbf{u}}]$ are *not* identical to $\tilde{\mathbf{u}} = \mathbf{Q}^T\mathbf{u}$ and $\mathbf{u} = \mathbf{Q}\tilde{\mathbf{u}}$, respectively. In (1.187)$_2$ and (1.188)$_2$, $[\tilde{\mathbf{u}}]$ and $[\mathbf{u}]$ are column matrices characterizing components of the *same* vector in two different coordinate systems, whereas $\tilde{\mathbf{u}}$ and $\mathbf{u}$ are *different* vectors.

Tensorial transformation law.   To determine the transformation laws for the Cartesian components of any second-order tensor $\mathbf{A}$, we describe its components along the two sets $\{\tilde{\mathbf{e}}_i\}$ and $\{\mathbf{e}_i\}$ of basis vectors, i.e.

$$\tilde{A}_{ij} = \tilde{\mathbf{e}}_i \cdot \mathbf{A}\tilde{\mathbf{e}}_j \qquad \text{in} \qquad \{\tilde{\mathbf{e}}_i\} \quad, \tag{1.189}$$

$$A_{ij} = \mathbf{e}_i \cdot \mathbf{A}\mathbf{e}_j \qquad \text{in} \qquad \{\mathbf{e}_i\} \quad. \tag{1.190}$$

Combining (1.189) with (1.182)$_1$ and using (1.183)$_1$ and (1.190), then the (rectangular Cartesian) components $\tilde{A}_{ij}$ and A_{ij} are related via the so-called **tensorial transformation law**

$$\begin{aligned} \tilde{A}_{ij} = \tilde{\mathbf{e}}_i \cdot \mathbf{A}\tilde{\mathbf{e}}_j &= (Q_{ki}\mathbf{e}_k) \cdot \mathbf{A}(Q_{mj}\mathbf{e}_m) \\ &= Q_{ki}Q_{mj}(\mathbf{e}_k \cdot \mathbf{A}\mathbf{e}_m) \\ &= Q_{ki}Q_{mj}A_{km} \qquad \text{or} \qquad [\tilde{\mathbf{A}}] = [\mathbf{Q}]^{\mathrm{T}}[\mathbf{A}][\mathbf{Q}] \quad. \end{aligned} \tag{1.191}$$

Transformation $[\tilde{\mathbf{A}}] = [\mathbf{Q}]^{\mathrm{T}}[\mathbf{A}][\mathbf{Q}]$ relates the *different* matrices $[\tilde{\mathbf{A}}]$ and $[\mathbf{A}]$, which have the components of the *same* tensor $\mathbf{A}$. In an analogous manner, we find that

$$A_{ij} = Q_{ik}Q_{jm}\tilde{A}_{km} \qquad \text{or} \qquad [\mathbf{A}] = [\mathbf{Q}][\tilde{\mathbf{A}}][\mathbf{Q}]^{\mathrm{T}} \quad. \tag{1.192}$$

In addition to previous considerations, the equations $[\tilde{\mathbf{A}}] = [\mathbf{Q}]^{\mathrm{T}}[\mathbf{A}][\mathbf{Q}]$ and $[\mathbf{A}] = [\mathbf{Q}][\tilde{\mathbf{A}}][\mathbf{Q}]^{\mathrm{T}}$ differ from the tensor equations $\tilde{\mathbf{A}} = \mathbf{Q}^{\mathrm{T}}\mathbf{A}\mathbf{Q}$ and $\mathbf{A} = \mathbf{Q}\tilde{\mathbf{A}}\mathbf{Q}^{\mathrm{T}}$, relating two *different* tensors, namely $\tilde{\mathbf{A}}$ and $\mathbf{A}$.

Finally, the 3^n components $A_{j_1 j_2 \dots j_n}$ of a tensor of order n (with n indices $j_1, j_2, \dots, j_n$) transform as

$$\tilde{A}_{i_1 i_2 \dots i_n} = Q_{j_1 i_1} Q_{j_2 i_2} \cdots Q_{j_n i_n} A_{j_1 j_2 \dots j_n} \quad. \tag{1.193}$$

This tensorial transformation law relates the different components $\tilde{A}_{i_1 i_2 \dots i_n}$ (along the directions $\tilde{\mathbf{e}}_1, \tilde{\mathbf{e}}_2, \tilde{\mathbf{e}}_3$) and $A_{j_1 j_2 \dots j_n}$ (along the directions $\mathbf{e}_1, \mathbf{e}_2, \mathbf{e}_3$) of the same tensor of order n.

Isotropic tensors.   A tensor $\mathbf{A}$ is said to be **isotropic** if its components are the same under arbitrary rotations of the basis vectors. The requirement is deduced from eq. (1.191) as

$$A_{ij} = Q_{ki}Q_{mj}A_{km} \qquad \text{or} \qquad [\mathbf{A}] = [\mathbf{Q}]^{\mathrm{T}}[\mathbf{A}][\mathbf{Q}] \quad. \tag{1.194}$$

Note that all scalars as well as zero tensors and unit tensors of all orders are *isotropic* tensors.

EXAMPLE 1.6   Show that the unit tensor, as given in (1.65), is an isotropic tensor of order two.

Solution.   In accord with manipulation (1.191)$_4$ we can show that the components of the unit tensor $\mathbf{I}$ are δ_{ij} in any coordinate system. Namely, using (1.61) and (1.184) we obtain

$$\tilde{\delta}_{ij} = \tilde{\mathbf{e}}_i \cdot \mathbf{I}\tilde{\mathbf{e}}_j = Q_{ki}Q_{mj}\delta_{km} = Q_{ki}Q_{kj} = \delta_{ij} \ , \tag{1.195}$$

which means that the components δ_{ij} do not change upon a *rotation* or *reflection* of axes. ∎

By multiplying both sides of $\tilde{\delta}_{ij} = \delta_{ij}$ with a scalar we can see that every scalar multiple of $\mathbf{I}$ is an isotropic tensor. Consequently, any spherical tensor in the form of

$$\mathbf{A} = \alpha\mathbf{I} \qquad \text{or} \qquad (\mathbf{A})_{ij} = \alpha\delta_{ij} = A_{ij} \tag{1.196}$$

is an isotropic tensor, where α denotes a scalar.

The permutation tensor (1.143) is an isotropic tensor of order three with respect to rotation of coordinate axes (proper orthogonal transformations). With respect to reflection it emerges that the permutation tensor is *not* isotropic.

It can be shown that the fourth-order unit tensors $\mathbb{I}$, $\bar{\mathbb{I}}$ and the tensor $\mathbf{I} \otimes \mathbf{I}$ with components $(\mathbf{I} \otimes \mathbf{I})_{ijkl} = \delta_{ij}\delta_{kl}$ are also isotropic tensors. The most general isotropic tensor of order four is of the form

$$\alpha(\mathbf{I} \otimes \mathbf{I}) + \beta\mathbb{I} + \gamma\bar{\mathbb{I}} \ , \tag{1.197}$$

where α, β, γ are scalars. The components are expressed according to definitions (1.161), (1.162) as $(\alpha(\mathbf{I} \otimes \mathbf{I}) + \beta\mathbb{I} + \gamma\bar{\mathbb{I}})_{ijkl} = \alpha\delta_{ij}\delta_{kl} + \beta\delta_{ik}\delta_{jl} + \gamma\delta_{il}\delta_{jk}$.

EXERCISES

1. Let a new right-handed Cartesian coordinate system be represented by the set $\{\tilde{\mathbf{e}}_i\}$ of basis vectors with transformation law $\tilde{\mathbf{e}}_2 = -\sin\theta\mathbf{e}_1 + \cos\theta\mathbf{e}_2$, $\tilde{\mathbf{e}}_3 = \mathbf{e}_3$. The origin of the new coordinate system coincides with the old origin.

    (a) Find $\tilde{\mathbf{e}}_1$ in terms of the old set $\{\mathbf{e}_i\}$ of basis vectors.

    (b) Find the orthogonal matrix $[\mathbf{Q}]$ and express the new coordinates in terms of the old one.

    (c) Express the vector $\mathbf{u} = -6\mathbf{e}_1 - 3\mathbf{e}_2 + \mathbf{e}_3$ in terms of the new set $\{\tilde{\mathbf{e}}_i\}$ of basis vectors.

2. Considering eqs. (1.189) and (1.190), show that $A_{ii} = \tilde{A}_{ii}$.

3. Assume that $\mathbf{A}$ is an isotropic tensor of order two, for example, $\alpha\mathbf{I}$. Show that the deviatoric part of $\mathbf{A}$ is always the zero tensor of order two, i.e.

$$\mathrm{dev}(\alpha\mathbf{I}) = \mathbf{O} \quad . \tag{1.198}$$

(In general, equations within exercises are not numbered. However, particular results may be given equation numbers, for later reference.)

1.6 General Bases

There are more general ways to introduce the notion of a vector and a tensor. A more general theory assumes the existence of a vector space and its so-called *dual space*. Thereby, a vector and a tensor of order n are introduced through a *linear* and a *multilinear function*, respectively. Within this broader framework we must distinguish between the scalar product (which is only defined between elements of a vector space and between elements of its dual) and the inner product. For a detailed explanation of the underlying theory the interested reader is referred to, for example, BOWEN and WANG [1976b] and OGDEN [1997, Section 1.4.3].

However, for the purpose we pursue in this text it is sufficiently general to work with a theory which assumes the existence of *Euclidean vector spaces* and which identifies one vector space with its dual so that the scalar product is identified with the inner product.

Thus far we have worked with Cartesian tensors which are resolved along a set of orthonormal basis vectors. However, tensor quantities can be projected onto more general bases. The aim of this section is to present the fundamental rules and standard results for these projections onto the basis of the aforementioned identification.

General basis vectors. We now resolve vectors and tensors along a fixed set $\{\mathbf{g}_i\}_{i\in\{1,2,3\}}$ of basis vectors. The elements $\mathbf{g}_1, \mathbf{g}_2, \mathbf{g}_3$ of this basis are called **covariant basis vectors**. They are assumed to be non-zero and non-parallel but they need not necessarily be orthogonal to each other or of unit length. The general basis vectors are linearly independent, requiring $(\mathbf{g}_1 \times \mathbf{g}_2) \cdot \mathbf{g}_3 \neq 0$ (the vectors do not lie in a plane).

We now introduce the so-called **reciprocal** (or **dual**) **basis** of $\{\mathbf{g}_i\}$, i.e. the fixed set $\{\mathbf{g}^i\}_{i\in\{1,2,3\}}$ (note that for a more general theory which assumes the existence of a vector space and its dual the reciprocal and dual bases are in different spaces). The elements $\mathbf{g}^1, \mathbf{g}^2, \mathbf{g}^3$ of the reciprocal basis are called **contravariant basis vectors** or **reciprocal basis vectors** and have *super*scripts rather than *sub*scripts. The two bases $\{\mathbf{g}_i\}$ and $\{\mathbf{g}^i\}$ satisfy the condition

$$\mathbf{g}^i \cdot \mathbf{g}_j = \delta^i_j \quad , \tag{1.199}$$

where δ^i_j is the (mixed) **Kronecker delta**. As in definition (1.21) $\delta^i_j = 1$ for $i = j$ and $\delta^i_j = 0$ for $i \neq j$.

Basically, eq. (1.199) means that each vector of a basis is orthogonal to the two vectors of the other basis whose indices are different. For example, the contravariant base vector $\mathbf{g}^1$ is orthogonal to the two covariant basis vectors $\mathbf{g}_2$ and $\mathbf{g}_3$, as illustrated in Figure 1.8.

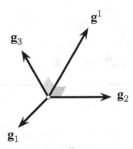

Figure 1.8   General basis $\{\mathbf{g}_i\}$, $\mathbf{g}^1$ is orthogonal to $\mathbf{g}_2$ and $\mathbf{g}_3$.

For subsequent use we introduce the useful definitions

$$\mathbf{g}^i \cdot \mathbf{g}^j = g^{ij} \quad , \qquad \mathbf{g}_i \cdot \mathbf{g}_j = g_{ij} \qquad (1.200)$$

of the quantities g^{ij} and g_{ij}. Since, from eq. (1.11), the dot product is commutative we may write $g^{ij} = g^{ji}$ and $g_{ij} = g_{ji}$.

The quantities g^{ii} and g_{ii} with the same indices represent the squares of the lengths of the associated basis vectors; for example, $g_{11} = |\mathbf{g}_1|^2$. The quantities g^{ij} and g_{ij} with different indices represent the product of the lengths of the associated basis vectors and the cosine of the angle θ between the basis vectors; for example, $g_{12} = |\mathbf{g}_1||\mathbf{g}_2|\cos\theta(\mathbf{g}_1, \mathbf{g}_2)$. Thus,

$$|\mathbf{g}_i| = (g_{ii})^{1/2} \quad , \qquad \cos\theta(\mathbf{g}_i, \mathbf{g}_j) = \frac{g_{ij}}{(g_{ii}g_{jj})^{1/2}} \qquad (1.201)$$

$(i, j = 1, 2, 3;$ no summation) determine the geometrical characteristics (the so-called **metric**) of a given basis. Hence, g^{ij} and g_{ij} are customarily referred to as the **metric coefficients**.

EXAMPLE 1.7   Show that the properties

$$\mathbf{g}^i = g^{ij}\mathbf{g}_j \quad , \qquad \mathbf{g}_i = g_{ij}\mathbf{g}^j \qquad (1.202)$$

hold.

Solution.   We tentatively write

$$\mathbf{g}^i = a^{ik}\mathbf{g}_k \quad , \tag{1.203}$$

where a^{ik} are unknown quantities. Taking the dot product of (1.203) with $\mathbf{g}^j$ and using relation (1.199), we obtain

$$\mathbf{g}^i \cdot \mathbf{g}^j = a^{ik}\mathbf{g}_k \cdot \mathbf{g}^j = a^{ik}\delta_k^j \quad . \tag{1.204}$$

This means that a^{ik} is only multiplied with the factor 1 if $j = k$ otherwise it is multiplied with zero. Hence, we may write $\mathbf{g}^i \cdot \mathbf{g}^j = a^{ij}$, which is the analogue of (1.200)$_1$. The proof is similar for property (1.202)$_2$.   ■

Taking the dot product of $\mathbf{g}^i$ with $\mathbf{g}_k$ and using (1.202) we obtain $\mathbf{g}^i \cdot \mathbf{g}_k = g^{ij}g_{kl}\mathbf{g}_j \cdot \mathbf{g}^l$. Finally, using (1.199) and replacing the subscript l on g_{kl} by a j, we arrive at the reciprocal property

$$g^{ij}g_{kj} = \delta_k^i \quad , \tag{1.205}$$

which means that the matrix with components g_{ij}, i.e. $[g_{ij}]$, is the inverse of the matrix with components g^{ij}, i.e. $[g^{ij}]$. Hence, we write $[g_{ij}] = [g^{ij}]^{-1}$, or explicitly

$$\begin{bmatrix} g_{11} & g_{12} & g_{13} \\ g_{21} & g_{22} & g_{23} \\ g_{31} & g_{32} & g_{33} \end{bmatrix} = \begin{bmatrix} g^{11} & g^{12} & g^{13} \\ g^{21} & g^{22} & g^{23} \\ g^{31} & g^{32} & g^{33} \end{bmatrix}^{-1} \quad . \tag{1.206}$$

The distinction between *sub*script and *super*script is essential for arbitrary bases. It is important to note that a *free* (or *live*) index, which is not summed over, such as i and j in eq. (1.199), must be either a subscript or a superscript on both sides of an equation. In the following we will sometimes use the term 'index' to denote either a subscript or a superscript. Only if the basis is orthonormal may we write $\mathbf{g}^i = \mathbf{g}_i$ and then no distinction between upper and lower indices is necessary. For this case we write all indices as subscripts, as in previous sections where we used a Cartesian basis.

Covariant and contravariant components of a vector.   Any vector $\mathbf{u}$ may be represented uniquely by a linear combination of the contravariant or covariant basis vectors. We write

$$\mathbf{u} = u_i\mathbf{g}^i = u^i\mathbf{g}_i \quad , \tag{1.207}$$

where u_i are the **covariant components** of $\mathbf{u}$ (with respect to the general basis $\{\mathbf{g}^i\}$), while u^i are the **contravariant components** of $\mathbf{u}$ (with respect to the general basis

$\{\mathbf{g}_i\}$). It is only in this theory, in which the reciprocal basis is identified with the dual basis, that the vector $\mathbf{u}$ has both covariant components u_i and contravariant components u^i. We shall call a vector with covariant components a **covariant vector** and a vector with contravariant components a **contravariant vector**.

By analogy with eq. (1.22)$_2$, the Kronecker delta δ_j^i has the useful (replacement) properties

$$\delta_j^i u_i = u_j \quad , \qquad \delta_i^j u^i = u^j \quad , \tag{1.208}$$

in which the index i on u is replaced by j. Note that summation can only take place over a *dummy* (or *summation*) index if one index is a subscript and the other is a superscript, such as index i in eq. (1.207). Hence, the j-th covariant and contravariant components of any vector $\mathbf{u}$ are the projections of $\mathbf{u}$ onto the bases $\{\mathbf{g}_j\}$ and $\{\mathbf{g}^j\}$, respectively. Taking the dot product of (1.207)$_1$ with $\mathbf{g}_j$ (and the dot product of (1.207)$_2$ with $\mathbf{g}^j$), we obtain, by means of properties (1.13), (1.199), (1.208)$_1$ (and (1.208)$_2$), the useful relations

$$\mathbf{u} \cdot \mathbf{g}_j = u_i \mathbf{g}^i \cdot \mathbf{g}_j = u_i \delta_j^i = u_j \quad , \tag{1.209}$$

$$\mathbf{u} \cdot \mathbf{g}^j = u^i \mathbf{g}_i \cdot \mathbf{g}^j = u^i \delta_i^j = u^j \quad , \tag{1.210}$$

which should be compared with eq. (1.23) for the case of a Cartesian basis.

The contravariant components of any vector $\mathbf{u}$ may also be expressed in terms of its covariant components, and vice versa. Thus, taking the dot product of $\mathbf{u} = u_j \mathbf{g}^j$ and $\mathbf{u} = u^j \mathbf{g}_j$ with $\mathbf{g}^i$ and $\mathbf{g}_i$, respectively, on both sides gives with definitions (1.200) and eqs. (1.210)$_3$ and (1.209)$_3$, i.e.

$$u^i = g^{ij} u_j \quad , \qquad u_i = g_{ij} u^j \quad , \tag{1.211}$$

which are properties analogous to those given in (1.202). Note that the new vector components formed depend on the original one. As indicated in eqs. (1.211), the set of six quantities g^{ij} (or g_{ij}) can be used as a useful operator to *raise* or *lower* an index on vector components. This process of raising or lowering indices can similarly be applied to tensor components (see Exercise 5 on p. 39).

Consider the two vectors $\mathbf{u}$ and $\mathbf{v}$ which may be represented by (1.207). Their dot product reads

$$\mathbf{u} \cdot \mathbf{v} = g^{ij} u_i v_j = g_{ij} u^i v^j = u^i v_i = u_i v^i \quad , \tag{1.212}$$

and the length $|\mathbf{u}|$ of the vector $\mathbf{u}$ is

$$|\mathbf{u}| = (g^{ij} u_i u_j)^{1/2} = (g_{ij} u^i u^j)^{1/2} = (u^i u_i)^{1/2} \quad , \tag{1.213}$$

since $|\mathbf{u}| = (\mathbf{u} \cdot \mathbf{u})^{1/2}$.

Covariant, contravariant and mixed components of a tensor. Similarly to the above, any second-order tensor $\mathbf{A}$ may be represented with respect to a basis consisting of tensor products of the $\mathbf{g}_i$ and the $\mathbf{g}^i$. Thus,

$$\mathbf{A} = A^{ij}\mathbf{g}_i \otimes \mathbf{g}_j = A_{ij}\mathbf{g}^i \otimes \mathbf{g}^j = A^i_{\cdot j}\mathbf{g}_i \otimes \mathbf{g}^j = A^{\cdot j}_i\mathbf{g}^i \otimes \mathbf{g}_j \ , \tag{1.214}$$

where A^{ij} and A_{ij} are the **contravariant** and **covariant components**, whereas $A^i_{\cdot j}$ and $A^{\cdot j}_i$ are the **mixed components** of the tensor $\mathbf{A}$. The dot in front of each of the indices of the mixed components indicates the order of occurrence of these indices. Thus, in $A^i_{\cdot j}$ the first index is 'contravariant' and the second 'covariant'. In the following we shall call a tensor with covariant (respectively contravariant, mixed) components a **covariant** (respectively **contravariant, mixed**) **tensor** (the connections between these terminologies are discussed in more detail by OGDEN [1997, Section 1.4.3]).

Note that the mixed components $A^i_{\cdot j}$ and $A^{\cdot i}_j$ of a tensor $\mathbf{A}$ are equal *if and only if* the tensor $\mathbf{A}$ is symmetric ($A^{ij} = A^{ji}$ and $A_{ij} = A_{ji}$). Instead of 3^2 components for a Cartesian second-order tensor we have now four sets of components of a general tensor of order two, i.e. 6^2 different contravariant, covariant and mixed components, in general. However, no more than 3^2 components are independent.

Moreover, equations such as

$$A_{ij}\delta^j_k = A_{ik} \ , \qquad A^{ij}\delta^k_j = A^{ik} \tag{1.215}$$

are analogous to (1.61).

Now, with rule (1.53) and properties (1.199) and (1.215), the projection of $\mathbf{A}$ onto the general basis $\{\mathbf{g}_i\}$ and its reciprocal $\{\mathbf{g}^i\}$ gives

$$A^{ij} = \mathbf{g}^i \cdot \mathbf{A}\mathbf{g}^j \ , \qquad A_{ij} = \mathbf{g}_i \cdot \mathbf{A}\mathbf{g}_j \ , \tag{1.216}$$

$$A^i_{\cdot j} = \mathbf{g}^i \cdot \mathbf{A}\mathbf{g}_j \ , \qquad A^{\cdot i}_j = \mathbf{g}_j \cdot \mathbf{A}\mathbf{g}^i \ . \tag{1.217}$$

The four sets of components follow in an analogous manner to that given in Example 1.2 on p. 11.

The tensor relations for Cartesian components introduced in previous sections may be generalized to arbitrary components by basically replacing the Kronecker delta with the metric coefficients and using the different types of indices in an appropriate way. If tensors $\mathbf{A}$ and $\mathbf{B}$ are referred to a general basis, then the properties

$$(\mathbf{A}\mathbf{B})_{ij} = A_{ik}B^k_{\cdot j} = A^{\cdot k}_i B_{kj} \ , \tag{1.218}$$

$$\mathbf{A} : \mathbf{B} = A_{ij}B^{ij} = A^{ij}B_{ij} = A^i_{\cdot j}B^{\cdot j}_i \ , \tag{1.219}$$

$$|\mathbf{A}| = (A_{ij}A^{ij})^{1/2} \ , \tag{1.220}$$

$$\mathrm{tr}\mathbf{A} = A^i_{\cdot i} = A^{\cdot i}_i = g^{ij}A_{ij} = g_{ij}A^{ij} \ , \tag{1.221}$$

$$\det\mathbf{A} = \det\begin{bmatrix} A^1_{\cdot 1} & A^1_{\cdot 2} & A^1_{\cdot 3} \\ A^2_{\cdot 1} & A^2_{\cdot 2} & A^2_{\cdot 3} \\ A^3_{\cdot 1} & A^3_{\cdot 2} & A^3_{\cdot 3} \end{bmatrix} = \det\begin{bmatrix} A^{\cdot 1}_1 & A^{\cdot 1}_2 & A^{\cdot 1}_3 \\ A^{\cdot 2}_1 & A^{\cdot 2}_2 & A^{\cdot 2}_3 \\ A^{\cdot 3}_1 & A^{\cdot 3}_2 & A^{\cdot 3}_3 \end{bmatrix} \tag{1.222}$$

hold.

EXAMPLE 1.8 Using the fact that $\mathbf{Iu} = \mathbf{uI} = \mathbf{u}$ holds for all vectors $\mathbf{u}$, show that the four sets of quantities

$$g^{ij} = (\mathbf{I})^{ij} \quad , \qquad g_{ij} = (\mathbf{I})_{ij} \quad , \qquad \delta^i_j = (\mathbf{I})^i_{.j} \quad , \qquad \delta^i_j = (\mathbf{I})^{.i}_j \quad , \qquad (1.223)$$

as defined in eqs. (1.200) and (1.199), are actually the components of the second-order unit tensor $\mathbf{I}$. Verify the representation

$$\mathbf{I} = g^{ij}\mathbf{g}_i \otimes \mathbf{g}_j = g_{ij}\mathbf{g}^i \otimes \mathbf{g}^j = \mathbf{g}_j \otimes \mathbf{g}^j = \mathbf{g}^j \otimes \mathbf{g}_j \quad , \qquad (1.224)$$

with the properties $g^{ij} = g^{ji}$ and $g_{ij} = g_{ji}$ of the metric coefficients.

Solution. Recall the form $(1.207)_2$ and eq. $(1.211)_1$ to obtain $\mathbf{u} = u^i\mathbf{g}_i = g^{ij}u_j\mathbf{g}_i$. Hence, use relation $(1.209)_3$ for the components u_j and property (1.53) to obtain

$$\mathbf{u} = g^{ij}(\mathbf{u} \cdot \mathbf{g}_j)\mathbf{g}_i = \underbrace{g^{ij}(\mathbf{g}_i \otimes \mathbf{g}_j)}_{\mathbf{I}}\mathbf{u} \quad , \qquad (1.225)$$

where $\mathbf{I} = g^{ij}\mathbf{g}_i \otimes \mathbf{g}_j$ is the unit tensor in the representation $(1.224)_1$ and $g^{ij} = (\mathbf{I})^{ij}$ are identified as the six *contravariant components* of $\mathbf{I}$ with respect to the basis $\{\mathbf{g}_i\}$.

In an analogous manner, starting from $\mathbf{u} = u_i\mathbf{g}^i$, we may obtain representation $(1.224)_2$ and the six *covariant components* $g_{ij} = (\mathbf{I})_{ij}$ of $\mathbf{I}$ with respect to the reciprocal basis $\{\mathbf{g}^i\}$. Since the metric coefficients g_{ij} and g^{ij} are components of $\mathbf{I}$, we also call $\mathbf{I}$ the **Euclidean metric tensor**.

In order to show the mixed component form we apply $u^i = \delta^i_j u^j$ to $\mathbf{u} = u^i\mathbf{g}_i$. Then, with eq.$(1.210)_3$ and rule (1.53), we obtain

$$\mathbf{u} = \delta^i_j(\mathbf{u} \cdot \mathbf{g}^j)\mathbf{g}_i = \delta^i_j\underbrace{(\mathbf{g}_i \otimes \mathbf{g}^j)}_{\mathbf{I}}\mathbf{u} = (\mathbf{g}_j \otimes \mathbf{g}^j)\mathbf{u} \quad , \qquad (1.226)$$

where the Kronecker delta $\delta^i_j = (\mathbf{I})^i_{.j}$ represents the *mixed components* of the metric tensor $\mathbf{I}$. Since $\mathbf{I}$ is a symmetric tensor we conclude that $(\mathbf{I})^i_{.j} = (\mathbf{I})^{.i}_j$. ∎

A generalization to higher-order tensors is a straightforward task. A general tensor of order n, having n indices $i_1, i_2, \ldots, i_n$, may be represented in the form

$$A^{i_1 i_2 \ldots i_n}\mathbf{g}_{i_1} \otimes \mathbf{g}_{i_2} \otimes \cdots \otimes \mathbf{g}_{i_n} \quad , \qquad A_{i_1 i_2 \ldots i_n}\mathbf{g}^{i_1} \otimes \mathbf{g}^{i_2} \otimes \cdots \otimes \mathbf{g}^{i_n} \quad , \qquad (1.227)$$

where only 3^n contravariant components $A^{i_1 i_2 \ldots i_n}$ or 3^n covariant components $A_{i_1 i_2 \ldots i_n}$ are involved. Now, a general tensor of order n may have mixed components of various kinds according to, for example,

$$A^{i_1 i_2 \ldots i_j}{}_{i_{j+1} \ldots i_{k-1}}{}^{i_k \ldots i_n}\mathbf{g}_{i_1} \otimes \cdots \otimes \mathbf{g}_{i_j} \otimes \mathbf{g}^{i_{j+1}} \otimes \cdots \otimes \mathbf{g}^{i_{k-1}} \otimes \mathbf{g}_{i_k} \otimes \cdots \otimes \mathbf{g}_{i_n} \quad . \qquad (1.228)$$

In a fixed set of orthonormal basis vectors the components given in (1.227) and (1.228) coincide (compare with the form (1.135)).

EXERCISES

1. Let the covariant basis vectors $\mathbf{g}_1$, $\mathbf{g}_2$ and a vector $\mathbf{u}$ be given by

$$\mathbf{g}_1 = \mathbf{e}_1 \ , \qquad \mathbf{g}_2 = \frac{1}{\sqrt{2}}(\mathbf{e}_1 + \mathbf{e}_2) \ , \qquad \mathbf{u} = \mathbf{g}_1 + 2\mathbf{g}_2 \ ,$$

where $\{\mathbf{e}_\alpha\}$, $\alpha = 1, 2$ is a set of orthonormal basis vectors.

Compute the contravariant basis vectors and $\mathbf{u} = \alpha \mathbf{g}^1 + \beta \mathbf{g}^2$. Draw a picture and interpret the geometrical situation.

2. Consider the two sets of covariant basis vectors

$$[\mathbf{g}_1] = \begin{bmatrix} 1 \\ 2 \\ 3 \end{bmatrix} \ , \qquad [\mathbf{g}_2] = \begin{bmatrix} -1 \\ 0 \\ -1 \end{bmatrix} \ , \qquad [\mathbf{g}_3] = \begin{bmatrix} -1 \\ 4 \\ 3 \end{bmatrix} \ ,$$

$$[\mathbf{g}_1] = \begin{bmatrix} 0 \\ 1 \\ 3 \end{bmatrix} \ , \qquad [\mathbf{g}_2] = \begin{bmatrix} -1 \\ 2 \\ 4 \end{bmatrix} \ , \qquad [\mathbf{g}_3] = \begin{bmatrix} 0 \\ 1 \\ 1 \end{bmatrix} \ .$$

Find the set of basis vectors which form a possible general basis. Compute the covariant and contravariant components of a vector joining the origin to the point $(1, 2, 1)$.

3. Let $\mathbf{g}_1$ and $\mathbf{g}_2$ be covariant basis vectors which are given as

$$[\mathbf{g}_1] = \begin{bmatrix} 1 \\ 0 \end{bmatrix} \ , \qquad [\mathbf{g}_2] = \begin{bmatrix} 1 \\ 1 \end{bmatrix} \ .$$

Consider the two vectors $\mathbf{u} = \mathbf{g}_1 + 3\mathbf{g}_2$ and $\mathbf{v} = -\mathbf{g}_1 + 2\mathbf{g}_2$ resolved along this basis and compute the scalar product $\mathbf{u} \cdot \mathbf{v}$ and the lengths $|\mathbf{u}|$, $|\mathbf{v}|$ using

   (a) the contravariant components of $\mathbf{u}$, $\mathbf{v}$ and the metric coefficients, and

   (b) the contravariant and covariant components only.

4. Let $\mathbf{A}$ be a second-order tensor and consider the linear transformation $\mathbf{v} = \mathbf{Au}$ that assigns a vector $\mathbf{v}$ to each vector $\mathbf{u}$. By using eqs. (1.207) and (1.214), show that the transformation may be represented by

$$v^i = A^{ij} u_j = A^i_{\cdot j} u^j \ , \qquad v_i = A_{ij} u^j = A_i^{\cdot j} u_j \ .$$

5. The contravariant, covariant and mixed components of a second-order tensor **A**, as represented by (1.214), are related to each other.

(a) Take (1.214) and multiply both sides by the basis vectors $\mathbf{g}^k$ (and $\mathbf{g}_k$). Using the rule (1.53) and properties (1.205) and (1.215), show that the relations between the various components of **A** are given by

$$A_{ij} = g_{ik}g_{jl}A^{kl} = g_{jk}A_i^{\cdot k} = g_{ik}A_{\cdot j}^k , \qquad (1.229)$$
$$A^{ij} = g^{ik}g^{jl}A_{kl} = g^{ik}A_k^{\cdot j} = g^{jk}A_{\cdot k}^i ,$$
$$A_i^{\cdot j} = g^{jk}A_{ik} = g_{ik}A^{kj} ,$$
$$A_{\cdot j}^i = g^{ik}A_{kj} = g_{jk}A^{ik} .$$

Note that the effect of multiplying the tensor components A^{ik}, A_{ik}, $A_{\cdot k}^i$, $A_i^{\cdot k}$ by the set of quantities g^{jk} (or g_{jk}) and summing on the index k is to *raise* or *lower* an index on the associated tensor components. The operations of raising and lowering indices have the same meaning as in (1.211).

(b) Consider the second-order tensor

$$\mathbf{A} = \mathbf{e}_1 \otimes \mathbf{e}_1 + 2\mathbf{e}_1 \otimes \mathbf{e}_2 - \mathbf{e}_1 \otimes \mathbf{e}_3$$
$$-\mathbf{e}_2 \otimes \mathbf{e}_1 + \mathbf{e}_2 \otimes \mathbf{e}_2 - 2\mathbf{e}_2 \otimes \mathbf{e}_3$$
$$+2\mathbf{e}_3 \otimes \mathbf{e}_1 + \mathbf{e}_3 \otimes \mathbf{e}_2 + 3\mathbf{e}_3 \otimes \mathbf{e}_3 ,$$

where $\mathbf{e}_1, \mathbf{e}_2, \mathbf{e}_3$ form an orthonormal basis. Consider a (new) basis represented by the covariant basis vectors

$$\mathbf{g}_1 = \mathbf{e}_1 , \qquad \mathbf{g}_2 = \mathbf{e}_1 + \mathbf{e}_2 , \qquad \mathbf{g}_3 = \mathbf{e}_1 + \mathbf{e}_2 + \mathbf{e}_3 .$$

From eqs.(1.216) and (1.217) derive the contravariant, covariant and mixed components of **A** with respect to the general basis $\{\mathbf{g}_i\}$ and its reciprocal $\{\mathbf{g}^i\}$. Check the results by applying relation (1.229).

6. Let an orthogonal tensor **Q** be represented by

$$\mathbf{Q} = Q_{\cdot j}^i \mathbf{g}_i \otimes \mathbf{g}^j = Q_i^{\cdot j} \mathbf{g}^i \otimes \mathbf{g}_j .$$

(a) Use the identity (1.81) to verify that $\mathbf{Q}^T = Q_{\cdot j}^i \mathbf{g}^j \otimes \mathbf{g}_i = Q_i^{\cdot j} \mathbf{g}_j \otimes \mathbf{g}^i$ and show that the orthogonality condition may be written as

$$Q_j^{\cdot k} Q_{\cdot k}^i = \delta_j^i , \qquad g_{kl} Q_i^{\cdot k} Q_j^{\cdot l} = g_{ij} .$$

(b) Show that under the transformation $\tilde{\mathbf{g}}_i = \mathbf{Q}\mathbf{g}_i$, $\mathbf{g}_i = \mathbf{Q}^T\tilde{\mathbf{g}}_i$, $i = 1, 2, 3$, the components of the Euclidean metric tensor **I** remain unchanged.

1.7   Scalar, Vector, Tensor Functions

A **tensor function** is a function whose arguments are *one* or *more* tensor variables and whose values are *scalars, vectors* or *tensors*. The functions $\Phi(\mathbf{B})$, $\mathbf{u}(\mathbf{B})$, and $\mathbf{A}(\mathbf{B})$ are examples of so-called **scalar-valued, vector-valued** and **tensor-valued tensor functions** of *one* tensor variable $\mathbf{B}$, respectively. In an analogous manner, $\Phi(\mathbf{u})$, $\mathbf{v}(\mathbf{u})$, and $\mathbf{A}(\mathbf{u})$ are **vector functions** of *one* vector variable $\mathbf{u}$ with the value of a scalar, vector and tensor, respectively.

Next, we consider **scalar functions** of *one* scalar variable, such as time t, in more detail. Let $\Phi = \Phi(t), \mathbf{u} = \mathbf{u}(t) = u_i(t)\mathbf{e}_i$ and $\mathbf{A} = \mathbf{A}(t) = A_{ij}(t)\mathbf{e}_i \otimes \mathbf{e}_j$ be scalar-valued, vector-valued and tensor-valued scalar functions with a set $\{\mathbf{e}_i\}$ of basis vectors assumed to be **fixed**. The components $u_i(t)$ and $A_{ij}(t)$ are assumed to be real-valued smooth functions of t varying over a certain interval.

Hence, the **first derivative** of $\mathbf{u}$ and $\mathbf{A}$ with respect to t (rate of change) denoted by $\dot{\mathbf{u}} = \mathrm{d}\mathbf{u}/\mathrm{d}t$ and $\dot{\mathbf{A}} = \mathrm{d}\mathbf{A}/\mathrm{d}t$, is given by the *first derivative* of their associated components. Since $\mathrm{d}\mathbf{e}_i/\mathrm{d}t = \mathbf{o}$, we write

$$\dot{\mathbf{u}} = \dot{u}_i(t)\mathbf{e}_i \quad , \qquad \dot{\mathbf{A}} = \dot{A}_{ij}(t)\mathbf{e}_i \otimes \mathbf{e}_j \quad . \tag{1.230}$$

The first derivative of the scalar function Φ is simply $\dot{\Phi} = \mathrm{d}\Phi(t)/\mathrm{d}t$.

In general, the n-th derivative of $\mathbf{u}$ and $\mathbf{A}$ (for any desired n) denoted by $\mathrm{d}^n\mathbf{u}/\mathrm{d}t^n$ and $\mathrm{d}^n\mathbf{A}/\mathrm{d}t^n$, is a vector-valued and tensor-valued function whose components are $\mathrm{d}^n[u_i(t)]/\mathrm{d}t^n$ and $\mathrm{d}^n[A_{ij}(t)]/\mathrm{d}t^n$, respectively. The n-th derivative of Φ is simply $\mathrm{d}^n\Phi(t)/\mathrm{d}t^n$. Note that if Φ, $\mathbf{u}$, $\mathbf{A}$ are constant quantities, then their derivatives are equal to zero.

By applying the usual rules of differentiation, we obtain the identities:

$$\overline{\mathbf{u} \pm \mathbf{v}} = \dot{\mathbf{u}} \pm \dot{\mathbf{v}} \quad , \tag{1.231}$$

$$\overline{\Phi\mathbf{u}} = \dot{\Phi}\mathbf{u} + \Phi\dot{\mathbf{u}} \quad , \tag{1.232}$$

$$\overline{\mathbf{u} \otimes \mathbf{v}} = \dot{\mathbf{u}} \otimes \mathbf{v} + \mathbf{u} \otimes \dot{\mathbf{v}} \quad , \tag{1.233}$$

$$\overline{\mathbf{A} \pm \mathbf{B}} = \dot{\mathbf{A}} \pm \dot{\mathbf{B}} \quad , \tag{1.234}$$

$$\overline{\mathbf{A}^{\mathrm{T}}} = \dot{\mathbf{A}}^{\mathrm{T}} \quad , \tag{1.235}$$

$$\overline{\mathrm{tr}\mathbf{A}} = \mathrm{tr}\dot{\mathbf{A}} \quad . \tag{1.236}$$

Here and elsewhere the overbars cover quantities to which the dot operations are applied.

EXAMPLE 1.9   Show the identity

$$\overline{\mathbf{A}^{-1}} = -\mathbf{A}^{-1}\dot{\mathbf{A}}\mathbf{A}^{-1} \quad . \tag{1.237}$$

Solution.   Starting from $\mathbf{A}^{-1}\mathbf{A} = \mathbf{I}$ we find by means of the chain rule that $\overline{\mathbf{A}^{-1}}\mathbf{A} + \mathbf{A}^{-1}\dot{\mathbf{A}} = \mathbf{O}$. Finally, by multiplying the relation with $\mathbf{A}^{-1}$ from the right-hand side we obtain the desired result (1.237).   ■

Gradient of a scalar-valued (tensor) function.   We consider a nonlinear and smooth scalar-valued (tensor) function $\Phi(\mathbf{A})$ of one second-order tensor variable $\mathbf{A}$. Thus, for each $\mathbf{A}$, $\Phi(\mathbf{A})$ results in a scalar.

The aim is to approximate the nonlinear function Φ at $\mathbf{A}$ by a linear function. First-order (Taylor's) expansion for Φ at $\mathbf{A}$ yields, using property $(1.93)_1$,

$$\Phi(\mathbf{A} + \mathrm{d}\mathbf{A}) = \Phi(\mathbf{A}) + \mathrm{d}\Phi + o(\mathrm{d}\mathbf{A}) \quad , \tag{1.238}$$

$$\mathrm{d}\Phi = \frac{\partial\Phi(\mathbf{A})}{\partial\mathbf{A}} : \mathrm{d}\mathbf{A} = \mathrm{tr}\left[\left(\frac{\partial\Phi(\mathbf{A})}{\partial\mathbf{A}}\right)^{\mathrm{T}}\mathrm{d}\mathbf{A}\right] \quad , \tag{1.239}$$

where $\mathrm{d}\Phi$ denotes the **total differential**. The remainder is characterized by $o(\mathrm{d}\mathbf{A})$, where $o(\bullet)$ denotes the **Landau order** symbol. It is a small **error** that tends to zero faster than $\mathrm{d}\mathbf{A} \rightarrow \mathbf{O}$, i.e.

$$\lim_{\mathrm{d}\mathbf{A} \rightarrow \mathbf{O}} \frac{o(\mathrm{d}\mathbf{A})}{|\mathrm{d}\mathbf{A}|} = \mathbf{O} \quad . \tag{1.240}$$

The term $\partial\Phi(\mathbf{A})/\partial\mathbf{A}$ (or $\mathrm{grad}\Phi(\mathbf{A})$) denotes the **gradient** (or **derivative**) of function Φ at $\mathbf{A}$, characterizing a *second-order tensor*. In order to make the derivation clear, the tensor variable to be derived is sometimes indicated in the literature by a corresponding index, for example, $\mathrm{grad}_{\mathbf{A}}\Phi(\mathbf{A})$.

Note that the **first variation** of Φ is developed in the same way as its total differential. In Chapter 8 the concept of linearization is treated in more detail.

EXAMPLE 1.10   If the second-order tensor $\mathbf{A}$ is invertible, establish the important relation

$$\frac{\partial\mathrm{det}\mathbf{A}}{\partial\mathbf{A}} = \mathrm{det}\mathbf{A}\,\mathbf{A}^{-\mathrm{T}} \quad . \tag{1.241}$$

Solution.   With (1.101) we find that

$$\det(\mathbf{A} + d\mathbf{A}) = \det[\mathbf{A}(\mathbf{I} + \mathbf{A}^{-1}d\mathbf{A})] = \det\mathbf{A}\det(\mathbf{I} + \mathbf{A}^{-1}d\mathbf{A}) \ . \tag{1.242}$$

By means of eq. (1.168), we conclude with $\mathbf{A}$ replaced by $\mathbf{A}^{-1}d\mathbf{A}$ and with $\lambda = -1$ that

$$\det(\mathbf{A}^{-1}d\mathbf{A} + \mathbf{I}) = 1 + I_{\mathbf{A}^{-1}d\mathbf{A}} + II_{\mathbf{A}^{-1}d\mathbf{A}} + III_{\mathbf{A}^{-1}d\mathbf{A}}$$
$$= 1 + \mathrm{tr}(\mathbf{A}^{-1}d\mathbf{A}) + o(d\mathbf{A}) \ . \tag{1.243}$$

In eq. (1.243) we have used the fact that in regard to (1.171) and (1.172) the second invariant is quadratic in $d\mathbf{A}$ and the third invariant is cubic in $d\mathbf{A}$. Therefore, both are suitably small and summarized by the remainder of order $o(d\mathbf{A})$.

Replacing $\Phi(\mathbf{A})$ in relations (1.238) and (1.239) by the scalar-valued function $\det\mathbf{A}$, we deduce that

$$\det(\mathbf{A} + d\mathbf{A}) = \det\mathbf{A} + \mathrm{tr}\left[\left(\frac{\partial\det\mathbf{A}}{\partial\mathbf{A}}\right)^{\mathrm{T}} d\mathbf{A}\right] + o(d\mathbf{A}) \ . \tag{1.244}$$

On the other hand we find by means of $(1.242)_2$, $(1.243)_2$ and property (1.92) that

$$\det(\mathbf{A} + d\mathbf{A}) = \det\mathbf{A}[1 + \mathrm{tr}(\mathbf{A}^{-1}d\mathbf{A})] + o(d\mathbf{A})$$
$$= \det\mathbf{A} + \mathrm{tr}(\det\mathbf{A}\mathbf{A}^{-1}d\mathbf{A}) + o(d\mathbf{A}) \ . \tag{1.245}$$

Using rule $(1.93)_1$ and property (1.83) we find by comparing (1.244) with $(1.245)_2$ that

$$\frac{\partial\det\mathbf{A}}{\partial\mathbf{A}} : d\mathbf{A} = (\det\mathbf{A}\mathbf{A}^{-1})^{\mathrm{T}} : d\mathbf{A} = \det\mathbf{A}\,\mathbf{A}^{-\mathrm{T}} : d\mathbf{A} \ , \tag{1.246}$$

which proves finally eq. (1.241), since this holds for all $d\mathbf{A}$.   ■

Gradient of a tensor-valued (tensor) function.   We consider a nonlinear and smooth tensor-valued (tensor) function $\mathbf{A}(\mathbf{B})$ of one second-order tensor variable $\mathbf{B}$. Thus, for each $\mathbf{B}$, $\mathbf{A}(\mathbf{B})$ gives the value of a second-order tensor.

In order to compute the total differential $d\mathbf{A}$ we consider first-order (Taylor's) expansion for $\mathbf{A}$ at $\mathbf{B}$. We obtain

$$\mathbf{A}(\mathbf{B} + d\mathbf{B}) = \mathbf{A}(\mathbf{B}) + d\mathbf{A} + o(d\mathbf{B}) \ , \tag{1.247}$$

$$d\mathbf{A} = \frac{\partial\mathbf{A}(\mathbf{B})}{\partial\mathbf{B}} : d\mathbf{B} \ . \tag{1.248}$$

The term $\partial\mathbf{A}(\mathbf{B})/\partial\mathbf{B}$ (or $\mathrm{grad}_{\mathbf{B}}\mathbf{A}(\mathbf{B})$) denotes the **gradient** (or **derivative**) of function $\mathbf{A}$ at $\mathbf{B}$, characterizing a *fourth-order tensor*. Note that $\partial\mathbf{A}/\partial\mathbf{A}$ and $\partial\mathbf{A}^{\mathrm{T}}/\partial\mathbf{A}$ produce the respective fourth-order unit tensors $\mathbb{I}$ and $\bar{\mathbb{I}}$, as given in eqs. (1.161), (1.162).

EXAMPLE 1.11 Assume that $\mathbf{A}$ is an invertible smooth tensor-valued function with property $\mathbf{A} = \mathbf{A}^{\mathrm{T}}$. Show that

$$\left(\frac{\partial \mathbf{A}^{-1}}{\partial \mathbf{A}}\right)_{ijkl} = -\frac{1}{2}(A_{ik}^{-1}A_{lj}^{-1} + A_{il}^{-1}A_{kj}^{-1}) \ . \tag{1.249}$$

Solution. Starting from the identity that $\partial(\mathbf{A}^{-1}\mathbf{A})/\partial \mathbf{A} = \mathbb{O}$ gives the *fourth-order zero tensor* $\mathbb{O}$, we continue in index notation and find by means of the product rule

$$\left(\frac{\partial \mathbf{A}^{-1}\mathbf{A}}{\partial \mathbf{A}}\right)_{ijkl} = \frac{\partial A_{im}^{-1}}{\partial A_{kl}}A_{mj} + A_{im}^{-1}\frac{\partial A_{mj}}{\partial A_{kl}} = 0 \tag{1.250}$$

and finally, after multiplication with $(\mathbf{A}^{-1})_{jn}$ from the right-hand side, and using property (1.61),

$$\frac{\partial A_{im}^{-1}}{\partial A_{kl}}A_{mj}A_{jn}^{-1} = -A_{im}^{-1}\frac{\partial A_{mj}}{\partial A_{kl}}A_{jn}^{-1} \ ,$$

$$\frac{\partial A_{im}^{-1}}{\partial A_{kl}}\delta_{mn} = -\frac{1}{2}(A_{im}^{-1}\delta_{mk}\delta_{jl}A_{jn}^{-1} + A_{im}^{-1}\delta_{ml}\delta_{jk}A_{jn}^{-1}) \ ,$$

$$\frac{\partial A_{in}^{-1}}{\partial A_{kl}} = -\frac{1}{2}(A_{ik}^{-1}A_{ln}^{-1} + A_{il}^{-1}A_{kn}^{-1}) \ . \tag{1.251}$$

A replacement of the free index n by j gives (1.249). Looking at the indices in (1.251) it is important to note that $\partial \mathbf{A}^{-1}/\partial \mathbf{A} \neq -\mathbf{A}^{-1} \otimes \mathbf{A}^{-1}$. ■

EXERCISES

1. Suppose that $\mathbf{Q} = \mathbf{Q}(t)$ is a given orthogonal tensor-valued function of a scalar, such as time t. Show that $\dot{\mathbf{Q}}\mathbf{Q}^{\mathrm{T}}$ is a skew tensor with the property $\dot{\mathbf{Q}}\mathbf{Q}^{\mathrm{T}} = -(\dot{\mathbf{Q}}\mathbf{Q}^{\mathrm{T}})^{\mathrm{T}}$.

2. If the tensor $\mathbf{A}(t)$ is invertible, prove the important relation

$$\overline{\det \mathbf{A}(t)} = \det \mathbf{A}(t)\mathrm{tr}[\mathbf{A}^{-1}(t)\dot{\mathbf{A}}(t)] = \det \mathbf{A}(t)\mathbf{A}(t)^{-\mathrm{T}} : \dot{\mathbf{A}}(t)$$

and check the result with the matrix representation of the tensor $\mathbf{A}(t)$ given by

$$[\mathbf{A}(t)] = \begin{bmatrix} 1/2 & 0 & 0 \\ t^{-\alpha} & 1 & 0 \\ 0 & 0 & 2 \end{bmatrix} \ ,$$

where α denotes a scalar.

3. Consider two smooth tensor-valued functions $\mathbf{A}$ and $\mathbf{B}$ of a tensor argument.

   (a) Show the properties

$$\frac{\partial \mathrm{tr}\mathbf{A}}{\partial \mathbf{A}} = \mathbf{I} \ , \qquad \frac{\partial \mathrm{tr}(\mathbf{A}^2)}{\partial \mathbf{A}} = 2\mathbf{A}^{\mathrm{T}} \ , \qquad \frac{\partial \mathrm{tr}(\mathbf{A}^3)}{\partial \mathbf{A}} = 3(\mathbf{A}^{\mathrm{T}})^2 \ , \quad (1.252)$$

$$\frac{\partial \mathbf{A}}{\partial \mathbf{A}} : \mathbf{B} = \mathbf{B} : \frac{\partial \mathbf{A}}{\partial \mathbf{A}} = \mathbf{B} \ ,$$

$$\frac{\partial \mathbf{A}^{\mathrm{T}}}{\partial \mathbf{A}} : \mathbf{B} = \mathbf{B} : \frac{\partial \mathbf{A}^{\mathrm{T}}}{\partial \mathbf{A}} = \mathbf{B}^{\mathrm{T}} \ ,$$

$$\frac{\partial \mathbf{A}^n}{\partial \mathbf{A}} : \mathbf{B} = \sum_{k=0}^{n-1} \mathbf{A}^k \mathbf{B} \mathbf{A}^{n-k-1} \ , \qquad (n \text{ is a positive integer}) \ .$$

   (b) If $\mathbf{A}$ is an *invertible* smooth tensor-valued function, show that

$$\frac{\partial \mathbf{A}^{-1}}{\partial \mathbf{A}} : \mathbf{B} = -\mathbf{A}^{-1}\mathbf{B}\mathbf{A}^{-1} \ , \qquad (1.253)$$

$$\frac{\partial \mathbf{A}^{-1}}{\partial \mathbf{A}} : \mathbf{A} \otimes \mathbf{A}^{-1} = -\mathbf{A}^{-1} \otimes \mathbf{A}^{-1} \ . \qquad (1.254)$$

4. Let Φ be a smooth scalar-valued function and $\mathbf{A}, \mathbf{B}$ smooth tensor-valued functions of a tensor variable $\mathbf{C}$. Obtain the properties

$$\frac{\partial(\mathbf{A}:\mathbf{B})}{\partial \mathbf{C}} = \mathbf{A} : \frac{\partial \mathbf{B}}{\partial \mathbf{C}} + \mathbf{B} : \frac{\partial \mathbf{A}}{\partial \mathbf{C}} \ , \qquad (1.255)$$

$$\frac{\partial(\Phi\mathbf{A})}{\partial \mathbf{C}} = \mathbf{A} \otimes \frac{\partial \Phi}{\partial \mathbf{C}} + \Phi\frac{\partial \mathbf{A}}{\partial \mathbf{C}} \ . \qquad (1.256)$$

5. Alternatively to (1.241), show that

$$\frac{\partial \det\mathbf{A}}{\partial \mathbf{A}} = (\mathbf{A}^{\mathrm{T}})^2 - I_1\mathbf{A}^{\mathrm{T}} + I_2\mathbf{I} \ . \qquad (1.257)$$

Hint: By means of property (1.92) and scalar invariants (1.170)–(1.172) take the trace of the *Cayley-Hamilton* equation (1.174) and derive it with respect to tensor $\mathbf{A}$, use result (1.252).

1.8   Gradients and Related Operators

In this section we consider scalar, vector and tensor-valued functions that assign a scalar, vector and tensor to each material point $\mathbf{x}$ varying over some region of definition.

We call a *tensor* **A** and a *vector* **u** whose components A_{ij} and u_i depend on three (Cartesian) coordinates x_i a **tensor field A(x)** and a **vector field u(x)** in space at a fixed time, respectively. A **scalar field** $\Phi(\mathbf{x})$ of a body is a function that assigns a *scalar* Φ to each material point **x**.

Gradient of a scalar field.   Subsequently, we consider a smooth *scalar field* $\Phi(\mathbf{x})$ (for example, a *temperature, density* or *mass field*). First-order (Taylor's) expansion for Φ at **x** yields by analogy with (1.238) and (1.239)

$$\Phi(\mathbf{x} + d\mathbf{x}) = \Phi(\mathbf{x}) + d\Phi + o(d\mathbf{x}) \quad , \tag{1.258}$$

$$d\Phi = \frac{\partial \Phi}{\partial \mathbf{x}} \cdot d\mathbf{x} \quad , \tag{1.259}$$

denoting the *total differential* of Φ. The remainder $o(d\mathbf{x})$ tends to zero faster than $d\mathbf{x} \to \mathbf{o}$. The total differential, characterized as the difference between the scalars at $\mathbf{x} + d\mathbf{x}$ and $\mathbf{x}$, may be written in index notation as

$$d\Phi = \frac{\partial \Phi}{\partial x_i} dx_i = \frac{\partial \Phi}{\partial x_1} dx_1 + \frac{\partial \Phi}{\partial x_2} dx_2 + \frac{\partial \Phi}{\partial x_3} dx_3 \quad . \tag{1.260}$$

In the literature the partial derivatives of Φ with respect to x_i is frequently denoted by the **subscript comma**, i.e. $\partial \Phi / \partial x_i = \Phi_{,i}$.

We introduce further the **vector operator** (or **Nabla operator**) ∇ of vector calculus, i.e.

$$\nabla(\bullet) = \frac{\partial(\bullet)}{\partial x_i} \mathbf{e}_i = \frac{\partial(\bullet)}{\partial x_1} \mathbf{e}_1 + \frac{\partial(\bullet)}{\partial x_2} \mathbf{e}_2 + \frac{\partial(\bullet)}{\partial x_3} \mathbf{e}_3 \quad . \tag{1.261}$$

Then (1.259) may be written in a different form, by means of (1.261)

$$d\Phi = \nabla \Phi \cdot d\mathbf{x} \quad , \tag{1.262}$$

$$\text{grad}\Phi = \nabla \Phi = \frac{\partial \Phi}{\partial x_i} \mathbf{e}_i = \frac{\partial \Phi}{\partial x_1} \mathbf{e}_1 + \frac{\partial \Phi}{\partial x_2} \mathbf{e}_2 + \frac{\partial \Phi}{\partial x_3} \mathbf{e}_3 \quad . \tag{1.263}$$

The sum over the partial derivatives of Φ multiplied by the appropriate basis vectors is a *vector field* $(\partial \Phi / \partial x_i) \mathbf{e}_i$, called the **gradient** (or **derivative**) of the scalar field $\Phi(\mathbf{x})$.

By analogy with (1.261), the dot product, cross product and tensor product of the vector operator ∇ with a smooth vector or tensor field $(\bullet)$ is governed by the rules

$$\nabla \cdot (\bullet) = \frac{\partial(\bullet)}{\partial x_i} \cdot \mathbf{e}_i \quad , \qquad \nabla \times (\bullet) = \mathbf{e}_i \times \frac{\partial(\bullet)}{\partial x_i} \quad , \qquad \nabla \otimes (\bullet) = \frac{\partial(\bullet)}{\partial x_i} \otimes \mathbf{e}_i \quad . \tag{1.264}$$

These relations turn out to be very useful for subsequent derivations.

Concept of directional derivative.   We consider $\Phi(\mathbf{x})$ describing a scalar field that varies throughout a three-dimensional space. We recall that $\Phi(\mathbf{x}) = \Phi(x_1, x_2, x_3) = $ const denotes a so-called **level surface** characterized by all points $\mathbf{x}$ resulting in the same value Φ. All neighboring points $\mathbf{x} + \mathrm{d}\mathbf{x}$ on a level surface in space are therefore governed by the relation $\mathrm{d}\Phi = 0$.

A normal to the surface is defined by three values, namely $\partial\Phi/\partial x_1$, $\partial\Phi/\partial x_2$, $\partial\Phi/\partial x_3$, which are the components of the vector $\mathrm{grad}\Phi$ introduced in (1.263). We conclude that $\mathrm{grad}\Phi = \partial\Phi/\partial\mathbf{x}$ is a vector perpendicular to the surface of constant Φ. If we introduce a unit vector $\mathbf{n}$ at a point $\mathbf{x}$ normal to the surface (directed along $\mathrm{grad}\Phi$), then, $\mathbf{n} = \mathrm{grad}\Phi/|\mathrm{grad}\Phi|$ (see Figure 1.9).

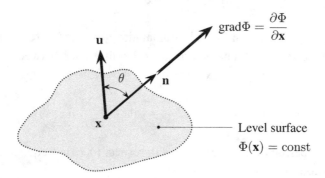

Figure 1.9 Level surface with unit normal vector $\mathbf{n}$ and an arbitrary vector $\mathbf{u}$ at $\mathbf{x}$.

We introduce further a vector $\mathbf{u}$ at the point $\mathbf{x}$ inclined at an angle θ to the direction of $\mathrm{grad}\Phi$. We call the dot product $\mathrm{grad}\Phi \cdot \mathbf{u}$ a **directional derivative** (or a **Gâteaux derivative**) of Φ at $\mathbf{x}$ in the direction of the vector $\mathbf{u}$. Note that for the definition of a directional derivative the normalized vector $\mathbf{u}$ is often used.

As we let $\mathbf{u}$ vary, the directional derivative $\mathrm{grad}\Phi \cdot \mathbf{u}$ takes its maximum (for a fixed point $\mathbf{x}$) when $\mathbf{u}$ approaches the direction of $\mathrm{grad}\Phi$ (compare with the expression for the dot product, i.e. (1.10)). For this particular situation $\mathbf{u}$ points in the direction of the normal vector $\mathbf{n}$ and $\cos\theta = 1$. Similarly, $\mathrm{grad}\Phi \cdot \mathbf{u}$ takes its minimum value (for a fixed point $\mathbf{x}$) when $\mathbf{u}$ approaches the opposite direction of the gradient (i.e. when $\cos\theta = -1$).

The directional derivative of Φ in the direction of $\mathbf{n}$ is called the **normal derivative**; we write

$$\mathrm{grad}\Phi \cdot \mathbf{n} = |\mathrm{grad}\Phi| \ . \tag{1.265}$$

For the case in which $\mathbf{u}$ is a unit vector the normal derivative represents the maximum of all the directional derivatives of Φ at $\mathbf{x}$.

For further consideration we briefly note the equivalent representation for the directional derivative of Φ at $\mathbf{x}$, namely

$$D_{\mathbf{u}}\Phi(\mathbf{x}) = \frac{\mathrm{d}}{\mathrm{d}\varepsilon}\Phi(\mathbf{x}+\varepsilon\mathbf{u})|_{\varepsilon=0} \ , \tag{1.266}$$

characterizing the rate of change of Φ along the (straight) line through $\mathbf{x}$ in the direction $\mathbf{u}$ (for a more detailed account see, for example, ABRAHAM et al. [1988] and MARSDEN and HUGHES [1994]). Usually $D(\bullet)$ is called the **Gâteaux operator** and ε is a scalar parameter. By means of the chain rule we immediately find from definition (1.266) that $D_{\mathbf{u}}\Phi(\mathbf{x}) = \mathrm{grad}\Phi\cdot\mathbf{u}$. Note that the directional derivative satisfies the usual rules of differentiation, i.e. the chain rule, product rule and so forth.

Later in the text we will see that the Lie time derivative, Section 2.8, and the variation and linearization of fields, Sections 8.1 and 8.4, are equivalent operations which are based on the powerful concept of the directional derivative.

EXAMPLE 1.12   Suppose that the scalar field

$$\Phi(\mathbf{x}) = x_1^2 + 3x_2x_3 \tag{1.267}$$

describes some physical quantity in space. Compute the directional derivative of Φ in the direction of the vector $\mathbf{u} = 1/\sqrt{3}(\mathbf{e}_1 + \mathbf{e}_2 + \mathbf{e}_3)$ at the point $\mathbf{x}$ with coordinates $(2,-1,0)$ and interpret the result.

Solution.   The parametric expression $\mathbf{x}+\varepsilon\mathbf{u}$ represents a line through the given point $\mathbf{x}$ with direction $\mathbf{u}$. We find the components of $\mathbf{x}+\varepsilon\mathbf{u}$ (for arbitrary, but fixed ε) as $(2+\varepsilon/\sqrt{3}, -1+\varepsilon/\sqrt{3}, \varepsilon/\sqrt{3})$.

Hence, according to definition (1.266) and given expression (1.267), $D_{\mathbf{u}}\Phi(2,-1,0)$ is the derivative of

$$\Phi(\mathbf{x}+\varepsilon\mathbf{u}) = \left(2+\frac{\varepsilon}{\sqrt{3}}\right)^2 + 3\left(-1+\frac{\varepsilon}{\sqrt{3}}\right)\frac{\varepsilon}{\sqrt{3}} = \frac{4}{3}\varepsilon^2 + \frac{\varepsilon}{\sqrt{3}} + 4 \tag{1.268}$$

with respect to the one variable ε, evaluated at zero. Therefore,

$$D_{\mathbf{u}}\Phi(2,-1,0) = \frac{\mathrm{d}}{\mathrm{d}\varepsilon}\left(\frac{4}{3}\varepsilon^2 + \frac{\varepsilon}{\sqrt{3}} + 4\right)\Bigg|_{\varepsilon=0} = \frac{1}{\sqrt{3}} \ . \tag{1.269}$$

As noted above the directional derivative may also be computed by the dot product $\mathrm{grad}\Phi\cdot\mathbf{u}$. In matrix notation we equivalently find from the given expression (1.267) that

$$D_{\mathbf{u}}\Phi(\mathbf{x}) = [\mathrm{grad}\Phi][\mathbf{u}] = \frac{1}{\sqrt{3}}\begin{bmatrix} 2x_1 \\ 3x_3 \\ 3x_2 \end{bmatrix}\begin{bmatrix} 1 \\ 1 \\ 1 \end{bmatrix} = \frac{1}{\sqrt{3}}(2x_1 + 3x_2 + 3x_3) \ , \tag{1.270}$$

which, evaluated at $x_1 = 2, x_2 = -1, x_3 = 0$, gives the desired result of $1/\sqrt{3}$.

Hence, if we are located at point $(2, -1, 0)$ and walk away in the given direction $\mathbf{u}$, we recognize an increase in some physical quantity, at a rate of $1/\sqrt{3}$ per distance, which is here $|\mathbf{u}| = 1$. ∎

Divergence of a vector field.   The introduced vector operator ∇ dotted into any smooth *vector field* $\mathbf{u}(\mathbf{x})$ (for example, a *force, velocity* or *acceleration field*) is called the **divergence** of $\mathbf{u}(\mathbf{x})$. It is a *scalar field* denoted by div$\mathbf{u}$ (or $\nabla \cdot \mathbf{u}$). With $(1.264)_1$ and eqs. (1.20), (1.21) and $(1.22)_2$ we find the important rule

$$\text{div}\mathbf{u} = \nabla \cdot \mathbf{u} = \frac{\partial u_j}{\partial x_i}\mathbf{e}_j \cdot \mathbf{e}_i = \frac{\partial u_j}{\partial x_i}\delta_{ji} = \frac{\partial u_i}{\partial x_i}$$
$$= \frac{\partial u_1}{\partial x_1} + \frac{\partial u_2}{\partial x_2} + \frac{\partial u_3}{\partial x_3} \ . \tag{1.271}$$

If div$\mathbf{u} = 0$, then the vector field $\mathbf{u}(\mathbf{x})$ is said to be **solenoidal** (or **divergence-free**).

Curl of a vector field.   The cross product of the vector operator ∇ with a smooth *vector field* $\mathbf{u}(\mathbf{x})$ is called the **curl** (or **rotation**) of $\mathbf{u}$. It is a *vector field* denoted by curl$\mathbf{u}$ (or $\nabla \times \mathbf{u}$, or rot$\mathbf{u}$). With $(1.264)_2$ and eqs. (1.20) and (1.35) we find that

$$\text{curl}\mathbf{u} = \nabla \times \mathbf{u} = \frac{\partial u_j}{\partial x_i}\mathbf{e}_i \times \mathbf{e}_j = \varepsilon_{ijk}\frac{\partial u_j}{\partial x_i}\mathbf{e}_k \ , \tag{1.272}$$

and with definition (1.33)

$$\text{curl}\mathbf{u} = \left(\frac{\partial u_3}{\partial x_2} - \frac{\partial u_2}{\partial x_3}\right)\mathbf{e}_1 + \left(\frac{\partial u_1}{\partial x_3} - \frac{\partial u_3}{\partial x_1}\right)\mathbf{e}_2 + \left(\frac{\partial u_2}{\partial x_1} - \frac{\partial u_1}{\partial x_2}\right)\mathbf{e}_3 \ . \tag{1.273}$$

If curl$\mathbf{u} = \mathbf{o}$, then the vector field $\mathbf{u}(\mathbf{x})$ is said to be **irrotational** (or **conservative**, or **curl-free**). We can show further that

$$\text{curl}\,\text{grad}\Phi = \mathbf{o} \ , \tag{1.274}$$
$$\text{div}\,\text{curl}\mathbf{u} = 0 \ . \tag{1.275}$$

We now consider any vector field $\mathbf{u}$ of the form $\mathbf{u} = \text{grad}\Phi$, where Φ is some scalar field, called the **potential** of $\mathbf{u}$. According to (1.274) it readily follows that the vector field $\mathbf{u}$ is automatically *irrotational*, since curl$\mathbf{u} = \mathbf{o}$.

EXAMPLE 1.13   Show that curl$\mathbf{u}$ is determined to be twice the axial vector ω of the skew part of tensor grad$\mathbf{u}$.

Solution.   We denote the skew part of grad$\mathbf{u}$ as $\mathbf{W}$, with components $W_{ji} = \partial u_j / \partial x_i$. With the index relation $W_{ji} = -W_{ij}$ for skew tensors and $(1.124)_2$, we deduce from $(1.272)_3$ that

$$\text{curl}\mathbf{u} = \varepsilon_{ijk} \frac{\partial u_j}{\partial x_i}\mathbf{e}_k = -\varepsilon_{ijk}W_{ij}\mathbf{e}_k = 2\boldsymbol{\omega} \quad . \quad \blacksquare \qquad (1.276)$$

Gradient of a vector field.   The **gradient** (or **derivative**) of a smooth *vector field* $\mathbf{u}(\mathbf{x})$ is defined to be a (second-order) *tensor field*. It is denoted by grad$\mathbf{u}$ (or $\nabla\mathbf{u}$, or $\partial\mathbf{u}/\partial\mathbf{x}$ or sometimes more explicitly by $\nabla \otimes \mathbf{u}$) and according to $(1.264)_3$ is given as

$$\text{grad}\mathbf{u} = \nabla \otimes \mathbf{u} = \frac{\partial u_i}{\partial x_j}\mathbf{e}_i \otimes \mathbf{e}_j \quad , \qquad (1.277)$$

with Cartesian components $(\text{grad}\mathbf{u})_{ij} = \partial u_i / \partial x_j$. In matrix notation this reads as

$$[\text{grad}\mathbf{u}] = \begin{bmatrix} \dfrac{\partial u_1}{\partial x_1} & \dfrac{\partial u_1}{\partial x_2} & \dfrac{\partial u_1}{\partial x_3} \\[2mm] \dfrac{\partial u_2}{\partial x_1} & \dfrac{\partial u_2}{\partial x_2} & \dfrac{\partial u_2}{\partial x_3} \\[2mm] \dfrac{\partial u_3}{\partial x_1} & \dfrac{\partial u_3}{\partial x_2} & \dfrac{\partial u_3}{\partial x_3} \end{bmatrix} \quad . \qquad (1.278)$$

It is easily seen from eq. (1.278) that the sum $\text{tr}(\text{grad}\mathbf{u})$ of the diagonal elements of the matrix is equivalent to the divergence of $\mathbf{u}$ characterized in eq. (1.271). Thus,

$$\text{tr}(\text{grad}\mathbf{u}) = \text{grad}\mathbf{u} : \mathbf{I} = \text{div}\mathbf{u} \quad . \qquad (1.279)$$

In addition, the **transposed gradient** of a smooth *vector field* $\mathbf{u}(\mathbf{x})$ of position $\mathbf{x}$ is denoted by $\text{grad}^{\text{T}}\mathbf{u}$ (or $\mathbf{u}\nabla$, or $(\partial\mathbf{u}/\partial\mathbf{x})^{\text{T}}$ or sometimes more explicitly by $\mathbf{u} \otimes \nabla$). It is given as

$$\text{grad}^{\text{T}}\mathbf{u} = \mathbf{u} \otimes \nabla = \mathbf{e}_i \otimes \frac{\partial\mathbf{u}}{\partial x_i} = \frac{\partial u_j}{\partial x_i}\mathbf{e}_i \otimes \mathbf{e}_j \quad , \qquad (1.280)$$

with Cartesian components $(\text{grad}^{\text{T}}\mathbf{u})_{ij} = \partial u_j / \partial x_i$.

Divergence and gradient of a (second-order) tensor field.   The vector operator ∇ dotted into any smooth (second-order) *tensor field* $\mathbf{A}(\mathbf{x})$ is a *vector field* denoted by $\text{div}\mathbf{A}$ (or $\nabla \cdot \mathbf{A}$), which is the **divergence** of $\mathbf{A}(\mathbf{x})$. With $(1.264)_1$, representation (1.59) and properties (1.53), (1.21) and (1.61) we have

$$\text{div}\mathbf{A} = \nabla \cdot \mathbf{A} = \frac{\partial A_{ik}}{\partial x_j}(\mathbf{e}_i \otimes \mathbf{e}_k) \cdot \mathbf{e}_j$$

$$= \frac{\partial A_{ik}}{\partial x_j}\delta_{kj}\mathbf{e}_i = \frac{\partial A_{ij}}{\partial x_j}\mathbf{e}_i \quad . \qquad (1.281)$$

Note that, especially in the mathematical community, the alternative definition $\mathrm{div}\mathbf{A} = \partial A_{ji}/\partial x_j \mathbf{e}_i$ of the divergence of a second-order tensor field $\mathbf{A}$ is frequently used.

The **gradient** (or **derivative**) of a smooth (second-order) *tensor field* $\mathbf{A}(\mathbf{x})$ is defined to be a (third-order) *tensor field*. It is denoted by $\mathrm{grad}\mathbf{A}$ (or $\nabla\mathbf{A}$, or $\partial\mathbf{A}/\partial\mathbf{x}$, or more explicitly by $\nabla \otimes \mathbf{A}$). Using $(1.264)_3$ and representation (1.59), we may write

$$\mathrm{grad}\mathbf{A} = \nabla \otimes \mathbf{A} = \frac{\partial A_{ij}}{\partial x_k}\mathbf{e}_i \otimes \mathbf{e}_j \otimes \mathbf{e}_k \quad . \tag{1.282}$$

By analogy with (1.279) we may show the equivalence of the trace of $\mathrm{grad}\mathbf{A}$ with the divergence of $\mathbf{A}$, i.e. eq. (1.281). We have the definition

$$\mathrm{tr}(\mathrm{grad}\mathbf{A}) = \mathrm{grad}\mathbf{A} : \mathbf{I} = \mathrm{div}\mathbf{A} \quad , \tag{1.283}$$

with Cartesian components $\mathrm{tr}(\mathrm{grad}\mathbf{A})_{ijk} = \delta_{jk}(\partial A_{ij}/\partial x_k) = \partial A_{ij}/\partial x_j$.

Laplacian and Hessian. The vector operator ∇ dotted into itself gives the **Laplacian** (or the **Laplacian operator**) ∇^2 (or $\nabla \cdot \nabla$, or in the literature sometimes denoted by Δ). With the rules $(1.261)_1$, $(1.264)_1$ and properties (1.21), $(1.22)_2$ we find that

$$\nabla^2(\bullet) = \nabla \cdot \nabla(\bullet) = \nabla \cdot \frac{\partial(\bullet)}{\partial x_i}\mathbf{e}_i = \frac{\partial^2(\bullet)}{\partial x_i \partial x_j}\mathbf{e}_i \cdot \mathbf{e}_j$$

$$= \frac{\partial^2(\bullet)}{\partial x_i \partial x_j}\delta_{ij} = \frac{\partial^2(\bullet)}{\partial x_i \partial x_i} \quad . \tag{1.284}$$

The Laplacian ∇^2 operated upon a scalar field Φ yields another scalar field. One example is **Laplace's equation** $\nabla^2\Phi = 0$. Another example is the inhomogeneous Laplace's equation often referred to as **Poisson's equation**, i.e. $\nabla^2\Phi = \Psi$. The differential equations of *Laplace* and *Poisson* occur in many fields of physics and engineering, for example, in electrostatics, in elasticity theory, in heat conduction problems of solids and so on.

The operator $\nabla\nabla$ (or more explicitly denoted by $\nabla \otimes \nabla$) is called the **Hessian**. With the rules $(1.261)_1$ and $(1.264)_3$ we obtain

$$\nabla\nabla(\bullet) = \nabla \otimes \nabla(\bullet) = \nabla \otimes \frac{\partial(\bullet)}{\partial x_i}\mathbf{e}_i = \frac{\partial^2(\bullet)}{\partial x_i \partial x_j}\mathbf{e}_i \otimes \mathbf{e}_j \tag{1.285}$$

for the Hessian. Note the properties

$$\mathrm{curl}\,\mathrm{curl}\mathbf{u} = \mathrm{grad}\,(\mathrm{div}\mathbf{u}) - \nabla^2\mathbf{u} \quad , \tag{1.286}$$

$$\nabla^2(\mathbf{u} \cdot \mathbf{v}) = \nabla^2\mathbf{u} \cdot \mathbf{v} + 2\mathrm{grad}\mathbf{u} \cdot \mathrm{grad}\mathbf{v} + \mathbf{u} \cdot \nabla^2\mathbf{v} \quad . \tag{1.287}$$

If a vector field $\mathbf{u}$ is both *solenoidal* ($\mathrm{div}\mathbf{u} = 0$) and *irrotational* ($\mathrm{curl}\mathbf{u} = \mathbf{o}$), then $\nabla^2\mathbf{u} = \mathbf{o}$ (see eq. (1.286)). For this case the vector field $\mathbf{u}(\mathbf{x})$ is said to be **harmonic**. If a scalar field Φ satisfies Laplace's equation $\nabla^2\Phi = 0$, then Φ is said to be **harmonic**.

EXERCISES

1. Consider a smooth scalar field $\Phi = x_1 x_2 x_3 - x_1$. Find a vector $\mathbf{n}$ of unit length normal to the level surface $\Phi = \text{const}$ passing through $(2, 0, 3)$.

2. Assume a force with magnitude F acting in the direction radially away from the origin at point $(2a, 3a, 2\sqrt{3}c)$ on the surface of a hyperboloid $\Phi(\mathbf{x}) = \Phi(x_1, x_2, x_3) = x_1^2/a^2 + x_2^2/a^2 - x_3^2/c^2 = 1$. Compute the components of the force located in the tangent plane to the surface at the point $(2a, 3a, 2\sqrt{3}c)$.

3. Consider a scalar field $\Phi(x_1, x_2) = e^{x_1}\cos(3x_1 - 2x_2)$. Compute the directional derivative of Φ in the direction of the line $x_1 = x_2/3 - 1$ at the point with coordinates $(0, 1)$, for increasing values of x_1.

4. If $\mathbf{u} = \mathbf{u}(\mathbf{x}) = x_1 x_2 x_3 \mathbf{e}_1 + x_1 x_2 \mathbf{e}_2 + x_1 \mathbf{e}_3$, determine $\mathrm{div}\,\mathbf{u}$, $\mathrm{curl}\,\mathbf{u}$, $\mathrm{grad}\,\mathbf{u}$, $\nabla^2\mathbf{u}$, respectively. Verify that (1.286) is satisfied.

5. Apply the operator ∇ to products of smooth scalar fields Φ, Ψ, vector fields $\mathbf{u}$, $\mathbf{v}$ and tensor fields $\mathbf{A}$, $\mathbf{B}$. Establish the important identities

$$\mathrm{div}(\Phi\mathbf{u}) = \Phi\,\mathrm{div}\mathbf{u} + \mathbf{u}\cdot\mathrm{grad}\Phi \quad , \tag{1.288}$$

$$\mathrm{div}(\Phi\mathbf{A}) = \Phi\,\mathrm{div}\mathbf{A} + \mathbf{A}\,\mathrm{grad}\Phi \quad , \tag{1.289}$$

$$\mathrm{div}(\mathbf{A}^{\mathrm{T}}\mathbf{u}) = \mathrm{div}\mathbf{A}\cdot\mathbf{u} + \mathbf{A} : \mathrm{grad}\mathbf{u} \quad , \tag{1.290}$$

$$\mathrm{div}(\mathbf{A}\mathbf{B}) = \mathrm{grad}\mathbf{A} : \mathbf{B} + \mathbf{A}\,\mathrm{div}\mathbf{B} \quad , \tag{1.291}$$

$$\mathrm{div}(\mathbf{u}\times\mathbf{v}) = \mathbf{v}\cdot\mathrm{curl}\mathbf{u} - \mathbf{u}\cdot\mathrm{curl}\mathbf{v} \quad ,$$

$$\mathrm{div}(\mathbf{u}\otimes\mathbf{v}) = (\mathrm{grad}\mathbf{u})\,\mathbf{v} + \mathbf{u}\,\mathrm{div}\mathbf{v} \quad , \tag{1.292}$$

$$\mathrm{grad}(\Phi\Psi) = (\mathrm{grad}\Phi)\,\Psi + \Phi\,\mathrm{grad}\Psi \quad ,$$

$$\mathrm{grad}(\Phi\mathbf{u}) = \mathbf{u}\otimes\mathrm{grad}\Phi + \Phi\,\mathrm{grad}\mathbf{u} \quad ,$$

$$\mathrm{grad}(\mathbf{u}\cdot\mathbf{v}) = (\mathrm{grad}^{\mathrm{T}}\mathbf{u})\,\mathbf{v} + (\mathrm{grad}^{\mathrm{T}}\mathbf{v})\,\mathbf{u} \quad , \tag{1.293}$$

$$\mathrm{curl}(\Phi\mathbf{u}) = \mathrm{grad}\Phi\times\mathbf{u} + \Phi\,\mathrm{curl}\mathbf{u} \quad ,$$

$$\mathrm{curl}(\mathbf{u}\times\mathbf{v}) = \mathbf{u}\,\mathrm{div}\mathbf{v} - \mathbf{v}\,\mathrm{div}\mathbf{u} + (\mathrm{grad}\mathbf{u})\,\mathbf{v} - (\mathrm{grad}\mathbf{v})\,\mathbf{u} \quad . \tag{1.294}$$

6. Consider the inverse square law defined by the particular vector field

$$\mathbf{u} = \alpha\frac{\mathbf{x}}{|\mathbf{x}|^3} \quad ,$$

with the positive constant α and $\mathbf{x} = x_i\mathbf{e}_i$. The region of $\mathbf{u}$ is the three-dimensional space, excluding the origin ($|\mathbf{x}| \neq 0$). Note that in fluid mechanics this relation characterizes the *velocity field* of a **point source**.

(a) Show that the vector field $\mathbf{u}(\mathbf{x})$ is harmonic.

(b) Derive a scalar field Φ whose gradient is $\mathbf{u}$. ($\Phi = -\alpha/|\mathbf{x}|$)

1.9   Integral Theorems

In the following we summarize some results of the integral theorems of *Gauss* and *Stokes* which are of essential importance in the field of continuum mechanics.

Divergence theorem.   Suppose $\mathbf{u}(\mathbf{x})$ and $\mathbf{A}(\mathbf{x})$ are any smooth vector and tensor fields defined on some convex three-dimensional region in physical space with volume v, and on a *closed* surface s bounding this volume (see Figure 1.10).

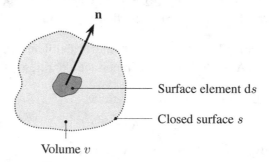

Figure 1.10 Volume v and bounded closed surface s (with infinitesimal surface element ds and associated unit normal vector $\mathbf{n}$).

Then, for $\mathbf{u}$ and $\mathbf{A}$ we have – without proof

$$\int_s \mathbf{u} \cdot \mathbf{n} \mathrm{d}s = \int_v \mathrm{div}\mathbf{u}\mathrm{d}v \qquad \text{or} \qquad \int_s u_i n_i \mathrm{d}s = \int_v \frac{\partial u_i}{\partial x_i}\mathrm{d}v \ , \qquad (1.295)$$

$$\int_s \mathbf{A}\mathbf{n}\mathrm{d}s = \int_v \mathrm{div}\mathbf{A}\mathrm{d}v \qquad \text{or} \qquad \int_s A_{ij}n_j \mathrm{d}s = \int_v \frac{\partial A_{ij}}{\partial x_j}\mathrm{d}v \qquad (1.296)$$

(given here in both symbolic and index notation), where $\mathbf{n}$ is the outward unit normal field acting along the surface s, dv and ds are infinitesimal volume and surface elements at $\mathbf{x}$, respectively. The important transformation of a surface integral into a volume integral (i.e. (1.295), (1.296)) is known as the **divergence theorem** (or **Gauss' divergence theorem**). Similar results hold for higher-order tensors.

The surface integral $\int_s \mathbf{u} \cdot \mathbf{n}\mathrm{d}s$ in the expression (1.295) is called the **(total) flux** (the name goes back to *Maxwell*) of $\mathbf{u}$ (or the **outward normal flux**) out of the total boundary surface s enclosing v.

By setting $\mathbf{A} = \Phi\mathbf{I}$ in theorem (1.296) and knowing that $\mathrm{div}(\Phi\mathbf{I}) = \mathrm{grad}\Phi$, recall identity (1.289), we consequently obtain from (1.296)

$$\int_s \Phi\mathbf{n}\mathrm{d}s = \int_v \mathrm{grad}\Phi\mathrm{d}v \qquad \text{or} \qquad \int_s \Phi n_i\mathrm{d}s = \int_v \frac{\partial\Phi}{\partial x_i}\mathrm{d}v \ , \qquad (1.297)$$

which is known as the **Green-Gauss-Ostrogradskiĭ theorem**.

Stokes' theorem.   For the sake of completeness the **theorem of Stokes** is briefly summarized. It relates a surface integral, which is valid over any *open* surface s, to a line integral around the bounding *closed* curve c in three-dimensional space.

We now introduce a **tangent vector** to c denoted by $\mathrm{d}\mathbf{x}$ (with components $\mathrm{d}x_i$) and an outward unit vector field $\mathbf{n}$ normal to s. Curve c has positive orientation relative to $\mathbf{n}$ in the sense shown in Figure 1.11.

The indicated circuit with the adjacent points $1, 2, 3$ ($1, 2$ on curve c and 3 an interior point of the surface s) induced by the orientation of c is related to the direction of $\mathbf{n}$ (i.e. a unit vector normal to s at point 3) by the right-hand screw-rule (see Figure 1.11).

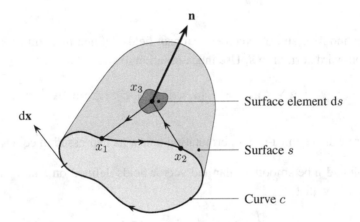

Figure 1.11 Open surface.

For a smooth vector field $\mathbf{u}$ defined on some region containing s, we have

$$\oint_c \mathbf{u} \cdot \mathrm{d}\mathbf{x} = \int_s \mathrm{curl}\mathbf{u} \cdot \mathbf{n}\mathrm{d}s \qquad \text{or} \qquad \oint_c u_k\mathrm{d}x_k = \int_s \varepsilon_{ijk}\frac{\partial u_k}{\partial x_j}n_i\mathrm{d}s \qquad (1.298)$$

(the proof is omitted).

Note that if the surface s is *closed* the integral on the left-hand side of eq. (1.298) reduces to zero, and then $\int_s \mathrm{curl}\mathbf{u} \cdot \mathbf{n}\mathrm{d}s = 0$. The line integral $\oint_c \mathbf{u} \cdot \mathrm{d}\mathbf{x}$ represents the **circulation** of $\mathbf{u}$ around the closed space curve c.

1. Let Φ, $\mathbf{u}$, and $\mathbf{A}$ be smooth scalar, vector and tensor fields defined in v and on s and let $\mathbf{n}$ be the outward unit normal field acting along the surface s. Show that

$$\int_s \Phi\mathbf{u} \cdot \mathbf{n}\mathrm{d}s = \int_v \mathrm{div}(\Phi\mathbf{u})\mathrm{d}v \quad , \tag{1.299}$$

$$\int_s \Phi\mathbf{A}\mathbf{n}\mathrm{d}s = \int_v \mathrm{div}(\Phi\mathbf{A})\mathrm{d}v \quad , \tag{1.300}$$

$$\int_s \mathbf{n} \times \mathbf{u}\mathrm{d}s = \int_v \mathrm{curl}\mathbf{u}\mathrm{d}v \quad ,$$

$$\int_s \mathbf{u} \otimes \mathbf{n}\mathrm{d}s = \int_v \mathrm{grad}\mathbf{u}\mathrm{d}v \quad ,$$

$$\int_s \mathbf{u} \cdot \mathbf{A}\mathbf{n}\mathrm{d}s = \int_v \mathrm{div}(\mathbf{A}^\mathrm{T}\mathbf{u})\mathrm{d}v \quad . \tag{1.301}$$

2. Let $\mathbf{u}$ and $\mathbf{A}$ be smooth vector and tensor fields defined in v and on s and let $\mathbf{n}$ be the outward normal to s. Use index notation to show that

$$\int_s \mathbf{u} \times \mathbf{A}\mathbf{n}\mathrm{d}s = \int_v [\mathbf{u} \times \mathrm{div}\mathbf{A} + \boldsymbol{\mathcal{E}} : (\mathrm{grad}\mathbf{u})\mathbf{A}^\mathrm{T}]\mathrm{d}v \quad , \tag{1.302}$$

where $\boldsymbol{\mathcal{E}}$ is the third-order permutation tensor as expressed in eq. (1.143).

3. Let Φ and $\mathbf{u}$ be smooth scalar and vector fields defined on s and c and let $\mathbf{n}$ act on s. Show that

$$\oint_c \Phi\mathrm{d}\mathbf{x} = \int_s \mathbf{n} \times \mathrm{grad}\Phi\mathrm{d}s \quad ,$$

$$\oint_c \mathbf{u} \times \mathrm{d}\mathbf{x} = \int_s [(\mathrm{div}\mathbf{u})\mathbf{n} - (\mathrm{grad}^\mathrm{T}\mathbf{u})\mathbf{n}]\mathrm{d}s \quad .$$

2 Kinematics

In the real world all physical objects are composed of **molecules** which are formed by atomic and subatomic particles. A **microscopic system** is studied by means of magnifying instruments such as a microscope. Microscopic studies are effective at the atomic level and very important in the exploration of a variety of physical phenomena. The atomistic point of view, however, is not a useful and adequate approach for common engineering applications. The microscopic approach is used in this text only briefly within Chapter 7.

We use the method of **continuum mechanics** as a powerful and effective tool to explain various physical phenomena successfully without detailed knowledge of the complexity of their internal (micro)structures. For example, think about *water* which is made up of billions of molecules. A good approximation is to treat water as a *continuous medium* characterized by certain field quantities which are associated with the internal structure, such as *density, temperature* and *velocity*. From the physical point of view this is an approximation in which the very large numbers of particles are replaced by a few quantities; we consider a **macroscopic system**. Hence, our primary interface with nature is through these quantities which represent averages over dimensions that are small enough to capture high gradients and to reflect some microstructural effects. Of course the predictions based on macroscopic studies are not exact but good enough for the design of machine elements in engineering.

The subject of continuum mechanics roughly comprises the following basic ingredients:

  (i) the study of motion and deformation (kinematics),

  (ii) the study of stress in a continuum (the concept of stress), and

  (iii) the mathematical description of the fundamental laws of physics governing the motion of a continuum (balance principles).

The aim of the following three Chapters 2-4 is to derive the essential equations within the basic fields (i)-(iii). Note that the provided results are applicable to all classes of materials, regardless of their internal physical structure. In order to explain the macroscopic behavior of physical objects, first of all we must understand the motion and deformation that cause stresses in a material (or are caused by stresses) arising from forces and moments. Thus, to study the motion and (finite) deformation of a continuum, i.e. kinematics, is essential and mainly the aim of the following chapter.

The reader who requires additional information on the subject of this chapter is referred to the monographs by, for example, MALVERN [1969], WANG and TRUES-DELL [1973], TRUESDELL [1977], GURTIN [1981a], CIARLET [1988], MARSDEN and HUGHES [1994], OGDEN [1997], CHADWICK [1999], TRUESDELL and NOLL [2004].

2.1   Configurations, and Motions of Continuum Bodies

Under an electron microscope we see the discontinuous atomic structure of matter. The molecules may be crystalline or randomly oriented. Between the particles there are large gaps. Theories considering the discrete structure of matter are **molecular** and **atomistic theories**. They are based on a discrete particle approach. For macroscopic systems such theories tend to become too complicated to yield the desired results and therefore do not meet our needs. However, under certain circumstances the microscopic approach is indispensable for the study of physical phenomena. A review of atomistic models is presented in the work of ORTIZ [1999, and references therein].

Notation of a particle and a continuum body.   Macroscopic systems usually can be described successfully with a **continuum approach (macroscopic approach)**. Such an approach leads to the **continuum theory**. The continuum theory has been developed independently of the molecular and atomistic theory and is meeting our needs. A fundamental assumption therein states that a body, denoted by $\mathcal{B}$, may be viewed as having a continuous (or at least a piecewise continuous) distribution of matter in space and time. The body is imagined as being a composition of a (continuous) set of **particles** (or **continuum particles** or **material points**), represented by $P \in \mathcal{B}$ as indicated in Figure 2.1.

It is important to note that the notion 'particle' (or 'continuum particle' or 'material point') refers to a part of a body and does not imply any association with the point mass of Newtonian mechanics or the discrete particle of the atomistic theory as mentioned above. A typical continuum particle is an accumulation of a large number of molecules, yet is small enough to be considered as a particle. The behavior of a continuum particle is a consequence of the collective behavior of all the molecules constituting that particle.

Hence, in a macroscopic study we are concerned with the mechanics of a body in which both **mass** and **volume** are continuous (or at least piecewise continuous) func-

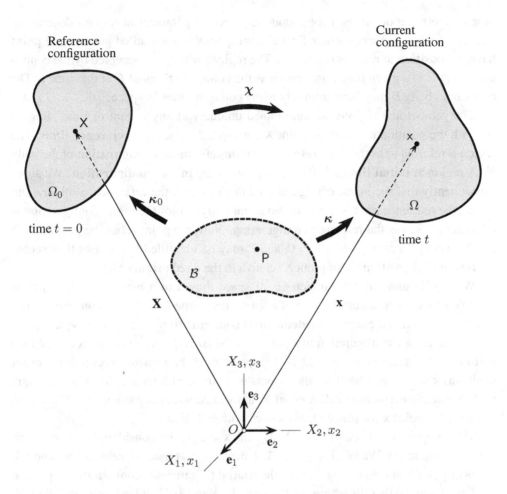

Figure 2.1 Configuration and motion of a continuum body.

tions of continuum particles. Such a body is called a **continuum body**, or just a **continuum**. A continuum is determined by **macroscopic quantities**. It has macroscopic dimensions that are much larger than the intermolecular spacings.

For the linkage of atomistic and continuum theories and for mixed atomistic/continuum computational schemes the reader is referred to ORTIZ [1999].

Configuration.   Consider a continuum body $\mathcal{B}$ with particle P $\in \mathcal{B}$ which is embedded in the three-dimensional Euclidean space at a given instant of **time** t, as indicated in Figure 2.1.

We introduce a **reference frame** of right-handed, rectangular coordinate axes at a **fixed** origin O with orthonormal basis vectors $\mathbf{e}_a$, $a = 1, 2, 3$, according to properties introduced in (1.18). As the continuum body $\mathcal{B}$ moves in space from one instant of

time to another it **occupies** a continuous sequence of geometrical regions denoted by
$\Omega_0, \ldots, \Omega$. Hence, every particle P of $\mathcal{B}$ corresponds to a so-called geometrical **point**
having a **position** in regions $\Omega_0, \ldots, \Omega$. The regions which are occupied by the continuum body $\mathcal{B}$ at a given time t are known as the **configurations** of $\mathcal{B}$ at that time t. The
continuum body $\mathcal{B}$ may have infinitely many configurations in space.

The geometrical regions are determined uniquely at any instant of time. Region
Ω_0 with the position of a typical point X corresponds to a fixed **reference time**. The
region is referred to as the fixed **reference** (or **undeformed**) **configuration** of the body
$\mathcal{B}$. A region at **initial time** $t = 0$ is referred to as the **initial configuration**. We agree
subsequently that the initial configuration coincides with the reference configuration;
hence, the reference time is at $t = 0$. Note that in dynamics the initial configuration is
often not chosen as the reference configuration. Now, the point X has the position of a
particle occupied by P $\in \mathcal{B}$ at $t = 0$, and P may be identified by the **position vector**
(or **referential position**) **X** of point X relative to the fixed origin O.

We now assume that the region Ω_0 of space moves to a new region Ω which is
occupied by the continuum body $\mathcal{B}$ at a subsequent time $t > 0$. The configuration of
$\mathcal{B}$ at t is its so-called **current** (or **deformed**) **configuration**. We relate a typical point
X of the reference configuration to a point x of the current configuration occupied by a
particle P $\in \mathcal{B}$ at times $t = 0$ and $t > 0$, respectively. The **position vector** (or **current
position**) **x** serves as a label for the associated point x with respect to the fixed origin
O. It is often convenient to call X **point X**, associated with the particle P $\in \mathcal{B}$ at $t = 0$,
and to call x **point x** (or **place x**), associated with P $\in \mathcal{B}$ at t.

The components of vectors $\mathbf{X} = X_A \mathbf{E}_A$ and $\mathbf{x} = x_a \mathbf{e}_a$ are considered as being along
the axes introduced. We label X_A, $A = 1, 2, 3$, as the **material** (or **referential**) **coordinates** of point X and x_a, $a = 1, 2, 3$, as the **spatial** (or **current**) **coordinates** of point x.
We have assumed that the origins of the sets $\{\mathbf{E}_A\}$ and $\{\mathbf{e}_a\}$ of basis vectors coincide.
In addition, we have agreed to have the same reference frame for the reference and
current configurations, which is why we set the basis $\{\mathbf{E}_A\}$ identical to $\{\mathbf{e}_a\}$. We use
just $\{\mathbf{e}_a\}$ in the following.

To denote scalar, vector and tensor quantities we use uppercase letters when they are
evaluated in the reference configuration, and lowercase letters for corresponding quantities in the current configuration. Sometimes we employ the index zero for quantities
acting in the reference configuration (for example, Ω_0, ρ_0, $\mathbf{a}_0$ ...). This convention is
often used in textbooks of continuum mechanics; however, it occasionally leads to con-
flicts with the general aim of Chapter 1 where we agreed to use lowercase, bold-face
letters for *vectors* and uppercase, bold-face letters for *second-order tensors*.

For example, we use **X** for the position vector of a point corresponding to the reference configuration despite the convention of Chapter 1 that uppercase, bold-face Latin
letters usually denote second-order tensors. A clear and unique notation is difficult to
establish, considering that certain quantities in continuum mechanics are characterized

by unequivocal symbols.

Whenever index notation is employed, as for the components of vectors and tensors, conventionally we use uppercase letters for the reference configuration (for example, $X_A, B_A \ldots$) and lowercase letters for the current configuration (for example, $x_a, b_a \ldots$).

Motion.   We assume that the map $\mathbf{X} = \kappa_0(\mathsf{P}, t)$ is a one-to-one correspondence between a particle $\mathsf{P} \in \mathcal{B}$ and the point $\mathbf{X} \in \Omega_0$ that $\mathcal{B}$ occupies at the given instant of time $t = 0$ (see Figure 2.1). Furthermore, let the map κ act on $\mathcal{B}$ to produce the region Ω at time t. The place $\mathbf{x} = \kappa(\mathsf{P}, t)$ that the particle P (evidently identified with $\mathbf{X}$ and t) occupies at t is described by (in symbolic and index notation)

$$\mathbf{x} = \kappa[\kappa_0^{-1}(\mathbf{X}, t)] = \chi(\mathbf{X}, t) \quad , \qquad x_a = \chi_a(X_1, X_2, X_3, t) \tag{2.1}$$

for all $\mathbf{X} \in \Omega_0$ and for all times t. In (2.1) χ is a vector field that specifies the place $\mathbf{x}$ of $\mathbf{X}$ for all fixed t, and is called the **motion** of the body $\mathcal{B}$. The motion χ is suitably regular and carries points $\mathbf{X}$ located at Ω_0 to places $\mathbf{x}$ in the current configuration Ω. We assume subsequently that χ possesses continuous derivatives with respect to position and time.

The parametric equation (2.1) determines successive positions $\mathbf{x}$ of a typical particle P in space. All successive points together form a curve in the Euclidean space which we call the **path line** (or **trajectory**) of the particle P.

The motion χ is assumed to be uniquely invertible. Consider $(\mathbf{x}, t)$, the position of point $\mathbf{X}$, which is associated with the place $\mathbf{x}$ at time t, is specified uniquely by (2.1) as

$$\mathbf{X} = \chi^{-1}(\mathbf{x}, t) \quad , \qquad X_A = \chi_A^{-1}(x_1, x_2, x_3, t) \quad , \tag{2.2}$$

with the **inverse motion** denoted by χ^{-1}, i.e. the inverse of the mapping χ, as defined in eq. (2.1). For a given time t, the inverse motion (2.2) carries points located at Ω to points in the reference configuration Ω_0. In (2.1) and (2.2) respectively the pairs $(\mathbf{X}, t)$ and $(\mathbf{x}, t)$ denote independent variables.

A motion χ of a body will generally change its shape, position and orientation. A continuum body which is able to change its shape is said to be **deformable**. By a **deformation** χ (or **inverse deformation** χ^{-1}) of a body we mean a motion (or inverse motion) of a body that is independent of time.

EXAMPLE 2.1   Let a deformed configuration of an initially rectangular region with lengths L_1 and L_2 be given as shown in Figure 2.2. The time-dependent angle θ is given by $\theta(t) = \omega t$, where ω denotes the *angular velocity*.

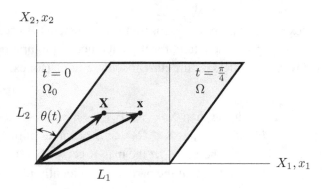

Figure 2.2 Deformed configuration of an initially rectangular region.

Determine the motion of a particle given by the position vector $\mathbf{X} \in \Omega_0$ and time t. In particular, determine the motion $\mathbf{x} = \chi(\mathbf{X}, t)$ at $(L_1/2, L_2/2, 1)$ and at time $t = \pi/4$, with $\omega = 1$.

Solution.   The reference configuration of the region Ω_0 of interest is characterized by (X_1, X_2, X_3) at $t = 0$, with $0 \leq X_1 \leq L_1$, $0 \leq X_2 \leq L_2$ and $0 \leq X_3 \leq 1$. The motion results in accord with (2.1)

$$\left.\begin{aligned}
x_1 &= \chi_1(X_1, X_2, X_3, t) = X_1 + c(t)X_2 = (L_1 + L_2)/2 \ , \\
x_2 &= \chi_2(X_1, X_2, X_3, t) = X_2 = L_2/2 \ , \\
x_3 &= \chi_3(X_1, X_2, X_3, t) = X_3 = 1 \ ,
\end{aligned}\right\} \qquad (2.3)$$

where $c(t) = \tan\theta(t)$ is a positive scalar. ∎

Material and spatial descriptions.   The so-called **material (or referential) description** is a characterization of the motion (or any other quantity) with respect to the material coordinates (X_1, X_2, X_3) and time t, given by eq. (2.1). In the material description attention is paid to a particle, and we observe what happens to the particle as it moves. Traditionally the material description is often referred to as the **Lagrangian description** (or **Lagrangian form**). Note that at $t = 0$ we have the consistency condition $\mathbf{X} = \mathbf{x}$, $X_A = x_a$.

The so-called **Eulerian (or spatial) description** (or **Eulerian form**), is a characterization of the motion (or any other quantity) with respect to the spatial coordinates (x_1, x_2, x_3) and time t, given by eq. (2.2). In the spatial description attention is paid to a point in space, and we study what happens at the point as time changes.

In fluid mechanics we quite often work in the Eulerian description in which we refer all relevant quantities to the position in space at time t. It is not useful to refer the quantities to the material coordinates X_A, $A = 1, 2, 3$, at $t = 0$, which are, in general, not known in fluid mechanics. However, in solid mechanics we use both types of description. Due to the fact that the constitutive behavior of solids is often given in terms of material coordinates we often prefer the Lagrangian description.

EXERCISE

1. Consider the two motions defined by the following sets of equations:

$$\mathbf{x} = \alpha t^3 \mathbf{e} + \mathbf{X} \ ,$$

$$x_1 = \alpha \ , \qquad x_2 = -\beta - \gamma e^{-\beta/\gamma} \cos\left(\frac{\delta + \omega t}{\gamma}\right) \ ,$$

$$x_3 = \delta + \gamma e^{-\beta/\gamma} \sin\left(\frac{\delta + \omega t}{\gamma}\right) \ ,$$

where α, β, γ, δ and ω are scalar constants and $\mathbf{e}$ denotes a fixed unit vector. Show that the path lines are straight lines and circles, respectively.

2.2   Displacement, Velocity, Acceleration Fields

Displacement field.   The vector field

$$\mathbf{U}(\mathbf{X}, t) = \mathbf{x}(\mathbf{X}, t) - \mathbf{X} \tag{2.4}$$

represents the **displacement field** of a typical particle and relates its position $\mathbf{X}$ in the undeformed configuration to its position $\mathbf{x}$ in the deformed configuration at time t (see Figure 2.3). Relation (2.4) holds for all particles. The displacement field $\mathbf{U}$ is a function of the referential position $\mathbf{X}$ and time t, which characterizes the *material description* (Lagrangian form) of the displacement field.

The displacement field in the *spatial description* (Eulerian form), denoted $\mathbf{u}$, is a function of the current position $\mathbf{x}$ and time t. We write

$$\mathbf{u}(\mathbf{x}, t) = \mathbf{x} - \mathbf{X}(\mathbf{x}, t) \ . \tag{2.5}$$

Representation (2.5) specifies the current position $\mathbf{x}$ of a particle at time t which results from its referential position $\mathbf{X}(\mathbf{x}, t)$ plus its displacement $\mathbf{u}(\mathbf{x}, t)$ from that position. Representation (2.4) expresses the displacement of a particle at time t in terms of its

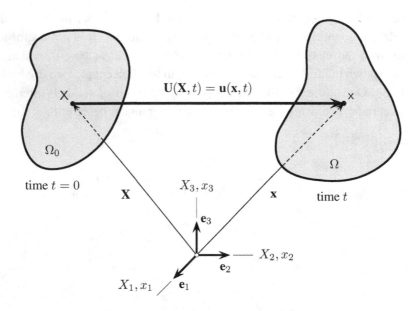

Figure 2.3 Displacement field **U** of a typical particle.

referential position **X**. The two descriptions are related by means of the motion $\mathbf{x} = \boldsymbol{\chi}(\mathbf{X}, t)$, namely

$$\mathbf{U}(\mathbf{X}, t) = \mathbf{U}[\boldsymbol{\chi}^{-1}(\mathbf{x}, t), t] = \mathbf{u}(\mathbf{x}, t) \quad . \tag{2.6}$$

Note that **U** and **u** have the same values. They represent functions of different arguments. The vectors **U** and **u** are referred to the material coordinates U_A and the spatial coordinates u_a, respectively. Since we agreed that the initial configuration coincides with the reference configuration, the displacements vanish in the reference configuration.

For a so-called **rigid-body translation** any particle moves an identical distance, with the same magnitude and the same direction at time t. For this case the displacement field is independent of **X** (it can be a function of time t).

Velocity and acceleration field.   In *solid mechanics* the motion and the deformation of a continuum body are, in general, described in terms of the displacement field. However, the primary field quantities in *fluid mechanics* describing the fundamental kinematic properties are the **velocity field** and the **acceleration field**.

The first and second derivatives of the motion $\boldsymbol{\chi}$ with respect to time t are performed by holding **X** fixed. We obtain

$$\mathbf{V}(\mathbf{X}, t) = \frac{\partial \boldsymbol{\chi}(\mathbf{X}, t)}{\partial t} \quad , \qquad \mathbf{A}(\mathbf{X}, t) = \frac{\partial \mathbf{V}(\mathbf{X}, t)}{\partial t} = \frac{\partial^2 \boldsymbol{\chi}(\mathbf{X}, t)}{\partial t^2} \quad , \tag{2.7}$$

where $\mathbf{V}(\mathbf{X}, t)$ and $\mathbf{A}(\mathbf{X}, t)$ denote the material description of the velocity field and the acceleration field, respectively. They are functions of the material coordinates (X_1, X_2, X_3) and time t representing the time rate of change of position and velocity of a particle $\mathbf{X}$ at time t.

Quite often we need a formulation in terms of the spatial coordinates (x_1, x_2, x_3) characterizing a fixed location in space at time t. As mentioned above, such a description is known as the Eulerian or spatial description in which now we like to define the velocity field and the acceleration field.

The two descriptions are transformed into each other by using the motion χ. In the spatial description, $\mathbf{X}$ may be expressed in terms of $\mathbf{x}$ and t. Hence, from (2.7) we conclude by analogy with eq. (2.6) that by change of variables

$$\mathbf{V}(\mathbf{X}, t) = \mathbf{V}[\chi^{-1}(\mathbf{x}, t), t] = \mathbf{v}(\mathbf{x}, t) \quad , \tag{2.8}$$

$$\mathbf{A}(\mathbf{X}, t) = \mathbf{A}[\chi^{-1}(\mathbf{x}, t), t] = \mathbf{a}(\mathbf{x}, t) \quad , \tag{2.9}$$

where $\mathbf{v}(\mathbf{x}, t)$ and $\mathbf{a}(\mathbf{x}, t)$ denote the spatial description of the velocity field and the acceleration field, respectively.

The **velocity components** and the **acceleration components** for $\mathbf{V}$ and $\mathbf{A}$ are denoted by V_A, A_A, while for $\mathbf{v}$ and $\mathbf{a}$ we write v_a, a_a, respectively.

EXAMPLE 2.2   A certain motion of a continuum body in the material description is given in the form

$$x_1 = e^t X_1 - e^{-t} X_2 \quad , \qquad x_2 = e^t X_1 + e^{-t} X_2 \quad , \qquad x_3 = X_3 \quad , \tag{2.10}$$

for $t > 0$. Find the velocity and acceleration components in terms of the material and spatial coordinates and time, i.e. $V_A = V_A(X_A, t)$, $A_A = A_A(X_A, t)$ and $v_a = v_a(x_a, t)$, $a_a = a_a(x_a, t)$, respectively.

Solution.   For the given motion $x_a = \chi_a(X_1, X_2, X_3, t)$, the velocity and acceleration components in terms of the material coordinates X_A, $A = 1, 2, 3$, and time t follow from eqs. (2.7) and have the forms

$$V_1 = e^t X_1 + e^{-t} X_2 \quad , \qquad V_2 = e^t X_1 - e^{-t} X_2 \quad , \qquad V_3 = 0 \quad , \tag{2.11}$$

$$A_1 = e^t X_1 - e^{-t} X_2 \quad , \qquad A_2 = e^t X_1 + e^{-t} X_2 \quad , \qquad A_3 = 0 \quad . \tag{2.12}$$

In order to find the velocity and acceleration components in terms of the spatial coordinates x_a, $a = 1, 2, 3$, and time t, we substitute simply the given motion into derived relations (2.11) and (2.12). Thus, according to eqs. (2.8) and (2.9) we find that

$$v_1 = x_2 \quad , \qquad v_2 = x_1 \quad , \qquad v_3 = 0 \quad , \tag{2.13}$$

$$a_1 = x_1 = v_2 \quad , \qquad a_2 = x_2 = v_1 \quad , \qquad a_3 = 0 \ . \quad \blacksquare \qquad (2.14)$$

EXERCISES

1. Consider the motion of a continuum body given by the equations

$$x_1 = X_1(1 + \alpha t^3) \quad , \qquad x_2 = X_2 \quad , \qquad x_3 = X_3 \quad , \qquad (2.15)$$

   where α is a constant. Determine the displacement, velocity and acceleration fields in each of the material and spatial descriptions.

2. A motion of a continuum body is defined by the velocity components

$$v_1 = \frac{3x_1}{1+t} \quad , \qquad v_2 = \frac{x_2}{1+t} \quad , \qquad v_3 = \frac{5x_3^2}{1+t} \quad . \qquad (2.16)$$

   Assume that the reference configuration of the continuum body is at $t = 0$, with the consistency condition $\mathbf{X} = \mathbf{x}$.

   (a) Derive the particle path, i.e. the motion $\mathbf{x} = \boldsymbol{\chi}(\mathbf{X}, t)$.

   (b) Compute the velocity components in terms of the material coordinates X_A and time t, i.e. V_A, $A = 1, 2, 3$, and the associated acceleration in the material and spatial descriptions.

2.3 Material, Spatial Derivatives

We introduce the terminology of a **material field** in which the independent variables are $(\mathbf{X}, t)$, i.e. the referential position $\mathbf{X}$ with material coordinates X_A, $A = 1, 2, 3$, and the time t. In a **spatial field** the independent variables are $(\mathbf{x}, t)$, i.e. the current position (the place) $\mathbf{x}$ with spatial coordinates x_a, $a = 1, 2, 3$, and the time t. In the following we denote a smooth material field and a spatial field of some physical scalar, vector or tensor quantity associated with the motion χ as $\mathcal{F} = \mathcal{F}(\mathbf{X}, t)$ and $f = f(\mathbf{x}, t)$, respectively.

Material time derivative of a material field.   Consider the **material time derivative** of a smooth material field $\mathcal{F}(\mathbf{X}, t)$ which is conventionally denoted by $\mathrm{D}\mathcal{F}(\mathbf{X}, t)/\mathrm{D}t$ or by a superposed dot, i.e. $\dot{\mathcal{F}}(\mathbf{X}, t)$. Other names for it in the literature are **total** or **substantial** time derivative.

It is the derivative of $\mathcal{F}$ with respect to time t (holding $\mathbf{X}$ fixed); thus, in symbolic and index notation,

$$\dot{\mathcal{F}}(\mathbf{X}, t) = \frac{\mathrm{D}\mathcal{F}(\mathbf{X}, t)}{\mathrm{D}t} = \left(\frac{\partial \mathcal{F}(\mathbf{X}, t)}{\partial t} \right)_{\mathbf{X}}$$

$$\text{or} \quad \dot{\mathcal{F}}(X_A, t) = \left(\frac{\partial \mathcal{F}(X_A, t)}{\partial t} \right)_{X_A} . \tag{2.17}$$

The subscript $\mathbf{X}$, (X_A), indicates the variable which is held constant while taking the partial differentiation of $\mathcal{F}$. In the following the subscripts are dropped to reduce the cumbersome notation. However, in cases where additional information is needed, they will be employed.

The material time derivative represents the rate at which the material field $\mathcal{F}$ changes with time, seen by an observer following the path line of a particle.

EXAMPLE 2.3   Determine the material description of the velocity field $\mathbf{V}$ and the acceleration field $\mathbf{A}$ of a particle at time t as the time rate of change of the displacement of that particle.

Solution.   By virtue of eqs. $(2.1)_2$ and (2.4) we may derive from (2.7)

$$\mathbf{V}(\mathbf{X}, t) = \left(\frac{\partial \mathbf{U}(\mathbf{X}, t)}{\partial t} \right)_{\mathbf{X}} = \dot{\mathbf{U}} , \qquad \mathbf{A}(\mathbf{X}, t) = \left(\frac{\partial \mathbf{V}(\mathbf{X}, t)}{\partial t} \right)_{\mathbf{X}} = \dot{\mathbf{V}} , \tag{2.18}$$

so that with $\dot{\mathbf{V}}(\mathbf{X}, t) = \partial^2 \mathbf{U}(\mathbf{X}, t)/\partial t^2 = \ddot{\mathbf{U}}$,

$$\mathbf{A} = \dot{\mathbf{V}} = \ddot{\mathbf{U}} \tag{2.19}$$

results. With the terminology introduced above, $\mathbf{V}(\mathbf{X}, t)$ and $\mathbf{A}(\mathbf{X}, t)$ denote the *material velocity field* and the *material acceleration field*, respectively.   ∎

For some studies following in this text we introduce the useful relationship between the material time derivative and the directional derivative. Considering $\mathcal{F} = \mathcal{F}(\mathbf{X}, t)$, the material time derivative of $\mathcal{F}$ is equal to the directional derivative of $\mathcal{F}$ in the direction of the velocity vector $\mathbf{v}$, we write

$$\frac{\mathrm{D}\mathcal{F}(\mathbf{X}, t)}{\mathrm{D}t} = D_{\mathbf{v}}\mathcal{F}(\mathbf{X}, t) , \tag{2.20}$$

which can be shown by applying the definition of the directional derivative, i.e. (1.266).

Material gradient of a material field.   The **material gradient** of a smooth material field $\mathcal{F}(\mathbf{X}, t)$, denoted by $\mathrm{Grad}\mathcal{F}(\mathbf{X}, t)$ (with uppercase G), is obtained by deriving $\mathcal{F}$ relative to the referential position $\mathbf{X}$, (X_A), at a fixed time t, i.e.

$$\mathrm{Grad}\mathcal{F}(\mathbf{X}, t) = \frac{\partial \mathcal{F}(\mathbf{X}, t)}{\partial \mathbf{X}} \quad . \tag{2.21}$$

Subsequently, the **material divergence** and the **material curl** of $\mathcal{F}$ are denoted by $\mathrm{Div}\mathcal{F}$ and $\mathrm{Curl}\mathcal{F}$ (with uppercase D and C), respectively. They are divergence and curl differential operators characterizing quantities with respect to the reference configuration.

Spatial time derivative and spatial gradient of a spatial field.   The **spatial time derivative** of a smooth spatial field $f(\mathbf{x}, t)$ is the derivative of f with respect to time t holding the current position $\mathbf{x}$ fixed. It is simply denoted by $\partial f(\mathbf{x}, t)/\partial t$ representing the rate at which f changes with time, seen by an observer currently stationed at $\mathbf{x}$. In the literature the spatial time derivative is also referred to as the **local time derivative**.

The **spatial gradient** of f, denoted by $\mathrm{grad}f$ (with lowercase g), is the derivative of f with respect to the current position $\mathbf{x}$, (x_a), at a fixed time t, i.e.

$$\mathrm{grad}f(\mathbf{x}, t) = \frac{\partial f(\mathbf{x}, t)}{\partial \mathbf{x}} \quad . \tag{2.22}$$

Subsequently, the **spatial divergence** and the **spatial curl** of f are denoted by $\mathrm{div}f$ and $\mathrm{curl}f$ (with lowercase d and c), respectively. They are differential operators characterizing quantities with respect to the current configuration.

Material time derivative of a spatial field.   Finally, we consider a very important operation, i.e. the material time derivative of a smooth spatial field $f(\mathbf{x}, t)$, which is denoted by $\mathrm{D}f(\mathbf{x}, t)/\mathrm{D}t$ or $\dot{f}(\mathbf{x}, t)$. It is the time derivative of f holding $\mathbf{X}$ fixed.

In order to evaluate $\dot{f}$ we map f to the material description as a first step, and then we take the material time derivative. The result we map back to the spatial description in order to obtain

$$\dot{f}(\mathbf{x}, t) = \frac{\mathrm{D}f(\mathbf{x}, t)}{\mathrm{D}t} = \left(\frac{\partial f[\boldsymbol{\chi}(\mathbf{X}, t), t]}{\partial t} \right)_{\mathbf{X} = \boldsymbol{\chi}^{-1}(\mathbf{x}, t)}. \tag{2.23}$$

Let Φ be a smooth spatial field which assigns a scalar $\Phi(\mathbf{x}, t)$ to each point $\mathbf{x}$ at t. By the chain rule we find from (2.23) that

$$\dot{\Phi}(\mathbf{x}, t) = \left(\frac{\partial \Phi(\mathbf{x}, t)}{\partial t} \right)_{\mathbf{x}} + \left(\frac{\partial \Phi(\mathbf{x}, t)}{\partial \mathbf{x}} \right)_{t} \cdot \left(\frac{\partial \boldsymbol{\chi}(\mathbf{X}, t)}{\partial t} \right)_{\mathbf{X} = \boldsymbol{\chi}^{-1}(\mathbf{x}, t)}, \tag{2.24}$$

and by means of eqs. $(2.7)_1$, (2.8) and notation (2.22),

$$\dot{\Phi}(\mathbf{x}, t) = \frac{\mathrm{D}\Phi(\mathbf{x}, t)}{\mathrm{D}t} = \frac{\partial \Phi(\mathbf{x}, t)}{\partial t} + \mathrm{grad}\Phi(\mathbf{x}, t) \cdot \mathbf{v}(\mathbf{x}, t) \quad ,$$

$$\text{or} \qquad \dot{\Phi} = \frac{\mathrm{D}\Phi}{\mathrm{D}t} = \frac{\partial \Phi}{\partial t} + \frac{\partial \Phi}{\partial x_b} v_b \quad . \tag{2.25}$$

Herein, the first term on the right-hand side of eq. (2.25) denotes the spatial (or local) time derivative of the spatial scalar field Φ, while the second term is called the **convective rate of change** of Φ, which describes the changing position of a particle.

By analogy with the above, the material time derivative of a smooth spatial field $\mathbf{v}(\mathbf{x}, t)$, which is vector-valued, is given by

$$\dot{\mathbf{v}}(\mathbf{x}, t) = \frac{D\mathbf{v}(\mathbf{x}, t)}{Dt} = \frac{\partial \mathbf{v}(\mathbf{x}, t)}{\partial t} + \mathrm{grad}\mathbf{v}(\mathbf{x}, t)\mathbf{v}(\mathbf{x}, t) \quad,$$

$$\text{or} \quad \dot{v}_a = \frac{Dv_a}{Dt} = \frac{\partial v_a}{\partial t} + \frac{\partial v_a}{\partial x_b}v_b \quad, \tag{2.26}$$

where $\mathbf{v} = \mathbf{v}(\mathbf{x}, t)$, $(v_a, a = 1, 2, 3)$, and its material time derivative $\dot{\mathbf{v}} = \dot{\mathbf{v}}(\mathbf{x}, t)$, $(\dot{v}_a, a = 1, 2, 3)$, may also be viewed as the spatial description of a velocity and acceleration field, referred to briefly as the *spatial velocity field* and the *spatial acceleration field*, respectively. The spatial acceleration field we often denote by $\mathbf{a} = \dot{\mathbf{v}}$, $(a_a = \dot{v}_a, a = 1, 2, 3)$.

Hence, we may write eq. $(2.26)_2$ as

$$\mathbf{a} = \frac{\partial \mathbf{v}}{\partial t} + (\mathrm{grad}\mathbf{v})\mathbf{v} \quad \text{or} \quad a_a = \frac{\partial v_a}{\partial t} + \frac{\partial v_a}{\partial x_b}v_b \tag{2.27}$$

(the arguments of the functions have been omitted for simplicity).

The first term on the right-hand side, namely $\partial\mathbf{v}/\partial t$, describes the **local acceleration** (local rate of change of the velocity field). The second term is *quadratically nonlinear* in the velocity field and describes the **convective acceleration field** (convective rate of change of the velocity field). Relation (2.27) turns out to be very useful, because the spatial acceleration field $\mathbf{a}(\mathbf{x}, t)$ may be determined from $\mathbf{v}(\mathbf{x}, t)$ without knowing the motion explicitly, which is an important question in applied fluid mechanics.

By virtue of $(2.7)_1$ and (2.8), the spatial velocity field $\mathbf{v}(\mathbf{x}, t)$ may be expressed as the material time derivative of the motion $\mathbf{x} = \boldsymbol{\chi}(\mathbf{X}, t)$, i.e.

$$\mathbf{v} = \dot{\mathbf{x}} = \frac{\partial \mathbf{x}}{\partial t} \quad \text{or} \quad v_a = \dot{x}_a = \frac{\partial x_a}{\partial t} \quad, \tag{2.28}$$

and from (2.5) we find that $\mathbf{v} = \dot{\mathbf{u}}$. The spatial acceleration field reads by analogy with relation (2.19) as

$$\mathbf{a} = \dot{\mathbf{v}} = \ddot{\mathbf{u}} \quad. \tag{2.29}$$

A very useful short-hand notation of the operator $D(\bullet)/Dt$ which acts on spatial fields reads, in symbolic and index notation,

$$\frac{D(\bullet)}{Dt} = \frac{\partial(\bullet)}{\partial t} + \mathrm{grad}(\bullet)\mathbf{v} \quad \text{or} \quad \frac{D(\bullet)}{Dt} = \frac{\partial(\bullet)}{\partial t} + \frac{\partial(\bullet)}{\partial x_a}v_a \quad, \tag{2.30}$$

which should be compared with eqs. (2.25) and (2.26). Relation (2.30) shows clearly
the relationship between the material and spatial time derivatives.

EXAMPLE 2.4   In the current configuration of a continuum body a certain physical
quantity is given in space and time according to

$$\Phi(\mathbf{x}, t) = -\frac{t^{-1}}{|\mathbf{x}|} \quad , \tag{2.31}$$

with place $\mathbf{x} = x_i \mathbf{e}_i$. The region of the spatial (scalar) field $\Phi = \Phi(\mathbf{x}, t)$ is a space
characterized by three variables corresponding to the coordinates and one to the time,
excluding the origin, i.e. $|\mathbf{x}| \neq 0$, $t \neq 0$. Suppose that an observer moves with velocity

$$\mathbf{v} = x_1 x_3 (t\, e^t)^{-1} \mathbf{e}_1 + x_2 x_3 (t\, e^t)^{-1} \mathbf{e}_2 + x_3^2 (t\, e^t)^{-1} \mathbf{e}_3 \quad . \tag{2.32}$$

Find the time rate of change of Φ as seen by the observer.

Solution.   In order to compute the material time derivative of the spatial field Φ
apply relation (2.25)$_2$. Therefore, we need to compute the spatial time derivative and
the spatial gradient of Φ. The spatial time derivative of Φ gives

$$\frac{\partial \Phi(\mathbf{x}, t)}{\partial t} = \frac{t^{-2}}{|\mathbf{x}|} \quad . \tag{2.33}$$

With definition (2.22) we find directly from (2.31) the spatial gradient of Φ, i.e.

$$\mathrm{grad}\Phi(\mathbf{x}, t) = \frac{\partial \Phi(\mathbf{x}, t)}{\partial \mathbf{x}} = t^{-1} \frac{\mathbf{x}}{|\mathbf{x}|^3} \quad . \tag{2.34}$$

Note that the spatial vector field $\mathrm{grad}\Phi$ is harmonic and describes a *point source* at a
certain time t (compare with Exercise 6 on p. 51).

By virtue of the derived solutions (2.33), (2.34) and the given velocity field (2.32)
we obtain finally

$$\dot{\Phi} = \frac{t^{-2}}{|\mathbf{x}|} + t^{-1} \frac{\mathbf{x}}{|\mathbf{x}|^3} \cdot \mathbf{v} = \frac{t^{-2}}{|\mathbf{x}|}(1 + x_3 e^{-t}) \quad , \tag{2.35}$$

which gives the desired solution (2.31), the material time derivative of Φ at $\mathbf{x}$ and at
time t.   ∎

Types of motions.   If the velocity field is independent of time, i.e. $\partial \mathbf{v}/\partial t = \mathbf{o}$
and $\mathbf{v} = \mathbf{v}(\mathbf{x})$, the associated motion is said to be **steady** and (2.27) reduces to $\mathbf{a} =$

(grad**v**)**v**. On the other hand, if the velocity field at each instant is independent of position **x**, i.e. grad**v** $= \mathbf{O}$ and $\mathbf{v} = \mathbf{v}(t)$, the associated motion is said to be **uniform** and (2.27) reduces to $\mathbf{a} = \partial\mathbf{v}/\partial t$. If the velocity field has components of the form $v_1 = v_1(x_1, x_2, t), v_2 = v_2(x_1, x_2, t), v_3 = 0$ the associated motion is called **plane**. A motion is **potential** if there exists a spatial velocity field $\mathbf{v} = \mathrm{grad}\Phi$, where the spatial scalar field Φ is called the **velocity potential**. A motion satisfying curl**v** $= \mathbf{o}$ is **irrotational**. By recalling relation (1.274) we conclude that if a motion is potential then it is automatically an irrotational motion.

EXERCISES

1. For a one-dimensional problem the displacement field $U = U(X, t)$ is given by the equation

$$U = ctX \ ,$$

   with c denoting a constant. The relation between the spatial and the material coordinates is given by $x = (1 + ct)X$.

   Derive the first and second derivatives of U with respect to time t in both the material and spatial descriptions.

2. Recall Example 2.2 on p. 63. The spatial components $a_1 = x_1, a_2 = x_2, a_3 = 0$, i.e. eq. (2.14), are derived directly from the acceleration in the material description, i.e. eq. (2.12), by means of given motion (2.10).

   Show that the spatial components a_a, $a = 1, 2, 3$, may also be derived from expression (2.27) using eq. (2.13), i.e. $v_1 = x_2, v_2 = x_1, v_3 = 0$.

3. In a certain region the spatial velocity components of $\mathbf{v} = \mathbf{v}(\mathbf{x}, t)$ are given as

$$v_1 = -\alpha(x_1^3 + x_1 x_2^2)e^{-\beta t} \ , \qquad v_2 = \alpha(x_1^2 x_2 + x_2^3)e^{-\beta t} \ , \qquad v_3 = 0 \ ,$$

   where $\alpha, \beta > 0$ are given constants. Find the components of the spatial acceleration field $\mathbf{a} = \mathbf{a}(\mathbf{x}, t)$ at point $(1, 0, 0)$ and time $t = 0$.

4. Consider a plane motion defined by the velocity components

$$v_1 = -\frac{\partial\Phi(x_1, x_2, t)}{\partial x_2} \ , \qquad v_2 = \frac{\partial\Phi(x_1, x_2, t)}{\partial x_1} \ , \qquad v_3 = 0 \ ,$$

   where Φ is a harmonic scalar field. Show that the plane motion is irrotational and find the spatial acceleration field.

2.4   Deformation Gradient

One goal of this section is to study the deformation (i.e. the changes of size and shape) of a continuum body occurring when moved from the reference configuration Ω_0 to some current configuration Ω. For simplicity, we often omit the arguments of the tensor quantities in subsequent considerations.

Deformation gradient.   As we know from Section 2.1, a typical point $X \in \Omega_0$ identified by the position vector $\mathbf{X}$ maps into the point $x \in \Omega$ with position vector $\mathbf{x}$. Now we want to know how curves and tangent vectors deform.

Consider a **material** (or **undeformed**) **curve** $\mathbf{X} = \mathbf{\Gamma}(\xi) \subset \Omega_0$, $(X_A = \Gamma_A(\xi))$, where ξ denotes a parametrization (see Figure 2.4). The material curve is associated with the reference configuration Ω_0 of the continuum body. Hence, the material curve is not a function of time. During a certain motion χ the material curve deforms into a **spatial** (or **deformed**) **curve** $\mathbf{x} = \boldsymbol{\gamma}(\xi, t) \subset \Omega$, $(x_a = \gamma_a(\xi, t))$, at time t.

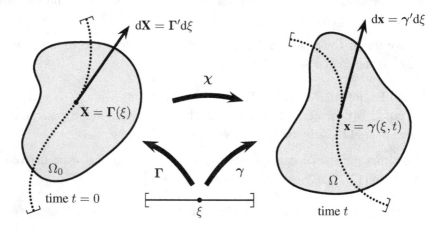

Figure 2.4 Deformation of a material curve $\mathbf{\Gamma} \subset \Omega_0$ into a spatial curve $\boldsymbol{\gamma} \subset \Omega$.

The spatial curve at a fixed time t is then defined by the parametric equation

$$\mathbf{x} = \boldsymbol{\gamma}(\xi, t) = \boldsymbol{\chi}(\mathbf{\Gamma}(\xi), t) \qquad \text{or} \qquad x_a = \gamma_a(\xi, t) = \chi_a(\Gamma_A(\xi), t)) \ . \quad (2.36)$$

We denote the **spatial tangent vector** to the spatial curve as $d\mathbf{x}$ and the **material tangent vector** to the material curve as $d\mathbf{X}$. They are defined by

$$d\mathbf{x} = \boldsymbol{\gamma}'(\xi, t)d\xi \ , \qquad d\mathbf{X} = \mathbf{\Gamma}'(\xi)d\xi \ , \qquad\qquad (2.37)$$

with the abbreviation $(\bullet)' = \partial(\bullet)/\partial\xi$. In the literature the tangent vectors $d\mathbf{x}$ and $d\mathbf{X}$, which are infinitesimal vector elements in the current and reference configuration (see

Figure 2.4), are often referred to as the **spatial** (or **deformed**) **line element** and the **material** (or **undeformed**) **line element**, respectively.

By using (2.36) and the chain rule we find that $\gamma'(\xi, t) = (\partial \chi(\mathbf{X}, t)/\partial \mathbf{X})\Gamma'(\xi)$. Hence, from eq. (2.37) we deduce the fundamental relation

$$\mathrm{d}\mathbf{x} = \mathbf{F}(\mathbf{X}, t)\mathrm{d}\mathbf{X} \qquad \text{or} \qquad \mathrm{d}x_a = F_{aA}\mathrm{d}X_A \quad , \tag{2.38}$$

where the definition

$$\mathbf{F}(\mathbf{X}, t) = \frac{\partial \chi(\mathbf{X}, t)}{\partial \mathbf{X}} = \mathrm{Grad}\mathbf{x}(\mathbf{X}, t)$$

$$\text{or} \qquad F_{aA} = \frac{\partial \chi_a}{\partial X_A} = \mathrm{Grad}_{X_A} x_a \tag{2.39}$$

is to be used. The quantity $\mathbf{F}$ is crucial in nonlinear continuum mechanics and is a primary measure of deformation, called the **deformation gradient**. In general, $\mathbf{F}$ has nine components for all t, and it characterizes the behavior of motion in the neighborhood of a point.

Expression (2.38) clearly defines a *linear* transformation which generates a vector $\mathrm{d}\mathbf{x}$ by the action of the second-order tensor $\mathbf{F}$ on the vector $\mathrm{d}\mathbf{X}$. Hence, eq. (2.38) serves as a transformation rule. Therefore, $\mathbf{F}$ is said to be a **two-point tensor** involving points in two distinct configurations. One index describes *spatial coordinates*, x_a, and the other *material coordinates*, X_A. In summary: material tangent vectors map (i.e. transform) into spatial tangent vectors via the deformation gradient.

We suppose that the derivative of the inverse motion χ^{-1} with respect to the current position $\mathbf{x}$ of a (material) point exists so that

$$\mathbf{F}^{-1}(\mathbf{x}, t) = \frac{\partial \chi^{-1}(\mathbf{x}, t)}{\partial \mathbf{x}} = \mathrm{grad}\mathbf{X}(\mathbf{x}, t)$$

$$\text{or} \qquad F_{Aa}^{-1} = \frac{\partial \chi_A^{-1}}{\partial x_a} = \mathrm{grad}_{x_a} X_A \quad , \tag{2.40}$$

where the tensor $\mathbf{F}^{-1}$ is the **inverse of the deformation gradient**. It carries the spatial line element $\mathrm{d}\mathbf{x}$ into the material line element $\mathrm{d}\mathbf{X}$ according to the (linear) transformation rule $\mathrm{d}\mathbf{X} = \mathbf{F}^{-1}(\mathbf{x}, t)\mathrm{d}\mathbf{x}$, or, in index notation, $\mathrm{d}X_A = F_{Aa}^{-1}\mathrm{d}x_a$.

Generally, the nonsingular (invertible, i.e. $\det\mathbf{F} \neq 0$) tensor $\mathbf{F}$ depends on $\mathbf{X}$ which denotes a so-called **inhomogeneous deformation**. A deformation of a body in question is said to be **homogeneous** if $\mathbf{F}$ does not depend on the space coordinates. The components F_{aA} depend only on time. Every part of a specimen deforms as the whole does. The associated motion is called **affine**. For a rigid-body translation for which the displacement field is independent of $\mathbf{X}$ we have $\mathbf{F} = \mathbf{I}$, $F_{aA} = \delta_{aA}$. However, if there is no motion we have $\mathbf{F} = \mathbf{I}$ and $\mathbf{x} = \mathbf{X}$.

EXAMPLE 2.5   For a two-dimensional problem the deformation is given by the explicit equations

$$x_1 = 4 - 2X_1 - X_2 \ , \qquad x_2 = 2 + \frac{3}{2}X_1 - \frac{1}{2}X_2 \ . \qquad (2.41)$$

Determine the matrix representation of the deformation gradient and its inverse and study the deformation of a unit square with region Ω_0 (see Figure 2.5). Consider a material line element $\mathbf{a}_0$ and a spatial line element $\mathbf{b}$ with unit lengths. Show how $\mathbf{a}_0$ deforms, and carry out the inverse operation with $\mathbf{b}$. For the components of $\mathbf{a}_0$ and $\mathbf{b}$ take $(1/\sqrt{2}, 1/\sqrt{2})$ and $(1, 0)$, respectively.

Solution.   By recalling definitions (2.39) and (2.40) we find after some simple algebra with the given deformation (2.41) that

$$[\mathbf{F}] = \begin{bmatrix} -2 & -1 \\ 3/2 & -1/2 \end{bmatrix} \ , \qquad [\mathbf{F}^{-1}] = \begin{bmatrix} -1/5 & 2/5 \\ -3/5 & -4/5 \end{bmatrix} \ . \qquad (2.42)$$

The given deformation χ carries the unit vector $\mathbf{a}_0$ into its new position characterized by $\mathbf{a}$ (see Figure 2.5). By means of $(2.42)_1$ and the given components of $\mathbf{a}_0$ we may specify the deformed position of $\mathbf{a}_0$, i.e.

$$[\mathbf{a}] = [\mathbf{F}][\mathbf{a}_0] = \begin{bmatrix} -2 & -1 \\ 3/2 & -1/2 \end{bmatrix} \begin{bmatrix} 1/\sqrt{2} \\ 1/\sqrt{2} \end{bmatrix}$$

$$= \frac{1}{\sqrt{2}} \begin{bmatrix} -3 \\ 1 \end{bmatrix} \ , \qquad |\mathbf{a}| = 5^{1/2} \ . \qquad (2.43)$$

The inverse deformation χ^{-1} carries the unit vector $\mathbf{b}$ into its undeformed position characterized by $\mathbf{b}_0$. By analogy with the above, using $(2.42)_2$ and the given components of $\mathbf{b}$, we may specify the undeformed position of $\mathbf{b}$, i.e.

$$[\mathbf{b}_0] = [\mathbf{F}^{-1}][\mathbf{b}] = \begin{bmatrix} -1/5 & 2/5 \\ -3/5 & -4/5 \end{bmatrix} \begin{bmatrix} 1 \\ 0 \end{bmatrix}$$

$$= -\frac{1}{5} \begin{bmatrix} 1 \\ 3 \end{bmatrix} \ , \qquad |\mathbf{b}_0| = \left(\frac{2}{5}\right)^{1/2} . \qquad (2.44)$$

Observe that during deformation χ the length of $\mathbf{a}_0$ increases from 1 up to $5^{1/2}$, while during the inverse map χ^{-1} the length of $\mathbf{b}$ decreases from 1 up to $(2/5)^{1/2}$ (see Figure 2.5).   ∎

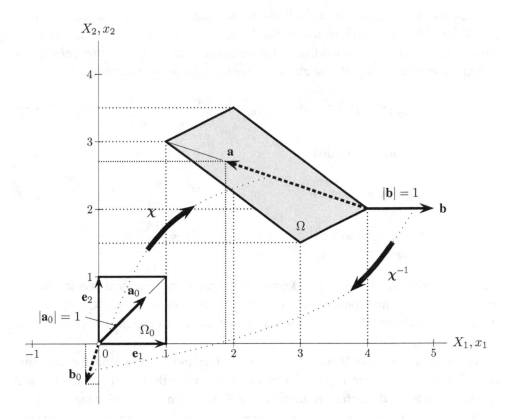

Figure 2.5 Deformation of a unit square showing the map χ of a material line element $\mathbf{a}_0$ into $\mathbf{a}$ and the inverse map χ^{-1} of a spatial line element $\mathbf{b}$ into $\mathbf{b}_0$.

Displacement gradient tensor. To combine the deformation gradient with the displacement vector we deduce from (2.4) and definition (2.39) that

$$\text{Grad}\mathbf{U} = \text{Grad}\mathbf{x}(\mathbf{X}, t) - \text{Grad}\mathbf{X}$$

$$= \mathbf{F}(\mathbf{X}, t) - \mathbf{I} \quad \text{or} \quad \frac{\partial U_a}{\partial X_A} = F_{aA} - \delta_{aA} \ . \tag{2.45}$$

The second-order tensor $\text{Grad}\mathbf{U}$ is called the **displacement gradient tensor** in the *material description*. From (2.5) we find using definition (2.40) that

$$\text{grad}\mathbf{u} = \text{grad}\mathbf{x} - \text{grad}\mathbf{X}(\mathbf{x}, t)$$

$$= \mathbf{I} - \mathbf{F}^{-1}(\mathbf{x}, t) \quad \text{or} \quad \frac{\partial u_A}{\partial x_a} = \delta_{Aa} - F_{Aa}^{-1} \ , \tag{2.46}$$

where the second-order tensor $\text{grad}\mathbf{u}$ is called the **displacement gradient tensor** in the *spatial description*.

Note the following relationships between the material gradient and material divergence, and the spatial gradient and spatial divergence of the smooth scalar, vector and tensor fields, Φ, $\mathbf{u}$ and $\mathbf{A}$, respectively. The fields are defined on the current configuration of a continuum body. By the chain rule we have the useful properties

$$\text{grad}\Phi = \mathbf{F}^{-T}\text{Grad}\Phi \qquad \text{or} \qquad \frac{\partial\Phi}{\partial x_a} = F^{-1}_{Aa}\frac{\partial\Phi}{\partial X_A} \quad , \tag{2.47}$$

$$\text{grad}\mathbf{u} = \text{Grad}\mathbf{u}\,\mathbf{F}^{-1} \qquad \text{or} \qquad \frac{\partial u_a}{\partial x_b} = \frac{\partial u_a}{\partial X_A}F^{-1}_{Ab} \quad , \tag{2.48}$$

$$\text{grad}\mathbf{A} = \text{Grad}\mathbf{A}\,\mathbf{F}^{-1} \qquad \text{or} \qquad \frac{\partial A_{ab}}{\partial x_c} = \frac{\partial A_{ab}}{\partial X_A}F^{-1}_{Ac} \quad , \tag{2.49}$$

$$\text{div}\mathbf{A} = \text{Grad}\mathbf{A} : \mathbf{F}^{-T} \qquad \text{or} \qquad \frac{\partial A_{ab}}{\partial x_b} = \frac{\partial A_{ab}}{\partial X_B}F^{-1}_{Bb} \quad . \tag{2.50}$$

Nanson's formula.   We already know that points, curves, tangent vectors, for example, $\mathbf{X}$, Γ, $d\mathbf{X}$, map onto points, curves, tangent vectors $\mathbf{x}$, γ, $d\mathbf{x}$, respectively. An arbitrary differential vector maps via the deformation gradient $\mathbf{F}$ (see the linear transformation(2.38)).

However, a unit vector $\mathbf{N}$ normal to an infinitesimal **material** (or **undeformed**) **surface element** dS *does not* map to a unit vector $\mathbf{n}$ normal to the associated infinitesimal **spatial** (or **deformed**) **surface element** ds via $\mathbf{F}$, as shown in the following.

We perform the change in volume between the reference and the current configuration at time t

$$dv = J(\mathbf{X}, t)dV \quad , \tag{2.51}$$

$$J(\mathbf{X}, t) = \det\mathbf{F}(\mathbf{X}, t) > 0 \quad , \tag{2.52}$$

in which J is the determinant of the deformation gradient $\mathbf{F}$, known as the **volume ratio** (or **Jacobian determinant**). In (2.51), dV and dv denote infinitesimal volume elements defined in the reference and current configurations called **material** (or **undeformed**) and **spatial** (or **deformed**) **volume elements**, respectively. Further, we assume that the volume is a continuous (or at least a piecewise continuous) function of continuum particles so that $dV = dX_1 dX_2 dX_3$ and $dv = dx_1 dx_2 dx_3$ (continuum idealization).

Since $\mathbf{F}$ is invertible we have $J(\mathbf{X}, t) = \det\mathbf{F}(\mathbf{X}, t) \neq 0$. Because of the impenetrability of matter, i.e. volume elements cannot have negative volumes, we reject $J(\mathbf{X}, t) < 0$ which mathematically is possible. Consequently, the volume ratio $J(\mathbf{X}, t) > 0$ must be greater than zero for all $\mathbf{X} \in \Omega_0$ and for all times t. The inverse of relation (2.52) follows with identity (1.109) as $J^{-1} = \det\mathbf{F}^{-1}(\mathbf{x}, t) > 0$, with $\mathbf{F}^{-1}$ introduced in eq. (2.40).

If there is no motion ($\mathbf{F} = \mathbf{I}$ and $\mathbf{x} = \mathbf{X}$), we obtain the consistency condition $J = 1$, since $\det\mathbf{F} = \det\mathbf{I} = 1$. However, a motion (or a deformation) with $J = 1$ (at every particle in every configuration and time t) is called **isochoric** or **volume-preserving**. It keeps the volume constant.

In order to compute the relationship between the unit vectors $\mathbf{n}$ and $\mathbf{N}$ consider an *arbitrary* material line element $d\mathbf{X}$, which maps to $d\mathbf{x}$ during a certain motion χ. We now express the infinitesimal volume element in the current configuration dv as a dot product. By means of (2.51) we have the following relation

$$dv = d\mathbf{s} \cdot d\mathbf{x} = J d\mathbf{S} \cdot d\mathbf{X} \quad , \tag{2.53}$$

with $d\mathbf{s} = ds\mathbf{n}$ and $d\mathbf{S} = dS\mathbf{N}$ denoting *vector elements* of infinitesimally small areas defined in the current and reference configurations, respectively.

With transformation (2.38) and identity (1.81) we may rewrite eq. (2.53) as

$$\underbrace{(\mathbf{F}^\mathrm{T}d\mathbf{s} - J d\mathbf{S})}_{\mathbf{0}} \cdot d\mathbf{X} = 0 \quad . \tag{2.54}$$

Since (2.54) holds for arbitrary material line elements $d\mathbf{X}$, we find that

$$d\mathbf{s} = J\mathbf{F}^{-\mathrm{T}}d\mathbf{S} \quad , \tag{2.55}$$

which shows how the vector elements of the infinitesimally small areas $d\mathbf{s}$ and $d\mathbf{S}$ on the current and reference configurations are related. Relationship (2.55) is well-known as **Nanson's formula**.

EXERCISES

1. Consider a parametric curve in space of the form $\mathbf{X} = \boldsymbol{\Gamma}(\xi)$, with

$$X_1 = \Gamma_1(\xi) = \cos\xi \quad , \qquad X_2 = \Gamma_2(\xi) = \sin\xi \quad , \qquad X_3 = \Gamma_3(\xi) = \xi \quad ,$$

   defining a helix. Find the length of the helix for $0 \le \xi \le \Pi$ ($\int |\boldsymbol{\Gamma}'(\xi)|d\xi$).

2. In a deformation of a three-dimensional problem, the displacement components of $\mathbf{u}$ are found to be

$$u_1 = x_1 - \frac{1}{4}x_2 \quad , \qquad u_2 = x_1 + 2x_2 \quad , \qquad u_3 = -3x_3 \quad .$$

   Compute the matrix representations of $\mathbf{F}^{-1}$ and $\mathbf{F}$ and deduce that the deformation is isochoric.

3. Let a continuum body undergo a homogeneous deformation which is defined by $\mathbf{x} = \chi(\mathbf{X}, t) = \mathbf{c}(t) + \mathbf{A}(t)\mathbf{X}$, where the components of the vector $\mathbf{c}$ and the tensor $\mathbf{A}$ are *constants* or *time-dependent* functions.

(a) Show that $[\mathbf{F}] = [\mathbf{A}]$ and interpret this result.

(b) Consider the components of vector $\mathbf{c}$ and tensor $\mathbf{A}$ given by

$$[\mathbf{c}] = \begin{bmatrix} \alpha \\ 2 \\ 0 \end{bmatrix} , \qquad [\mathbf{A}] = \begin{bmatrix} \sqrt{2} & 0 & 0 \\ 0 & 2\beta & 0 \\ 0 & 0 & \sqrt{3}/2\,\gamma \end{bmatrix} ,$$

where α, β, γ are constants. Show that a particle that lies on a spherical surface in the current configuration initially lies on the surface of an ellipsoid.

(c) Consider (b) and determine how a unit normal $\mathbf{N}$ of an infinitesimal small area dS in the reference configuration deforms. Take $\mathbf{N} = 1/\sqrt{3}\,(\mathbf{e}_1 - \mathbf{e}_2 + \mathbf{e}_3)$.

4. Let $\mathbf{A}$ be a Cartesian tensor of order two. Show by means of the analogue of definition $(1.93)_3$ and properties (2.49), (1.283) that

$$\mathrm{Grad}\mathbf{A} : \mathbf{F}^{-\mathrm{T}} = \mathrm{tr}(\mathrm{grad}\mathbf{A}) = \mathrm{div}\mathbf{A} \ . \tag{2.56}$$

2.5 Strain Tensors

We have learnt from the preceding section that the deformation gradient is the fundamental kinematic (second-order) tensor in finite deformation kinematics that characterizes changes of material elements during motion. The aim of this section is to determine these changes in the form of (second-order) strain tensors related to either the reference or the current configuration.

Note that unlike displacements, which are measurable quantities, strains are based on a *concept* that is introduced to simplify analyses. Therefore, numerous definitions and names of strain tensors have been proposed in the literature. We discuss (and compare) the most common definitions of strain tensors established in nonlinear continuum mechanics.

Material strain tensors.   We compute the change in length between two **neighboring** points $\mathbf{X}$ and $\mathbf{Y}$, located in region Ω_0, occurring during a motion (see Figure 2.6). By neighboring we mean that $\mathbf{X}$ is 'close' to $\mathbf{Y}$.

The geometry in the *reference configuration* is given by

$$\mathbf{Y} = \mathbf{Y} + (\mathbf{X} - \mathbf{X}) = \mathbf{X} + |\mathbf{Y} - \mathbf{X}|\frac{\mathbf{Y} - \mathbf{X}}{|\mathbf{Y} - \mathbf{X}|} = \mathbf{X} + d\mathbf{X} \ , \tag{2.57}$$

$$d\mathbf{X} = d\varepsilon\,\mathbf{a}_0 \quad \text{and} \quad d\varepsilon = |\mathbf{Y} - \mathbf{X}| \ , \quad \mathbf{a}_0 = \frac{\mathbf{Y} - \mathbf{X}}{|\mathbf{Y} - \mathbf{X}|} \ . \tag{2.58}$$

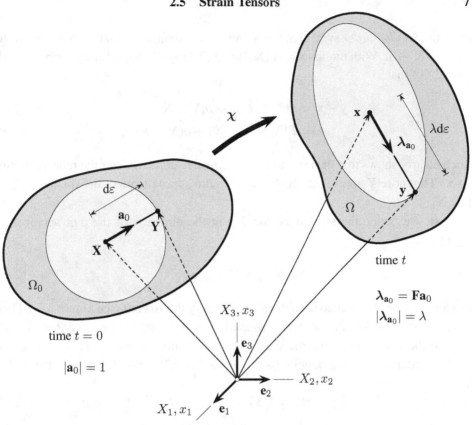

Figure 2.6 Deformation of a material line element with length $d\varepsilon$ into a spatial line element with length $\lambda d\varepsilon$.

We denote the (material) length of the material line element $d\mathbf{X} = \mathbf{Y} - \mathbf{X}$ by $d\varepsilon$. It is the distance between the neighboring points $\mathbf{X} \in \Omega_0$ and $\mathbf{Y} \in \Omega_0$, i.e. $d\varepsilon = |\mathbf{Y} - \mathbf{X}|$ with $d\varepsilon/|\mathbf{X}| \ll 1$ ($|\mathbf{X}| \neq 0$). The unit vector $\mathbf{a}_0$, $|\mathbf{a}_0| = 1$, at the referential position $\mathbf{X}$ describes the direction of the material line element (which may be imagined as a **fiber**), as illustrated in Figure 2.6. Hence, additionally, we find using $(2.58)_1$ that

$$d\mathbf{X} \cdot d\mathbf{X} = d\varepsilon \mathbf{a}_0 \cdot d\varepsilon \mathbf{a}_0 = d\varepsilon^2 \quad . \tag{2.59}$$

Note that the vector quantities $d\mathbf{X}$ and $\mathbf{a}_0$ are naturally associated with the *reference configuration* of the body.

Certain motions transform the two neighboring points $\mathbf{X}$ and $\mathbf{Y}$ into their displaced positions $\mathbf{x} = \chi(\mathbf{X}, t)$ and $\mathbf{y} = \chi(\mathbf{Y}, t)$ of region Ω, respectively. We now ask how close is $\mathbf{x}$ to $\mathbf{y}$. Using Taylor's expansion according to (1.238), (1.239), $\mathbf{y}$ may be expressed by means of (2.57), (2.58) and the deformation gradient (2.39), as

$$\begin{aligned} \mathbf{y} = \chi(\mathbf{Y}, t) &= \chi(\mathbf{X} + d\varepsilon \mathbf{a}_0, t) \\ &= \chi(\mathbf{X}, t) + d\varepsilon \mathbf{F}(\mathbf{X}, t)\mathbf{a}_0 + o(\mathbf{Y} - \mathbf{X}) \quad , \end{aligned} \tag{2.60}$$

where the *Landau order* symbol $o(\mathbf{Y}-\mathbf{X})$ refers to a small error that tends to zero faster than $\mathbf{Y} - \mathbf{X} \rightarrow \mathbf{o}$. With motion $\mathbf{x} = \chi(\mathbf{X}, t)$ and $(2.58)_2$, $(2.58)_3$ it follows subsequently from (2.60) that

$$\begin{aligned} \mathbf{y} - \mathbf{x} &= \mathrm{d}\varepsilon \mathbf{F}(\mathbf{X}, t)\mathbf{a}_0 + o(\mathbf{Y} - \mathbf{X}) \\ &= \mathbf{F}(\mathbf{X}, t)(\mathbf{Y} - \mathbf{X}) + o(\mathbf{Y} - \mathbf{X}) \quad, \end{aligned} \tag{2.61}$$

which clearly shows that the term $\mathbf{F}(\mathbf{Y} - \mathbf{X})$ *linearly approximates* the relative motion $\mathbf{y} - \mathbf{x}$. The more $\mathbf{Y}$ approaches $\mathbf{X}$ the better is the approximation, the smaller is $\mathrm{d}\varepsilon = |\mathbf{Y} - \mathbf{X}|$.

Next, we define the **stretch vector** $\boldsymbol{\lambda}_{\mathbf{a}_0}$ in the direction of the unit vector $\mathbf{a}_0$ at $\mathbf{X} \in \Omega_0$, i.e.

$$\boldsymbol{\lambda}_{\mathbf{a}_0}(\mathbf{X}, t) = \mathbf{F}(\mathbf{X}, t)\mathbf{a}_0 \quad, \tag{2.62}$$

with length $\lambda = |\boldsymbol{\lambda}_{\mathbf{a}_0}|$ called **stretch ratio** or simply the **stretch** (see Figure 2.6). Then, the length of a spatial line element (originally in the direction of $\mathbf{a}_0$), i.e. the distance between the two neighboring places $\mathbf{x}$ and $\mathbf{y}$, is obtained from $(2.61)_1$ by neglecting terms of order $\mathrm{d}\varepsilon^2$. Using definitions $(1.15)_1$ and (2.62) we find with $\mathbf{y} - \mathbf{x} = \mathrm{d}\varepsilon \mathbf{F}\mathbf{a}_0$ that

$$|\mathbf{y} - \mathbf{x}| = [(\mathbf{y} - \mathbf{x}) \cdot (\mathbf{y} - \mathbf{x})]^{1/2} = (\boldsymbol{\lambda}_{\mathbf{a}_0} \cdot \boldsymbol{\lambda}_{\mathbf{a}_0})^{1/2}\mathrm{d}\varepsilon = \lambda \mathrm{d}\varepsilon \quad. \tag{2.63}$$

In summary: a material line element $\mathrm{d}\mathbf{X}$ at $\mathbf{X}$ with length $\mathrm{d}\varepsilon$ at time $t = 0$ becomes the length $\lambda \mathrm{d}\varepsilon$ at time t. The stretch λ is a measure of how much the unit vector $\mathbf{a}_0$ has stretched. We say that a line element is **extended**, **unstretched** or **compressed** according to $\lambda > 1$, $\lambda = 1$ or $\lambda < 1$, respectively.

With definitions $(1.15)_2$, (2.62) and property (1.81), the square of λ is computed according to

$$\begin{aligned} \lambda^2 &= \boldsymbol{\lambda}_{\mathbf{a}_0} \cdot \boldsymbol{\lambda}_{\mathbf{a}_0} = \mathbf{F}\mathbf{a}_0 \cdot \mathbf{F}\mathbf{a}_0 \\ &= \mathbf{a}_0 \cdot \mathbf{F}^{\mathrm{T}}\mathbf{F}\mathbf{a}_0 = \mathbf{a}_0 \cdot \mathbf{C}\mathbf{a}_0 \quad, \end{aligned} \tag{2.64}$$

$$\mathbf{C} = \mathbf{F}^{\mathrm{T}}\mathbf{F} \qquad \text{or} \qquad C_{AB} = F_{aA}F_{aB} \quad, \tag{2.65}$$

where we have introduced the **right Cauchy-Green tensor** $\mathbf{C}$ as an important strain measure in material coordinates ($\mathbf{F}$ is on the *right*). Frequently in the literature $\mathbf{C}$ is referred to as the **Green deformation tensor**. From (2.65) we learn that to determine the stretch of a fiber one only needs the direction $\mathbf{a}_0$ at a point $\mathbf{X} \in \Omega_0$ and the second-order tensor $\mathbf{C}$.

Note that $\mathbf{C}$ is *symmetric* and *positive definite* at each $\mathbf{X} \in \Omega_0$. Thus,

$$\mathbf{C} = \mathbf{F}^{\mathrm{T}}\mathbf{F} = (\mathbf{F}^{\mathrm{T}}\mathbf{F})^{\mathrm{T}} = \mathbf{C}^{\mathrm{T}} \qquad \text{and} \qquad \mathbf{u} \cdot \mathbf{C}\mathbf{u} > 0 \quad \text{for all} \quad \mathbf{u} \neq \mathbf{o} \quad. \tag{2.66}$$

Consequently, given the nine components F_{aA}, it is easy to compute the six components $C_{AB} = C_{BA}$ via (2.65), but given C_{AB} it is impossible to compute the nine components F_{aA}. With definition (2.65) and eqs. (1.101) and (2.52) we find that

$$\det\mathbf{C} = (\det\mathbf{F})^2 = J^2 > 0 \quad . \tag{2.67}$$

The so-called **Piola deformation tensor**, denoted by $\mathbf{B}$, is defined by the inverse of the right Cauchy-Green tensor, i.e. $\mathbf{B} = \mathbf{C}^{-1}$, with $\mathbf{C}^{-1} = (\mathbf{F}^{\mathrm{T}}\mathbf{F})^{-1} = \mathbf{F}^{-1}\mathbf{F}^{-\mathrm{T}}$.

As a further strain measure we define the change in the squared lengths, i.e. $(\lambda d\varepsilon)^2 - d\varepsilon^2$. With $(2.64)_3$, the use of the unit tensor $\mathbf{I}$ and eq. $(2.59)_2$ we have

$$\frac{1}{2}[(\lambda d\varepsilon)^2 - d\varepsilon^2] = \frac{1}{2}[(d\varepsilon\mathbf{a}_0) \cdot \mathbf{F}^{\mathrm{T}}\mathbf{F}(d\varepsilon\mathbf{a}_0) - d\varepsilon^2] = d\mathbf{X} \cdot \mathbf{E} d\mathbf{X} \quad , \tag{2.68}$$

$$\mathbf{E} = \frac{1}{2}(\mathbf{F}^{\mathrm{T}}\mathbf{F} - \mathbf{I}) \quad \text{or} \quad E_{AB} = \frac{1}{2}(F_{aA}F_{aB} - \delta_{AB}) \quad , \tag{2.69}$$

where the introduced normalization factor $1/2$ will be evident within the linear theory. This expression describes a strain measure in the direction of $\mathbf{a}_0$ at point $\mathbf{X} \in \Omega_0$. In (2.69) we have introduced the commonly used strain tensor $\mathbf{E}$, which is known as the **Green-Lagrange strain tensor**. Since $\mathbf{I}$ and $\mathbf{C}$ are symmetric we deduce from (2.69) that $\mathbf{E} = \mathbf{E}^{\mathrm{T}}$ also.

So far the introduced strain tensors operate solely on the material vectors $\mathbf{a}_0, \mathbf{X}$. Thus, $\mathbf{C}, \mathbf{E}$ and their inverse are also referred to as **material strain tensors**.

Spatial strain tensors.   In order to relate strain measures to quantities which are associated with the current configuration we continue with arguments entirely similar to those just used.

The geometry in the *current configuration* is given by

$$\mathbf{y} = \mathbf{y} + (\mathbf{x} - \mathbf{x}) = \mathbf{x} + |\mathbf{y} - \mathbf{x}|\frac{\mathbf{y} - \mathbf{x}}{|\mathbf{y} - \mathbf{x}|} = \mathbf{x} + d\mathbf{x} \quad , \tag{2.70}$$

$$d\mathbf{x} = d\tilde{\varepsilon}\mathbf{a} \quad \text{and} \quad d\tilde{\varepsilon} = |\mathbf{y} - \mathbf{x}| \quad , \quad \mathbf{a} = \frac{\mathbf{y} - \mathbf{x}}{|\mathbf{y} - \mathbf{x}|} \quad . \tag{2.71}$$

The (spatial) length of the spatial line element $d\mathbf{x} = \mathbf{y} - \mathbf{x}$ is given by $d\tilde{\varepsilon} = |\mathbf{y} - \mathbf{x}|$, with $d\tilde{\varepsilon}/|\mathbf{x}| \ll 1$ ($|\mathbf{x}| \neq 0$). The unit vector $\mathbf{a}$, $|\mathbf{a}| = 1$, acts at the current position $\mathbf{x}$ and points in the direction of the spatial line element, which is the direction of $\boldsymbol{\lambda}_{\mathbf{a}_0}$ (see Figure 2.7). Since $|\mathbf{a}| = 1$, we find using $(2.71)_1$ that

$$d\mathbf{x} \cdot d\mathbf{x} = d\tilde{\varepsilon}\mathbf{a} \cdot d\tilde{\varepsilon}\mathbf{a} = d\tilde{\varepsilon}^2 \quad . \tag{2.72}$$

The vector $\mathbf{a}$ may be viewed as a spatial element characterizing the direction of a fiber in the current configuration.

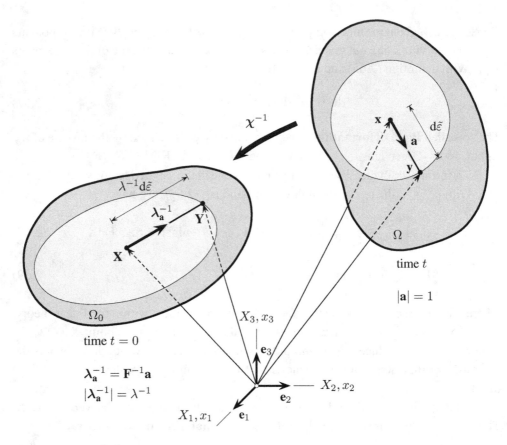

Figure 2.7 Deformation of a spatial line element with length $d\tilde{\varepsilon}$ into a material line element with length $\lambda^{-1}d\tilde{\varepsilon}$.

In order to perform the relationship between $\boldsymbol{\lambda}_{\mathbf{a}_0}$ and $\mathbf{a}$ we recall eqs. $(2.61)_1$, $(2.63)_3$ and definition $(2.71)_3$ to give $\mathbf{Fa}_0 = \lambda\mathbf{a}$. Hence, from (2.62) we deduce that

$$\boldsymbol{\lambda}_{\mathbf{a}_0} = \lambda\mathbf{a} \quad . \tag{2.73}$$

Note that the vector quantities $\mathbf{a}$ and $d\mathbf{x}$ are naturally associated with the *current configuration* of the body.

Using Taylor's expansion, the associated position vector $\mathbf{Y} \in \Omega_0$, which is described by the inverse motion $\boldsymbol{\chi}^{-1}(\mathbf{y}, t)$, may be expressed by means of (2.70), (2.71) and the inverse of the deformation gradient (2.40). By the chain rule we find, by analogy with eq. (2.60), that

$$\begin{aligned}
\mathbf{Y} &= \boldsymbol{\chi}^{-1}(\mathbf{y}, t) = \boldsymbol{\chi}^{-1}(\mathbf{x} + d\tilde{\varepsilon}\mathbf{a}, t) \\
&= \boldsymbol{\chi}^{-1}(\mathbf{x}, t) + d\tilde{\varepsilon}\mathbf{F}^{-1}(\mathbf{x}, t)\mathbf{a} + o(\mathbf{y} - \mathbf{x}) \quad .
\end{aligned} \tag{2.74}$$

In an analogous way to (2.62), we now define the **stretch vector** $\boldsymbol{\lambda}_\mathbf{a}$ in the direction of the unit vector $\mathbf{a}$ at $\mathbf{x} \in \Omega$, i.e.

$$\boldsymbol{\lambda}_\mathbf{a}^{-1}(\mathbf{x}, t) = \mathbf{F}^{-1}(\mathbf{x}, t)\mathbf{a} \quad . \tag{2.75}$$

The length of a material line element (originally in the direction of $\mathbf{a}$) is obtained from $(2.74)_3$ by neglecting terms of order $\mathrm{d}\tilde{\varepsilon}^2$. With definitions $(1.15)_1$ and (2.75), we obtain, by means of $\mathbf{Y} - \mathbf{X} = \mathrm{d}\tilde{\varepsilon}\mathbf{F}^{-1}\mathbf{a}$ that

$$|\mathbf{Y} - \mathbf{X}| = [(\mathbf{Y} - \mathbf{X}) \cdot (\mathbf{Y} - \mathbf{X})]^{1/2} = (\boldsymbol{\lambda}_\mathbf{a}^{-1} \cdot \boldsymbol{\lambda}_\mathbf{a}^{-1})^{1/2}\mathrm{d}\tilde{\varepsilon} = \lambda^{-1}\mathrm{d}\tilde{\varepsilon} \quad , \tag{2.76}$$

where the length of the inverse stretch vector $\boldsymbol{\lambda}_\mathbf{a}^{-1}$ is the **inverse stretch ratio** λ^{-1} (or simply the **inverse stretch**) (see Figure 2.7).

The square of λ^{-1} follows with definitions $(1.15)_2$, (2.75) and property (1.81) as

$$\begin{aligned} \lambda^{-2} &= \boldsymbol{\lambda}_\mathbf{a}^{-1} \cdot \boldsymbol{\lambda}_\mathbf{a}^{-1} = \mathbf{F}^{-1}\mathbf{a} \cdot \mathbf{F}^{-1}\mathbf{a} \\ &= \mathbf{a} \cdot \mathbf{F}^{-\mathrm{T}}\mathbf{F}^{-1}\mathbf{a} = \mathbf{a} \cdot \mathbf{b}^{-1}\mathbf{a} \quad , \end{aligned} \tag{2.77}$$

$$\mathbf{b}^{-1} = \mathbf{F}^{-\mathrm{T}}\mathbf{F}^{-1} \quad \text{or} \quad b_{ab}^{-1} = F_{Aa}^{-1}F_{Ab}^{-1} \quad . \tag{2.78}$$

The strain tensor $\mathbf{b}^{-1}$ is the inverse of the **left Cauchy-Green tensor b**, which is defined by

$$\mathbf{b} = \mathbf{FF}^{\mathrm{T}} \quad \text{or} \quad b_{ab} = F_{aA}F_{bA} \tag{2.79}$$

($\mathbf{F}$ is on the *left*). In the literature the left Cauchy-Green tensor $\mathbf{b}$ is sometimes referred to as the **Finger deformation tensor**. It is an important strain measure in terms of spatial coordinates. The left Cauchy-Green tensor is *symmetric* and *positive definite* at each $\mathbf{x} \in \Omega$,

$$\mathbf{b} = \mathbf{FF}^{\mathrm{T}} = (\mathbf{FF}^{\mathrm{T}})^{\mathrm{T}} = \mathbf{b}^{\mathrm{T}} \quad \text{and} \quad \mathbf{u} \cdot \mathbf{bu} > 0 \quad \text{for all} \quad \mathbf{u} \neq \mathbf{o} \quad . \tag{2.80}$$

With definition (2.79) and (1.101), (2.52) we find consequently that

$$\det \mathbf{b} = (\det \mathbf{F})^2 = J^2 > 0 \quad . \tag{2.81}$$

As a last important strain measure we define the change in the squared lengths, i.e. $\mathrm{d}\tilde{\varepsilon}^2 - (\lambda^{-1}\mathrm{d}\tilde{\varepsilon})^2$. With $(2.77)_3$, the use of the unit tensor $\mathbf{I}$ and eq. $(2.72)_2$ we find a relation expressed solely through quantities in Ω. Thus,

$$\frac{1}{2}[\mathrm{d}\tilde{\varepsilon}^2 - (\lambda^{-1}\mathrm{d}\tilde{\varepsilon})^2] = \frac{1}{2}[\mathrm{d}\tilde{\varepsilon}^2 - (\mathrm{d}\tilde{\varepsilon}\mathbf{a}) \cdot \mathbf{F}^{-\mathrm{T}}\mathbf{F}^{-1}(\mathrm{d}\tilde{\varepsilon}\mathbf{a})] = \mathrm{d}\mathbf{x} \cdot \mathbf{e}\mathrm{d}\mathbf{x} \quad , \tag{2.82}$$

$$\mathbf{e} = \frac{1}{2}(\mathbf{I} - \mathbf{F}^{-\mathrm{T}}\mathbf{F}^{-1}) \quad \text{or} \quad e_{ab} = \frac{1}{2}(\delta_{ab} - F_{Ca}^{-1}F_{Cb}^{-1}) \quad . \tag{2.83}$$

Relation (2.83) describes a strain measure in the direction of $\mathbf{a}$ at place $\mathbf{x} \in \Omega$. We have introduced the commonly used *symmetric* strain tensor $\mathbf{e}$, which is well-known as the **Euler-Almansi strain tensor**.

Since the strain tensors $\mathbf{b}, \mathbf{e}$ and their inverse operate on the spatial vectors $\mathbf{a}, \mathbf{x}$, we call them **spatial strain tensors**.

We now consider the case $\mathbf{C} = \mathbf{F}^{\mathrm{T}}\mathbf{F} = \mathbf{I}$. From the relation $(2.58)_2$ we know that the distance between any two neighboring points $\mathbf{X}$ and $\mathbf{Y}$ located in the reference configuration is $d\varepsilon = |\mathbf{Y} - \mathbf{X}|$. On the other hand the distance between the associated neighboring points $\mathbf{x}$ and $\mathbf{y}$ located in the current configuration is given via $(2.63)_3$, i.e. $\lambda d\varepsilon = |\mathbf{y} - \mathbf{x}|$. For $\mathbf{C} = \mathbf{I}$ we conclude from $(2.64)_4$ that the line element is unstretched, i.e. $\lambda = 1$, and consequently $d\varepsilon = |\mathbf{y} - \mathbf{x}| = |\mathbf{Y} - \mathbf{X}|$. Hence, the distance between any two points is unchanged during such a motion. This means that there is *no* relative motion of points under χ. Since $\mathbf{C} = \mathbf{I}$ we conclude additionally from relation (2.69) that the strain tensor $\mathbf{E}$ vanishes identically, which means that the body does not change its size and shape (no changes in distances and angles).

This particular motion, which preserves the distance between any pair of points of a continuum body, is called a **rigid-body motion** and is dealt with in more detail in Section 5.2. Hence, a rigid-body motion induces no strains and consequently no stresses. A body which is only able to undergo a rigid-body motion is said to be a **rigid body**. The idealization that a body is rigid is often considered in engineering dynamics.

Push-forward, pull-back operation.   As already seen, vector and tensor-valued quantities may be resolved along triads of basis vectors belonging to either the reference or the current configuration. Additionally, there are two-point tensors which are associated with both configurations, one example being the deformation gradient (2.39). The transformations between material and spatial quantities are typically called a **push-forward operation** and a **pull-back operation** (familiar in differential geometry) and are denoted by short-hand $\chi_*(\bullet)$ and $\chi_*^{-1}(\bullet)$, respectively. In the literature the pull-back operation is often written as $\chi^*(\bullet)$.

In particular, a *push-forward* is an operation which transforms a vector or tensor-valued quantity based on the reference configuration to the current configuration. Since the Euler-Almansi strain tensor $\mathbf{e}$ is defined with respect to spatial coordinates we can compute it as a push-forward of the Green-Lagrange strain tensor $\mathbf{E}$, which is given in terms of material coordinates. From eq. (2.83) we conclude, using definition (2.69), that

$$
\begin{aligned}
\mathbf{e} &= \frac{1}{2}(\mathbf{I} - \mathbf{F}^{-\mathrm{T}}\mathbf{F}^{-1}) = \mathbf{F}^{-\mathrm{T}}[\frac{1}{2}\mathbf{F}^{\mathrm{T}}(\mathbf{I} - \mathbf{F}^{-\mathrm{T}}\mathbf{F}^{-1})\mathbf{F}]\mathbf{F}^{-1} \\
&= \mathbf{F}^{-\mathrm{T}}[\frac{1}{2}(\mathbf{F}^{\mathrm{T}}\mathbf{F} - \mathbf{I})]\mathbf{F}^{-1} = \mathbf{F}^{-\mathrm{T}}\mathbf{E}\mathbf{F}^{-1} \\
&= \chi_*(\mathbf{E}) \ .
\end{aligned}
\tag{2.84}
$$

A *pull-back* is an inverse operation, which transforms a vector or tensor-valued quantity based on the current configuration to the reference configuration. Similarly to the above, the pull-back of **e** is

$$\mathbf{E} = \frac{1}{2}(\mathbf{F}^T\mathbf{F} - \mathbf{I}) = \mathbf{F}^T[\frac{1}{2}\mathbf{F}^{-T}(\mathbf{F}^T\mathbf{F} - \mathbf{I})\mathbf{F}^{-1}]\mathbf{F}$$
$$= \mathbf{F}^T[\frac{1}{2}(\mathbf{I} - \mathbf{F}^{-T}\mathbf{F}^{-1})]\mathbf{F} = \mathbf{F}^T\mathbf{e}\mathbf{F}$$
$$= \boldsymbol{\chi}_*^{-1}(\mathbf{e}) \quad . \tag{2.85}$$

As can be seen from eqs. (2.84) and (2.85) the transformations are based on multi-plications by one description of the deformation gradient, i.e. $\mathbf{F}, \mathbf{F}^{-1}, \mathbf{F}^T, \mathbf{F}^{-T}$. Which form of the deformation gradient we have to take depends on the tensor to be trans-formed.

Following MARSDEN and HUGHES [1994] we indicate *covariant tensors* by $(\bullet)^\flat$ and *contravariant tensors* by $(\bullet)^\sharp$ (for the notions 'covariant' and 'contravariant' the reader is referred to Section 1.6). The push-forward and pull-back operations on *co-variant* second-order tensors (such as $\mathbf{E}^\flat$, $\mathbf{C}^\flat$, $\mathbf{e}^\flat$, $(\mathbf{b}^{-1})^\flat$) are according to

$$\boldsymbol{\chi}_*(\bullet)^\flat = \mathbf{F}^{-T}(\bullet)^\flat\mathbf{F}^{-1} \quad , \qquad \boldsymbol{\chi}_*^{-1}(\bullet)^\flat = \mathbf{F}^T(\bullet)^\flat\mathbf{F} \quad . \tag{2.86}$$

An example was given previously in eqs. (2.84) and (2.85), i.e. $\mathbf{e} = \boldsymbol{\chi}_*(\mathbf{E}^\flat)$ and $\mathbf{E} = \boldsymbol{\chi}_*^{-1}(\mathbf{e}^\flat)$, which provide the relationships between the material and spatial quantities, i.e. $\mathbf{e} = \mathbf{F}^{-T}\mathbf{E}\mathbf{F}^{-1}$ and $\mathbf{E} = \mathbf{F}^T\mathbf{e}\mathbf{F}$, respectively.

However, the push-forward and pull-back operations on *contravariant* second-order tensors (such as $(\mathbf{C}^{-1})^\sharp$, $\mathbf{b}^\sharp$ and most of the common stress tensors) are according to

$$\boldsymbol{\chi}_*(\bullet)^\sharp = \mathbf{F}(\bullet)^\sharp\mathbf{F}^T \quad , \qquad \boldsymbol{\chi}_*^{-1}(\bullet)^\sharp = \mathbf{F}^{-1}(\bullet)^\sharp\mathbf{F}^{-T} \quad . \tag{2.87}$$

In the following chapters we use *covariant strain tensors* in combination with *con-travariant stress tensors*.

For completeness we write down the push-forward and pull-back operations on *covariant vectors*, i.e.

$$\boldsymbol{\chi}_*(\bullet)^\flat = \mathbf{F}^{-T}(\bullet)^\flat \quad , \qquad \boldsymbol{\chi}_*^{-1}(\bullet)^\flat = \mathbf{F}^T(\bullet)^\flat \quad , \tag{2.88}$$

and of *contravariant vectors*, i.e.

$$\boldsymbol{\chi}_*(\bullet)^\sharp = \mathbf{F}(\bullet)^\sharp \quad , \qquad \boldsymbol{\chi}_*^{-1}(\bullet)^\sharp = \mathbf{F}^{-1}(\bullet)^\sharp \quad . \tag{2.89}$$

Finally we provide the so-called **Piola transformation** of a spatial vector field $\mathbf{u} = \mathbf{u}(\mathbf{x}, t)$, with components u_a, i.e.

$$\mathbf{U} = J\boldsymbol{\chi}_*^{-1}(\mathbf{u}) \quad , \tag{2.90}$$

where $\mathbf{U} = \mathbf{U}(\mathbf{X}, t)$ denotes a material vector field with the components U_A. The transformation from $\mathbf{u}$ to $\mathbf{U}$ involves the pull-back of $\mathbf{u}$ scaled by the volume ratio J. The inverse of eq. (2.90), i.e. $\mathbf{u} = J^{-1}\chi_*(\mathbf{U})$, involves the push-forward operation on $\mathbf{U}$ here scaled by the inverse of the volume ratio, i.e. J^{-1}.

For a more complete source on the underlying concept the mathematically oriented reader is referred to the book by MARSDEN and HUGHES [1994, and references therein].

EXERCISES

1. For a given material point express conditions on the right Cauchy-Green tensor
   C which ensure that

   (a) no stretch occurs in a specified direction $\mathbf{a}_0$ of a fiber,

   (b) no change in the angle between a pair of specified directions $(\mathbf{a}_{01}, \mathbf{a}_{02})$ takes
       place, and

   (c) no change occurs in an infinitesimal surface element $\mathrm{d}s$ placed in a plane
       perpendicular to a given direction (set $\mathrm{d}s = \mathrm{d}S$).

2. Let a body undergo a homogeneous deformation defined by

$$x_1 = \alpha X_1 \ , \qquad x_2 = -(\beta X_2 + \gamma X_3) \ , \qquad x_3 = \gamma X_2 - \beta X_3 \ ,$$

   where α, β, γ are constants.

   (a) Determine the components of the material and spatial strain tensors **C**, **E**
       and **b**, **e**, respectively.

   (b) Take the values $\beta = -\cos\theta$ and $\gamma = -\sin\theta$ and show that if $\alpha = 1$ the
       strains are zero. Explain that for $\alpha = 1$ the map corresponds to a rotation
       of magnitude θ about the X_1-axis.

3. If the deformation of a body is defined by

$$x_1 = x_1(X_1, X_2) \ , \qquad x_2 = x_2(X_1, X_2) \ , \qquad x_3 = X_3 \ ,$$

   the body is said to be in a state of **plane strain**. Deformations and strains occur
   only in planes $x_3 = \mathrm{const}$ and do not depend on the x_3-coordinate. Strains in the
   x_3-direction are zero. Determine the matrix representations of tensors **F**, **E**, **e**.

4. Using relations $(2.45)_2$ and $(2.46)_2$ show that the strain tensors $\mathbf{E}$ and $\mathbf{e}$ may be expressed in terms of the displacement gradient tensor according to

$$\mathbf{E} = \frac{1}{2}(\text{Grad}^T\mathbf{U} + \text{Grad}\,\mathbf{U}) + \frac{1}{2}\text{Grad}^T\mathbf{U}\,\text{Grad}\,\mathbf{U}\ , \qquad (2.91)$$

$$\mathbf{e} = \frac{1}{2}(\text{grad}^T\mathbf{u} + \text{grad}\,\mathbf{u}) - \frac{1}{2}\text{grad}^T\mathbf{u}\,\text{grad}\,\mathbf{u}\ , \qquad (2.92)$$

or in index notation, in terms of material and spatial coordinates, as

$$E_{AB} = \frac{1}{2}\left(\frac{\partial U_B}{\partial X_A} + \frac{\partial U_A}{\partial X_B}\right) + \frac{1}{2}\frac{\partial U_C}{\partial X_A}\frac{\partial U_C}{\partial X_B}\ ,$$

$$e_{ab} = \frac{1}{2}\left(\frac{\partial u_b}{\partial x_a} + \frac{\partial u_a}{\partial x_b}\right) - \frac{1}{2}\frac{\partial u_c}{\partial x_a}\frac{\partial u_c}{\partial x_b}\ ,$$

with implied summations on C and c, respectively.

2.6   Rotation, Stretch Tensors

In the following we decompose a *local* motion, characterized by the nonsingular (invertible) tensor $\mathbf{F}(\mathbf{X}, t)$, into a pure **stretch** and a pure **rotation**. As above, the arguments of the tensors are omitted, for convenience. However, they will be employed when additional information is needed.

Polar decomposition.   At each point $\mathbf{X} \in \Omega_0$ and each time t, we have the following **unique polar decomposition** of the deformation gradient $\mathbf{F}$:

$$\mathbf{F} = \mathbf{R}\mathbf{U} = \mathbf{v}\mathbf{R} \qquad \text{or} \qquad F_{aA} = R_{aB}U_{BA} = v_{ab}R_{bA}\ , \qquad (2.93)$$

$$\mathbf{R}^T\mathbf{R} = \mathbf{I}\ , \qquad \mathbf{U} = \mathbf{U}^T\ , \qquad \mathbf{v} = \mathbf{v}^T\ . \qquad (2.94)$$

This is a fundamental theorem in continuum mechanics.

In (2.93) $\mathbf{U}$ and $\mathbf{v}$ define *unique, positive definite, symmetric* tensors, which we call the **right** (or **material**) **stretch tensor** and the **left** (or **spatial**) **stretch tensor**, respectively. They measure local stretching (or contraction) along their mutually orthogonal eigenvectors, that is a change of *local shape*. The right stretch tensor $\mathbf{U}$ (with components U_{AB}) is defined with respect to the reference configuration while the left stretch tensor $\mathbf{v}$ (with components v_{ab}) acts on the current configuration. Note that in this text the symbol $\mathbf{U}$ also stands for the material displacement field and $\mathbf{v}$ for the spatial velocity field, respectively.

The positive definite and symmetric tensors $\mathbf{U}$ and $\mathbf{v}$ are introduced, so that

$$\mathbf{U}^2 = \mathbf{U}\mathbf{U} = \mathbf{C} \qquad \text{and} \qquad \mathbf{v}^2 = \mathbf{v}\mathbf{v} = \mathbf{b}\ , \qquad (2.95)$$

which is based on the **square-root theorem**; see, for example, GURTIN [1981a, p. 13]. Consider the relation $\det\mathbf{C} = \det\mathbf{b} = J^2$ we deduce with (2.95) and rule (1.101) the important property

$$\det\mathbf{U} = \det\mathbf{v} = J > 0 \quad . \tag{2.96}$$

The *unique* $\mathbf{R}$ is a *proper orthogonal* tensor, with $\det\mathbf{R} = 1$, called the **rotation tensor**. It measures the local rotation that is a change of *local orientation*. A **rigid-body rotation** about a fixed origin is then characterized *if and only if* $\mathbf{U} = \mathbf{v} = \mathbf{I}$, so that $\mathbf{F} = \mathbf{R}$. Hence, each material line element $d\mathbf{X}$ is rotated into a unique spatial line element $d\mathbf{x}$ (and vice versa) according to (2.38), i.e.

$$d\mathbf{x} = \mathbf{R}(\mathbf{X}, t)d\mathbf{X} \qquad \text{or} \qquad dx_a = R_{aA}dX_A \quad . \tag{2.97}$$

On the other hand if $\mathbf{R} = \mathbf{I}$, the deformation is called **pure stretch**, for which $\mathbf{F} = \mathbf{U} = \mathbf{v}$. Using (2.97), the pointwise polar decomposition (2.93) may be written as

$$d\mathbf{x} = \mathbf{R}\,(\mathbf{U}d\mathbf{X}) \qquad \text{or} \qquad dx_a = R_{aA}\,(U_{AB}dX_B) \quad , \tag{2.98}$$

$$d\mathbf{x} = \mathbf{v}\,(\mathbf{R}d\mathbf{X}) \qquad \text{or} \qquad dx_a = v_{ab}\,(R_{bA}dX_A) \quad . \tag{2.99}$$

Relation (2.98) describes a pure stretch of $d\mathbf{X}$ by the material tensor $\mathbf{U}$, (U_{AB}), followed by a pure rotation performed by the rotation tensor $\mathbf{R}$, (R_{aA}). The rotation tensor transforms the material vector $\mathbf{U}d\mathbf{X}$ into the spatial vector $d\mathbf{x}$. However, relation (2.99) describes a pure rotation of $d\mathbf{X}$ performed by the *same* rotation tensor $\mathbf{R}$, (R_{aA}), which transforms the material vector $d\mathbf{X}$ into the spatial vector $\mathbf{R}d\mathbf{X}$. This rotation is followed by a pure stretch of $\mathbf{R}d\mathbf{X}$ with the spatial tensor $\mathbf{v}$, (v_{ab}), which gives the spatial vector $d\mathbf{x}$. In both cases the rotation tensor $\mathbf{R}$ maps between the reference and the current configuration and, therefore, $\mathbf{R}$ is a two-point tensor like $\mathbf{F}$.

Relation $\mathbf{F} = \mathbf{R}\mathbf{U}$ is also known as the unique **right** polar decomposition, while $\mathbf{F} = \mathbf{v}\mathbf{R}$ is referred to as the unique **left** polar decomposition.

EXAMPLE 2.6   Show that $\mathbf{R}$ is *proper orthogonal*, i.e. $\mathbf{R}^T\mathbf{R} = \mathbf{I}$ and $\det\mathbf{R} = 1$. Show further that the multiplicative decomposition $\mathbf{F} = \mathbf{R}\mathbf{U}$ is *unique*.

Solution.   Using $\mathbf{R} = \mathbf{F}\mathbf{U}^{-1}$ and eqs. (2.95)$_1$ and (2.94)$_2$ we find that

$$\mathbf{R}^T\mathbf{R} = (\mathbf{F}\mathbf{U}^{-1})^T(\mathbf{F}\mathbf{U}^{-1}) = \mathbf{U}^{-T}\mathbf{F}^T\mathbf{F}\mathbf{U}^{-1} = \mathbf{U}^{-T}\mathbf{U}^2\mathbf{U}^{-1}$$
$$= (\mathbf{U}\mathbf{U}^{-1})^T\mathbf{U}\mathbf{U}^{-1} = \mathbf{I} \quad . \tag{2.100}$$

In addition, computing $\det\mathbf{R} = \det(\mathbf{F}\mathbf{U}^{-1})$ we deduce with rule (1.101) and relations (2.52) and (2.96) that

$$\det\mathbf{R} = \det\mathbf{F}\frac{1}{\det\mathbf{U}} = 1 \quad . \tag{2.101}$$

In order to show uniqueness of the polar decomposition we suppose that there exist positive definite, second-order tensors $\overline{\mathbf{R}}$ and $\overline{\mathbf{U}}$, so that

$$\mathbf{F} = \mathbf{RU} = \overline{\mathbf{R}}\,\overline{\mathbf{U}} \ , \tag{2.102}$$

$$\overline{\mathbf{R}}^{\mathrm{T}}\overline{\mathbf{R}} = \mathbf{I} \qquad \text{and} \qquad \overline{\mathbf{U}} = \overline{\mathbf{U}}^{\mathrm{T}} \ . \tag{2.103}$$

Then, with $\mathbf{C} = \mathbf{F}^{\mathrm{T}}\mathbf{F}$ and eqs. $(2.102)_2$ and (2.103) it follows that

$$\mathbf{C} = \mathbf{F}^{\mathrm{T}}\mathbf{F} = \overline{\mathbf{U}}^{\mathrm{T}}\overline{\mathbf{R}}^{\mathrm{T}}\overline{\mathbf{R}}\,\overline{\mathbf{U}} = \overline{\mathbf{U}}^2 \ , \tag{2.104}$$

which means using $(2.95)_1$ that $\overline{\mathbf{U}} = \mathbf{U}$, since $\mathbf{C}$ has a unique square root, and hence $\overline{\mathbf{R}} = \mathbf{R}$. The proof is similar for $\mathbf{F} = \mathbf{vR}$. ∎

We discuss further a physical interpretation of the right stretch tensor $\mathbf{U}$. Look in the direction of $\mathbf{a}_0$ at point $\mathbf{X} \in \Omega_0$, with $|\mathbf{a}_0| = 1$. According to (2.62) we find using $\mathbf{F} = \mathbf{RU}$ that the stretch vector $\boldsymbol{\lambda}_{\mathbf{a}_0}$ may be expressed as

$$\boldsymbol{\lambda}_{\mathbf{a}_0} = \mathbf{RUa}_0 \ . \tag{2.105}$$

Then, the square of the stretch ratio $\lambda = (\boldsymbol{\lambda}_{\mathbf{a}_0} \cdot \boldsymbol{\lambda}_{\mathbf{a}_0})^{1/2}$ may be computed with (2.105), identities (1.81), (1.84) and relation $\mathbf{R}^{\mathrm{T}}\mathbf{R} = \mathbf{I}$. Thus, we have

$$\begin{aligned} \lambda^2 &= \mathbf{RUa}_0 \cdot \mathbf{RUa}_0 = \mathbf{a}_0 \cdot \mathbf{U}^{\mathrm{T}}(\mathbf{R}^{\mathrm{T}}\mathbf{R})\mathbf{Ua}_0 \\ &= \mathbf{Ua}_0 \cdot \mathbf{Ua}_0 = |\mathbf{Ua}_0|^2 \ . \end{aligned} \tag{2.106}$$

Note that the length of $\boldsymbol{\lambda}_{\mathbf{a}_0}$ acting at $\mathbf{X} \in \Omega_0$ along the direction of the unit vector $\mathbf{a}_0$ depends only on a state of pure stretch, i.e. $\mathbf{U}$; therefore we call $\mathbf{U}$ the *stretch tensor*.

Another form of λ^2 may be obtained from $(2.106)_2$ using $(2.94)_2$, i.e.

$$\lambda^2 = \mathbf{a}_0 \cdot \mathbf{U}^{\mathrm{T}}\mathbf{Ua}_0 = \mathbf{a}_0 \cdot \mathbf{U}^2\mathbf{a}_0 \ . \tag{2.107}$$

Consequently, the right Cauchy-Green tensor $\mathbf{C}$ (recall eq. $(2.95)_1$) and the Green-Lagrange strain tensor $\mathbf{E} = (\mathbf{C} - \mathbf{I})/2$ do *not* include information about the rotation, that is experienced by a particle during motion. The deformation gradient $\mathbf{F}$ contains more information than strains do in general (rotation *and* strain-like information).

Finally we find the relation between $\mathbf{v}$ and $\mathbf{U}$, and between $\mathbf{b}$ and $\mathbf{C}$. By means of the polar decomposition (2.93) we have

$$\mathbf{v} = \mathbf{FR}^{\mathrm{T}} = \mathbf{RUR}^{\mathrm{T}} \qquad \text{and} \qquad \mathbf{v}^2 = \mathbf{RUR}^{\mathrm{T}}\mathbf{RUR}^{\mathrm{T}} = \mathbf{RU}^2\mathbf{R}^{\mathrm{T}} \ . \tag{2.108}$$

Consequently, with $\mathbf{v}^2 = \mathbf{R}\mathbf{U}^2\mathbf{R}^{\mathrm{T}}$ and eq. (2.95) we find an important relation between the left and right Cauchy-Green tensors, namely

$$\mathbf{b} = \mathbf{v}^2 = \mathbf{R}\mathbf{U}^2\mathbf{R}^{\mathrm{T}} = \mathbf{R}\mathbf{C}\mathbf{R}^{\mathrm{T}} \quad . \tag{2.109}$$

A general formula for strain measures was introduced by SETH [1964] and HILL [1968, 1970, 1978]. The generalized strain measures in the Lagrangian and Eulerian descriptions are defined as

$$\left. \begin{array}{llll} \dfrac{1}{n}(\mathbf{U}^n - \mathbf{I}) \quad , & \dfrac{1}{n}(\mathbf{v}^n - \mathbf{I}) \quad , & \text{if} \quad n \neq 0 \quad , \\[2mm] \ln\mathbf{U} \quad , & \ln\mathbf{v} \quad , & \text{if} \quad n = 0 \quad , \end{array} \right\} \tag{2.110}$$

where n is a real number (not necessarily an integer).

For the special cases $n = 0$ and $n = 1$ we obtain strain tensors associated with the names *Hencky* and *Biot*, respectively, while for $n = 2$ and $n = -2$ we obtain the *Green-Lagrange strain tensor* $(\mathbf{U}^2 - \mathbf{I})/2$ (or also $(\mathbf{v}^2 - \mathbf{I})/2$) and the *Euler-Almansi strain tensor* $(\mathbf{I} - \mathbf{U}^{-2})/2$ (or also $(\mathbf{I} - \mathbf{v}^{-2})/2$), which we have introduced in (2.69) and (2.83), respectively. For more details see, for example, MAN and GUO [1993], and for more discussion of the logarithmic strain tensor, see, for example, HOGER [1987].

The Hencky strain tensor in the material and spatial form, i.e. $\ln\mathbf{U}$ and $\ln\mathbf{v}$, is of particular interest in nonlinear constitutive theories. Because of the logarithmic functions, these tensors are decomposed **additively** into so-called *volumetric* and *isochoric* parts (for a more detailed exposition of the decomposition procedure of strain measures see, for example, p. 228).

EXAMPLE 2.7   Assume a certain two-dimensional motion which is given in the form of the deformation gradient $\mathbf{F}(\mathbf{X}, t) = \partial\boldsymbol{\chi}(\mathbf{X}, t)/\partial\mathbf{X}$. All tensor quantities are defined with respect to the orthonormal basis $\{\mathbf{e}_\alpha\}$, $\alpha = 1, 2$.

Show that for the case of two dimensions the polar decomposition $\mathbf{F} = \mathbf{R}\mathbf{U}$ may be given by the closed-form expression

$$\mathbf{U} = [I_1(\mathbf{C}) + 2J(\mathbf{C})]^{-1/2}[\mathbf{C} + J(\mathbf{C})\mathbf{I}] \tag{2.111}$$

for the right stretch tensor, and by the rotation tensor $\mathbf{R}$ which follows by means of the inverse of expression (2.111), i.e. $\mathbf{R} = \mathbf{F}\mathbf{U}^{-1}$. In eq. (2.111), $I_1(\mathbf{C}) = C_{11} + C_{22}$ and $J(\mathbf{C}) = (C_{11}C_{22})^{1/2}$ characterize the first scalar invariant of the right Cauchy-Green tensor $\mathbf{C}$ and the volume ratio, respectively.

Solution.   In two dimensions, the characteristic polynomial of $\mathbf{U}$ gives a quadratic equation in λ, i.e.

$$\lambda_\alpha^2 - i_1(\mathbf{U})\lambda_\alpha + i_2(\mathbf{U}) = 0 \quad , \qquad \alpha = 1, 2 \quad , \tag{2.112}$$

with the principal scalar invariants $i_1(\mathbf{U}) = \mathrm{tr}\mathbf{U} = U_{11} + U_{22}$ and $i_2(\mathbf{U}) = \det\mathbf{U} = U_{11}U_{22} - U_{12}^2$.

Knowing from the *Cayley-Hamilton* equation (1.174) that any (second-order) tensor satisfies its own characteristic equation, we may write instead of (2.112), $\mathbf{U}^2 - i_1(\mathbf{U})\mathbf{U} + i_2(\mathbf{U})\mathbf{I} = \mathbf{O}$. By use of $\mathbf{U}^2 = \mathbf{C}$ and rearranging we may write the explicit expression for $\mathbf{U}$, i.e.

$$\mathbf{U} = \frac{1}{i_1(\mathbf{U})}[\mathbf{C} + i_2(\mathbf{U})\mathbf{I}] \quad , \tag{2.113}$$

which coincides with the relation obtained in, for example, HOGER and CARLSON [1984] and TING [1985].

On taking the trace of the last equation (2.113) and knowing that $\mathrm{tr}\mathbf{C} = I_1(\mathbf{C})$ and $\mathrm{tr}\mathbf{I} = 2$ in two dimensions we are able to express the first invariant of $\mathbf{U}$ in terms of the first invariant of $\mathbf{C}$. Thus,

$$i_1(\mathbf{U}) = [I_1(\mathbf{C}) + 2i_2(\mathbf{U})]^{1/2} \quad . \tag{2.114}$$

In addition, using the relation $\det\mathbf{C} = \det(\mathbf{U}\mathbf{U}) = (\det\mathbf{U})^2$, where the property (1.101) is applied, and recalling $(2.67)_2$ we conclude that

$$i_2(\mathbf{U}) = \det\mathbf{U} = (\det\mathbf{C})^{1/2} = J(\mathbf{C}) \quad . \tag{2.115}$$

Using eqs. (2.114), $(2.115)_3$ we find finally from (2.113) the desired result (2.111). ∎

Eigenvalues and eigenvectors of strain tensors.   We introduce the *mutually orthogonal* and *normalized* set of eigenvectors $\{\hat{\mathbf{N}}_a\}$ and their corresponding eigenvalues λ_a, $a = 1, 2, 3$, of the material tensor $\mathbf{U}$ as

$$\mathbf{U}\hat{\mathbf{N}}_a = \lambda_a\hat{\mathbf{N}}_a \qquad \text{with} \qquad |\hat{\mathbf{N}}_a| = 1 \ , \qquad a = 1, 2, 3 \ . \tag{2.116}$$

Furthermore, after combining $(2.95)_1$ with (2.116) we obtain the eigenvalue problem for $\mathbf{C}$, i.e.

$$\mathbf{C}\hat{\mathbf{N}}_a = \mathbf{U}^2\hat{\mathbf{N}}_a = \lambda_a^2\hat{\mathbf{N}}_a \ , \qquad a = 1, 2, 3 \ . \tag{2.117}$$

Clearly $\mathbf{U}$ and $\mathbf{C}$ have the **same** orthonormal eigenvectors, i.e. the set $\{\hat{\mathbf{N}}_a\}$, called the **principal referential directions** (or **principal referential axes**). However, the corresponding positive and real eigenvalues differ. The eigenvalues of the symmetric tensor $\mathbf{U}$ are λ_a, $a = 1, 2, 3$, called the **principal stretches**, while for the symmetric tensor $\mathbf{C}$ we find the squares of the principal stretches denoted by λ_a^2.

Hence, with $\mathbf{R}^T\mathbf{R} = \mathbf{I}$ and $\mathbf{v} = \mathbf{R}\mathbf{U}\mathbf{R}^T$ we obtain from (2.116) the eigenvalue problem for $\mathbf{v}$, i.e.

$$\mathbf{v}(\mathbf{R}\hat{\mathbf{N}}_a) = \mathbf{R}\mathbf{U}\mathbf{R}^T(\mathbf{R}\hat{\mathbf{N}}_a)$$
$$= \mathbf{R}\mathbf{U}\hat{\mathbf{N}}_a = \lambda_a(\mathbf{R}\hat{\mathbf{N}}_a) \ , \qquad a = 1,2,3 \ . \tag{2.118}$$

Combining $(2.95)_2$ with (2.118) we obtain, by analogy with (2.117), the eigenvalue problem for $\mathbf{b}$, i.e.

$$\mathbf{b}(\mathbf{R}\hat{\mathbf{N}}_a) = \mathbf{v}^2(\mathbf{R}\hat{\mathbf{N}}_a) = \lambda_a^2(\mathbf{R}\hat{\mathbf{N}}_a) \ , \qquad a = 1,2,3 \ , \tag{2.119}$$

which means that the two tensors $\mathbf{v}$ and $\mathbf{b}$ have the same eigenvectors $\mathbf{R}\hat{\mathbf{N}}_{(a)}$, while their positive and real eigenvalues are λ_a and λ_a^2, respectively.

The eigenvectors of $\mathbf{v}$ and $\mathbf{b}$ are those of $\mathbf{U}$ and $\mathbf{C}$ rotated with $\mathbf{R}$. We conclude that the principal referential directions $\hat{\mathbf{N}}_a$ transform onto the **principal spatial directions** (or **principal spatial axes**) $\hat{\mathbf{n}}_a$ (which are *mutually orthogonal* and *normalized* eigenvectors of $\mathbf{v}$ and $\mathbf{b}$) via

$$\hat{\mathbf{n}}_a = \mathbf{R}\hat{\mathbf{N}}_a \qquad \text{with} \qquad |\hat{\mathbf{n}}_a| = 1 \ , \qquad a = 1,2,3 \ . \tag{2.120}$$

This means that the two-point tensor $\mathbf{R}$ rotates the material vectors $\hat{\mathbf{N}}_a$ into the spatial vectors $\hat{\mathbf{n}}_a$. So the principal directions of $\mathbf{v}$ and $\mathbf{b}$ may be obtained by rotating the corresponding principal directions of $\mathbf{U}$ and $\mathbf{C}$ by $\mathbf{R}$. Relation (2.120) will prove useful in a moment.

We summarize the four introduced (symmetric) strain tensors in the convenient form of their *spectral decomposition*, for $\lambda_1 \neq \lambda_2 \neq \lambda_3 \neq \lambda_1$, i.e.

$$\mathbf{U}^2 = \mathbf{C} = \sum_{a=1}^{3} \lambda_a^2 \hat{\mathbf{N}}_a \otimes \hat{\mathbf{N}}_a \ , \tag{2.121}$$

$$\mathbf{v}^2 = \mathbf{b} = \sum_{a=1}^{3} \lambda_a^2 \hat{\mathbf{n}}_a \otimes \hat{\mathbf{n}}_a \ . \tag{2.122}$$

Note that the eigenvalues and the eigenvectors depend on the point and time t.

In order to express the two-point tensors $\mathbf{F}$ and $\mathbf{R}$ in terms of principal stretches and principal directions we employ the polar decomposition $\mathbf{F} = \mathbf{R}\mathbf{U}$ and eq. (2.121). Using $(2.120)_1$ we find the deformation gradient in the form

$$\mathbf{F} = \mathbf{R}\sum_{a=1}^{3} \lambda_a \hat{\mathbf{N}}_a \otimes \hat{\mathbf{N}}_a = \sum_{a=1}^{3} \lambda_a(\mathbf{R}\hat{\mathbf{N}}_a) \otimes \hat{\mathbf{N}}_a = \sum_{a=1}^{3} \lambda_a \hat{\mathbf{n}}_a \otimes \hat{\mathbf{N}}_a \ . \tag{2.123}$$

Knowing that the unit tensor $\mathbf{I}$ may be expressed as $\hat{\mathbf{N}}_a \otimes \hat{\mathbf{N}}_a$ (see relation (1.65)) we find by means of (2.120)$_1$ that

$$\mathbf{R} = \mathbf{R}\mathbf{I} = (\mathbf{R}\hat{\mathbf{N}}_a) \otimes \hat{\mathbf{N}}_a = \sum_{a=1}^{3} \hat{\mathbf{n}}_a \otimes \hat{\mathbf{N}}_a \quad . \tag{2.124}$$

Note that both tensors involve principal directions in the reference and current configuration, which emphasizes the two-point character of these tensors. Since tensors $\mathbf{F}$ and $\mathbf{R}$ are, in general, non-symmetric, the representations (2.123) and (2.124) may not be viewed as spectral decompositions in the sense introduced on p. 25. Hence, it is clear that the principal stretches λ_a, $a = 1, 2, 3$, in eq. (2.123)$_3$ may *not* be interpreted as the eigenvalues of the deformation gradient $\mathbf{F}$.

EXAMPLE 2.8   Suppose that the spectral decomposition of the right Cauchy-Green tensor $\mathbf{C}$, i.e. eq. (2.121), is given. Since the principal stretches are functions of $\mathbf{C}$ we may write $\lambda_a = \lambda_a(\mathbf{C})$. For later use we introduce the following relations

$$\frac{\partial \lambda_a^2}{\partial \mathbf{C}} = \hat{\mathbf{N}}_a \otimes \hat{\mathbf{N}}_a \quad (a = 1, 2, 3) \qquad \text{for} \qquad \lambda_1 \neq \lambda_2 \neq \lambda_3 \neq \lambda_1 \quad , \tag{2.125}$$

$$\left.\begin{aligned}\frac{\partial \lambda_1^2}{\partial \mathbf{C}} &= \mathbf{I} - \hat{\mathbf{N}}_3 \otimes \hat{\mathbf{N}}_3 \\[1em] \frac{\partial \lambda_3^2}{\partial \mathbf{C}} &= \hat{\mathbf{N}}_3 \otimes \hat{\mathbf{N}}_3\end{aligned}\right\} \qquad \text{for} \qquad \lambda_1 = \lambda_2 \neq \lambda_3 \quad , \tag{2.126}$$

$$\frac{\partial \lambda^2}{\partial \mathbf{C}} = \sum_{a=1}^{3} \hat{\mathbf{N}}_a \otimes \hat{\mathbf{N}}_a = \mathbf{I} \qquad \text{for} \qquad \lambda_1 = \lambda_2 = \lambda_3 = \lambda \tag{2.127}$$

(compare also with SIMO and TAYLOR [1991a] or SALEEB et al. [1992]). Prove this set of equations for these distinct cases.

Solution.   Firstly we consider the case $\lambda_a \neq \lambda_b$ for $a \neq b$ and compute the total differential of the spectral decomposition of $\mathbf{C}$ (eq. (2.121)$_1$)

$$d\mathbf{C} = \sum_{a=1}^{3} [2\lambda_a d\lambda_a \hat{\mathbf{N}}_a \otimes \hat{\mathbf{N}}_a + \lambda_a^2 (d\hat{\mathbf{N}}_a \otimes \hat{\mathbf{N}}_a + \hat{\mathbf{N}}_a \otimes d\hat{\mathbf{N}}_a)] \quad . \tag{2.128}$$

Recall that the three vectors $\hat{\mathbf{N}}_a$, with $|\hat{\mathbf{N}}_a| = 1$, form an orthonormal basis ($\hat{\mathbf{N}}_a \cdot \hat{\mathbf{N}}_b = \delta_{ab}$), so that $\hat{\mathbf{N}}_a \cdot d\hat{\mathbf{N}}_a = 0$ (the change of a vector with constant length is always orthogonal to the vector itself). If we pre- and postmultiply eq. (2.128) with $\hat{\mathbf{N}}_a$ we find that

$$\hat{\mathbf{N}}_a \cdot d\mathbf{C}\hat{\mathbf{N}}_a = 2\lambda_a d\lambda_a \quad , \qquad a = 1, 2, 3 \quad . \tag{2.129}$$

Applying property (1.96) to the left-hand side of (2.129) we may write $\hat{\mathbf{N}}_a \cdot d\mathbf{C}\hat{\mathbf{N}}_a =$

$dC : \hat{\mathbf{N}}_a \otimes \hat{\mathbf{N}}_a$, $a = 1, 2, 3$, and by means of identity $dC = (\partial \mathbf{C}/\partial \lambda_a)d\lambda_a$ we have

$$\frac{\partial \mathbf{C}}{\partial \lambda_a}d\lambda_a : \hat{\mathbf{N}}_a \otimes \hat{\mathbf{N}}_a = 2\lambda_a d\lambda_a \ ,$$

$$\frac{1}{2\lambda_a}\frac{\partial \mathbf{C}}{\partial \lambda_a} : \hat{\mathbf{N}}_a \otimes \hat{\mathbf{N}}_a = 1 \ , \qquad a = 1, 2, 3 \ . \tag{2.130}$$

Knowing that the value 1 may be written as the contraction $(\partial \mathbf{C}/\partial \lambda_a) : (\partial \lambda_a/\partial \mathbf{C})$, the last equation implies $\partial \lambda_a/\partial \mathbf{C} = (2\lambda_a)^{-1}\hat{\mathbf{N}}_a \otimes \hat{\mathbf{N}}_a$. With help of the chain rule we obtain finally the basic relation (2.125).

The other two cases are straightforward results, compare with the general considerations on the spectral decomposition of a tensor, in particular, relations (1.178) and (1.179). ∎

Uniform and biaxial deformation, pure shear.   Consider an extension or compression of a rod (with uniform cross-section) in the direction of the x_1-axis up to the stretch λ_1. The associated relation $x_1 = \lambda_1 X_1$ defines a **uniform deformation** along the x_1-axis. For $\lambda_1 > 1$ we call the deformation a **uniform extension** (or in the literature sometimes referred to as the **uniaxial extension**), however, for $\lambda_1 < 1$ we call it a **uniform compression** (or **uniaxial compression**). Uniform extensions or compressions in all three directions follow the relations

$$x_1 = \lambda_1 X_1 \ , \qquad x_2 = \lambda_2 X_2 \ , \qquad x_3 = \lambda_3 X_3 \ . \tag{2.131}$$

If two stretches, for example, λ_1, λ_2, can be chosen arbitrarily and the third stretch λ_3 is determined by the condition $J = \lambda_1\lambda_2\lambda_3 = 1$ (this type of deformation keeps the volume constant) the associated mode of deformation is often referred to as **biaxial** (although there are changes of lengths in all three directions). A typical biaxial deformation is characterized in the following form

$$x_1 = \lambda_1 X_1 \ , \qquad x_2 = \lambda_2 X_2 \ , \qquad x_3 = \frac{1}{\lambda_1\lambda_2}X_3 \ . \tag{2.132}$$

An **equibiaxial deformation** is defined as the mode of deformation in which $\lambda_1 = \lambda_2 = \lambda$ (and $\lambda_3 = \lambda^{-2}$ provided it is isochoric).

If $\lambda_1 = \lambda_2 = \lambda_3$, then the continuum body undergoes a so-called **uniform dilation**, i.e. a *uniform expansion* or *uniform contraction* in all three directions.

Next, we consider a thin sheet of material which is fixed along a parallel pair of edges normal to the $\mathbf{e}_2$ direction. An extension with the principal stretch λ_1 in the plane normal to these edges results in a *plane deformation* (state of plane strain) of the form

$$x_1 = \lambda_1 X_1 \ , \qquad x_2 = X_2 \ , \qquad x_3 = \frac{1}{\lambda_1}X_3 \tag{2.133}$$

(provided that the plane deformation is isochoric), known as **pure shear** (or **strip-biaxial extension**). The associated principal stretches are λ_1, $\lambda_2 = 1$ and $\lambda_3 = \lambda_1^{-1}$, while $\mathbf{R} = \mathbf{I}$. The area remains constant in the plane spanned by the orthogonal eigen-vectors $\hat{\mathbf{n}}_1$ and $\hat{\mathbf{n}}_3$. Experimental results on a thin sheet of rubber under pure shear were first published by RIVLIN and SAUNDERS [1951], see also the book by TRELOAR [2005, Chapter 5].

The volume of a continuum body under biaxial or pure shear deformations according to kinematic relations (2.132) and (2.133) remains constant, as we will see later in the text.

EXERCISES

1. Use (2.108) to show that $\mathbf{v}^n = \mathbf{R}\mathbf{U}^n\mathbf{R}^T$, where n is an integer.

2. The motion presented by eq. (2.3) may be given in symbolic notation as $\mathbf{x} = \chi(\mathbf{X}, t) = \mathbf{X} + c(t)(\mathbf{e}_2 \cdot \mathbf{X})\mathbf{e}_1$, $\mathbf{e}_1$ and $\mathbf{e}_2$ denoting orthogonal unit vectors fixed in space, with the components $(1, 0, 0)$ and $(0, 1, 0)$, respectively, and the parameter $c(t) = \tan\theta(t) > 0$. The motion described causes a so-called **simple shear** deformation (also known as a **uniform shear** deformation), where the planes $x_2 = \text{const}$ are the **shear planes** and the direction along x_1 is the **shear direction**. The angle $\theta(t)$ is a measure of the amount of shear.

   (a) Compute the matrix representations of tensors $\mathbf{F}$, $\mathbf{C}$, $\mathbf{b}$, $\mathbf{C}^{-1}$, $\mathbf{b}^{-1}$.

   (b) Show that $\mathbf{F}$, $\mathbf{C}$ and $\mathbf{b}$ may be expressed as

   $$\mathbf{F} = \mathbf{I} + c\mathbf{e}_1 \otimes \mathbf{e}_2 \ , \qquad \mathbf{C} = \mathbf{I} + c^2\mathbf{e}_2 \otimes \mathbf{e}_2 + c(\mathbf{e}_1 \otimes \mathbf{e}_2 + \mathbf{e}_2 \otimes \mathbf{e}_1) \ ,$$

   $$\mathbf{b} = \mathbf{I} + c^2\mathbf{e}_1 \otimes \mathbf{e}_1 + c(\mathbf{e}_1 \otimes \mathbf{e}_2 + \mathbf{e}_2 \otimes \mathbf{e}_1) \ .$$

   (c) Show that a simple shear deformation is isochoric.

   (d) Compute the principal stretches λ_a, $a = 1, 2, 3$, and the three principal referential directions $\hat{\mathbf{N}}_a$, i.e. the normalized eigenvectors of $\mathbf{C}$.

   (e) Compute the angle between the principal referential directions $\hat{\mathbf{N}}_a$ and $\mathbf{e}_1$, $\mathbf{e}_2$, $\mathbf{e}_3 = \mathbf{e}_1 \times \mathbf{e}_2$ in terms of $c(t)$.

   (f) Compute the three principal spatial directions $\hat{\mathbf{n}}_a$, i.e. the normalized eigenvectors of $\mathbf{b}$.

   Hint: The principal directions $\hat{\mathbf{n}}_a$ are simply obtained from $\hat{\mathbf{N}}_a$ by interchanging $\mathbf{e}_1$ and $\mathbf{e}_2$.

   (g) By using results (d) and (f) determine the spectral decompositions of the Cauchy-Green tensors $\mathbf{C} = \mathbf{U}^2$, $\mathbf{b} = \mathbf{v}^2$, i.e. (2.121), (2.122), and the representations of the deformation gradient $\mathbf{F}$ and the rotation tensor $\mathbf{R}$, as derived in eqs. $(2.123)_3$ and $(2.124)_3$, respectively.

(h) From (g) we have basically obtained the polar decomposition $\mathbf{F} = \mathbf{RU}$ via the spectral decomposition of $\mathbf{C}$. For the particular deformation of simple shear, compare (and check) your results for the right stretch tensor $\mathbf{U}$ and the rotation tensor $\mathbf{R}$ with those obtained from the closed-form expression (2.111) and $\mathbf{R} = \mathbf{FU}^{-1}$.

3. Recall the representation $(2.123)_3$, the polar decomposition $\mathbf{F} = \mathbf{RU}$ and $(2.120)_1$ and (2.121). Establish the following expressions relating the two sets of principal directions $\{\hat{\mathbf{N}}_a\}$ and $\{\hat{\mathbf{n}}_a\}$, $a = 1, 2, 3$, i.e.

$$\mathbf{F}\hat{\mathbf{N}}_a = \lambda_a \hat{\mathbf{n}}_a \quad , \qquad \mathbf{F}^{-\mathrm{T}}\hat{\mathbf{N}}_a = \lambda_a^{-1}\hat{\mathbf{n}}_a \quad , \tag{2.134}$$

$$\mathbf{F}^{-1}\hat{\mathbf{n}}_a = \lambda_a^{-1}\hat{\mathbf{N}}_a \quad , \qquad \mathbf{F}^{\mathrm{T}}\hat{\mathbf{n}}_a = \lambda_a \hat{\mathbf{N}}_a \quad .$$

4. A body undergoes uniform extension in all three directions according to (2.131). Find the matrix representations of tensors $\mathbf{F}, \mathbf{R}, \mathbf{U}, \mathbf{E}, \mathbf{e}$.

5. Consider the deformation gradient $\mathbf{F}$ in the form of a 3×3 matrix with $\det\mathbf{F} > 0$. Write a computer program in order to determine the right stretch tensor $\mathbf{U}$ and the rotation tensor $\mathbf{R}$.

Start the procedure by computing $\mathbf{C} = \mathbf{F}^{\mathrm{T}}\mathbf{F}$ and the eigenvalues and eigenvectors of $\mathbf{C}$. Then use the spectral decomposition (2.121) for the tensor $\mathbf{U}$, and by means of $\mathbf{R} = \mathbf{FU}^{-1}$ finally find $\mathbf{R}$.

6. Show that the second-order tensors $\hat{\mathbf{N}}_a \otimes \hat{\mathbf{N}}_a$ and $\hat{\mathbf{n}}_a \otimes \hat{\mathbf{n}}_a$ occurring in eqs. (2.121) and (2.122) may be obtained in the closed form, for $\lambda_1 \neq \lambda_2 \neq \lambda_3 \neq \lambda_1$, as

$$\hat{\mathbf{N}}_a \otimes \hat{\mathbf{N}}_a = \lambda_a^2 \frac{\mathbf{C} - (I_1 - \lambda_a^2)\mathbf{I} + I_3\lambda_a^{-2}\mathbf{C}^{-1}}{D_a} \quad , \tag{2.135}$$

$$\hat{\mathbf{n}}_a \otimes \hat{\mathbf{n}}_a = \frac{\mathbf{b}^2 - (I_1 - \lambda_a^2)\mathbf{b} + I_3\lambda_a^{-2}\mathbf{I}}{D_a} \quad , \tag{2.136}$$

($a = 1, 2, 3$; no summation) where the scalar D_a must be nonzero, with $D_a = 2\lambda_a^4 - I_1\lambda_a^2 + I_3\lambda_a^{-2}$. The first and third invariants of $\mathbf{C}$, and also of $\mathbf{b}$, are I_1 and I_3, respectively. The closed form expressions for $\hat{\mathbf{N}}_a \otimes \hat{\mathbf{N}}_a$ and $\hat{\mathbf{n}}_a \otimes \hat{\mathbf{n}}_a$ circumvent the explicit computation of the eigenvectors.

Hint: Relations (2.135) and (2.136) follow from the *Rivlin-Ericksen representation theorem* (compare with p. 201 of this text), see BOWEN and WANG [1976b], MORMAN [1986] for an analytical treatment and inter alia SIMO and TAYLOR [1991a] and MIEHE [1994] for a numerical treatment.

2.7 Rates of Deformation Tensors

Within this section we want to study how some of the tensor fields introduced above change with time, by knowing the motion $\mathbf{x} = \boldsymbol{\chi}(\mathbf{X}, t)$. In particular, we study the rate at which changes of shape, position and orientation of a continuum body occur.

Spatial and material velocity gradient. The derivative of a spatial velocity field $\mathbf{v}(\mathbf{x}, t)$ with respect to the spatial coordinates is defined by

$$\mathbf{l}(\mathbf{x}, t) = \frac{\partial \mathbf{v}(\mathbf{x}, t)}{\partial \mathbf{x}} = \mathrm{grad}\mathbf{v}(\mathbf{x}, t) \qquad \text{or} \qquad l_{ab} = \frac{\partial v_a}{\partial x_b} \quad , \qquad (2.137)$$

here given in symbolic and index notation. The spatial field $\mathbf{l}$, in general a non-symmetric second-order tensor, is known as the **spatial velocity gradient** commonly used in both solid and fluid mechanics.

The material time derivative of the deformation gradient $\mathbf{F}$ gives, with definition $(2.39)_1$ and relation $(2.7)_1$,

$$\dot{\mathbf{F}}(\mathbf{X}, t) = \frac{\partial}{\partial t}\left(\frac{\partial \boldsymbol{\chi}(\mathbf{X}, t)}{\partial \mathbf{X}}\right) = \frac{\partial}{\partial \mathbf{X}}\left(\frac{\partial \boldsymbol{\chi}(\mathbf{X}, t)}{\partial t}\right)$$

$$= \frac{\partial \mathbf{V}(\mathbf{X}, t)}{\partial \mathbf{X}} = \mathrm{Grad}\mathbf{V}(\mathbf{X}, t) \qquad \text{or} \qquad \dot{F}_{aA} = \frac{\partial v_a}{\partial X_A} \quad , \qquad (2.138)$$

where the time rate of change of the deformation gradient, i.e. $\dot{\mathbf{F}} = \mathrm{D}\mathbf{F}/\mathrm{D}t$, is equal to $\mathrm{Grad}\mathbf{V}$, called the **material velocity gradient**. In eq. (2.138) we have used, additionally, the property that the material time derivative commutes with the material gradient.

Another valuable derivation of the time rate of change of the deformation gradient is by means of the directional derivative introduced in eq. (1.266). At a given time t, we compute the directional derivative of $\mathbf{F}$ in the direction of the velocity vector $\mathbf{v}$ at position $\mathbf{x}$. We obtain

$$D_{\mathbf{v}}\mathbf{F}(\mathbf{X}, t) = \frac{\mathrm{d}}{\mathrm{d}\varepsilon}\mathrm{Grad}(\mathbf{x} + \varepsilon\mathbf{v})|_{\varepsilon=0} = \frac{\mathrm{d}}{\mathrm{d}\varepsilon}\left(\frac{\partial \mathbf{x}}{\partial \mathbf{X}} + \varepsilon\frac{\partial \mathbf{v}}{\partial \mathbf{X}}\right)\Big|_{\varepsilon=0} = \frac{\partial \mathbf{v}}{\partial \mathbf{X}} \quad . \qquad (2.139)$$

On comparing eq. $(2.138)_4$ with $(2.139)_3$ (by means of $(2.8)_2$), we find finally that

$$\frac{\mathrm{D}\mathbf{F}(\mathbf{X}, t)}{\mathrm{D}t} = D_{\mathbf{v}}\mathbf{F}(\mathbf{X}, t) \quad , \qquad (2.140)$$

which is merely an application of relationship (2.20).

The spatial velocity gradient $\mathbf{l} = \mathrm{grad}\mathbf{v}$ may be expressed through the material velocity gradient $\dot{\mathbf{F}} = \mathrm{Grad}\mathbf{V}$. With definition $(2.137)_1$, identities $(2.8)_2$, $(2.7)_1$, the chain rule and the definitions of $\mathbf{F}$ and $\mathbf{F}^{-1}$, we find the useful relation

$$\begin{aligned}
\mathbf{l} = \frac{\partial \mathbf{v}}{\partial \mathbf{x}} &= \frac{\partial \dot{\boldsymbol{\chi}}(\mathbf{X}, t)}{\partial \mathbf{X}} \frac{\partial \mathbf{X}}{\partial \mathbf{x}} \\
&= \frac{\partial}{\partial t} \left(\frac{\partial \boldsymbol{\chi}(\mathbf{X}, t)}{\partial \mathbf{X}} \right) \mathbf{F}^{-1} = \dot{\mathbf{F}} \mathbf{F}^{-1} \qquad \text{or} \qquad \dot{\mathbf{F}} = \mathbf{l}\mathbf{F} \ .
\end{aligned} \qquad (2.141)$$

This may be written in index notation as

$$\begin{aligned}
l_{ab} = \frac{\partial v_a}{\partial x_b} &= \frac{\partial \dot{\chi}_a}{\partial X_A} \frac{\partial X_A}{\partial x_b} \\
&= \frac{\partial}{\partial t} \left(\frac{\partial \chi_a}{\partial X_A} \right) F_{Ab}^{-1} = \dot{F}_{aA} F_{Ab}^{-1} \qquad \text{or} \qquad \dot{F}_{aA} = l_{ab} F_{bA} \ .
\end{aligned} \qquad (2.142)$$

EXAMPLE 2.9   Obtain the relationship which expresses the rate of change of a unit vector $\mathbf{a}$ (characterizing the direction of a fiber in the deformed configuration, as introduced in $(2.71)_3$, with reference to Figure 2.7) in terms of the spatial velocity gradient $\mathbf{l}$ in the form

$$\dot{\mathbf{a}} = \mathbf{l}\mathbf{a} - \frac{\dot{\lambda}}{\lambda} \mathbf{a} \ . \qquad (2.143)$$

Solution.   The combination of eqs. (2.62) and (2.73) gives $\boldsymbol{\lambda}_{\mathbf{a}_0} = \mathbf{F}\mathbf{a}_0 = \lambda \mathbf{a}$. Taking the material time derivative, we find, by means of the product rule, that

$$\dot{\mathbf{F}}\mathbf{a}_0 = \dot{\lambda}\mathbf{a} + \lambda \dot{\mathbf{a}} \ . \qquad (2.144)$$

Hence, using $\dot{\mathbf{F}} = \mathbf{l}\mathbf{F}$ and applying $\mathbf{F}\mathbf{a}_0 = \lambda\mathbf{a}$ once more, we obtain eq. (2.143) after some straightforward algebra. ∎

Next, we compute the material time derivative of $\mathbf{F}^{-1}$ and $\mathbf{F}^{-T}$ which will be useful later in the text. Starting from $\mathbf{F}^{-1}\mathbf{F} = \mathbf{I}$ we deduce, using the product rule and relation $\mathbf{l} = \dot{\mathbf{F}}\mathbf{F}^{-1}$ (see $(2.141)_4$), that

$$\overline{\mathbf{F}^{-1}\mathbf{F}} = -\mathbf{F}^{-1}\dot{\mathbf{F}}$$

$$\dot{\overline{\mathbf{F}^{-1}}} = -\mathbf{F}^{-1}\dot{\mathbf{F}}\mathbf{F}^{-1} = -\mathbf{F}^{-1}\mathbf{l} \qquad \text{or} \qquad \dot{\overline{F_{Aa}^{-1}}} = -F_{Ab}^{-1} l_{ba} \qquad (2.145)$$

(compare also with relation (1.237)). Consequently, from eq. $(2.145)_2$ we obtain $\mathbf{l} = -\mathbf{F}\dot{\overline{\mathbf{F}^{-1}}}$. However, starting from $\mathbf{F}^{-T}\mathbf{F}^{T} = \mathbf{I}$ we find that

$$\overline{\mathbf{F}^{-T}\mathbf{F}^{T}} = -\mathbf{F}^{-T}\dot{\overline{\mathbf{F}^{T}}}$$

$$\dot{\overline{\mathbf{F}^{-T}}} = -\mathbf{F}^{-T}\dot{\overline{\mathbf{F}^{T}}}\mathbf{F}^{-T} = -\mathbf{l}^{T}\mathbf{F}^{-T} \ , \qquad (2.146)$$

which is the analogue of (2.145). The overbar covers the quantity to which the time differentiation is applied.

We now additively decompose the spatial velocity gradient $\mathbf{l}$ according to

$$\mathbf{l}(\mathbf{x}, t) = \mathbf{d}(\mathbf{x}, t) + \mathbf{w}(\mathbf{x}, t) \quad , \tag{2.147}$$

where we have defined

$$\mathbf{d} = \frac{1}{2}(\mathbf{l} + \mathbf{l}^{\mathrm{T}}) = \frac{1}{2}(\mathrm{grad}\mathbf{v} + \mathrm{grad}^{\mathrm{T}}\mathbf{v}) = \mathbf{d}^{\mathrm{T}} \quad , \tag{2.148}$$

$$\mathbf{w} = \frac{1}{2}(\mathbf{l} - \mathbf{l}^{\mathrm{T}}) = \frac{1}{2}(\mathrm{grad}\mathbf{v} - \mathrm{grad}^{\mathrm{T}}\mathbf{v}) = -\mathbf{w}^{\mathrm{T}} \quad , \tag{2.149}$$

or, in index notation, as

$$d_{ab} = \frac{1}{2}\left(\frac{\partial v_a}{\partial x_b} + \frac{\partial v_b}{\partial x_a}\right) = d_{ba} \quad , \tag{2.150}$$

$$w_{ab} = \frac{1}{2}\left(\frac{\partial v_a}{\partial x_b} - \frac{\partial v_b}{\partial x_a}\right) = -w_{ba} \quad , \tag{2.151}$$

with $\mathbf{l}^{\mathrm{T}} = \mathrm{grad}^{\mathrm{T}}\mathbf{v}$ and the associated components $l_{ba} = \partial v_b / \partial x_a$.

In eqs. (2.148) and (2.149) the symmetric part of the spatial velocity gradient $\mathbf{l}$, i.e. the **rate of deformation tensor** (or in the literature sometimes referred to as the **rate of strain tensor**), and the antisymmetric (skew) part of $\mathbf{l}$, i.e. the **spin tensor** (or sometimes called the **rate of rotation tensor** or **vorticity tensor**), are denoted by $\mathbf{d}$ and $\mathbf{w}$, respectively. Both $\mathbf{d}$ and $\mathbf{w}$ are spatial fields involving only quantities acting on the current configuration. We also observe that the components d_{ab} and w_{ab} are *linear* in the velocity components v_a. Note that the spatial tensors $\mathbf{l}, \mathbf{d}, \mathbf{w}$ are viewed as covariant second-order tensors.

EXAMPLE 2.10 Express the spatial acceleration field $\mathbf{a}(\mathbf{x}, t)$, as introduced in (2.27), in terms of the spin tensor $\mathbf{w}(\mathbf{x}, t)$.

Solution. Expanding (2.27) with $(1/2)\mathrm{grad}(\mathbf{v}^2) - (\mathrm{grad}^{\mathrm{T}}\mathbf{v})\mathbf{v} = \mathbf{o}$ (which follows directly from the identity (1.293) by setting $\mathbf{u} = \mathbf{v}$), we obtain

$$\mathbf{a} = \frac{\partial \mathbf{v}}{\partial t} + (\mathrm{grad}\mathbf{v})\mathbf{v} + \frac{1}{2}\mathrm{grad}(\mathbf{v}^2) - (\mathrm{grad}^{\mathrm{T}}\mathbf{v})\mathbf{v} \quad . \tag{2.152}$$

With definition $(2.149)_2$ we end up with an essential relation

$$\mathbf{a} = \frac{\partial \mathbf{v}}{\partial t} + \frac{1}{2}\mathrm{grad}(\mathbf{v}^2) + 2\mathbf{w}\mathbf{v} \tag{2.153}$$

for the spatial acceleration. ∎

The antisymmetric spin tensor **w** may also be represented by its axial vector ω, according to

$$\mathbf{wv} = \omega \times \mathbf{v} \tag{2.154}$$

(compare with eq. (1.118)), for any vector **v**. By recalling eq. (1.276) and property $w_{ab} = -w_{ba} = -\partial v_b/\partial x_a$, we have $2\omega = -\varepsilon_{abc}w_{ab}\mathbf{e}_c = \varepsilon_{abc}\partial v_b/\partial x_a\mathbf{e}_c$. Thus,

$$2\omega = \text{curl}\mathbf{v} \ , \tag{2.155}$$

where 2ω is referred to as the **vorticity vector** (or **spin**), which is of fundamental importance in fluid mechanics. The spatial vector field $\omega = (\text{curl}\mathbf{v})/2$ is called the **angular velocity vector**. For an irrotational motion, i.e. $\text{curl}\mathbf{v} = \mathbf{o}$, the vorticity vector 2ω and (therefore) the spin tensor **w** vanish at each point.

Note that by means of eq. (2.154) and the vorticity vector, as given in (2.155), the term $2\mathbf{wv}$ in relation (2.153) may be expressed as

$$2\mathbf{wv} = (\text{curl}\mathbf{v}) \times \mathbf{v} \ . \tag{2.156}$$

EXAMPLE 2.11   Assume a certain motion wherein the spatial velocity field $\mathbf{v}(\mathbf{x}, t)$ is the gradient of a given spatial scalar field $\Phi(\mathbf{x}, t)$, known as the potential of **v** ($\mathbf{v} = \text{grad}\Phi$).

Compute the spatial acceleration field $\mathbf{a}(\mathbf{x}, t)$ and show that **a** may also be derived from that potential Φ.

Solution.   Knowing that every potential motion is an irrotational motion, i.e. $\text{curl}\mathbf{v} = \mathbf{o}$ (or equivalently $\mathbf{w} = \mathbf{o}$, see eq. (2.156)), we find that the third term in (2.153) vanishes. Consequently we have

$$\mathbf{a} = \frac{\partial \mathbf{v}}{\partial t} + \frac{1}{2}\text{grad}(\mathbf{v}^2) = \frac{\partial}{\partial t}\left(\frac{\partial \Phi}{\partial \mathbf{x}}\right) + \frac{1}{2}\frac{\partial}{\partial \mathbf{x}}\left(\frac{\partial \Phi}{\partial \mathbf{x}} \cdot \frac{\partial \Phi}{\partial \mathbf{x}}\right) \ , \tag{2.157}$$

and with the property that the spatial time derivative commutes with the spatial gradient,

$$\mathbf{a} = \frac{\partial}{\partial \mathbf{x}}\left(\frac{\partial \Phi}{\partial t} + \frac{1}{2}\frac{\partial \Phi}{\partial \mathbf{x}} \cdot \frac{\partial \Phi}{\partial \mathbf{x}}\right) = \text{grad}\left(\frac{\partial \Phi}{\partial t} + \frac{1}{2}(\text{grad}\Phi)^2\right) \ . \tag{2.158}$$

Relation (2.158) shows that if **v** is the gradient of a potential Φ, then **a** may be derived from Φ.   ∎

Spatial velocity gradient in terms of $\dot{\mathbf{U}}$.   We determine a relationship between the spatial velocity gradient $\mathbf{l}$ and the material time derivative of the right stretch tensor, $\dot{\mathbf{U}}$, which is given in terms of material coordinates. By means of $\mathbf{l} = \dot{\mathbf{F}}\mathbf{F}^{-1}$, the polar decomposition $\mathbf{F} = \mathbf{R}\mathbf{U}$, the product rule of differentiation and the property $\mathbf{R}^{\mathrm{T}}\mathbf{R} = \mathbf{I}$, we conclude that

$$\mathbf{l} = (\dot{\mathbf{R}}\mathbf{U})\mathbf{F}^{-1} + (\mathbf{R}\dot{\mathbf{U}})\mathbf{F}^{-1} = \dot{\mathbf{R}}\mathbf{R}^{\mathrm{T}} + \mathbf{R}(\dot{\mathbf{U}}\mathbf{U}^{-1})\mathbf{R}^{\mathrm{T}} \ , \tag{2.159}$$

where $\dot{\mathbf{R}}$ characterizes the time rate of change of the proper orthogonal rotation tensor $\mathbf{R}$. From $\mathbf{R}\mathbf{R}^{\mathrm{T}} = \mathbf{I}$ we deduce directly upon differentiation that $\dot{\mathbf{R}}\mathbf{R}^{\mathrm{T}} + (\dot{\mathbf{R}}\mathbf{R}^{\mathrm{T}})^{\mathrm{T}} = \mathbf{O}$. Thus,

$$\dot{\mathbf{R}}\mathbf{R}^{\mathrm{T}} = -(\dot{\mathbf{R}}\mathbf{R}^{\mathrm{T}})^{\mathrm{T}} \ , \tag{2.160}$$

which clearly shows that $\dot{\mathbf{R}}\mathbf{R}^{\mathrm{T}}$ is a skew tensor.

In order to express the rate of deformation tensor $\mathbf{d}$ and the spin tensor $\mathbf{w}$ in terms of $\dot{\mathbf{U}}$ we adopt the additive decomposition of the spatial velocity gradient (2.147). Hence, using $(2.159)_2$ and property (2.160), the definitions $(2.148)_1$ and $(2.149)_1$ give

$$\mathbf{d} = \mathbf{R}\operatorname{sym}(\dot{\mathbf{U}}\mathbf{U}^{-1})\mathbf{R}^{\mathrm{T}} \ , \qquad \mathbf{w} = \dot{\mathbf{R}}\mathbf{R}^{\mathrm{T}} + \mathbf{R}\operatorname{skew}(\dot{\mathbf{U}}\mathbf{U}^{-1})\mathbf{R}^{\mathrm{T}} \ . \tag{2.161}$$

These relations show that $\mathbf{d}$ is not a pure rate of strain and $\mathbf{w}$ is not a pure rate of rotation.

EXAMPLE 2.12   Show that for a rigid-body rotation (motion) the rate of deformation tensor $\mathbf{d}$ vanishes and the identity $\mathbf{w} = \dot{\mathbf{R}}\mathbf{R}^{\mathrm{T}}$ holds.

Solution.   For a rigid-body rotation (motion) we know that $\mathbf{F} = \mathbf{R}$, which implies $\mathbf{U} = \mathbf{I}$ and $\dot{\mathbf{U}} = \mathbf{O}$ for all $\mathbf{X}$. Hence, we find from $(2.161)_1$ that $\mathbf{d}$ vanishes, meaning that $\mathbf{l} = \mathbf{w}$. On the other hand, from $(2.161)_2$ we conclude that

$$\mathbf{w} = \dot{\mathbf{R}}\mathbf{R}^{\mathrm{T}} \ , \tag{2.162}$$

which means that for the case of a rigid-body rotation the spin tensor $\mathbf{w}$ coincides with the skew tensor $\dot{\mathbf{R}}\mathbf{R}^{\mathrm{T}}$. ∎

EXAMPLE 2.13   Suppose a rigid body is rotating about an axis. The rotation is characterized by the angular velocity vector $\boldsymbol{\omega} = \boldsymbol{\omega}(t)$, as depicted in Figure 2.8.

An arbitrary point x of the body with current position $\mathbf{x} \in \Omega$ is moving around a circle with the spatial velocity $\mathbf{v}(\mathbf{x}, t)$ relative to a fixed point O. Show that the velocity

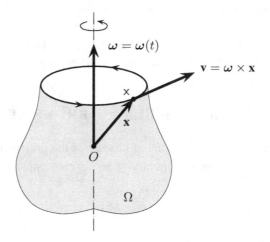

Figure 2.8 Velocity **v** of a particle relative to a fixed point O.

of the point x may be expressed as

$$v(\mathbf{x}, t) = \mathbf{w}(t)\mathbf{x} = \boldsymbol{\omega}(t) \times \mathbf{x} \quad , \tag{2.163}$$

where **w** is the time-dependent antisymmetric spin tensor.

Solution.   The rotation of the rigid body may be described by the linear transformation $\mathbf{x} = \mathbf{R}(t)\mathbf{X}$ (compare with eq. (2.97)), with the proper orthogonal rotation tensor **R** and the referential position **X** of the point. The material time derivative of **x** yields, by means of the product rule, eq. $(2.28)_1$ and $\mathbf{X} = \mathbf{R}^{\mathrm{T}}(t)\mathbf{x}$,

$$v(\mathbf{x}, t) = \dot{\mathbf{R}}(t)\mathbf{X} = \dot{\mathbf{R}}(t)\mathbf{R}^{\mathrm{T}}(t)\mathbf{x} \quad . \tag{2.164}$$

For a rigid-body rotation we know from the last example that the skew tensor $\dot{\mathbf{R}}\mathbf{R}^{\mathrm{T}}$ coincides with the spin tensor **w** (see eq. (2.162)), which gives the desired expression $(2.163)_1$. Relation $v(\mathbf{x}, t) = \boldsymbol{\omega}(t) \times \mathbf{x}$ is well-known from rigid-body dynamics.

In conclusion, relation (2.163) shows that the spin tensor **w** is associated with the angular velocity of a rotating rigid body characterized by the vector $\boldsymbol{\omega}$. In fact, $\boldsymbol{\omega}$ is simply the axial vector of the skew tensor **w**. Eq. (2.163) represents a nice physical interpretation of relations (1.118) and (2.154).   ∎

Material time derivatives of some strain tensors.   Our present starting point involves material strain tensors and their derivatives with respect to time t. In particular, we compute the material time derivative of the Green-Lagrange strain tensor, $\dot{\mathbf{E}}$, the right Cauchy-Green tensor, $\dot{\mathbf{C}}$, and introduce the rotated rate of deformation tensor $\mathbf{D_R}$.

From the definition of the Green-Lagrange strain tensor (2.69), and with the product rule, eq. $(2.141)_4$ and the rate of deformation tensor $\mathbf{d}$ introduced in $(2.148)_1$, we find that

$$\dot{\mathbf{E}} = \frac{1}{2}(\overline{\dot{\mathbf{F}^T \mathbf{F}}} + \mathbf{F}^T \dot{\mathbf{F}}) = \frac{1}{2}(\mathbf{F}^T \mathbf{l}^T \mathbf{F} + \mathbf{F}^T \mathbf{l} \mathbf{F})$$

$$= \mathbf{F}^T \frac{1}{2}(\mathbf{l}^T + \mathbf{l})\mathbf{F} = \mathbf{F}^T \mathbf{d} \mathbf{F} \quad . \tag{2.165}$$

The material time derivative of tensor $\mathbf{E}$ is also known as the **material strain rate tensor** $\dot{\mathbf{E}}$. As can be seen from $(2.165)_4$, it is simply the pull-back of the covariant rate of deformation tensor $\mathbf{d}$, which we may write as $\dot{\mathbf{E}} = \chi_*^{-1}(\mathbf{d}^\flat) = \mathbf{F}^T \mathbf{d} \mathbf{F}$ (see rule $(2.86)_2$).

Now we may show that the directional derivative of the Green-Lagrange strain tensor $\mathbf{E}$ in the direction of $\mathbf{v}$ equals the material strain rate tensor $\dot{\mathbf{E}}$, as given in (2.165). Using eq. (2.140) and the common properties of the product rule, we find using definition (2.69) that

$$D_\mathbf{v}\mathbf{E} = \frac{1}{2}(D_\mathbf{v}\mathbf{F}^T \mathbf{F} + \mathbf{F}^T D_\mathbf{v}\mathbf{F}) = \frac{1}{2}(\overline{\dot{\mathbf{F}^T} \mathbf{F}} + \mathbf{F}^T \dot{\mathbf{F}}) \quad , \tag{2.166}$$

and with $(2.165)_1$,

$$\frac{D\mathbf{E}(\mathbf{X}, t)}{Dt} = D_\mathbf{v}\mathbf{E}(\mathbf{X}, t) \quad . \tag{2.167}$$

The resulting relation (2.167) may also be found immediately from (2.20).

The time rate of change of the right Cauchy-Green tensor $\mathbf{C}$ follows from definitions (2.65), (2.69) as $\dot{\mathbf{C}} = 2\dot{\mathbf{E}}$. With eq. $(2.165)_4$ we obtain

$$\dot{\mathbf{C}} = 2\dot{\mathbf{E}} = 2\mathbf{F}^T \mathbf{d} \mathbf{F} \quad . \tag{2.168}$$

By replacing $\mathbf{F}$ with the rotation tensor $\mathbf{R}$ in $(2.165)_4$ we obtain $\mathbf{D_R} = \mathbf{R}^T \mathbf{d} \mathbf{R}$, which is known as the **rotated rate of deformation tensor** $\mathbf{D_R}$. Using the polar decomposition $\mathbf{F} = \mathbf{R}\mathbf{U}$ and the symmetry $\mathbf{U}^T = \mathbf{U}$ we find from $(2.168)_2$ that $\dot{\mathbf{C}} = 2\mathbf{U}(\mathbf{R}^T \mathbf{d} \mathbf{R})\mathbf{U} = 2\mathbf{U}\mathbf{D_R}\mathbf{U}$. Finally, we have an alternative expression for $\mathbf{D_R}$, i.e.

$$\mathbf{D_R} = \mathbf{R}^T \mathbf{d} \mathbf{R} = \frac{1}{2}\mathbf{C}^{-1/2}\dot{\mathbf{C}}\mathbf{C}^{-1/2} \quad , \tag{2.169}$$

where $\mathbf{C}^{-1/2} = \mathbf{U}^{-1}$ is in accordance with $(2.95)_1$.

In the following we consider spatial strain tensors and compute a relationship between the material time derivative of the left Cauchy-Green tensor, $\dot{\mathbf{b}}$, and the spatial tensors $\mathbf{l}$ and $\mathbf{b}$. Recall the definition (2.80) of $\mathbf{b}$ and use the product rule in order to obtain

$$\dot{\mathbf{b}} = \overline{\dot{\mathbf{F}}\mathbf{F}^{\mathrm{T}}} = \dot{\mathbf{F}}\mathbf{F}^{\mathrm{T}} + \mathbf{F}\overline{\dot{\mathbf{F}^{\mathrm{T}}}}$$
$$= (\dot{\mathbf{F}}\mathbf{F}^{-1})\mathbf{F}\mathbf{F}^{\mathrm{T}} + (\mathbf{F}\mathbf{F}^{\mathrm{T}})\mathbf{F}^{-\mathrm{T}}\overline{\dot{\mathbf{F}^{\mathrm{T}}}} \quad . \tag{2.170}$$

Then, we find the important relationship

$$\dot{\mathbf{b}} = \mathbf{l}\mathbf{b} + \mathbf{b}\mathbf{l}^{\mathrm{T}} \quad , \tag{2.171}$$

where the definition $(2.141)_4$ of the spatial velocity gradient $\mathbf{l}$ is to be used.

EXAMPLE 2.14   Show the useful relation

$$\dot{\mathbf{e}} = \mathbf{d} - \mathbf{l}^{\mathrm{T}}\mathbf{e} - \mathbf{e}\mathbf{l} \tag{2.172}$$

for the material time derivative of the Euler-Almansi strain tensor $\mathbf{e}$.

Solution.   Recall the definition of the spatial strain tensor $\mathbf{e}$, i.e. eq. (2.83). Using the product rule we obtain

$$\dot{\mathbf{e}} = -\frac{1}{2}\overline{\dot{\mathbf{F}^{-\mathrm{T}}}\mathbf{F}^{-1}} = -\frac{1}{2}(\overline{\dot{\mathbf{F}^{-\mathrm{T}}}}\mathbf{F}^{-1} + \mathbf{F}^{-\mathrm{T}}\overline{\dot{\mathbf{F}^{-1}}}) \quad . \tag{2.173}$$

Hence, with the derived relations $(2.146)_2$ and $(2.145)_2$ and the definition of the Euler-Almansi strain tensor (2.83), we obtain

$$\dot{\mathbf{e}} = -\frac{1}{2}(-\mathbf{l}^{\mathrm{T}}\mathbf{F}^{-\mathrm{T}}\mathbf{F}^{-1} - \mathbf{F}^{-\mathrm{T}}\mathbf{F}^{-1}\mathbf{l})$$
$$= \frac{1}{2}[\mathbf{l}^{\mathrm{T}}(\mathbf{I} - 2\mathbf{e}) + (\mathbf{I} - 2\mathbf{e})\mathbf{l}]$$
$$= \mathbf{d} - \mathbf{l}^{\mathrm{T}}\mathbf{e} - \mathbf{e}\mathbf{l} \quad , \tag{2.174}$$

where the definition $(2.148)_1$ of the rate of deformation tensor $\mathbf{d}$ is to be used. ∎

Material time derivatives of spatial line, surface and volume elements.   We consider first spatial and material line elements $d\mathbf{x} \in \Omega$ and $d\mathbf{X} \in \Omega_0$, as introduced in (2.37), and compute the material time derivative of $d\mathbf{x}$. We know from (2.38) that line elements map via the deformation gradient according to $d\mathbf{x} = \mathbf{F}(\mathbf{X}, t)d\mathbf{X}$. By means of the product rule and relation $(2.141)_4$ we find that

$$\overline{\dot{d\mathbf{x}}} = \dot{\mathbf{F}}d\mathbf{X} = \dot{\mathbf{F}}\mathbf{F}^{-1}d\mathbf{x} = \mathbf{l}d\mathbf{x} \qquad \text{or} \qquad \overline{\dot{dx_a}} = \frac{\partial v_a}{\partial x_b}dx_b \quad . \tag{2.175}$$

As can be seen, $\mathbf{l} = \mathrm{grad}\mathbf{v}$ is a spatial tensor field transforming $d\mathbf{x}$ into $\overline{\dot{d\mathbf{x}}}$.

Before proceeding it is necessary to provide the relation for the material time derivative of the volume ratio $J = \det \mathbf{F} > 0$. Using the chain rule we obtain simply $\dot{J} = \partial J/\partial \mathbf{F} : \dot{\mathbf{F}}$. We just need to specify the term $\partial J/\partial \mathbf{F}$ which results from relation (1.241) by taking $\mathbf{F}$ instead of $\mathbf{A}$. Hence, we may write

$$\frac{\partial J}{\partial \mathbf{F}} = J\mathbf{F}^{-\mathrm{T}} \ . \tag{2.176}$$

Consequently, with relation (2.176), $\dot{\mathbf{F}} = \mathbf{l}\mathbf{F}$ and properties (1.95), (1.94), (1.279) we find expressions for the material time derivative of the scalar field J, namely

$$\begin{aligned} \dot{J} &= J\mathbf{F}^{-\mathrm{T}} : \dot{\mathbf{F}} = J\mathbf{F}^{-\mathrm{T}} : \mathbf{l}\mathbf{F} \\ &= J\mathbf{F}^{-\mathrm{T}}\mathbf{F}^{\mathrm{T}} : \mathbf{l} = J\mathbf{I} : \mathrm{grad}\mathbf{v} \\ &= J\mathrm{tr}(\mathrm{grad}\mathbf{v}) = J\mathrm{div}\mathbf{v} \qquad \text{or} \qquad \dot{J} = J\frac{\partial v_a}{\partial x_a} \ . \end{aligned} \tag{2.177}$$

Using the additive split $\mathbf{l} = \mathrm{grad}\mathbf{v} = \mathbf{d} + \mathbf{w}$ and knowing that the trace of a skew tensor is zero, we deduce from $(2.177)_5$ an important alternative expression for $\dot{J}$, namely

$$\dot{J} = J\mathrm{tr}\mathbf{d} \qquad \text{or} \qquad \dot{J} = Jd_{aa} \ . \tag{2.178}$$

By recalling relationship (2.20), the material time derivative of J may also be evaluated as the directional derivative of J in the direction of the velocity vector $\mathbf{v}$. Thus, we also may write $\mathrm{D}J(\mathbf{X}, t)/\mathrm{D}t = D_{\mathbf{v}}J(\mathbf{X}, t)$.

A motion with $J = 1$ we called *isochoric* (keeping the volume constant, $\mathrm{d}v = \mathrm{const}$). From eqs. (2.177) and (2.178) we may deduce alternative expressions for $J = 1$ or $\mathrm{d}v = \mathrm{const}$, namely, $\dot{J} = 0$, $\mathbf{F}^{-\mathrm{T}} : \dot{\mathbf{F}} = 0$, $\mathrm{div}\mathbf{v} = 0$ or $\mathrm{tr}\mathbf{d} = 0$. In summary, the following six statements characterize necessary and sufficient conditions for an isochoric motion and are equivalent to one another:

$$\left.\begin{array}{ccc} J = 1 \ , & \mathrm{d}v = \mathrm{const} \ , & \dot{J} = 0 \ , \\[2mm] \mathbf{F}^{-\mathrm{T}} : \dot{\mathbf{F}} = 0 \ , & \mathrm{div}\mathbf{v} = 0 \ , & \mathrm{tr}\mathbf{d} = 0 \ . \end{array}\right\} \tag{2.179}$$

A continuum body is said to be **incompressible** if every motion it undergoes is isochoric. Consequently, for every motion of an incompressible body each of the conditions in (2.179) holds. The condition $\mathbf{F}^{-\mathrm{T}} : \dot{\mathbf{F}} = 0$ is essential in the treatment of motions of incompressible solids, while $\mathrm{div}\mathbf{v} = 0$, $\mathrm{tr}\mathbf{d} = 0$ is of fundamental importance in fluid mechanics. If the deformation behavior of a continuum body is restricted, we say that the body is subjected to so-called **internal constraints**. In particular, the conditions formulated in (2.179) characterize the most important internal constraint known as the **internal kinematic constraint**, or more precisely the **incompressibility constraint**.

Now we perform the material time derivative of a vector element $d\mathbf{s}$ of infinitesimally small area defined in the current configuration. Applying Nanson's formula (2.55), i.e. $d\mathbf{s} = J\mathbf{F}^{-T}d\mathbf{S}$, the product rule and eqs. $(2.177)_6$ and $(2.146)_2$, we obtain

$$\dot{\overline{d\mathbf{s}}} = (\dot{J}\mathbf{F}^{-T} + J\dot{\overline{\mathbf{F}^{-T}}})d\mathbf{S} = (\mathrm{div}\,\mathbf{v}\,\mathbf{I} - \mathbf{l}^T)J\mathbf{F}^{-T}d\mathbf{S}$$
$$= \mathrm{div}\,\mathbf{v}\,d\mathbf{s} - \mathbf{l}^T d\mathbf{s} \quad , \tag{2.180}$$

with the second-order unit tensor $\mathbf{I}$ (note that the spatial velocity gradient $\mathbf{l}$ has a similar symbol as $\mathbf{I}$). In index notation eq. $(2.180)_3$ reads as

$$\dot{\overline{ds_a}} = \frac{\partial v_b}{\partial x_b}ds_a - \frac{\partial v_b}{\partial x_a}ds_b \quad . \tag{2.181}$$

Expression $(2.180)_3$ (and (2.181)) represent the relation between the rate of change of the infinitesimal spatial vector area in terms of $d\mathbf{s}$, the divergence of the spatial velocity field $\mathbf{v}$ and the transpose of the spatial velocity gradient $\mathbf{l}$.

Finally, the material time derivative of the spatial volume element $dv = JdV$ gives by means of the product rule

$$\dot{\overline{dv}} = \dot{J}dV = \mathrm{div}\,\mathbf{v}\,dv \qquad \text{or} \qquad \dot{\overline{dv}} = \frac{\partial v_a}{\partial x_a}dv \quad , \tag{2.182}$$

where the relation $\dot{J} = J\mathrm{div}\,\mathbf{v}$ is to be used.

EXERCISES

1. Show that in a moving continuum the spatial velocity field $\mathbf{v}$ and the spatial vorticity field $2\boldsymbol{\omega}$ are related to their material time derivatives by the identity

$$\mathrm{curl}\dot{\mathbf{v}} = 2\dot{\boldsymbol{\omega}} + 2\boldsymbol{\omega}\mathrm{div}\,\mathbf{v} - (\mathrm{grad}\mathbf{v})2\boldsymbol{\omega} \quad . \tag{2.183}$$

   Hint: Take the curl of relation (2.153) by considering eqs. (2.155) and (2.156), and then use property (1.294) with (1.275).

2. Consider a deformed fiber characterized by the vector $\mathbf{a}$, with $|\mathbf{a}| = 1$. Show that the material time derivative of the logarithmic stretch ratio λ at a particle along that direction $\mathbf{a}$ is given by

$$\dot{\overline{\ln\lambda}} = \mathbf{a} \cdot \mathbf{da} \quad . \tag{2.184}$$

   Project the tensor $\mathbf{d}$ onto the orthonormal basis vectors $\mathbf{e}_a$ and give a physical interpretation of the rate of deformation tensor $\mathbf{d}$. In particular, discuss the diagonal components d_{aa} and the off-diagonal components d_{ab} $(a \neq b)$, $a, b = 1, 2, 3$, of the matrix $[\mathbf{d}]$.

   Hint: Compute the dot product of vector $\mathbf{a}$ and the vectors of equation (2.143) and use the fact that $\dot{\mathbf{a}} \cdot \mathbf{a} = 0$ (since $\mathbf{a} \cdot \mathbf{a} = 1$), and $\mathbf{a} \cdot \mathbf{wa} = \mathbf{a} \cdot (\boldsymbol{\omega} \times \mathbf{a}) = 0$ (since $\mathbf{u} \cdot (\mathbf{u} \times \mathbf{v}) = 0$, which should be compared with eq. (1.31)).

3. By combining eq. (2.143) with eq. (2.184) show that the material time derivative of a unit vector **a** may be expressed as

$$\dot{\mathbf{a}} = \mathbf{l}\mathbf{a} - (\mathbf{a} \cdot \mathbf{d}\mathbf{a})\mathbf{a} \quad . \tag{2.185}$$

We now assume that α_a, $a = 1, 2, 3$; denote the three eigenvalues of **d** and $\hat{\mathbf{n}}_a$ its three associated normalized eigenvectors. For the particular case in which **a** is an eigenvector of the rate of deformation tensor **d**, show by means of (2.185) and decomposition $\mathbf{l} = \mathbf{d} + \mathbf{w}$, that

$$\dot{\hat{\mathbf{n}}}_a = \mathbf{w}\hat{\mathbf{n}}_a = \boldsymbol{\omega} \times \hat{\mathbf{n}}_a \quad , \qquad a = 1, 2, 3 \quad , \tag{2.186}$$

with the eigenvalue problem $\mathbf{d}\hat{\mathbf{n}}_a = \alpha_a\hat{\mathbf{n}}_a$ ($a = 1, 2, 3$; no summation), where $\alpha_a = \overline{\dot{\ln \lambda_a}}$.

Obviously, the spin tensor **w** is a measure for the rate of change of the eigenvectors of **d**, which gives a physical interpretation of the spin tensor **w**.

4. Recall Exercise 2, p. 93, with the motion $\mathbf{x} = \mathbf{X} + c(t)(\mathbf{e}_2 \cdot \mathbf{X})\mathbf{e}_1$, the orthogonal unit vectors $\mathbf{e}_1$, $\mathbf{e}_2$, $\mathbf{e}_3 = \mathbf{e}_1 \times \mathbf{e}_2$ and the parameter $c(t) = \tan\theta(t) > 0$.

    (a) Show that the spatial velocity field $\mathbf{v}(\mathbf{x}, t)$ may be given by $\mathbf{v} = \dot{c}(t)(\mathbf{e}_2 \cdot \mathbf{x})\mathbf{e}_1$, where $\dot{c}(t) = \overline{\dot{\tan\theta(t)}}$ is called the **shear rate** (for typical ranges of shear rates of some specific materials see, for example, BARNES et al. [1989]).

    (b) Based on result (a) compute the rate of deformation tensor $\mathbf{d}(\mathbf{x}, t)$, the spin tensor $\mathbf{w}(\mathbf{x}, t)$ and the angular velocity vector $\boldsymbol{\omega}(\mathbf{x}, t)$, i.e. the axial vector of the spin tensor, in terms of $\mathbf{e}_1$, $\mathbf{e}_2$ and $\mathbf{e}_3$.

    (c) Show that the rate of deformation tensor may be expressed in its spectral form

$$\mathbf{d} = \sum_{a=1}^{3} \alpha_a\hat{\mathbf{n}}_a \otimes \hat{\mathbf{n}}_a \quad ,$$

    and hence compute the three eigenvalues of **d**, i.e. α_a, and the three normalized eigenvectors of **d**, i.e. the set $\{\hat{\mathbf{n}}_a\}$.

5. Consider the representation of the deformation gradient (compare with (2.123)$_2$):

$$\mathbf{F}(\mathbf{X}, t) = \sum_{a=1}^{3} \lambda_a(t)(\mathbf{R}(t)\hat{\mathbf{N}}_a) \otimes \hat{\mathbf{N}}_a \quad , \tag{2.187}$$

where the principal referential directions $\hat{\mathbf{N}}_a$ (orthonormal eigenvectors) are assumed not to change in time.

    (a) Compute the rate of deformation tensor $\mathbf{d}(\mathbf{x}, t)$, the spin tensor $\mathbf{w}(\mathbf{x}, t)$ and the angular velocity vector $\boldsymbol{\omega}(\mathbf{x}, t)$.

(b) Based on representation (2.187), find $\dot{\mathbf{U}}\mathbf{U}^{-1}$ and show that the antisymmetric tensors $\mathbf{w}$ and $\dot{\mathbf{R}}\mathbf{R}^{\mathrm{T}}$ coincide.

6. Show that in a plane motion, i.e. $v_1 = v_1(x_1, x_2, t), v_2 = v_2(x_1, x_2, t), v_3 = 0$,

$$\mathbf{w}\mathbf{d} + \mathbf{d}\mathbf{w} = (\mathrm{div}\mathbf{v})\mathbf{w} \quad .$$

2.8　Lie Time Derivatives

Consider a spatial field $f = f(\mathbf{x}, t)$ characterizing some physical scalar, vector or tensor quantity in space and time (for the relevant notation recall Section 2.3). In the following we compute the change of f relative to a vector field $\mathbf{v}$ which is commonly known as the **Lie time derivative** of f, denoted by $\mathcal{L}_\mathbf{v}(f)$.

The Lie time derivative of a spatial field f is obtained using the following concept:

(i) compute the *pull-back* operation of f to the reference configuration; as a result we obtain the associated material field $\mathcal{F}(\mathbf{X}, t) = \chi_*^{-1}(f(\mathbf{x}, t))$;

(ii) take the material time derivative of $\mathcal{F}$, i.e. $\dot{\mathcal{F}}$, and

(iii) carry out the *push-forward* operation of the result to the current configuration.

This technique is simply summarized as

$$\mathcal{L}_\mathbf{v}(f) = \chi_* \left(\frac{\mathrm{D}}{\mathrm{D}t} \chi_*^{-1}(f) \right) = \chi_*(\dot{\mathcal{F}}) \quad . \tag{2.188}$$

Since the material time derivative can be obtained from the directional derivative according to relation (2.20) we may apply the concept of directional derivative to eq. (2.188). Consequently, we express the Lie time derivative of $f = f(\mathbf{x}, t)$ as the directional derivative (see also WRIGGERS [1988]). Hence, eq. (2.188) reads equivalently as

$$\mathcal{L}_\mathbf{v}(f) = \chi_* \left(D_\mathbf{v}\chi_*^{-1}(f) \right) = \chi_*(D_\mathbf{v}\mathcal{F}) \quad . \tag{2.189}$$

In summary: the Lie time derivative of the spatial field f is the push-forward of the directional derivative of the associated material field $\mathcal{F} = \chi_*^{-1}(f)$ in the direction of the vector $\mathbf{v}$, identified as the *velocity vector*.

By recalling definition (1.266), we may specify the directional derivative of $\mathcal{F}$ at the reference configuration in the direction of the velocity vector $\mathbf{v}$ as

$$D_\mathbf{v}\mathcal{F} = \frac{\mathrm{d}}{\mathrm{d}\varepsilon}\mathcal{F}(\mathbf{X} + \varepsilon\mathbf{v})|_{\varepsilon=0} \quad . \tag{2.190}$$

If $\Phi = \Phi(\mathbf{x}, t)$ is a function that assigns a *scalar* Φ to each point $\mathbf{x}$ at time t, the Lie time derivative of Φ coincides with the material time derivative of Φ. Thus, $\mathcal{L}_{\mathbf{v}}(\Phi)$ $= \dot{\Phi}(\mathbf{x}, t)$.

The concept of Lie time derivatives occurs throughout constitutive theories in computing stress rates (see Section 5.3). In addition, it is a powerful concept used in the process of linearization and variation. In fact, it emerges that the variation or the linearization of a spatial field is formally equivalent to the Lie time derivative according to the rule (2.189) (see Chapter 8 for more details).

EXAMPLE 2.15 Show that the Lie time derivative of the Euler-Almansi strain tensor $\mathbf{e}$ is the rate of deformation tensor $\mathbf{d}$, i.e.

$$\mathcal{L}_{\mathbf{v}}(\mathbf{e}) = \mathbf{d} \quad . \tag{2.191}$$

Derive this result by using first the rule (2.188) and then the concept of directional derivative, in particular eqs. (2.189), (2.190).

Solution. Recall the pull-back operation of the covariant Euler-Almansi strain tensor $\mathbf{e}$ which gives the Green-Lagrange strain tensor $\mathbf{E} = \mathbf{F}^{\mathrm{T}}\mathbf{e}\mathbf{F}$, i.e. eq. (2.85)$_4$. Then recall that the push-forward operation on the material strain rate tensor $\dot{\mathbf{E}}$ is the rate of deformation tensor $\mathbf{d} = \mathbf{F}^{-\mathrm{T}}\dot{\mathbf{E}}\mathbf{F}^{-1}$, i.e. eq. (2.165)$_4$.

According to the steps expressed in (2.188) we find the desired result

$$\mathcal{L}_{\mathbf{v}}(\mathbf{e}^\flat) = \mathbf{F}^{-\mathrm{T}} \left(\frac{\mathrm{D}(\mathbf{F}^{\mathrm{T}}\mathbf{e}\mathbf{F})}{\mathrm{D}t} \right) \mathbf{F}^{-1} = \mathbf{F}^{-\mathrm{T}} \frac{\mathrm{D}\mathbf{E}}{\mathrm{D}t} \mathbf{F}^{-1}$$

$$= \mathbf{F}^{-\mathrm{T}}\dot{\mathbf{E}}\mathbf{F}^{-1} = \mathbf{d} \quad . \tag{2.192}$$

Another way to obtain eq. (2.191) is through relations (2.189) and (2.190). Eq. (2.189) gives

$$\mathcal{L}_{\mathbf{v}}(\mathbf{e}) = \mathbf{F}^{-\mathrm{T}}(D_{\mathbf{v}}(\mathbf{F}^{\mathrm{T}}\mathbf{e}\mathbf{F}))\mathbf{F}^{-1} = \mathbf{F}^{-\mathrm{T}}(D_{\mathbf{v}}\mathbf{E})\mathbf{F}^{-1} \quad . \tag{2.193}$$

With reference to eq. (2.190) we must determine the directional derivative of the Green-Lagrange strain tensor $\mathbf{E}$ in the direction of $\mathbf{v}$, which is, by reference to (2.167), the material strain rate tensor $\dot{\mathbf{E}}$. Finally, from eq. (2.193)$_2$, we find that $\mathcal{L}_{\mathbf{v}}(\mathbf{e}) = \mathbf{F}^{-\mathrm{T}}(D_{\mathbf{v}}\mathbf{E})\mathbf{F}^{-1} = \mathbf{F}^{-\mathrm{T}}\dot{\mathbf{E}}\mathbf{F}^{-1} = \mathbf{d}$. ∎

EXERCISES

1. Establish the properties

$$\mathcal{L}_{\mathbf{v}}(\mathbf{b}^\sharp) = \mathbf{O} \quad , \qquad \mathcal{L}_{\mathbf{v}}(\mathbf{b}^{-1})^\flat = \mathbf{O} \quad .$$

2. Consider a contravariant spatial vector field **u** with its material time derivative
   $\dot{\mathbf{u}} = \partial\mathbf{u}/\partial t + (\mathrm{grad}\,\mathbf{u})\mathbf{v}$ (see eq. (2.30)).

   Recall relations (2.89) and (2.145)$_2$ and show that the Lie time derivative of **u** is,
   according to the rule (2.188),

$$\pounds_{\mathbf{v}}(\mathbf{u}^{\sharp}) = \frac{\partial\mathbf{u}}{\partial t} + (\mathrm{grad}\,\mathbf{u})\mathbf{v} - \mathbf{l}\mathbf{u} \quad . \tag{2.194}$$

3 The Concept of Stress

In the previous chapter some kinematical aspects of the motion and deformation of a continuum body were discussed. Motion and deformation give rise to interactions between the material and neighboring material in the interior part of the body. One consequence of these interactions is stress, which has physical dimension force per unit of area. The notion of stress, which is responsible for the deformation of materials, is crucial in continuum mechanics.

In particular, we consider a deformable body during a finite motion. For that body we introduce the *concept* of stress and discuss the properties of traction vectors and stress tensors in different descriptions. We continue with a brief review of the determination of extremal stress values, noting that this topic has already been studied extensively in numerous other books on mechanics.

Since the current configuration of many problems, especially those involving solids, is not known, it is therefore not convenient to work with stress tensors which are expressed in terms of spatial coordinates. For some cases it is more convenient to work with stress tensors that are referred to the reference configuration or an intermediate configuration. Some important forms of alternative stress measures, which are associated with the names of *Piola, Kirchhoff, Biot, Green, Naghdi* or *Mandel*, are provided.

3.1 Traction Vectors, and Stress Tensors

Here we introduce surface tractions which are related to either the current or the reference configuration. They may be expressed in terms of unique stress fields acting on a normal vector to a plane surface. Stress components such as normal and shear stresses are introduced.

Surface tractions. We focus attention on a deformable continuum body $\mathcal{B}$ occupying an arbitrary region Ω of physical space with boundary surface $\partial\Omega$ at time t, as shown in Figure 3.1.

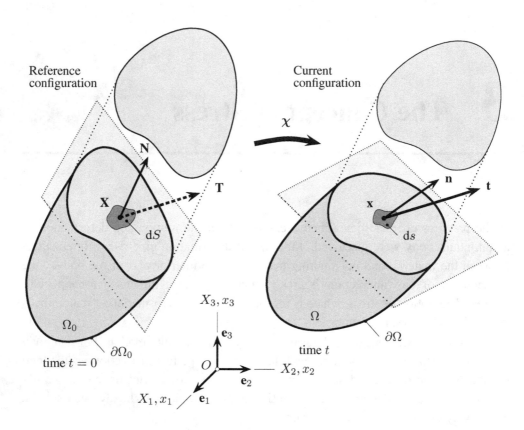

Figure 3.1 Traction vectors acting on infinitesimal surface elements with outward unit normals.

We postulate that arbitrary forces act on parts or the whole of the boundary surface (called *external forces*), and on an (imaginary) surface within the interior of that body (called *internal forces*) in some distributed manner.

Let the body now be cut by a plane surface which passes any given point $\mathbf{x} \in \Omega$ with spatial coordinates x_a at time t. As illustrated in Figure 3.1, the plane surface separates the deformable body into two portions. We focus attention on that part of the *free* (continuum) body lying on the tail of a unit vector $\mathbf{n}$ at $\mathbf{x}$, directed along the outward normal to an infinitesimal spatial surface element $\mathrm{d}s \in \partial\Omega$. Since we consider interaction of the two portions, forces are transmitted across the (internal) plane surface. We denote an infinitesimal **resultant (actual) force** acting on a surface element as $\mathrm{d}\mathbf{f}$. For purposes which will be made clear in Section 4.3 we omit distributed so-called **resultant couples** not occurring in the classical formulation of continuum mechanics. For a couple we can think of a pure torque.

Initially, before motion occurred, the continuum body $\mathcal{B}$ was in the reference (undeformed) configuration at the reference time $t = 0$ and has occupied the region Ω_0 of physical space with boundary surface $\partial\Omega_0$. The quantities $\mathbf{x}$ (with spatial coordinates x_a), $\mathrm{d}s$ and $\mathbf{n}$ which are associated with the current configuration of the body are denoted by $\mathbf{X}$ (with material coordinates X_A), $\mathrm{d}S$ and $\mathbf{N}$ when they are referred to the reference configuration.

According to Figure 3.1 we claim that for every surface element

$$\mathrm{d}\mathbf{f} = \mathbf{t}\mathrm{d}s = \mathbf{T}\mathrm{d}S \quad , \tag{3.1}$$

$$\mathbf{t} = \mathbf{t}(\mathbf{x}, t, \mathbf{n}) \quad , \qquad \mathbf{T} = \mathbf{T}(\mathbf{X}, t, \mathbf{N}) \quad . \tag{3.2}$$

Here, $\mathbf{t}$ represents the **Cauchy** (or **true**) **traction vector** (force measured per unit surface area defined in the *current* configuration), exerted on $\mathrm{d}s$ with outward normal $\mathbf{n}$. The vector $\mathbf{T}$ represents the **first Piola-Kirchhoff** (or **nominal**) **traction vector** (force measured per unit surface area defined in the *reference* configuration), and points in the *same* direction as the Cauchy traction vector $\mathbf{t}$. The (pseudo) traction vector $\mathbf{T}$ does not describe the actual intensity. It acts on the region Ω and is, in contrast to the Cauchy traction vector $\mathbf{t}$, a function of the referential position $\mathbf{X}$ and the outward normal $\mathbf{N}$ to the boundary surface $\partial\Omega_0$. This circumstance is indicated in Figure 3.1 in the form of a dashed line for $\mathbf{T}$. Relationship (3.2) is **Cauchy's postulate**.

The vectors $\mathbf{t}$ and $\mathbf{T}$ that act across the surface elements $\mathrm{d}s$ and $\mathrm{d}S$ with respective normals $\mathbf{n}$ and $\mathbf{N}$ are referred to as **surface tractions** or in some texts as **contact forces**, **stress vectors** or just **loads**. Typical surface tractions are contact and friction forces or are caused by liquids or gases, for example, water or wind.

Cauchy's stress theorem.   There exist unique second-order tensor fields $\boldsymbol{\sigma}$ and $\mathbf{P}$ so that

$$\left. \begin{array}{llll} \mathbf{t}(\mathbf{x}, t, \mathbf{n}) = \boldsymbol{\sigma}(\mathbf{x}, t)\mathbf{n} & \text{or} & t_a = \sigma_{ab}n_b \quad , \\[2mm] \mathbf{T}(\mathbf{X}, t, \mathbf{N}) = \mathbf{P}(\mathbf{X}, t)\mathbf{N} & \text{or} & T_a = P_{aA}N_A \end{array} \right\} \tag{3.3}$$

(the proof is omitted), where $\boldsymbol{\sigma}$ denotes a *symmetric* spatial tensor field called the **Cauchy** (or **true**) **stress tensor** (or simply the **Cauchy stress**), while $\mathbf{P}$ characterizes a tensor field called the **first Piola-Kirchhoff stress tensor** (or simply the **Piola stress**). The transpose of $\mathbf{P}$ is frequently called the **nominal stress tensor**. The index notation in relation (3.3) reveals that $\mathbf{P}$, like $\mathbf{F}$, is a two-point tensor in which one index describes *spatial coordinates* x_a, and the other *material coordinates* X_A. In Section 4.3, p. 147, we establish that the Cauchy stress tensor $\boldsymbol{\sigma}$ is symmetric, under the assumption that resultant couples are neglected.

Relation (3.3), which combines the surface traction with the stress tensor, is one of the most important axioms in continuum mechanics and is known as **Cauchy's stress theorem** (or in the literature sometimes referred to as **Cauchy's law**). Basically it states that if traction vectors such as $\mathbf{t}$ or $\mathbf{T}$ depend on the outward unit normals $\mathbf{n}$ or $\mathbf{N}$, then they must be **linear** in $\mathbf{n}$ or $\mathbf{N}$, respectively.

An immediate consequence of (3.3) is the following relationship between $\mathbf{t}, \mathbf{T}$ and the corresponding normal vectors, i.e.

$$\mathbf{t}(\mathbf{x}, t, \mathbf{n}) = -\mathbf{t}(\mathbf{x}, t, -\mathbf{n}) \qquad \text{or} \qquad \mathbf{T}(\mathbf{X}, t, \mathbf{N}) = -\mathbf{T}(\mathbf{X}, t, -\mathbf{N}) \qquad (3.4)$$

for all unit vectors $\mathbf{n}$ and $\mathbf{N}$. This is known as **Newton's (third) law of action** and **reaction** (see Figure 3.2).

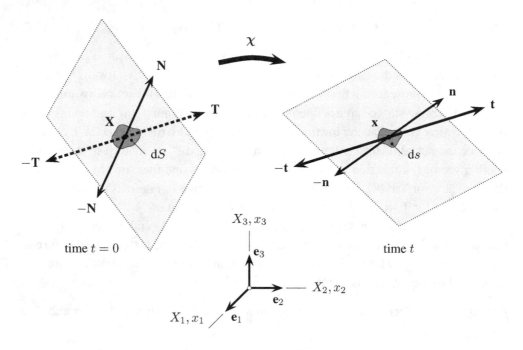

Figure 3.2 Newton's (third) law of action and reaction.

To write Cauchy's stress theorem, for example $(3.3)_1$, in the more convenient matrix notation which is useful for computational purposes, we have

$$[\mathbf{t}] = [\boldsymbol{\sigma}][\mathbf{n}] \quad , \qquad (3.5)$$

$$[\mathbf{t}] = \begin{bmatrix} t_1 \\ t_2 \\ t_3 \end{bmatrix} \quad , \qquad [\boldsymbol{\sigma}] = \begin{bmatrix} \sigma_{11} & \sigma_{12} & \sigma_{13} \\ \sigma_{21} & \sigma_{22} & \sigma_{23} \\ \sigma_{31} & \sigma_{32} & \sigma_{33} \end{bmatrix} \quad , \qquad [\mathbf{n}] = \begin{bmatrix} n_1 \\ n_2 \\ n_3 \end{bmatrix} \quad , \qquad (3.6)$$

where $[\boldsymbol{\sigma}]$ is usually called the **(Cauchy) stress matrix**.

Finally, we find the relation between the Cauchy stress tensor $\boldsymbol{\sigma}$ and the first Piola-Kirchhoff stress tensor $\mathbf{P}$. From eq. (3.1) we obtain with eqs. (3.3) and (3.2) the important transformation

$$\mathbf{t}(\mathbf{x}, t, \mathbf{n}) \mathrm{d}s = \mathbf{T}(\mathbf{X}, t, \mathbf{N}) \mathrm{d}S \quad ,$$

$$\boldsymbol{\sigma}(\mathbf{x}, t)\mathbf{n}\mathrm{d}s = \mathbf{P}(\mathbf{X}, t)\mathbf{N}\mathrm{d}S \quad . \tag{3.7}$$

Using Nanson's formula, i.e. eq. (2.55), $\mathbf{P}$ may be written in the form

$$\mathbf{P} = J\boldsymbol{\sigma}\mathbf{F}^{-\mathrm{T}} \qquad \text{or} \qquad P_{aA} = J\sigma_{ab}F_{Ab}^{-1} \quad . \tag{3.8}$$

The passage from $\boldsymbol{\sigma}$ to $\mathbf{P}$ and back is known as the *Piola transformation* (compare with the definition on p. 83). Strictly speaking, in order to obtain the two-point tensor $\mathbf{P}$, with components P_{aA}, we have performed a Piola transformation on the second index of tensor $\boldsymbol{\sigma}$, with components σ_{ab}.

For convenience, we omit subsequently the arguments of the tensor quantities. The explicit expression for the symmetric Cauchy stress tensor results as the inverse of relation (3.8), i.e.

$$\boldsymbol{\sigma} = J^{-1}\mathbf{P}\mathbf{F}^{\mathrm{T}} = \boldsymbol{\sigma}^{\mathrm{T}} \qquad \text{or} \qquad \sigma_{ab} = J^{-1}P_{aA}F_{bA} = \sigma_{ba} \quad , \tag{3.9}$$

which necessarily implies

$$\mathbf{P}\mathbf{F}^{\mathrm{T}} = \mathbf{F}\mathbf{P}^{\mathrm{T}} \quad . \tag{3.10}$$

Consequently, the second-order tensor $\mathbf{P}$ is, in general, *not* symmetric and has nine independent components P_{aA}.

EXAMPLE 3.1   A deformation of a body is described by

$$x_1 = -6X_2 \quad , \qquad x_2 = \frac{1}{2}X_1 \quad , \qquad x_3 = \frac{1}{3}X_3 \quad . \tag{3.11}$$

The Cauchy stress tensor for a certain point of a body is given by its matrix representation as

$$[\boldsymbol{\sigma}] = \begin{bmatrix} 0 & 0 & 0 \\ 0 & 50 & 0 \\ 0 & 0 & 0 \end{bmatrix} \text{kN/cm}^2 \quad . \tag{3.12}$$

Determine the Cauchy traction vector $\mathbf{t}$ and the first Piola-Kirchhoff traction vector $\mathbf{T}$ acting on a plane, which is characterized by the outward unit normal $\mathbf{n} = \mathbf{e}_2$ in the *current* configuration.

Solution.   From the given deformation (3.11) we find the components of the deformation gradient and its inverse as

$$[\mathbf{F}] = \begin{bmatrix} 0 & -6 & 0 \\ 1/2 & 0 & 0 \\ 0 & 0 & 1/3 \end{bmatrix} \quad , \qquad [\mathbf{F}^{-1}] = \begin{bmatrix} 0 & 2 & 0 \\ -1/6 & 0 & 0 \\ 0 & 0 & 3 \end{bmatrix} \quad , \tag{3.13}$$

while $\det \mathbf{F} = J = 1$. The components of the first Piola-Kirchhoff stress tensor read, according to (3.8), as

$$[\mathbf{P}] = J[\boldsymbol{\sigma}][\mathbf{F}^{-\text{T}}] = \begin{bmatrix} 0 & 0 & 0 \\ 100 & 0 & 0 \\ 0 & 0 & 0 \end{bmatrix} \text{kN/cm}^2 \quad . \tag{3.14}$$

In order to find the outward unit normal $\mathbf{N}$ in the reference configuration we recall Nanson's formula $\mathbf{N}\mathrm{d}S = J^{-1}\mathbf{F}^{\text{T}}\mathbf{n}\mathrm{d}s$. Hence, with the transpose of matrix $[\mathbf{F}]$, $J = 1$ and knowing that $\mathbf{n} = \mathbf{e}_2$ we find that

$$\mathbf{N}\mathrm{d}S = \frac{1}{2}\mathbf{e}_1\mathrm{d}s \quad , \tag{3.15}$$

thus, $\mathbf{N} = \mathbf{e}_1$. Finally, using Cauchy's stress theorem,

$$[\mathbf{t}] = [\boldsymbol{\sigma}][\mathbf{n}] = \begin{bmatrix} 0 \\ 50 \\ 0 \end{bmatrix} \text{kN/cm}^2 \quad , \qquad [\mathbf{T}] = [\mathbf{P}][\mathbf{N}] = \begin{bmatrix} 0 \\ 100 \\ 0 \end{bmatrix} \text{kN/cm}^2 \quad , \tag{3.16}$$

i.e. $\mathbf{t} = 50\mathbf{e}_2$ and $\mathbf{T} = 100\mathbf{e}_2$, respectively. As can be seen, $\mathbf{t}$ and $\mathbf{T}$ have the same direction. The magnitude of $\mathbf{T}$ is twice that of $\mathbf{t}$, because, in view of (3.15), the deformed area is half the undeformed area.   ■

Stress components.   We project the unique Cauchy stress tensor $\boldsymbol{\sigma}$ along an orthonormal set $\{\mathbf{e}_a\}$ of basis vectors; then, according to (1.62), we find that

$$\mathbf{e}_a \cdot \boldsymbol{\sigma}\mathbf{e}_b = \mathbf{e}_a \cdot \mathbf{t}_{\mathbf{e}_b} = \sigma_{ab} \qquad \text{with} \qquad \mathbf{t}_{\mathbf{e}_b} = \boldsymbol{\sigma}\mathbf{e}_b \quad . \tag{3.17}$$

In view of Cauchy's stress theorem (3.3)$_1$, $\mathbf{t}_{\mathbf{e}_b} = \boldsymbol{\sigma}\mathbf{e}_b$, $b = 1, 2, 3$, characterize the three Cauchy traction vectors acting on surface elements whose outward normals point in the directions $\mathbf{e}_1, \mathbf{e}_2, \mathbf{e}_3$, respectively. We use the notation $\mathbf{t}_{\mathbf{e}_a} = \mathbf{t}(\mathbf{x}, t, \mathbf{e}_a)$ to indicate explicitly the dependence of the traction vector $\mathbf{t}$ on the basis vectors $\mathbf{e}_a$.

In matrix notation, the columns of $[\boldsymbol{\sigma}]$ can be identified as the components of traction vectors acting on planes perpendicular to $\mathbf{e}_1, \mathbf{e}_2, \mathbf{e}_3$, respectively. We may write

$$\mathbf{t}_{\mathbf{e}_1} = \boldsymbol{\sigma}\mathbf{e}_1 = \sigma_{11}\mathbf{e}_1 + \sigma_{21}\mathbf{e}_2 + \sigma_{31}\mathbf{e}_3 \quad , \tag{3.18}$$

$$\mathbf{t}_{\mathbf{e}_2} = \boldsymbol{\sigma}\mathbf{e}_2 = \sigma_{12}\mathbf{e}_1 + \sigma_{22}\mathbf{e}_2 + \sigma_{32}\mathbf{e}_3 \quad , \tag{3.19}$$

$$\mathbf{t}_{\mathbf{e}_3} = \boldsymbol{\sigma}\mathbf{e}_3 = \sigma_{13}\mathbf{e}_1 + \sigma_{23}\mathbf{e}_2 + \sigma_{33}\mathbf{e}_3 \tag{3.20}$$

(see the three faces of a cube illustrated in Figure 3.3), characterizing the state of stress at a certain point.

The traction vectors on any surface element are determined uniquely by the set of given quantities σ_{ab}, called the **stress components** of the Cauchy stress tensor $\boldsymbol{\sigma}$. Since the Cauchy stress tensor is symmetric we have six independent stress components acting at a certain point of a body, with $\sigma_{12} = \sigma_{21}$, $\sigma_{13} = \sigma_{31}$, $\sigma_{23} = \sigma_{32}$. For each stress component σ_{ab} we adopt the mathematically logical convention that the first index characterizes the *component* of the vector $\mathbf{t}$ at a point $\mathbf{x}$ in the direction of the associated base vector $\mathbf{e}_a$, and the second index characterizes the *plane* that $\mathbf{t}$ is acting on. The plane is described by its unit normal, i.e. in our case the direction of the base vector $\mathbf{e}_b$ (see TRUESDELL and NOLL [2004]).

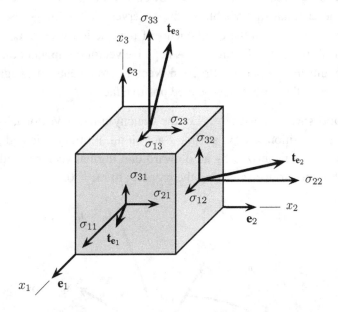

Figure 3.3 Positive stress components of the traction vectors $\mathbf{t}_{\mathbf{e}_a}$ acting on the faces of a cube.

It is important to note that some authors (in particular, those in the engineering community) reverse this convention by identifying the first index with the plane and the second index with the vector component (see MALVERN [1969]).

EXAMPLE 3.2   Project the Cauchy traction vector $\mathbf{t}_{\mathbf{e}_1}$ onto $\mathbf{e}_2$ and interpret the result.

Solution.   Recall relation $(3.18)_1$ and use the analogue of representation (1.59), we have

$$\mathbf{e}_2 \cdot \mathbf{t}_{\mathbf{e}_1} = \mathbf{e}_2 \cdot \boldsymbol{\sigma}\mathbf{e}_1 = \mathbf{e}_2 \cdot \sigma_{ab}(\mathbf{e}_a \otimes \mathbf{e}_b)\mathbf{e}_1 \quad . \tag{3.21}$$

With rule (1.53) and repeated applications of eqs. (1.21) and (1.61) to eq. $(3.21)_2$, we obtain

$$\mathbf{e}_2 \cdot \mathbf{t}_{\mathbf{e}_1} = \delta_{1b}\sigma_{ab}\mathbf{e}_2 \cdot \mathbf{e}_a = \delta_{2a}\delta_{b1}\sigma_{ab} = \sigma_{21} \quad . \tag{3.22}$$

Note that the quantity σ_{21} is the component of $\mathbf{t}_{\mathbf{e}_1}$ in the direction of $\mathbf{e}_2$ (compare with Figure 3.3).   ■

In order to introduce a sign convention for the stress components look at Figure 3.3. We denote those faces of a cube seen by the observer as *positive*. The so-called *negative* faces of the cube are not visible to the observer. When the stress components σ_{ab} are positive scalars, then the components of the traction vectors $\mathbf{t}_{\mathbf{e}_a}$ are defined to be positive, as illustrated in Figure 3.3. A negative vector component on positive faces of the cube points in the opposite direction. Vector components on a negative face are just the opposites of those on the associated positive face.

Normal and shear stresses.   Let the Cauchy traction vector $\mathbf{t}_{\mathbf{n}} = \mathbf{t}(\mathbf{x}, t, \mathbf{n})$ for a given current position $\mathbf{x} \in \Omega$ at time t act on an arbitrary oriented surface element. The surface is characterized by an outward unit normal vector $\mathbf{n}$ and a unit vector $\mathbf{m}$ embedded in the surface satisfying the property $\mathbf{m} \cdot \mathbf{n} = 0$.

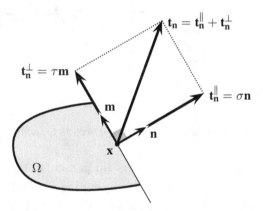

Figure 3.4 Normal and shear stresses at current position $\mathbf{x}$.

According to Figure 3.4, $\mathbf{t}_{\mathbf{n}}$ may be resolved into the sum of a vector along the normal $\mathbf{n}$ to the plane, denoted by $\mathbf{t}_{\mathbf{n}}^{\parallel}$, and a vector perpendicular to $\mathbf{n}$, denoted by $\mathbf{t}_{\mathbf{n}}^{\perp}$. With the rule (1.53) we find that

$$\mathbf{t}_{\mathbf{n}}^{\parallel} = (\mathbf{n} \cdot \mathbf{t}_{\mathbf{n}})\mathbf{n} = \mathbf{P}_{\mathbf{n}}^{\parallel}\mathbf{t}_{\mathbf{n}} \quad , \tag{3.23}$$

$$\mathbf{t}_{\mathbf{n}}^{\perp} = (\mathbf{m} \cdot \mathbf{t}_{\mathbf{n}})\mathbf{m} = \mathbf{P}_{\mathbf{n}}^{\perp}\mathbf{t}_{\mathbf{n}} \tag{3.24}$$

(compare with the equivalent equations (1.125) and (1.126)), where $\mathbf{P}_{\mathbf{n}}^{\|}$ and $\mathbf{P}_{\mathbf{n}}^{\perp}$ are projection tensors of order two defined by

$$\mathbf{P}_{\mathbf{n}}^{\|} = \mathbf{n} \otimes \mathbf{n} \quad , \qquad \mathbf{P}_{\mathbf{n}}^{\perp} = \mathbf{m} \otimes \mathbf{m} = \mathbf{I} - \mathbf{n} \otimes \mathbf{n} \quad , \tag{3.25}$$

with properties according to (1.127)–(1.130).

The lengths of $\mathbf{t}_{\mathbf{n}}^{\|}$ and $\mathbf{t}_{\mathbf{n}}^{\perp}$ are called the **normal stress** σ and **shear stress** (or **tangential stress**) τ acting on the surface element considered, respectively. Using eq. $(3.3)_1$ the explicit expressions for σ and τ are

$$\sigma = \mathbf{n} \cdot \mathbf{t}_{\mathbf{n}} = \mathbf{n} \cdot \boldsymbol{\sigma}\mathbf{n} \qquad \text{or} \qquad \sigma = n_a \sigma_{ab} n_b \quad , \tag{3.26}$$

$$\tau = \mathbf{m} \cdot \mathbf{t}_{\mathbf{n}} = \mathbf{m} \cdot \boldsymbol{\sigma}\mathbf{n} \qquad \text{or} \qquad \tau = m_a \sigma_{ab} n_b \quad . \tag{3.27}$$

If $\sigma > 0$, normal stresses are said to be **tensile stresses** while negative normal stresses ($\sigma < 0$) are known as **compressive stresses**. Traditionally, in soil mechanics the compressive stresses are characterized as positive and the tensile stresses as negative. It is important to note that a tensile stress is a fundamental different type of loading than a compression stress. However, the sign of a shear stress has no intrinsic physical relevance, the type of loading is the same.

From Figure 3.4 we find the useful geometrical relation

$$\mathbf{t}_{\mathbf{n}} = \mathbf{t}_{\mathbf{n}}^{\|} + \mathbf{t}_{\mathbf{n}}^{\perp} \qquad \text{with} \qquad \mathbf{t}_{\mathbf{n}}^{\|} = \sigma\mathbf{n} \quad , \qquad \mathbf{t}_{\mathbf{n}}^{\perp} = \tau\mathbf{m} \quad , \tag{3.28}$$

and hence

$$|\mathbf{t}_{\mathbf{n}}|^2 = \sigma^2 + \tau^2 \qquad \text{or} \qquad t_{n_a} t_{n_a} = \sigma^2 + \tau^2 \quad . \tag{3.29}$$

Consider the schematic diagram of Figure 3.3. Each component σ_{aa} ($a = 1, 2, 3$; no summation), i.e. $\sigma_{11}, \sigma_{22}, \sigma_{33}$, is a normal stress acting in the direction of the unit vector normal to a surface element. These stresses are the diagonal elements of the stress matrix $[\boldsymbol{\sigma}]$ given in eq. $(3.6)_2$. Each remaining component σ_{ab} ($a \neq b$), i.e. $\sigma_{12}, \sigma_{13}, \ldots$, is a shear stress acting tangentially to a surface element. Shear stresses are the off-diagonal elements of the stress matrix $[\boldsymbol{\sigma}]$.

EXERCISES

1. Consider Cauchy traction vectors $\mathbf{t}_{\mathbf{n}}$ and $\mathbf{t}_{\tilde{\mathbf{n}}}$ on two infinitesimal surface elements at $\mathbf{x}$ with normals $\mathbf{n}$ and $\tilde{\mathbf{n}}$. Show that

$$\mathbf{n} \cdot \mathbf{t}_{\tilde{\mathbf{n}}} = \tilde{\mathbf{n}} \cdot \mathbf{t}_{\mathbf{n}} \tag{3.30}$$

if and only if the associated stress tensor $\boldsymbol{\sigma}$ at $\mathbf{x}$ is symmetric.

2. At a certain point of a deformable body the Cauchy stress components are given with respect to a x_1, x_2, x_3-coordinate system as

$$[\boldsymbol{\sigma}] = \begin{bmatrix} 1 & 4 & -2 \\ 4 & 0 & 0 \\ -2 & 0 & 3 \end{bmatrix} \text{kN/m}^2 \ .$$

(a) Find the components of the Cauchy traction vector $\mathbf{t_n}$ and the length of the vector along the normal to the plane that passes through this point and that is parallel to the plane $2x_1 + 3x_2 + x_3 = 5$.

(b) Find the length of $\mathbf{t_n}$ and the angle that $\mathbf{t_n}$ makes with the normal to the plane.

(c) Find the stress components $\tilde{\sigma}_{ab}$ along a *new* orthonormal basis $\{\tilde{\mathbf{e}}_a\}$, $a = 1, 2, 3$, given by the following transformation

$$\tilde{\mathbf{e}}_2 = \frac{1}{\sqrt{2}}(\mathbf{e}_1 - \mathbf{e}_3) \ , \qquad \tilde{\mathbf{e}}_3 = \frac{1}{3}(2\mathbf{e}_1 - \mathbf{e}_2 + 2\mathbf{e}_3)$$

(compare with Section 1.5), where $\{\mathbf{e}_a\}$ characterizes the *old* orthonormal basis.

3. Consider the Cauchy stress matrix in the form

$$[\boldsymbol{\sigma}] = \begin{bmatrix} 5x_2x_3 & 3x_2^2 & 0 \\ 3x_2^2 & 0 & -x_1 \\ 0 & -x_1 & 0 \end{bmatrix} \text{kN/cm}^2 \qquad (3.31)$$

and find the components of the Cauchy traction vector $\mathbf{t_n}$ at the point with coordinates $(1/2, \sqrt{3}/2, -1)$ on the surface $x_1^2 + x_2^2 + x_3 = 0$.

4. At a certain point of a body the components of the Cauchy stress tensor are given by

$$[\boldsymbol{\sigma}] = \begin{bmatrix} 2 & 5 & 3 \\ 5 & 1 & 4 \\ 3 & 4 & 3 \end{bmatrix} \text{kN/cm}^2 \ .$$

(a) Find the components of the Cauchy traction vector $\mathbf{t_n}$ at the point on the plane whose normal has direction ratios $3 : 1 : -2$.

(b) Find the normal and shear components of $\mathbf{t}$ on that plane.

3.2 Extremal Stress Values

Normal and shear stresses vary in magnitude, direction and location. The designer of a structure must prove that a certain material will not fail as a result of these stresses. Magnitudes, directions and locations of extremal stress values must be identified. We present briefly the standard results on extremal stress values used in practical engineering.

Maximum and minimum normal stresses.   Consider the Cauchy traction vector $\mathbf{t}(\mathbf{x}, t, \mathbf{n}) = \boldsymbol{\sigma}(\mathbf{x}, t)\mathbf{n}$ on an arbitrary oriented surface element through any point $\mathbf{x} \in \Omega$ at time t. First of all, we wish to find the unit vector $\mathbf{n}$ at $\mathbf{x}$, indicating the direction of *maximum* and *minimum* values of *normal stresses* σ.

In order to obtain these maximum and minimum (or so-called extremal) values of σ we may apply the **Lagrange-multiplier method** and claim the stationary position of a functional $\mathcal{L}$ to be

$$\mathcal{L}(\mathbf{n}, \lambda^*) = \mathbf{n} \cdot \boldsymbol{\sigma}\mathbf{n} - \lambda^*(|\mathbf{n}|^2 - 1) \quad, \tag{3.32}$$

or, in index notation,

$$\mathcal{L}(n_a, \lambda^*) = n_a \sigma_{ab} n_b - \lambda^*(n_a n_a - 1) \quad. \tag{3.33}$$

Here, $|\mathbf{n}|^2 - 1 = 0$ ($n_a n_a - 1 = 0$) characterizes the **constraint (auxiliary) condition** and λ^* the **Lagrange multiplier**.

For the stationary position of $\mathcal{L}$, the derivatives $\partial\mathcal{L}/\partial n_a$ and $\partial\mathcal{L}/\partial\lambda^*$ must vanish, i.e., with respect to (3.33),

$$\frac{\partial\mathcal{L}}{\partial n_c} = \sigma_{ab}(\delta_{bc}n_a + \delta_{ac}n_b) - \lambda^*(2n_a\delta_{ac})$$
$$= 2(\sigma_{ca}n_a - \lambda^* n_c) = 0 \quad, \tag{3.34}$$

$$\frac{\partial\mathcal{L}}{\partial\lambda^*} = n_a n_a - 1 = 0 \quad. \tag{3.35}$$

We have used $\partial n_a/\partial n_b = \delta_{ab}$, property (1.22)$_2$ and the fact that the stress tensor $\boldsymbol{\sigma}$ is symmetric. In symbolic notation (3.34)$_2$ and (3.35) read as

$$\boldsymbol{\sigma}\mathbf{n} - \lambda^*\mathbf{n} = \mathbf{o} \quad, \qquad |\mathbf{n}| = 1 \quad. \tag{3.36}$$

Since this is an eigenvalue problem the general results and rules of Section 1.4 hold. The Lagrange multiplier λ^* may be identified as an eigenvalue. Adopting the relevant notation of Section 1.4, we must solve homogeneous algebraic equations for the unknown *eigenvalues* λ_a^*, $a = 1, 2, 3$, and the unknown *eigenvectors* $\hat{\mathbf{n}}_a$, $a = 1, 2, 3$, in the form

$$(\boldsymbol{\sigma} - \lambda_a^*\mathbf{I})\hat{\mathbf{n}}_a = \mathbf{o} \quad, \qquad (a = 1, 2, 3; \text{ no summation}) \tag{3.37}$$

(compare with eq. (1.166)). In order to obtain the eigenvalues we must solve the characteristic polynomial of $\boldsymbol{\sigma}$, which is cubic in $\lambda_a^\star$, i.e.

$$\lambda_a^{\star 3} - I_1 \lambda_a^{\star 2} + I_2 \lambda_a^\star - I_3 = 0 \ , \qquad a = 1, 2, 3 \ , \tag{3.38}$$

with the three principal stress invariants I_a of tensor $\boldsymbol{\sigma}$ (compare with eqs. (1.170)–(1.172)), i.e.

$$I_1(\boldsymbol{\sigma}) = \mathrm{tr}\boldsymbol{\sigma} = \lambda_1^\star + \lambda_2^\star + \lambda_3^\star \ , \tag{3.39}$$

$$I_2(\boldsymbol{\sigma}) = \frac{1}{2}\left[(\mathrm{tr}\boldsymbol{\sigma})^2 - \mathrm{tr}(\boldsymbol{\sigma}^2)\right] = \lambda_1^\star \lambda_2^\star + \lambda_1^\star \lambda_3^\star + \lambda_2^\star \lambda_3^\star \ , \tag{3.40}$$

$$I_3(\boldsymbol{\sigma}) = \mathrm{det}\boldsymbol{\sigma} = \lambda_1^\star \lambda_2^\star \lambda_3^\star \ . \tag{3.41}$$

The three eigenvalues $\lambda_a^\star$ are here called **principal normal stresses** and typically are denoted by σ_a, $a = 1, 2, 3$. The principal values σ_a include both the maximum and minimum normal stresses among all planes passing through a given $\mathbf{x}$.

The corresponding three orthonormal eigenvectors $\hat{\mathbf{n}}_a$, which are then characterized through relation (3.37), are the **principal directions** of $\boldsymbol{\sigma}$. The normal stress is stationary along these principal directions. Their related normal planes are known as **principal planes**. Since the stress tensor $\boldsymbol{\sigma}$ is symmetric the set of eigenvectors form a mutually orthogonal basis. Shear stress components vanish at normal (principal) planes, since obviously all eigenvectors are normal to their respective principal planes.

We may represent $\boldsymbol{\sigma}$ in the spectral form

$$\boldsymbol{\sigma}(\mathbf{x}, t) = \sum_{a=1}^{3} \sigma_a \hat{\mathbf{n}}_a \otimes \hat{\mathbf{n}}_a \tag{3.42}$$

(see Section 1.4), satisfying the eigenvalue problem $\boldsymbol{\sigma}\hat{\mathbf{n}}_a = \sigma_a\hat{\mathbf{n}}_a$, $a = 1, 2, 3$. Note that only for *isotropic* materials do the introduced three principal directions $\hat{\mathbf{n}}_a$ of $\boldsymbol{\sigma}$ coincide with the principal directions, as defined in Section 2.6. For a brief explanation the reader is referred to Section 5.4, in particular to eq. (5.88), and also to Section 6.2.

Maximum and minimum shear stresses.   The next goal is to find the direction of the unit vector $\mathbf{n}$ at $\mathbf{x}$ that gives the *maximum* and *minimum* values of *shear stresses* τ.

In the following we choose the eigenvectors $\hat{\mathbf{n}}_a$, $a = 1, 2, 3$, of $\boldsymbol{\sigma}$ as a possible set of basis vectors. Then, according to the spectral decomposition (3.42), all non-diagonal components of the matrix $[\boldsymbol{\sigma}]$ vanish. The components $\sigma_1, \sigma_2, \sigma_3$ form the only entries of the stress matrix, and $[\boldsymbol{\sigma}]$ is 'diagonalized'.

The Cauchy traction vector $\mathbf{t}_\mathbf{n}$ then appears in the simple form

$$\mathbf{t}_\mathbf{n} = \boldsymbol{\sigma}\mathbf{n} = \sigma_1 n_1 \hat{\mathbf{n}}_1 + \sigma_2 n_2 \hat{\mathbf{n}}_2 + \sigma_3 n_3 \hat{\mathbf{n}}_3 \ , \tag{3.43}$$

where $\mathbf{n} = n_1\hat{\mathbf{n}}_1 + n_2\hat{\mathbf{n}}_2 + n_3\hat{\mathbf{n}}_3$ denotes the unit vector normal to an arbitrary oriented surface element. The normal stress and shear stress on that surface element follow from eqs. $(3.26)_1$, (3.29) and have the forms

$$\sigma = \mathbf{n} \cdot \mathbf{t_n} = \sigma_1 n_1^2 + \sigma_2 n_2^2 + \sigma_3 n_3^2 \ , \tag{3.44}$$

$$\tau^2 = |\mathbf{t_n}|^2 - \sigma^2$$
$$= \sigma_1^2 n_1^2 + \sigma_2^2 n_2^2 + \sigma_3^2 n_3^2 - (\sigma_1 n_1^2 + \sigma_2 n_2^2 + \sigma_3 n_3^2)^2 \ , \tag{3.45}$$

where the relations $\mathbf{n} = n_a \hat{\mathbf{n}}_a$ and $(3.43)_2$ with the property $\hat{\mathbf{n}}_a \cdot \hat{\mathbf{n}}_b = \delta_{ab}$ (according to eq. (1.21)) are to be used.

With the constraint (auxiliary) condition $|\mathbf{n}|^2 - 1 = 0$ $(n_a n_a - 1 = 0)$ we are now able to eliminate n_3 from eq. $(3.45)_2$. Since the principal stresses $\sigma_1, \sigma_2, \sigma_3$ are known, τ^2 is a function of n_1 and n_2 only. In order to obtain the extremal values of τ^2 we differentiate τ^2 with respect to n_1 and n_2 and identify with zero:

$$\left. \begin{aligned} \frac{\partial \tau^2}{\partial n_1} &= 2n_1(\sigma_1 - \sigma_3) \left\{ \sigma_1 - \sigma_3 - 2[(\sigma_1 - \sigma_3)n_1^2 + (\sigma_2 - \sigma_3)n_2^2] \right\} = 0 \ , \\[2mm] \frac{\partial \tau^2}{\partial n_2} &= 2n_2(\sigma_2 - \sigma_3) \left\{ \sigma_2 - \sigma_3 - 2[(\sigma_1 - \sigma_3)n_1^2 + (\sigma_2 - \sigma_3)n_2^2] \right\} = 0 \ . \end{aligned} \right\} \tag{3.46}$$

Yet we must find the solutions for the stationary shear stress directions, i.e. for the complete set of principal directions. The first (obvious) solution of (3.46) is obtained by taking $n_1 = n_2 = 0$ and $n_3 = \pm 1$. If we consider further the two cases $n_1 = n_3 = 0$ and $n_2 = n_3 = 0$ as the associated component pairs, we find that $n_2 = \pm 1$ and $n_1 = \pm 1$, respectively. In summary, with $\mathbf{n} = n_a \hat{\mathbf{n}}_a$ we obtain the three respective unit normal vectors which are

$$\mathbf{n} = \pm \hat{\mathbf{n}}_3 \ , \qquad \mathbf{n} = \pm \hat{\mathbf{n}}_2 \ , \qquad \mathbf{n} = \pm \hat{\mathbf{n}}_1 \ , \tag{3.47}$$

corresponding to the set $\{\hat{\mathbf{n}}_a\}$ of principal directions of $\boldsymbol{\sigma}$. Obviously, this first solution describes the principal stress planes upon which τ is minimized (in fact $\tau = 0$), as easily concluded from (3.45).

The second solution of (3.46) may be found by assuming $n_1 = 0$, which gives, from $(3.46)_2$, the component $n_2 = \pm 1/\sqrt{2}$. The condition $|\mathbf{n}| = 1$ leads to $n_3 = \pm 1/\sqrt{2}$. Doing so for $n_2 = 0$ we find $n_1 = n_3 = \pm 1/\sqrt{2}$, while for $n_3 = 0$ we obtain the components $n_1 = n_2 = \pm 1/\sqrt{2}$. These results substituted back into (3.45) give the associated extremal values τ^2. In summary,

$$\mathbf{n} = \pm \frac{1}{\sqrt{2}} \mathbf{n}_2 \pm \frac{1}{\sqrt{2}} \mathbf{n}_3 \ , \qquad \tau^2 = \frac{1}{4}(\sigma_2 - \sigma_3)^2 \ , \tag{3.48}$$

$$\mathbf{n} = \pm \frac{1}{\sqrt{2}} \mathbf{n}_1 \pm \frac{1}{\sqrt{2}} \mathbf{n}_3 \ , \qquad \tau^2 = \frac{1}{4}(\sigma_1 - \sigma_3)^2 \ , \tag{3.49}$$

$$\mathbf{n} = \pm \frac{1}{\sqrt{2}} \mathbf{n}_1 \pm \frac{1}{\sqrt{2}} \mathbf{n}_2 \ , \qquad \tau^2 = \frac{1}{4}(\sigma_1 - \sigma_2)^2 \ . \tag{3.50}$$

Consequently, the maximum magnitude of the shear stress denoted by $\tau_{\max}$ is given by the largest of the three values of $(3.48)_2$–$(3.50)_2$. We obtain

$$\tau_{\max} = \frac{1}{2}|\sigma_{\max} - \sigma_{\min}| \ , \tag{3.51}$$

where $\sigma_{\max}$ and $\sigma_{\min}$ denote the maximum and the minimum magnitudes of principal stresses, respectively. It is important to note that the maximum shear stress acts on a plane that is shifted about an angle of $\pm 45°$ to the principal plane in which the maximum and minimum principal stresses act (compare (3.47) with $(3.48)_1$–$(3.50)_1$).

In addition, it can be shown that the normal stress σ which is associated with $\tau_{\max}$ has the value $\sigma = |\sigma_{\max} + \sigma_{\min}|/2$.

EXERCISES

1. A Cauchy stress tensor $\boldsymbol{\sigma}$, whose components σ_{ab} depend on the coordinates x_1, x_2, x_3, is given in the form of its matrix representation

$$[\boldsymbol{\sigma}] = \begin{bmatrix} 0 & 0 & \alpha x_2 \\ 0 & 0 & -\beta x_3 \\ \alpha x_2 & -\beta x_3 & 0 \end{bmatrix} \text{kN/cm}^2 \ ,$$

where α and β are constants. For a certain point $\mathbf{x}$, given by its coordinates $(0, \beta^2, \alpha)$, find

   (a) the three principal stress invariants of tensor $\boldsymbol{\sigma}$.

   (b) Compute the principal stress components and the associated principal directions of stress. Verify that the principal directions of stress are mutually orthogonal.

   (c) Compute the maximum magnitude of shear stress and the plane on which it acts.

2. The Cauchy stress matrix is given as

$$[\boldsymbol{\sigma}] = \begin{bmatrix} 7 & 0 & 14 \\ 0 & 8 & 0 \\ 14 & 0 & -4 \end{bmatrix} \text{kN/cm}^2 \ .$$

   (a) Compute the principal stress components and the associated principal directions.

   (b) Compute the maximum shear stress and the plane on which this maximum shear stress acts.

3. At a given current position $\mathbf{x} \in \Omega$ consider a plane whose unit normal $\mathbf{n}$ makes equal angles with each of the three principal directions $\hat{\mathbf{n}}_a$ of stress. Such a plane is typically called an **octahedral plane** characterized by the unit normal vector $\mathbf{n} = n_a \hat{\mathbf{n}}_a$, $a = 1, 2, 3$, with components $n_1 = n_2 = n_3 = 1/\sqrt{3}$.

    (a) With reference to eqs. $(3.26)_1$ and (3.29) verify that the normal stress and the shear stress on the octahedral plane are

$$\sigma_{\text{oct}} = \frac{1}{3}(\sigma_1 + \sigma_2 + \sigma_3) \ ,$$

$$\tau_{\text{oct}} = \frac{1}{3}[(\sigma_{11} - \sigma_{22})^2 + (\sigma_{22} - \sigma_{33})^2 + (\sigma_{33} - \sigma_{11})^2$$
$$+ 6(\sigma_{12}^2 + \sigma_{23}^2 + \sigma_{31})^2]^{1/2}$$
$$= \frac{1}{3}[(\sigma_1 - \sigma_2)^2 + (\sigma_2 - \sigma_3)^2 + (\sigma_3 - \sigma_1)^2]^{1/2} \ ,$$

    where σ_1, σ_2, σ_3 are the principal stresses. The normal and shear stress on the octahedral plane are typically denoted by σ_{oct} and τ_{oct}, respectively.

    (b) Find the traction vector, the normal and shear components on the octahedral plane for a certain point at which the principal stresses are $\sigma_1 = 2\,\text{kN/cm}^2$, $\sigma_2 = \alpha$, $\sigma_3 = 11\,\text{kN/cm}^2$, where α is a constant. Determine α so that τ_{oct} is the maximum shear stress.

3.3 Examples of States of Stress

Since a Cauchy traction vector at given $(\mathbf{x}, t)$ is defined for each surface element represented by the outward unit normal $\mathbf{n}$, we may find infinitely many traction vectors. In this section we show some important cases.

State of stress. The set of pairs $\{(\mathbf{t}, \mathbf{n})\}$ at a given point is called the **state of stress**, completely determined by the stress tensor $\boldsymbol{\sigma}$ at this given point and time t according to Cauchy's stress theorem. Since $\mathbf{t}$ is a linear combination of $\mathbf{n}$ in three dimensions, three linear independent pairs $(\mathbf{t}, \mathbf{n})$ form a complete basis. The state of stress is said to be **homogeneous** if the stress tensor does not depend on space coordinates at each time t.

We now give some important examples for different states of stress.

(i) A **pure normal stress** state at a certain point is given by the stress tensor

$$\boldsymbol{\sigma} = \sigma(\mathbf{n} \otimes \mathbf{n}) \qquad \text{or} \qquad \sigma_{ab} = \sigma n_a n_b \tag{3.52}$$

(see Figure 3.5(a)).

Post-multiplying (3.52) with the unit vector $\mathbf{n}$ we find using rule (1.53) and $(3.28)_2$ that

$$\boldsymbol{\sigma}\mathbf{n} = \sigma(\mathbf{n} \otimes \mathbf{n})\mathbf{n}$$
$$= \sigma \underbrace{(\mathbf{n} \cdot \mathbf{n})}_{1} \mathbf{n} = \sigma\mathbf{n} = \mathbf{t}_{\mathbf{n}}^{\parallel} \ . \tag{3.53}$$

Evidently, $\mathbf{t}_{\mathbf{n}}^{\parallel}$ is along (or opposite) $\mathbf{n}$ (compare with Figure 3.4). This stress state is characterized by a normal stress σ, the shear stress is zero. The stress σ characterizes either **pure tension** (if $\sigma > 0$) or **pure compression** (if $\sigma < 0$).

Consider a rectangular Cartesian coordinate system. If $\sigma_{11} = \sigma = \text{const}$ and all other stress components are identically zero, then we have either **uniform tension** or **uniform compression** in the x_1-direction. In the literature these stress states are often referred to as **uniaxial tension** and **uniaxial compression**, respectively. This may be imagined as the stress in a rod (with uniform cross-section) generated by forces applied to its plane ends in the x_1-direction, which is one principal direction of stress. A uniform tension or compression of a uniform cylindrical rod leads to a deformation, which we had already called *uniform extension* or *compression* (see p. 92).

A three-dimensional stress state in which all shear stress components vanish, i.e. $\sigma_{ab} = 0, a \neq b$, is said to be **triaxial**, while in a **biaxial stress state** (which is associated with biaxial deformations) we have a pair of non-vanishing normal stresses, i.e. $\sigma_1, \sigma_2; \sigma_2, \sigma_3; \sigma_1, \sigma_3$. However, in an **equibiaxial stress state** (which is associated with equibiaxial deformations) the non-vanishing normal stresses have the same value.

(ii) A **pure shear stress** state at a certain point is given by the stress tensor

$$\boldsymbol{\sigma} = \tau(\mathbf{n} \otimes \mathbf{m} + \mathbf{m} \otimes \mathbf{n}) \qquad \text{or} \qquad \sigma_{ab} = \tau(n_a m_b + m_a n_b) \tag{3.54}$$

(see Figure 3.5(b)). Here, the stress τ is related to the directions of the unit vectors $\mathbf{n}$ and $\mathbf{m}$, with property $\mathbf{m} \cdot \mathbf{n} = 0$.

Post-multiplying (3.54) with the unit vector $\mathbf{n}$ we obtain with rule (1.53) and $(3.28)_3$

$$\boldsymbol{\sigma}\mathbf{n} = \tau(\mathbf{n} \otimes \mathbf{m} + \mathbf{m} \otimes \mathbf{n})\mathbf{n}$$
$$= \tau[\underbrace{(\mathbf{m} \cdot \mathbf{n})}_{0}\mathbf{n} + \underbrace{(\mathbf{n} \cdot \mathbf{n})}_{1}\mathbf{m}] = \tau\mathbf{m} = \mathbf{t}_{\mathbf{m}}^{\perp} \ . \tag{3.55}$$

Evidently, $\mathbf{t}_{\mathbf{m}}^{\perp}$ is tangential to the surface and along (or opposite) $\mathbf{m}$, i.e. perpendicular to $\mathbf{n}$ (compare with Figure 3.4). This stress state is characterized by a **pure shear stress** τ (or **pure tangential stress**) and the normal stress is zero.

Consider a rectangular Cartesian coordinate system. If $\sigma_{12} = \sigma_{21} = \tau = \text{const}$ and all other stress components are identically zero, then this stress state is characterized by a **uniform shear stress**. A state of uniform shear stress leads to a uniform (or simple) shear deformation, as illustrated in Figure 2.2 (see also Exercise 2 on p. 93).

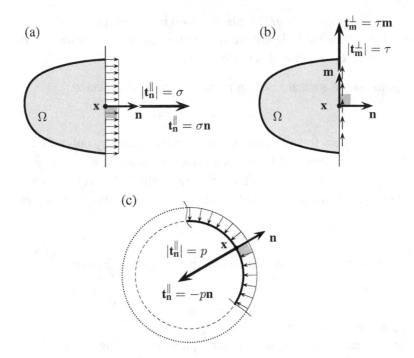

Figure 3.5 Examples of stress states.

(iii) A **hydrostatic stress** state at a certain point is given by the stress tensor

$$\boldsymbol{\sigma} = -p\mathbf{I} \qquad \text{or} \qquad \sigma_{ab} = -p\delta_{ab} \qquad (3.56)$$

(see Figure 3.5(c)). Within a rectangular Cartesian coordinate system that means that we have normal stresses $\sigma_{11} = \sigma_{22} = \sigma_{33} = -p$ and no shear stresses on *any* plane containing this point, i.e. $\sigma_{12} = \sigma_{23} = \sigma_{31} = 0$.

Post-multiplying (3.56) with the unit vector $\mathbf{n}$, we obtain, using (3.28)$_2$,

$$\boldsymbol{\sigma}\mathbf{n} = -(p\mathbf{I})\mathbf{n} = -p\mathbf{n} = \mathbf{t}_n^{\|} \quad . \qquad (3.57)$$

Evidently, $\mathbf{t}_n^{\|}$ is opposite to the normal vector $\mathbf{n}$. The relationship $\boldsymbol{\sigma}\mathbf{n} = -p\mathbf{n}$ holds only if $\mathbf{n}$ is a principal direction of $\boldsymbol{\sigma}$. Note that any three mutually orthogonal directions may be regarded as principal directions (compare also with eq. (1.179)). This stress state is characterized by a scalar p, known as the **hydrostatic pressure**. In general, the hydrostatic pressure is a scalar function of time t, and in the literature is often introduced with the opposite sign.

One example of stress state (3.56) is in an (elastic) fluid (without motion) that is not able to sustain shear stresses. Taking the trace of (3.56), we obtain

$$p = -\frac{1}{3}\text{tr}\boldsymbol{\sigma} \qquad \text{or} \qquad p = -\frac{1}{3}\sigma_{aa} \quad , \qquad (3.58)$$

indicating that the pressure of the (elastic) fluid is a **mean pressure**.

The three stress fields described for a certain point **x** at t (with σ constant) correspond to the states, as illustrated in Figure 3.5.

(iv) A **plane stress** state at a certain point is given by the relation

$$\sigma_{13} = \sigma_{23} = \sigma_{33} \equiv 0 \ , \tag{3.59}$$

and $\sigma_{11}, \sigma_{22}, \sigma_{12}$ are functions of the coordinates x_1 and x_2 only. Consequently, the x_3-direction, as represented by $\hat{\mathbf{n}}_3$, is a principal direction of stress with a zero corresponding principal stress σ_{33}. The other two principal directions acting in a plane normal to $\hat{\mathbf{n}}_3$ are inclined at an angle θ with the x_1 and x_2 direction, where

$$\tan 2\theta = \frac{2\sigma_{12}}{\sigma_{11} - \sigma_{22}} \ . \tag{3.60}$$

The corresponding maximum and minimum stresses are given by

$$\frac{1}{2}(\sigma_{11} + \sigma_{22}) \pm [\frac{1}{4}(\sigma_{11} - \sigma_{22})^2 + \sigma_{12}^2]^{1/2} \ , \tag{3.61}$$

defining a biaxial stress state.

The maximum shear stress $\tau_{\max}$ for a plane stress state will be the largest of the three values of $(3.48)_2$–$(3.50)_2$ since $\sigma_3 = 0$. The planes of extremal shear stresses form angles of $\pm 45°$ with the planes of the principal stresses.

A plane stress state occurs at any unloaded surface in a continuum body and is of practical interest.

EXERCISES

1. For the case of plane stress, show that eq. (3.38) reduces to eq. (3.61).

2. Assume a plane stress state in a rectangular cube bounded by the planes $x_1 = \pm a, x_2 = \pm b, x_3 = \pm c$. The state of stress at a point with coordinates x_1, x_2, x_3 in the cube is given by

$$[\sigma] = \begin{bmatrix} \alpha(x_1 - x_2) & \beta x_1^2 x_2 & 0 \\ \beta x_1^2 x_2 & -\alpha(x_1 - x_2) & 0 \\ 0 & 0 & 0 \end{bmatrix} \text{ kN/cm}^2 \ ,$$

where α, β are constants. For a point with coordinates $(a/2, -b/2, 0)$, determine

 (a) the principal normal stresses and the associated principal directions,

 (b) the planes, characterized by the unit normal **n**, that give the maximum and minimum shear stresses and the magnitude of the extremal shear stress.

 (c) Find the total Cauchy traction vector on each face of this rectangular cube.

3.4   Alternative Stress Tensors

Numerous definitions and names of stress tensors have been proposed in the literature. Each definition has advantages and disadvantages. In the following we discuss stress tensors used for practical nonlinear analyses. Most of their components do not have a direct physical interpretation.

Often it is convenient to work with the so-called **Kirchhoff stress tensor** τ, which differs from the Cauchy stress tensor by the volume ratio J. It is a contravariant spatial tensor field parameterized by spatial coordinates, and is defined by

$$\tau = J\sigma \qquad \text{or} \qquad \tau_{ab} = J\sigma_{ab} \ . \tag{3.62}$$

We introduce further the **second Piola-Kirchhoff stress tensor S** which does not admit a physical interpretation in terms of surface tractions. The contravariant material tensor field is symmetric and parameterized by material coordinates. Therefore, it often represents a very useful stress measure in computational mechanics and in the formulation of constitutive equations, in particular, for solids, as we will see in Chapter 6.

The second Piola-Kirchhoff stress tensor is obtained by the pull-back operation on the contravariant spatial tensor field $\tau^{\sharp}$ by the motion χ, which is, according to $(2.87)_2$,

$$\mathbf{S} = \chi_*^{-1}(\tau^{\sharp}) = \mathbf{F}^{-1}\tau\mathbf{F}^{-T} \qquad \text{or} \qquad S_{AB} = F_{Aa}^{-1}F_{Bb}^{-1}\tau_{ab} \ . \tag{3.63}$$

Hence, the Kirchhoff stress tensor is the push-forward of **S**, i.e., using $(2.87)_1$,

$$\tau = \chi_*(\mathbf{S}^{\sharp}) = \mathbf{F}\mathbf{S}\mathbf{F}^{T} \qquad \text{or} \qquad \tau_{ab} = F_{aA}F_{bB}S_{AB} \ . \tag{3.64}$$

Using eqs. $(3.63)_2$, (3.62) and (3.8) we obtain the Piola transformation relating the two stress fields **S** and σ, i.e.

$$\mathbf{S} = J\mathbf{F}^{-1}\sigma\mathbf{F}^{-T} = \mathbf{F}^{-1}\mathbf{P} = \mathbf{S}^{T}$$

$$\text{or} \qquad S_{AB} = JF_{Aa}^{-1}F_{Bb}^{-1}\sigma_{ab} = F_{Aa}^{-1}P_{aB} = S_{BA} \ , \tag{3.65}$$

with its inverse,

$$\sigma = J^{-1}\mathbf{F}\mathbf{S}\mathbf{F}^{T} \qquad \text{or} \qquad \sigma_{ab} = J^{-1}F_{aA}F_{bB}S_{AB} \ . \tag{3.66}$$

From eq. (3.65) we find a fundamental relationship between the first Piola-Kirchhoff stress tensor **P** introduced in (3.8) and the symmetric second Piola-Kirchhoff stress tensor **S**, i.e.

$$\mathbf{P} = \mathbf{F}\mathbf{S} \qquad \text{or} \qquad P_{aA} = F_{aB}S_{BA} \ . \tag{3.67}$$

In addition to the four stress tensors given above, we introduce another important quantity. It is a material stress tensor formally defined as

$$\mathbf{T}_B = \mathbf{R}^T \mathbf{P} \qquad \text{or} \qquad T_{B\,AB} = R_{aA} P_{aB} \ . \tag{3.68}$$

The non-symmetric tensor $\mathbf{T}_B$, which is not in general positive definite, is known as the **Biot stress tensor** (BIOT [1965]). With (3.67) and the polar decomposition $\mathbf{F} = \mathbf{RU}$ we deduce from (3.68) that $\mathbf{T}_B = \mathbf{R}^T(\mathbf{FS}) = \mathbf{US}$. Herein, $\mathbf{R}$ and $\mathbf{U}$ denote the rotation tensor (with $\det\mathbf{R} = 1$) and the (positive definite) symmetric right stretch tensor, respectively. They are according to the polar decomposition of the deformation gradient $\mathbf{F}$ (see Section 2.6).

Multiplying eq. (3.68) by $\mathbf{R}$ from the left-hand side, we obtain the polar decomposition

$$\mathbf{P} = \mathbf{R}\mathbf{T}_B \qquad \text{or} \qquad P_{aA} = R_{aB} T_{B\,BA} \tag{3.69}$$

for the first Piola-Kirchhoff stress tensor, which is in analogy with that for $\mathbf{F}$. Since the Biot stress tensor $\mathbf{T}_B$ is not positive definite this decomposition is *not* unique, in general.

Other examples of stress tensors are the symmetric so-called **corotated Cauchy stress tensor** σ_u, as introduced by *Green* and *Naghdi*, and the **Mandel stress tensor** Σ which is in general not symmetric. These tensors are defined with respect to an *intermediate* configuration. In order to obtain the corotated Cauchy stress tensor take relation (3.66) and use the symmetric right stretch tensor $\mathbf{U}$ instead of $\mathbf{F}$. Then apply relation $(3.65)_1$ and the polar decomposition $\mathbf{F} = \mathbf{RU}$ to find

$$\sigma_u = J^{-1}\mathbf{USU} = (\mathbf{UF}^{-1})\sigma(\mathbf{F}^{-T}\mathbf{U}) = \mathbf{R}^T\sigma\mathbf{R} \ . \tag{3.70}$$

The Mandel stress tensor is defined to be

$$\Sigma = \mathbf{CS} \ , \tag{3.71}$$

which is often used to describe inelastic (plastic) materials.

EXERCISES

1. Consider an infinitesimal **resultant (pseudo) force** $d\mathbf{f}_B = (\mathbf{T}_B\mathbf{N})dS$, where $\mathbf{T}_B\mathbf{N}$ denotes the **Biot traction vector**. Using (3.68) and the linear transformations $(3.3)_2$ and $(3.1)_2$, verify that

$$d\mathbf{f}_B = \mathbf{R}^T d\mathbf{f} \ ,$$

which shows that $d\mathbf{f}_B$ only differs from $d\mathbf{f}$ (introduced in (3.1)) by the rotation tensor $\mathbf{R}^T$.

2. The Cauchy stress components at a certain point are given with respect to a
x_1, x_2-coordinate system with the associated set $\{\mathbf{e}_\alpha\}$, $\alpha = 1, 2$, of orthonormal
basis vectors as

$$[\sigma] = \begin{bmatrix} 2 & 1 \\ 1 & -3 \end{bmatrix} \text{kN/m}^2 \quad .$$

Assume a rotation of $\{\mathbf{e}_\alpha\}$ into the *new* set $\{\tilde{\mathbf{e}}_\alpha\}$ of orthonormal basis vectors
according to

$$\tilde{\mathbf{e}}_1 = \frac{1}{3}(-\mathbf{e}_1 + 2\sqrt{2}\mathbf{e}_2) \quad .$$

Compute the components of the corotated Cauchy stress tensor σ_u.

4 Balance Principles

In this chapter we provide the classical *balance principles* and discuss some of their important consequences. The fundamental balance principles, i.e. conservation of mass, the momentum balance principles and balance of energy, are valid in all branches of continuum mechanics. They are applicable to any particular material and must be satisfied for all times.

We also discuss another fundamental set of laws that are expressed as *inequalities*, such as the second law of thermodynamics. The section devoted to continuum thermodynamics specifically addresses balance of energy and the entropy inequality principle. Finally, the structure of principles is summarized as the master balance (inequality) principle.

4.1 Conservation of Mass

Every continuum body B possesses **mass**, denoted by m. It is a fundamental physical property commonly defined to be a measure of the amount of a material contained in the body B. In order to perform a macroscopic study we assume that mass is continuously (or at least piecewise continuously) distributed over an arbitrary region Ω (of physical space) with boundary surface $\partial\Omega$ at time t. The mass is a scalar measure (a *positive* number) which is *invariant* during a motion. We exclude concentrated masses such as those used in classical Newtonian mechanics.

Closed and open systems. We define a **system** as a **quantity of mass** or a particular **collection of matter** in space. The complement of a system, i.e. the mass or region outside the system, we call the **surroundings**, while the surface that separates the system from its surroundings we call the **boundary** or **wall** of the system (see Figure 4.1).

A **closed system** (or **control mass**) consists of a **fixed** amount of *mass* in a properly selected region Ω in space with boundary surface $\partial\Omega$ which depends on time t (see Figure 4.1). No *mass* can cross (enter or leave) its boundary, but *energy*, in the form of *work* or *heat*, can cross the boundary. The volume of a closed system does not have

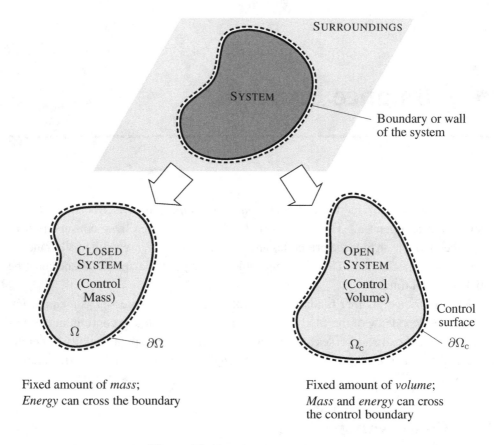

Figure 4.1 Closed and open systems.

to be fixed. If even energy does not interact between the system and its surroundings, then we say that the boundary is **insulated**. Such a system is called **mechanically** and **thermally isolated**, which is an idealization for a physical system. There always exist electromagnetic and other types of forces which permeate the space. Note that no physical system is truly isolated.

An **open system** (or **control volume**) consists of a **fixed** amount of *volume* of a properly selected region Ω_c which is independent of time t (see Figure 4.1). The enclosing boundary of a control volume, over which both *mass* and *energy* can cross (enter or leave), is called a **control surface**, which we denote by $\partial \Omega_c$.

Conservation of mass.   In non-relativistic physics mass cannot be produced or destroyed. It is assumed that during a motion there are neither mass sources (reservoirs that supply mass) nor mass sinks (reservoirs that absorb mass), so that the mass m of a body is a **conserved quantity**. Hence, if a particle has a certain mass in the reference configuration it must stay the same during a motion. Considering a closed system, obviously that holds for the total mass too. We write

$$m(\Omega_0) = m(\Omega) > 0 \quad , \tag{4.1}$$

which holds for all times t. Relation (4.1) is a statement of a fundamental mechanical law known as the **conservation of mass**. The boundary surfaces in the reference and the configurations, with volume V and v, are denoted by Ω_0 and Ω, respectively. Note that the mass m is independent of the motion and of the region occupied by the body. Hence, the material time derivative of the mass m gives

$$\frac{\mathrm{D}}{\mathrm{D}t}m(\Omega_0) = \frac{\mathrm{D}}{\mathrm{D}t}m(\Omega) = 0 \quad . \tag{4.2}$$

The differential form of eq. (4.1) reads

$$\mathrm{d}m(\mathbf{X}) = \mathrm{d}m(\mathbf{x}, t) > 0 \quad , \tag{4.3}$$

with the infinitesimal **mass element** $\mathrm{d}m$.

The mass of Ω_0 and Ω is characterized by continuous (or at least piecewise continuous) scalar fields, i.e. $\rho_0 = \rho_0(\mathbf{X}) > 0$ and $\rho = \rho(\mathbf{x}, t) > 0$, respectively. They denote physical properties of the same particle. Property ρ_0 is called the **reference mass density** (or just **density**) and ρ is called the **spatial mass density** during a motion $\mathbf{x} = \boldsymbol{\chi}(\mathbf{X}, t)$. The spatial mass density, also known as the **density in the motion**, depends on place $\mathbf{x} \in \Omega$ and time t throughout the body. Note that ρ_0 is *time-independent* and intrinsically associated with the reference configuration of the body. Hence, ρ_0 depends only on the position $\mathbf{X}$ chosen in configuration Ω_0. If the density does not depend on $\mathbf{X} \in \Omega_0$, i.e. if $\mathrm{Grad}\rho_0 = \mathbf{o}$, the configuration is said to be **homogeneous**.

The mass densities at the points $\mathbf{X}$ and $\mathbf{x}$ are defined by the limit

$$\rho_0(\mathbf{X}) = \lim_{\Delta V(\Omega_0) \to 0} \frac{\Delta m(\Omega_0)}{\Delta V(\Omega_0)} \quad , \qquad \rho(\mathbf{x}, t) = \lim_{\Delta v(\Omega) \to 0} \frac{\Delta m(\Omega)}{\Delta v(\Omega)} \quad , \tag{4.4}$$

where Δm denotes a continuous function of incremental mass of an incremental volume element in the reference and the current configurations, which we have denoted by ΔV and Δv, respectively.

Note that $\Delta V(\Omega_0)$, $\Delta v(\Omega)$, actually must not tend to zero since then the limit of ρ_0, ρ, would show a discrete distribution according to the atomistic structure of matter. Therefore, to obtain representative averages, $\Delta V(\Omega_0)$ and $\Delta v(\Omega)$ must be large in terms of an atomistic scale and small in terms of a length scale of a certain physical problem. Usually the ratio of the length scale of a physical problem and the length scale of an incremental volume element $\Delta V(\Omega_0)$ and $\Delta v(\Omega)$ is of the order 10^3 or more.

In the differential form eq. (4.4) reads

$$dm(\mathbf{X}) = \rho_0(\mathbf{X})dV \ , \qquad dm(\mathbf{x}, t) = \rho(\mathbf{x}, t)dv \ , \qquad (4.5)$$

with the standard infinitesimal volume elements dV and dv defined in the reference and the current configurations, respectively.

Substituting eq. (4.5) into (4.3) we obtain

$$\rho_0(\mathbf{X})dV = \rho(\mathbf{x}, t)dv > 0 \ , \qquad (4.6)$$

which means that volume increases when density decreases. By integrating the infinitesimal mass over the entire region, we find the total mass m of that region. Hence, an alternative expression for (4.6) reads

$$m = \int_{\Omega_0} \rho_0(\mathbf{X})dV = \int_{\Omega} \rho(\mathbf{x}, t)dv = \text{const} > 0 \qquad (4.7)$$

for all times t, which implies the rate form

$$\dot{m} = \frac{Dm}{Dt} = \frac{D}{Dt} \int_{\Omega} \rho(\mathbf{x}, t)dv = 0 \ . \qquad (4.8)$$

Hence, conservation of mass requires that the material time derivative of m is zero for all regions Ω of a body which change with time (mass remains unchanged during the motion of Ω).

An equation which holds at every point of a continuum and for all times, for example, eq. (4.6), is referred to as the **local** (or **differential**) **form** of that equation (local means pointwise). An equation in which physical quantities over a certain region of space are integrated is referred to as the **global** (or **integral**) **form** of that equation; see, for example, eq .(4.7). Consequently we may say that (4.6) is the local form and (4.7) is the global form of conservation of mass.

In general, local forms are ideally suited for approximation techniques such as the finite difference method while global forms are the best to start with when the finite element method is employed.

Continuity mass equation.   We want to find a relationship between the reference mass density $\rho_0(\mathbf{X}) \in \Omega_0$ and the spatial mass density $\rho(\mathbf{x}, t) \in \Omega$.

By recalling eqs. (2.51) and (2.52), i.e. $dv = J(\mathbf{X}, t)dV$, $J = \det\mathbf{F}(\mathbf{X}, t) > 0$, we may change the variable of integration in eq. (4.7) from $\mathbf{x} = \chi(\mathbf{X}, t)$ to $\mathbf{X}$ and we obtain the identity

$$\int_{\Omega_0} [\rho_0(\mathbf{X}) - \rho(\chi(\mathbf{X}, t), t)J(\mathbf{X}, t)]dV = 0 \ . \qquad (4.9)$$

By assuming that V is an *arbitrary* volume of region Ω_0, we conclude that the integrand in (4.9) must vanish everywhere. Hence,

$$\rho_0(\mathbf{X}) = \rho(\chi(\mathbf{X}, t), t) J(\mathbf{X}, t) \quad , \tag{4.10}$$

holds for all $\mathbf{X} \in \Omega_0$. It represents the **continuity mass equation** (continuity stands for constancy of mass) in the material (or Lagrangian) description which is the appropriate description in *solid mechanics*.

Since the reference mass density ρ_0 is independent of time we find simply from (4.10) that

$$\frac{\partial \rho_0(\mathbf{X})}{\partial t} = \dot{\rho}_0(\mathbf{X}) = 0 \quad , \tag{4.11}$$

which is the rate form of (4.10) in the material description.

EXAMPLE 4.1   Show how the spatial mass density $\rho = \rho(\mathbf{x}, t)$ changes with time. In particular, derive the rate form of continuity mass equation in the spatial (or Eulerian) description, which is (expressed in terms of the velocity components)

$$\dot{\rho}(\mathbf{x}, t) + \rho(\mathbf{x}, t) \operatorname{div} \mathbf{v}(\mathbf{x}, t) = 0 \qquad \text{or} \qquad \dot{\rho} + \rho \frac{\partial v_a}{\partial x_a} = 0 \quad , \tag{4.12}$$

or in the two equivalent forms

$$\frac{\partial \rho(\mathbf{x}, t)}{\partial t} + \operatorname{grad}\rho(\mathbf{x}, t) \cdot \mathbf{v}(\mathbf{x}, t) + \rho(\mathbf{x}, t) \operatorname{div} \mathbf{v}(\mathbf{x}, t) = 0 \quad , \tag{4.13}$$

$$\frac{\partial \rho(\mathbf{x}, t)}{\partial t} + \operatorname{div}(\rho(\mathbf{x}, t) \mathbf{v}(\mathbf{x}, t)) = 0 \tag{4.14}$$

for all $\mathbf{x} \in \Omega$ and for all times t.

Solution.   Since $\dot{\rho}_0 = 0$ we obtain from (4.10) that

$$\frac{\mathrm{D}}{\mathrm{D}t}(\rho J) = \dot{\overline{\rho J}} = 0 \tag{4.15}$$

(for simplicity written without arguments of the scalar quantities). In order to express eq. (4.15) in terms of the spatial velocity components we find using the product rule and $\dot{J} = J \operatorname{div} \mathbf{v}$, i.e. eq. (2.177)$_6$, that

$$\dot{\overline{\rho J}} = \dot{\rho} J + \rho \dot{J} = J(\dot{\rho} + \rho \operatorname{div} \mathbf{v}) = 0 \quad , \tag{4.16}$$

where the material time derivative of the spatial density function ρ is, having regard to (2.25), given by the explicit expression

$$\dot{\rho} = \frac{\mathrm{D}\rho}{\mathrm{D}t} = \frac{\partial \rho}{\partial t} + \operatorname{grad}\rho \cdot \mathbf{v} \quad . \tag{4.17}$$

Since $J > 0$ we deduce from $(4.16)_2$ the desired result (4.12), which is the corresponding local form of $(4.8)_2$. With the material time derivative of the spatial density function
(4.17) and by means of identity (1.288) we may obtain from (4.12) the two equivalent
forms (4.13) and (4.14).   ∎

If the density of a continuum body is constant at any particle, then from relation
(4.12) we find with $\dot{\rho} = 0$ a kinematical restriction which characterizes an isochoric
(volume-preserving) motion, i.e. $\mathrm{div}\,\mathbf{v} = 0$ (compare also with eqs. $(2.179)_5$). A continuum body is incompressible if every motion it undergoes satisfies $\dot{\rho} = 0$.

The rate forms of continuity mass equations (4.12)–(4.14) show how the spatial
mass density ρ changes as time changes. They represent the continuity mass equation
in the spatial description, which is the appropriate description in *fluid dynamics*.

Conservation of mass for an open system.   Sometimes we work with an open
system given by a region Ω_c and boundary control surface $\partial\Omega_c$.

At a certain time t a control volume contains the mass $m(t) = \int_{\Omega_c} \rho(\mathbf{x}, t)dv$. Since
the region of integration Ω_c does not depend on t, integration and differentiation commute and we may write

$$\dot{m}(t) = \frac{D}{Dt} \int_{\Omega_c} \rho(\mathbf{x}, t)dv = \int_{\Omega_c} \frac{\partial\rho(\mathbf{x}, t)}{\partial t}dv \quad . \tag{4.18}$$

Applying the divergence theorem for a fixed amount of volume Ω_c we have using
(1.299)

$$\int_{\Omega_c} \mathrm{div}(\rho(\mathbf{x}, t)\mathbf{v}(\mathbf{x}, t))dv = \int_{\partial\Omega_c} \rho(\mathbf{x}, t)\mathbf{v}(\mathbf{x}, t) \cdot \mathbf{n}ds \quad , \tag{4.19}$$

where $\mathbf{n}$ denotes the outward unit vector field perpendicular to the boundary control
surface $\partial\Omega_c$. The term $\int_{\partial\Omega_c} \rho\mathbf{v} \cdot \mathbf{n}ds$ determines the flux of $\rho\mathbf{v}$ out of Ω_c across $\partial\Omega_c$.

Integrating the continuity mass equation in the form of (4.14) over a certain region
Ω_c and using eqs. (4.18) and (4.19), we obtain the **conservation of mass** for a **control
volume** in the global form, i.e.

$$\frac{D}{Dt} \int_{\Omega_c} \rho(\mathbf{x}, t)dv = - \int_{\partial\Omega_c} \rho(\mathbf{x}, t)\mathbf{v}(\mathbf{x}, t) \cdot \mathbf{n}ds \quad . \tag{4.20}$$

Relation (4.20) asserts that the material time derivative of the mass inside a control
volume Ω_c is equal to the flux of $\rho\mathbf{v}$ entering Ω_c across $\partial\Omega_c$. The global form (4.20) is
widely used in fluid dynamics.

EXERCISES

1. A velocity field of a plane motion has components of the form

$$v_1 = (\alpha x_1 - \beta x_2)t \ , \qquad v_2 = \beta x_1 - \alpha x_2 \ , \qquad v_3 = 0 \ ,$$

where α and β are positive constants. Assume that the spatial mass density ρ is independent of the current position $\mathbf{x}$ so that $\mathrm{grad}\rho = \mathbf{o}$.

(a) Express ρ so that the continuity mass equation is satisfied.

(b) Find a condition for which the given motion is isochoric.

2. Two motions of a continuum body are given in the form

$$\mathbf{x} = (1 + \alpha(t)t)\mathbf{X} \ , \qquad \mathbf{x} = \mathbf{X} + \alpha(t)t^2[(\mathbf{e}_1 \otimes \mathbf{e}_2) - (\mathbf{e}_2 \otimes \mathbf{e}_1)]\mathbf{X} \ ,$$

with the scalar function $\alpha(t)$ and the set $\{\mathbf{e}_a\}$, $a - 1, 2, 3$, of orthogonal unit vectors.

Find expressions for the spatial mass density ρ in terms of ρ_0 so that the continuity mass equation is satisfied.

3. Consider a spatial scalar field $\Phi = \Phi(\mathbf{x}, t)$ and a spatial vector field $\mathbf{u} = \mathbf{u}(\mathbf{x}, t)$. Use the rate form of the continuity mass equation and obtain the identities

$$\rho\frac{D\Phi}{Dt} = \frac{\partial(\rho\Phi)}{\partial t} + \mathrm{div}(\rho\Phi\mathbf{v}) \ , \qquad \rho\frac{D\mathbf{u}}{Dt} = \frac{\partial(\rho\mathbf{u})}{\partial t} + \mathrm{div}(\rho\mathbf{u} \otimes \mathbf{v}) \ , \quad (4.21)$$

where $\mathbf{v}$ denotes the spatial velocity field.

4. An irrotational motion of an incompressible continuum is given by the spatial velocity field $\mathbf{v} = \mathrm{grad}\Phi$. Show that for this case the scalar field Φ is harmonic.

5. By means of continuity mass equation (4.12) and identity (2.183) show that the vorticity vector $2\boldsymbol{\omega}$ is related to the spatial mass density ρ and to the spatial velocity vector $\mathbf{v}$ by

$$\rho\frac{D}{Dt}\left(\frac{2\boldsymbol{\omega}}{\rho}\right) = \mathrm{curl}\frac{D\mathbf{v}}{Dt} + (\mathrm{grad}\mathbf{v})2\boldsymbol{\omega} \ ,$$

which is known as the **Beltrami vorticity equation**.

4.2   Reynolds' Transport Theorem

Suppose we have a spatial scalar field $\Phi = \Phi(\mathbf{x}, t)$ describing some physical quantity (for example, *mass, internal energy, entropy, heat* or *entropy sources*) of a particle in space per unit *volume* at time t.

Assume Φ to be smooth, so that it is continuously differentiable. Hence, the present status of a continuum body in some three-dimensional region Ω with volume v at given time t may be characterized by the scalar-valued function

$$I(t) = \int_\Omega \Phi(\mathbf{x}, t) dv \quad . \tag{4.22}$$

The aim is now to compute the material time derivative of the volume integral $I(t)$. Since the region of integration Ω depends on time t, integration and time differentiation do not commute. Therefore, as a first step $I(t)$ must be transformed to the reference configuration. By changing variables using the motion $\mathbf{x} = \chi(\mathbf{X}, t)$ and the relation $dv = J(\mathbf{X}, t) dV$ we find the time rate of change of $I(t)$ to be

$$\dot{I}(t) = \frac{D}{Dt} \int_\Omega \Phi(\mathbf{x}, t) dv = \frac{D}{Dt} \int_{\Omega_0} \Phi(\chi(\mathbf{X}, t), t) J(\mathbf{X}, t) dV \quad . \tag{4.23}$$

Since the region of integration is now time-independent, integration and differentiation commute. Hence, as a second step, from $(4.23)_2$ we obtain, using the product rule of differentiation,

$$\frac{D}{Dt} \int_\Omega \Phi(\mathbf{x}, t) dv = \int_{\Omega_0} \left[\dot{\Phi}(\chi(\mathbf{X}, t), t) J(\mathbf{X}, t) + \Phi(\chi(\mathbf{X}, t), t) \dot{J}(\mathbf{X}, t) \right] dV \quad , \tag{4.24}$$

where $\dot{\Phi}$ denotes the material time derivative of the spatial scalar field Φ according to relation (2.25). In a last step we undo the change of variables and convert the volume integral back to the current configuration. By means of eq. $(2.177)_6$, $dv = J(\mathbf{X}, t) dV$ and motion $\mathbf{x} = \chi(\mathbf{X}, t)$, we find finally that

$$\frac{D}{Dt} \int_\Omega \Phi(\mathbf{x}, t) dv = \int_{\Omega_0} [\dot{\Phi}(\chi(\mathbf{X}, t), t) + \Phi(\chi(\mathbf{X}, t), t) \frac{\dot{J}(\mathbf{X}, t)}{J(\mathbf{X}, t)}] J(\mathbf{X}, t) dV$$

$$= \int_\Omega [\dot{\Phi}(\mathbf{x}, t) + \Phi(\mathbf{x}, t) \frac{\dot{J}(\mathbf{X}, t)}{J(\mathbf{X}, t)}] dv$$

$$= \int_\Omega [\dot{\Phi}(\mathbf{x}, t) + \Phi(\mathbf{x}, t) \mathrm{div} \mathbf{v}(\mathbf{x}, t)] dv \quad . \tag{4.25}$$

In the following the arguments of the tensor quantities are dropped in order to simplify the notation. However, in cases where additional information is needed, they will be employed. Hence, relation $(4.25)_3$ reads as

$$\frac{D}{Dt} \int_\Omega \Phi dv = \int_\Omega (\dot\Phi + \Phi \mathrm{div} \mathbf{v}) dv \quad , \tag{4.26}$$

where we have assumed smoothness of the spatial velocity field $\mathbf{v}$.

Other forms of the time rate of change of the integral (4.22) result from (4.26) by means of the material time derivative of the spatial scalar field Φ in the form of eq. (2.25), and the product rule (1.288), i.e.

$$\frac{D}{Dt} \int_\Omega \Phi dv = \int_\Omega (\frac{\partial \Phi}{\partial t} + \mathrm{grad}\Phi \cdot \mathbf{v} + \Phi \mathrm{div}\mathbf{v}) dv$$

$$= \int_\Omega (\mathrm{div}(\Phi \mathbf{v}) + \frac{\partial \Phi}{\partial t}) dv \quad , \tag{4.27}$$

and finally, using the divergence theorem according to (1.299),

$$\frac{D}{Dt} \int_\Omega \Phi dv = \int_{\partial \Omega} \Phi \mathbf{v} \cdot \mathbf{n} ds + \int_\Omega \frac{\partial \Phi}{\partial t} dv \quad . \tag{4.28}$$

The first term on the right-hand side of eq. (4.28) characterizes the **rate of transport** (or the outward normal *flux*) of $\Phi \mathbf{v}$ across the surface $\partial \Omega$ out of region Ω, which is assumed to be fixed. This contribution arises from the moving region. The second term denotes the *local time rate of change* of the spatial scalar field Φ within region Ω. In (4.28) $\mathbf{n}$ denotes the outward unit normal field acting along $\partial \Omega$. Relation (4.28) is referred to as **Reynolds' transport theorem**.

Another very useful relationship is obtained by considering the scalar-valued function

$$\bar{I}(t) = \int_\Omega \rho(\mathbf{x}, t) \Psi(\mathbf{x}, t) dv \quad , \tag{4.29}$$

which is deduced from eq. (4.22) by changing Φ into $\rho \Psi$, where Ψ denotes a smooth spatial scalar field describing some physical quantity of a particle in space per unit *mass* at time t.

With reference to eq. (4.26) the time rate of change of $\bar{I}(t)$ is then given as

$$\frac{D}{Dt} \int_\Omega \rho \Psi dv = \int_\Omega (\overline{\dot{\rho \Psi}} + \rho \Psi \mathrm{div}\mathbf{v}) dv \quad . \tag{4.30}$$

We now apply the product rule of differentiation and the rate form (4.12) of continuity mass equation to the first term of the integrand in (4.30). We find $\overline{\rho \Psi} = \rho \dot{\Psi} + \dot{\rho} \Psi = \rho \dot{\Psi} - \rho \Psi \mathrm{div} \mathbf{v}$. As a consequence of the continuity mass equation, we obtain

$$\frac{\mathrm{D}}{\mathrm{D}t} \int_\Omega \rho(\mathbf{x}, t) \Psi(\mathbf{x}, t) \mathrm{d}v = \int_\Omega \rho(\mathbf{x}, t) \dot{\Psi}(\mathbf{x}, t) \mathrm{d}v \quad . \tag{4.31}$$

Assume $\Psi = 1$ so that the volume integral $\bar{I}(t)$ in (4.29) represents mass within region Ω (see eq. (4.7)). For that case, eq. (4.31) reduces to $(4.8)_2$, since $\dot{\Psi} = 0$.

EXAMPLE 4.2   Show that the important relation (4.31) may also be obtained by changing the scalar-valued integral (4.29) over space into an integral over mass.

Solution.   Using the infinitesimal mass element $\mathrm{d}m = \rho \mathrm{d}v$ (see eq. $(4.5)_2$) we find that

$$\frac{\mathrm{D}}{\mathrm{D}t} \int_\Omega \Psi(\mathbf{x}, t)(\rho(\mathbf{x}, t)\mathrm{d}v) = \int_\Omega \frac{\mathrm{D}\Psi(\mathbf{x}, t)}{\mathrm{D}t} \mathrm{d}m = \int_\Omega \rho(\mathbf{x}, t) \dot{\Psi}(\mathbf{x}, t) \mathrm{d}v \quad , \tag{4.32}$$

since the infinitesimal mass element is not affected by the material time derivative.   ∎

EXERCISES

1. Derive the rate form of continuity mass equation (4.12) in the spatial description, i.e. $\dot{\rho} + \rho \mathrm{div} \mathbf{v} = 0$, by applying relation (4.26) and using eq. $(4.8)_2$.

2. Essentially the same statements as derived in eqs. (4.26), $(4.27)_2$, (4.28) and (4.31) result for the material time derivative of a vector-valued volume integral (or tensor-valued volume integral) by regarding either Φ or Ψ as a smooth (continuously differentiable) spatial vector field $\mathbf{u} = \mathbf{u}(\mathbf{x}, t)$.

    Show that

$$\frac{\mathrm{D}}{\mathrm{D}t} \int_\Omega \mathbf{u}\mathrm{d}v = \int_\Omega (\dot{\mathbf{u}} + \mathbf{u}\mathrm{div}\mathbf{v})\mathrm{d}v = \int_\Omega [\frac{\partial \mathbf{u}}{\partial t} + \mathrm{div}(\mathbf{u} \otimes \mathbf{v})]\mathrm{d}v$$

$$= \int_{\partial\Omega} \mathbf{u}(\mathbf{v} \cdot \mathbf{n})\mathrm{d}s + \int_\Omega \frac{\partial \mathbf{u}}{\partial t}\mathrm{d}v \quad , \tag{4.33}$$

$$\frac{\mathrm{D}}{\mathrm{D}t} \int_\Omega \rho\mathbf{u}\mathrm{d}v = \int_\Omega \rho\dot{\mathbf{u}}\mathrm{d}v \quad , \tag{4.34}$$

    where the material time derivative of $\mathbf{u}$ is $\dot{\mathbf{u}} = \partial \mathbf{u}/\partial t + (\mathrm{grad}\mathbf{u})\mathbf{v}$. The vector $\mathbf{v}$ denotes the spatial velocity.

4.3 Momentum Balance Principles

In this section we describe balance of linear and angular momentum for a closed and an open system, essential in continuum mechanics. These principles are valid for the whole or arbitrary parts of a continuum body $\mathcal{B}$. In addition, we derive Cauchy's first equation of motion and show the symmetry of the Cauchy stress tensor.

Balance of linear and angular momentum in spatial and material description. Consider a continuum body $\mathcal{B}$ with a set of particles occupying an arbitrary region Ω with boundary surface $\partial\Omega$ at time t.

We consider a *closed system* with a given motion $\mathbf{x} = \chi(\mathbf{X}, t)$, spatial mass density $\rho = \rho(\mathbf{x}, t)$ and spatial velocity field $\mathbf{v} = \mathbf{v}(\mathbf{x}, t)$. We define the total **linear momentum L** (or in the literature sometimes called **translational momentum**) by the vector-valued function

$$\mathbf{L}(t) = \int_\Omega \rho(\mathbf{x}, t)\mathbf{v}(\mathbf{x}, t)\mathrm{d}v = \int_{\Omega_0} \rho_0(\mathbf{X})\mathbf{V}(\mathbf{X}, t)\mathrm{d}V \quad , \tag{4.35}$$

and the total **angular momentum J** relative to a fixed point (characterized by the position vector $\mathbf{x}_0$) as

$$\mathbf{J}(t) = \int_\Omega \mathbf{r} \times \rho(\mathbf{x}, t)\mathbf{v}(\mathbf{x}, t)\mathrm{d}v = \int_{\Omega_0} \mathbf{r} \times \rho_0(\mathbf{X})\mathbf{V}(\mathbf{X}, t)\mathrm{d}V \quad . \tag{4.36}$$

We used the identity (2.8), conservation of mass in the form of $\rho\mathrm{d}v = \rho_0\mathrm{d}V$ and the definition of the position vector $\mathbf{r}$, i.e.

$$\mathbf{r}(\mathbf{x}) = \mathbf{x} - \mathbf{x}_0 = \chi(\mathbf{X}, t) - \mathbf{x}_0 \quad . \tag{4.37}$$

In the literature the angular momentum $\mathbf{J}$ is often referred to as the **moment of momentum** or the **rotational momentum**.

Momentum equations (4.35) and (4.36) are formulated with respect to the current and reference configurations with associated quantities $\rho, \mathbf{v}, \mathrm{d}v$ and $\rho_0, \mathbf{V}, \mathrm{d}V$, respectively. **Linear momentum** and **angular momentum** per unit current and reference volume are defined as the products $\rho\mathbf{v}, \rho_0\mathbf{V}$ and $\mathbf{r} \times \rho\mathbf{v}, \mathbf{r} \times \rho_0\mathbf{V}$, respectively. To avoid congestion we often omit the arguments of the tensors for much of the remainder of this section.

The material time derivatives of linear and angular momentum $(4.35)_1$ and $(4.36)_1$ of the particles which fill an arbitrary region Ω result in fundamental axioms called **momentum balance principles** for a continuum body. We postulate the **balance of linear momentum** as

$$\dot{\mathbf{L}}(t) = \frac{\mathrm{D}}{\mathrm{D}t}\int_\Omega \rho\mathbf{v}\mathrm{d}v = \frac{\mathrm{D}}{\mathrm{D}t}\int_{\Omega_0} \rho_0\mathbf{V}\mathrm{d}V = \mathbf{F}(t) \quad , \tag{4.38}$$

and the **balance of angular momentum** (or **balance of moment of momentum** or **balance of rotational momentum**) as

$$\mathbf{\dot{J}}(t) = \frac{\mathrm{D}}{\mathrm{D}t} \int_{\Omega} \mathbf{r} \times \rho \mathbf{v} \mathrm{d}v = \frac{\mathrm{D}}{\mathrm{D}t} \int_{\Omega_0} \mathbf{r} \times \rho_0 \mathbf{V} \mathrm{d}V = \mathbf{M}(t) \quad, \tag{4.39}$$

which are given in both the spatial and material descriptions.

In relations (4.38) and (4.39), $\mathbf{F}(t)$ and $\mathbf{M}(t)$ are vector-valued functions. They characterize respectively the **resultant force** and the **resultant moment** (or **resultant torque**), i.e. the moment of $\mathbf{F}$ about $\mathbf{x}_0$. The momentum balance principles are generalizations of Newton's first and second principle of motion to the context of continuum mechanics, as introduced by *Cauchy* and *Euler*. The contributions to linear momentum $\mathbf{L}$ and angular momentum $\mathbf{J}$ of a body are due to *external sources*, i.e. $\mathbf{F}$ and $\mathbf{M}$, respectively. If the external sources vanish linear and angular momentum of the body are said to be conserved.

By virtue of relation (4.34) we may rewrite the balance principles as

$$\mathbf{\dot{L}}(t) = \int_{\Omega} \rho \mathbf{\dot{v}} \mathrm{d}v = \int_{\Omega_0} \rho_0 \mathbf{\dot{V}} \mathrm{d}V = \mathbf{F}(t) \quad, \tag{4.40}$$

$$\mathbf{\dot{J}}(t) = \int_{\Omega} \mathbf{r} \times \rho \mathbf{\dot{v}} \mathrm{d}v = \int_{\Omega_0} \mathbf{r} \times \rho_0 \mathbf{\dot{V}} \mathrm{d}V = \mathbf{M}(t) \quad. \tag{4.41}$$

Here, we have used the relation $\overline{\mathbf{r} \times \mathbf{v}} = \mathbf{r} \times \mathbf{\dot{v}}$, since $\mathbf{\dot{r}} = \mathbf{\dot{x}} = \mathbf{v}$ according to $(4.37)_1$ and $(2.28)_1$, and consequently $\mathbf{\dot{r}} \times \mathbf{v} = \mathbf{v} \times \mathbf{v} = \mathbf{o}$. The spatial and material acceleration fields are characterized by $\mathbf{\dot{v}}$ and $\mathbf{\dot{V}}$ (compare with Section 2.3). The so-called **inertia forces** per unit current and reference volume are denoted by $\rho \mathbf{\dot{v}}$ and $\rho_0 \mathbf{\dot{V}}$, respectively.

In the following we define the structure of forces acting on a continuum body. Consider a boundary surface $\partial\Omega$ of an arbitrary region Ω which is subjected to the Cauchy traction vector $\mathbf{t} = \mathbf{t}(\mathbf{x}, t, \mathbf{n})$ (force measured per unit current surface area of $\partial\Omega$), as introduced in Section 3.1. The unit vector $\mathbf{n}$ is the outward normal to an infinitesimal surface element $\mathrm{d}s$ of $\partial\Omega$. Furthermore, let $\mathbf{b} = \mathbf{b}(\mathbf{x}, t)$ denote a spatial vector field called the **body force**. It is defined per unit current volume of region Ω acting on a particle, as illustrated in Figure 4.2. Note that the symbol $\mathbf{b}$ should not be confused with the strain tensor introduced in eq. (2.79).

A body force is, for example, self-weight or gravity loading per unit volume, i.e. $\mathbf{b} = \rho\mathbf{g}$ with the spatial mass density ρ and the (constant) **gravitational acceleration** $\mathbf{g}$.

Hence, the resultant force $\mathbf{F}$ and the resultant moment $\mathbf{M}$ (about a point $\mathbf{x}_0$) on the body in the current configuration have the additive forms

$$\mathbf{F}(t) = \int_{\partial\Omega} \mathbf{t} \mathrm{d}s + \int_{\Omega} \mathbf{b} \mathrm{d}v \quad, \tag{4.42}$$

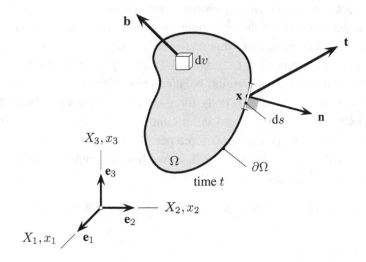

Figure 4.2 Structure of forces acting on the current configuration.

$$\mathbf{M}(t) = \int_{\partial\Omega} \mathbf{r} \times \mathbf{t}\,\mathrm{d}s + \int_{\Omega} \mathbf{r} \times \mathbf{b}\,\mathrm{d}v \quad . \tag{4.43}$$

Finally, by virtue of eqs. (4.38) and (4.39) the global forms of balance of linear momentum and balance of angular momentum may be given in the spatial description as

$$\frac{\mathrm{D}}{\mathrm{D}t} \int_{\Omega} \rho\mathbf{v}\,\mathrm{d}v = \int_{\partial\Omega} \mathbf{t}\,\mathrm{d}s + \int_{\Omega} \mathbf{b}\,\mathrm{d}v \quad , \tag{4.44}$$

$$\frac{\mathrm{D}}{\mathrm{D}t} \int_{\Omega} \mathbf{r} \times \rho\mathbf{v}\,\mathrm{d}v = \int_{\partial\Omega} \mathbf{r} \times \mathbf{t}\,\mathrm{d}s + \int_{\Omega} \mathbf{r} \times \mathbf{b}\,\mathrm{d}v \quad . \tag{4.45}$$

These equations are fundamental in continuum mechanics.

For the balance of angular momentum (4.45) we have assumed the restriction that distributed resultant couples are neglected. If we consider resultant couples throughout a body in motion, then the balance of angular momentum (4.45) reads as

$$\frac{\mathrm{D}}{\mathrm{D}t} \int_{\Omega} (\mathbf{r} \times \rho\mathbf{v} + \mathbf{p})\mathrm{d}v = \int_{\partial\Omega} (\mathbf{r} \times \mathbf{t} + \mathbf{m})\mathrm{d}s + \int_{\Omega} (\mathbf{r} \times \mathbf{b} + \mathbf{c})\mathrm{d}v \quad . \tag{4.46}$$

Here, **m** is the distributed assigned **coupled traction vector** per unit current area acting on the boundary surface $\partial\Omega$ while **c** is the distributed assigned **body couple** (or also called **body torque**) per unit volume acting within the volume of region Ω. The **spin angular momentum** (or **intrinsic angular momentum**) per unit current volume is denoted by **p**. A continuum without distributed resultant couples is called **non-polar**. If any couple acts on parts of the continuum we say that the continuum is **polar**. Polar continua are not considered in this text. For a detailed study of polar continuum mechanics see, for example, TRUESDELL and TOUPIN [1960] or MALVERN [1969].

In order to express the momentum balance principles in terms of material coordinates we introduce the (pseudo) body force called the **reference body force** $\mathbf{B} = \mathbf{B}(\mathbf{X}, t)$. It acts on the region Ω and is, in contrast to the body force $\mathbf{b}$, referred to the reference position $\mathbf{X}$ and measures force per unit reference volume. With volume change $dv = JdV$ and motion $\mathbf{x} = \chi(\mathbf{X}, t)$ we find the transformation of the body force terms of eqs. (4.44) and (4.45) in the form

$$\int_\Omega \mathbf{b}(\mathbf{x}, t)dv = \int_{\Omega_0} \mathbf{b}(\chi(\mathbf{X}, t), t)J(\mathbf{X}, t)dV = \int_{\Omega_0} \mathbf{B}(\mathbf{X}, t)dV \quad, \tag{4.47}$$

or in the local form as

$$\mathbf{B}(\mathbf{X}, t) = J(\mathbf{X}, t)\mathbf{b}(\mathbf{x}, t) \qquad \text{or} \qquad B_a = Jb_a \quad. \tag{4.48}$$

Using the first Piola-Kirchhoff traction vector $\mathbf{T} = \mathbf{T}(\mathbf{X}, t, \mathbf{N})$ introduced in (3.1), relations (4.38), (4.39), (4.48) and $dv = JdV$, we conclude from (4.44) and (4.45) that

$$\frac{D}{Dt}\int_{\Omega_0} \rho_0 \mathbf{V}dV = \int_{\partial\Omega_0} \mathbf{T}dS + \int_{\Omega_0} \mathbf{B}dV \quad, \tag{4.49}$$

$$\frac{D}{Dt}\int_{\Omega_0} \mathbf{r} \times \rho_0 \mathbf{V}dV = \int_{\partial\Omega_0} \mathbf{r} \times \mathbf{T}dS + \int_{\Omega_0} \mathbf{r} \times \mathbf{B}dV \quad, \tag{4.50}$$

which are the global forms of balance of linear momentum and balance of angular momentum, respectively, in the material description.

Equation of motion in spatial and material description.   A necessary and sufficient condition that the momentum balance principles (4.44) and (4.45) are satisfied is the existence of a spatial tensor field $\boldsymbol{\sigma}$ so that $\mathbf{t}(\mathbf{x}, t, \mathbf{n}) = \boldsymbol{\sigma}(\mathbf{x}, t)\mathbf{n}$ (see eq. (3.3)$_1$).

By computing the integral form of Cauchy's stress theorem (3.3)$_1$ and by using divergence theorem (1.296), which converts the surface integral into a volume integral, we find that

$$\int_{\partial\Omega} \mathbf{t}(\mathbf{x}, t, \mathbf{n})ds = \int_{\partial\Omega} \boldsymbol{\sigma}(\mathbf{x}, t)\mathbf{n}ds = \int_\Omega \mathrm{div}\boldsymbol{\sigma}(\mathbf{x}, t)dv \quad, \tag{4.51}$$

where $\boldsymbol{\sigma}$ is the symmetric Cauchy stress tensor. By substituting this result into the balance of linear momentum (4.44), and using (4.38), (4.40), we may show that $\boldsymbol{\sigma}$ satisfies the **Cauchy's first equation of motion**

$$\int_\Omega (\mathrm{div}\boldsymbol{\sigma} + \mathbf{b} - \rho\dot{\mathbf{v}})dv = \mathbf{o} \quad, \tag{4.52}$$

here presented in the global form. This relation is supposed to hold for any volume v. Hence, we may deduce Cauchy's first equation of motion in the local form, i.e.

$$\text{div}\boldsymbol{\sigma} + \mathbf{b} = \rho\dot{\mathbf{v}} \qquad \text{or} \qquad \frac{\partial\sigma_{ab}}{\partial x_b} + b_a = \rho\dot{v}_a \ , \tag{4.53}$$

for each point $\mathbf{x}$ of v and for all times t.

The differential equation of motion is here presented with respect to the current configuration. Note that the material time derivative of the spatial velocity field $\mathbf{v}$ is, according to (2.26), given as $\dot{\mathbf{v}} = \partial\mathbf{v}/\partial t + (\text{grad}\mathbf{v})\mathbf{v}$ or, in terms of the spin tensor $\mathbf{w}$, according to (2.153), given as $\dot{\mathbf{v}} = \partial\mathbf{v}/\partial t + (1/2)\text{grad}(\mathbf{v}^2) + 2\mathbf{w}\mathbf{v}$.

Generally, relation (4.53) is nonlinear in the displacement field $\mathbf{u}$. The nonlinearities are implicitly present due to *geometric* sources, i.e. the kinematics of motion of the body, and *material* sources, i.e. the material itself – the Cauchy stress $\boldsymbol{\sigma}$ may, in general, depend on $\mathbf{u}$.

If the acceleration is assumed to be zero for all $\mathbf{x} \in \Omega$ (note that a constant velocity field is not excluded), eq. (4.53) becomes

$$\text{div}\boldsymbol{\sigma} + \mathbf{b} = \mathbf{o} \qquad \text{or} \qquad \frac{\partial\sigma_{ab}}{\partial x_b} + b_a = 0 \ , \tag{4.54}$$

which is referred to as **Cauchy's equation of equilibrium** in elastostatics. A spatial stress field satisfying $\text{div}\boldsymbol{\sigma} = \mathbf{o}$ is called **self-equilibrated**.

EXAMPLE 4.3 Show that Cauchy's first equation of motion may also be written in the important equivalent form

$$\frac{\partial(\rho\mathbf{v})}{\partial t} = \text{div}(\boldsymbol{\sigma} - \rho\mathbf{v} \otimes \mathbf{v}) + \mathbf{b} \ . \tag{4.55}$$

Solution. In order to reformulate the term $\rho\dot{\mathbf{v}}$ on the right-hand side of eq. (4.53) we apply (2.26), the product rule and the continuity mass equation in the form of (4.14) to obtain

$$\rho\dot{\mathbf{v}} = \rho\frac{\partial\mathbf{v}}{\partial t} + \rho(\text{grad}\mathbf{v})\mathbf{v} = \frac{\partial(\rho\mathbf{v})}{\partial t} - \frac{\partial\rho}{\partial t}\mathbf{v} + (\text{grad}\mathbf{v})(\rho\mathbf{v})$$

$$= \frac{\partial(\rho\mathbf{v})}{\partial t} + \mathbf{v}\text{div}(\rho\mathbf{v}) + (\text{grad}\mathbf{v})(\rho\mathbf{v}) \ . \tag{4.56}$$

With (1.292) we find finally the identity

$$\rho\dot{\mathbf{v}} = \frac{\partial(\rho\mathbf{v})}{\partial t} + \text{div}(\rho\mathbf{v} \otimes \mathbf{v}) \tag{4.57}$$

(which also comes from eq. $(4.21)_2$ with $\mathbf{u}$ replaced by $\mathbf{v}$). Substituting (4.57) into (4.53) gives the desired result. ■

For solid bodies, it is sometimes more convenient to work with the material description. Hence, we rearrange the equations of motion (4.52) and (4.53) in terms of quantities which are referred to the reference configuration. To begin with, we introduce the **Piola identity**, which is

$$\text{Div}(J\mathbf{F}^{-\text{T}}) = \mathbf{o} \qquad \text{or} \qquad \frac{\partial(JF_{Aa}^{-1})}{\partial X_A} = 0 \ . \tag{4.58}$$

Proof.   To prove the important identity (4.58) we pick any region Ω_0 of a continuum body with boundary surface $\partial\Omega_0$ and apply the divergence theorem twice. With (1.300), Nanson's formula (2.55) and (1.296), we obtain simply that

$$\int_{\Omega_0} \text{Div}(J\mathbf{F}^{-\text{T}})\text{d}V = \int_{\partial\Omega_0} J\mathbf{F}^{-\text{T}}\mathbf{N}\text{d}S = \int_{\partial\Omega} \mathbf{n}\text{d}s$$

$$= \int_{\partial\Omega} \mathbf{I}\mathbf{n}\text{d}s = \int_{\Omega} \underbrace{\text{div}\mathbf{I}}_{\mathbf{0}}\text{d}v = \mathbf{o} \ . \quad \blacksquare \tag{4.59}$$

Now with the Piola transformation (3.8), and identities (1.291) and (4.58) we may find the divergence of the first Piola-Kirchhoff stress tensor $\mathbf{P}$ with respect to the material coordinates, i.e.

$$\text{Div}\mathbf{P} = \text{Div}(J\boldsymbol{\sigma}\mathbf{F}^{-\text{T}}) = \text{Div}[\boldsymbol{\sigma}(J\mathbf{F}^{-\text{T}})]$$

$$= J\text{Grad}\boldsymbol{\sigma} : \mathbf{F}^{-\text{T}} + \boldsymbol{\sigma}\underbrace{\text{Div}(J\mathbf{F}^{-\text{T}})}_{\mathbf{0}} \ . \tag{4.60}$$

By recalling relation (2.50) we obtain the transformation

$$\text{Div}\mathbf{P} = J\text{div}\boldsymbol{\sigma} \qquad \text{or} \qquad \frac{\partial P_{aB}}{\partial X_B} = J\frac{\partial \sigma_{ab}}{\partial x_b} \ . \tag{4.61}$$

Combining this result and eqs. (4.47)$_2$ and (4.40) with Cauchy's first equation of motion (4.52) we obtain, after a change of variables and use of $\text{d}v = J\text{d}V$,

$$\int_{\Omega_0} (\text{Div}\mathbf{P} + \mathbf{B} - \rho_0\dot{\mathbf{V}})\text{d}V = \mathbf{o} \ . \tag{4.62}$$

It is the global form of the equation of motion in the reference configuration. Since the volume v (and therefore V) is arbitrary, we obtain the associated local form, i.e.

$$\text{Div}\mathbf{P} + \mathbf{B} = \rho_0\dot{\mathbf{V}} \qquad \text{or} \qquad \frac{\partial P_{aA}}{\partial X_A} + B_a = \rho_0\dot{V}_a \ , \tag{4.63}$$

in which the independent variables are $(\mathbf{X}, t)$.

The equilibrium counterpart of (4.63) is given simply by setting $\dot{\mathbf{V}}$, $(\dot{V}_a)$, equal to zero.

Symmetry of the Cauchy stress tensor. The Cauchy stress tensor σ is symmetric, as can be seen from the global form of balance of angular momentum (4.45) as follows. Knowing Cauchy's stress theorem $(3.3)_1$ and the divergence theorem, as given in (1.302), we are able to convert the first term on the right-hand side of eq. (4.45) to a volume integral according to

$$\int_{\partial\Omega} \mathbf{r} \times \mathbf{t} \mathrm{d}s = \int_{\partial\Omega} \mathbf{r} \times \sigma \mathbf{n} \mathrm{d}s = \int_{\Omega} (\mathbf{r} \times \mathrm{div}\sigma + \mathcal{E} : \sigma^{\mathrm{T}}) \mathrm{d}v \ , \tag{4.64}$$

where $\mathcal{E}$ denotes the third-order permutation tensor introduced in eq. (1.143). With (4.64) and relations (4.39), (4.41) we are now able to rewrite (4.45) as

$$\int_{\Omega} \mathbf{r} \times (\rho\dot{\mathbf{v}} - \mathbf{b} - \mathrm{div}\sigma) \mathrm{d}v = \int_{\Omega} \mathcal{E} : \sigma^{\mathrm{T}} \mathrm{d}v \ . \tag{4.65}$$

Using the equation of motion (4.53) and the fact that the current volume v is arbitrary, we conclude that

$$\mathcal{E} : \sigma^{\mathrm{T}} = \mathbf{o} \qquad \text{or} \qquad \varepsilon_{abc}\sigma_{cb} = 0 \ , \tag{4.66}$$

which holds at each point $\mathbf{x}$ of the region and for all times t. The double contraction $\mathcal{E} : \sigma^{\mathrm{T}}$ gives a vector with components $\varepsilon_{abc}\sigma_{cb}$, which must be zero. We see that

$$\sigma_{32} - \sigma_{23} = 0 \ , \qquad \sigma_{13} - \sigma_{31} = 0 \ , \qquad \sigma_{21} - \sigma_{12} = 0 \ . \tag{4.67}$$

This relation is satisfied, *if and only if* the Cauchy stress tensor σ is *symmetric*, i.e.

$$\sigma = \sigma^{\mathrm{T}} \qquad \text{or} \qquad \sigma_{ab} = \sigma_{ba} \ . \tag{4.68}$$

The crucial result (4.68) is a local consequence of the balance of angular momentum (4.45), often referred to as **Cauchy's second equation of motion**. From eqs. (3.62) and $(3.65)_1$ we deduce that the Kirchhoff stress tensor τ and the second Piola-Kirchhoff stress tensor $\mathbf{S}$ are also symmetric. However, from eq. (3.67) we find that the first Piola-Kirchhoff stress tensor $\mathbf{P}$ is, in general, not symmetric as indicated in (3.10).

Note that for a polar continuum (resultant couples are not zero) the symmetry property does not hold any longer ($\sigma \neq \sigma^{\mathrm{T}}$) and therefore eq. (4.68) may also be viewed as a constitutive equation (compare with Exercise 5 on p. 152).

System of forces and flow. The Cauchy traction vector $\mathbf{t} = \mathbf{t}(\mathbf{x}, t, \mathbf{n})$ and the body force $\mathbf{b} = \mathbf{b}(\mathbf{x}, t)$ acting on $\partial\Omega$ and Ω (see eq. (4.42)) during a motion χ consistent with the momentum balance principles form a so-called **system of forces**. To each system of forces $(\mathbf{t}, \mathbf{b})$ corresponds exactly one symmetric stress tensor field $\sigma = \sigma(\mathbf{x}, t)$ satisfying the equation of motion (4.53), i.e. $\mathrm{div}\sigma + \mathbf{b} = \rho\dot{\mathbf{v}}$, and Cauchy's stress theorem $(3.3)_1$, i.e. $\mathbf{t} = \sigma\mathbf{n}$.

In other words, given the Cauchy stress tensor $\sigma = \sigma(\mathbf{x}, t)$ (that is defined to be a smooth function of $\mathbf{x}$) and a motion $\mathbf{x} = \chi(\mathbf{X}, t)$, then the system of forces $(\mathbf{t}, \mathbf{b})$ may be uniquely determined. The Cauchy traction vector $\mathbf{t}$ follows from Cauchy's stress theorem $(3.3)_1$, while the body force $\mathbf{b}$ is obtained from the equation of motion (4.53). The spatial mass density ρ therein results from the continuity mass equation. The pair (σ, χ) defines a so-called **dynamical process**, or just a **process**, for convenience.

EXAMPLE 4.4 A dynamical process (σ, χ) is given by the Cauchy stress tensor σ in the form of the matrix

$$[\sigma] = \begin{bmatrix} x_1^2 & \alpha x_2 x_3^2 & 0 \\ \alpha x_2 x_3^2 & x_2^2 & 0 \\ 0 & 0 & \beta x_1^3 \end{bmatrix}, \tag{4.69}$$

where α and β are scalar constants, and by a motion according to Example 2.2 (see eq. (2.10)), i.e. $x_1 = e^t X_1 - e^{-t} X_2$, $x_2 = e^t X_1 + e^{-t} X_2$, $x_3 = X_3$ for $t > 0$.

Find the system of forces (in matrix notation) so that Cauchy's first equation of motion (4.53) and continuity mass equation in the form of eq. (4.10) are satisfied. The Cauchy traction vector $\mathbf{t}$ is assumed to act at a point $\mathbf{x}$ of a plane tangential to a sphere, given by equation $\Phi = x_1^2 + x_2^2 + x_3^2$.

Solution.   In order to determine the components of the traction vector $\mathbf{t}$ from Cauchy's stress theorem we must find the unit vector $\mathbf{n}$ normal to the surface Φ, which is $\mathbf{n} = \mathrm{grad}\Phi/|\mathrm{grad}\Phi| = \mathbf{x}/|\mathbf{x}|$ (see Figure 1.9).

Hence, the components of $\mathbf{t}$ take on the values

$$[\mathbf{t}] = [\sigma][\mathbf{n}] = \frac{1}{|\mathbf{x}|} \begin{bmatrix} x_1^2 & \alpha x_2 x_3^2 & 0 \\ \alpha x_2 x_3^2 & x_2^2 & 0 \\ 0 & 0 & \beta x_1^3 \end{bmatrix} \begin{bmatrix} x_1 \\ x_2 \\ x_3 \end{bmatrix}$$

$$= \frac{1}{|\mathbf{x}|} \begin{bmatrix} x_1^3 + \alpha x_2^2 x_3^2 \\ \alpha x_1 x_2 x_3^2 + x_2^3 \\ \beta x_1^3 x_3 \end{bmatrix}. \tag{4.70}$$

In order to determine the components of the body force vector $\mathbf{b}$, we rewrite Cauchy's equation of motion so that $[\mathbf{b}] = \rho[\dot{\mathbf{v}}] - [\mathrm{div}\sigma]$, which requires the computation of ρ, $[\dot{\mathbf{v}}]$ and $[\mathrm{div}\sigma]$.

The spatial mass density ρ follows from the inverse of relation (4.10), i.e. $\rho = J^{-1}\rho_0$, where the volume ratio J is for the given motion (2.10) equal to 2. The components of the spatial acceleration field are $a_1 = \dot{v}_1 = x_1, a_2 = \dot{v}_2 = x_2, a_3 = \dot{v}_3 = 0$ (compare with Example 2.2, eq. (2.14)) and the divergence of the Cauchy stress tensor gives a vector with components $(2x_1 + \alpha x_3^2, 2x_2, 0)$.

Finally, the components of **b** are given uniquely as

$$[\mathbf{b}] = \frac{\rho_0}{2} \begin{bmatrix} x_1 \\ x_2 \\ 0 \end{bmatrix} - \begin{bmatrix} 2x_1 + \alpha x_3^2 \\ 2x_2 \\ 0 \end{bmatrix} , \qquad (4.71)$$

which together with $(4.70)_3$ determine the system of forces. ∎

We call a **flow** a set of quantities, such as the velocity field **v**, the spatial mass density ρ and the symmetric tensor field $\boldsymbol{\sigma}$, i.e. $(\mathbf{v}, \rho, \boldsymbol{\sigma})$, which is associated with the system of forces $(\mathbf{t}, \mathbf{b})$. To each system of forces consistent with the momentum balance principles (4.44) and (4.45) there corresponds exactly one flow and vice versa.

We call a flow $(\mathbf{v}, \rho, \boldsymbol{\sigma})$ **steady** if the associated spatial quantities are independent of time, i.e. $\partial \mathbf{v} / \partial t = \mathbf{o}$, $\partial \rho / \partial t = 0$, $\partial \boldsymbol{\sigma} / \partial t = \mathbf{O}$ and consequently $\mathbf{v} = \mathbf{v}(\mathbf{x})$, $\rho = \rho(\mathbf{x})$, $\boldsymbol{\sigma} = \boldsymbol{\sigma}(\mathbf{x})$. Since, for a steady flow, $\partial \mathbf{v} / \partial t = \mathbf{o}$, we have with reference to eq. (2.153)

$$\mathbf{v} \cdot \dot{\mathbf{v}} = \mathbf{v} \cdot \frac{1}{2} \operatorname{grad}(\mathbf{v}^2) , \qquad (4.72)$$

where the relation $\mathbf{v} \cdot \mathbf{w}\,\mathbf{v} = 0$ is to be used (since **w** is skew). Hence, the equation of motion (4.53) for a steady flow has the form

$$\mathbf{v} \cdot \operatorname{div}\boldsymbol{\sigma} + \mathbf{v} \cdot \mathbf{b} = \rho \mathbf{v} \cdot \frac{1}{2} \operatorname{grad}(\mathbf{v}^2) . \qquad (4.73)$$

By analogy with the types of motion introduced on p. 68, a flow is said to be a **potential flow** if the velocity field is $\mathbf{v} = \operatorname{grad}\Phi$, with a spatial scalar field Φ. Then the spatial acceleration field $\mathbf{a} = \dot{\mathbf{v}}$ has the form of $(2.158)_2$ and the equation of motion (4.53) takes on the form

$$\operatorname{div}\boldsymbol{\sigma} + \mathbf{b} = \rho \operatorname{grad} \left(\frac{\partial \Phi}{\partial t} + \frac{1}{2}(\operatorname{grad}\Phi)^2 \right) . \qquad (4.74)$$

A flow satisfying $\operatorname{curl}\mathbf{v} = \mathbf{o}$ is **irrotational**. By recalling relation (1.274) we conclude that potential flows are irrotational. If the flow is steady and irrotational we find that the spatial acceleration field $\dot{\mathbf{v}}$ on the right-hand side of the equation of motion (4.53) is simply

$$\dot{\mathbf{v}} = \frac{1}{2} \operatorname{grad}(\mathbf{v}^2) , \qquad (4.75)$$

since $\partial \mathbf{v} / \partial t = \mathbf{o}$ and $\operatorname{curl}\mathbf{v} = \mathbf{o}$.

Balance of linear and angular momentum for an open system.   We set up finally the momentum balance principles for an *open system*, i.e. a selected region in space with control volume Ω_c and enclosing control surface $\partial \Omega_c$, independent of time. By integrating the equation of motion (4.53) over Ω_c we find that

$$\int_{\Omega_c} \mathrm{div}\boldsymbol{\sigma}\, dv + \int_{\Omega_c} \mathbf{b}\, dv = \int_{\Omega_c} \rho\dot{\mathbf{v}}\, dv \quad . \tag{4.76}$$

With divergence theorem (1.296) and Cauchy's stress theorem (3.3)$_1$, the first term in eq. (4.76) yields, by analogy with (4.51),

$$\int_{\Omega_c} \mathrm{div}\boldsymbol{\sigma}\, dv = \int_{\partial\Omega_c} \boldsymbol{\sigma}\mathbf{n}\, ds = \int_{\partial\Omega_c} \mathbf{t}\, ds \quad . \tag{4.77}$$

The term on the right-hand side of eq. (4.76) follows by integration over the control volume Ω_c

$$\int_{\Omega_c} \rho\dot{\mathbf{v}}\, dv = \int_{\partial\Omega_c} (\rho\mathbf{v} \otimes \mathbf{v})\mathbf{n}\, ds + \int_{\Omega_c} \frac{\partial(\rho\mathbf{v})}{\partial t}\, dv$$

$$= \int_{\partial\Omega_c} (\mathbf{v} \cdot \mathbf{n})\rho\mathbf{v}\, ds + \frac{\mathrm{D}}{\mathrm{D}t} \int_{\Omega_c} \rho\mathbf{v}\, dv \quad . \tag{4.78}$$

We have used eq. (4.57), the analogue of the divergence theorem (1.300), the fact that the volume is time-independent and rule (1.53)$_1$. Consequently, by substituting eqs. (4.77)$_2$ and (4.78)$_2$ into (4.76) we arrive finally at the balance of linear momentum for a control volume in the form

$$\frac{\mathrm{D}}{\mathrm{D}t} \int_{\Omega_c} \rho\mathbf{v}\, dv = \int_{\partial\Omega_c} [\mathbf{t} - (\mathbf{v} \cdot \mathbf{n})\rho\mathbf{v}]\, ds + \int_{\Omega_c} \mathbf{b}\, dv \quad . \tag{4.79}$$

It states that the rate at which $\rho\mathbf{v}$ (linear momentum) changes in the control volume Ω_c equals the traction vector acting on the boundary control surface $\partial\Omega_c$ of Ω_c and the flux of $\rho\mathbf{v}$ entering across $\partial\Omega_c$ plus the body force $\mathbf{b}$ acting on Ω_c.

In an analogous manner we can show the balance of angular momentum for a control volume, i.e.

$$\frac{\mathrm{D}}{\mathrm{D}t} \int_{\Omega_c} \mathbf{r} \times \rho\mathbf{v}\, dv = \int_{\partial\Omega_c} \mathbf{r} \times (\mathbf{t} - (\mathbf{v} \cdot \mathbf{n})\rho\mathbf{v})\, ds + \int_{\Omega_c} \mathbf{r} \times \mathbf{b}\, dv \quad , \tag{4.80}$$

which has a similar interpretation.

EXERCISES

1. Consider a certain point $\mathbf{g}$ in the current configuration which we refer to as the **mass center** of an arbitrary region Ω of a body $\mathcal{B}$ with mass m. It is defined

(independently of the choice of the point $\mathbf{x}_0$) as

$$\mathbf{g}(t) = \frac{1}{m(\Omega)} \int_{\Omega} \rho(\mathbf{x}, t)\mathbf{x}\mathrm{d}v \quad .$$

Considering conservation of mass, $m(\Omega_0) = m(\Omega)$, this relation may be expressed by change of variables as $1/m(\Omega_0) \int_{\Omega_0} \rho_0(\mathbf{X})\chi(\mathbf{X}, t)\mathrm{d}V$.

  (a) Differentiate $\mathbf{g}$ with respect to time and show that the average spatial velocity $\dot{\mathbf{g}}$ and the average spatial acceleration $\ddot{\mathbf{g}}$ of Ω are given by

$$\dot{\mathbf{g}}(t) = \frac{1}{m(\Omega)} \int_{\Omega} \rho(\mathbf{x}, t)\mathbf{v}(\mathbf{x}, t)\mathrm{d}v \quad , \qquad \ddot{\mathbf{g}}(t) = \frac{1}{m(\Omega)} \int_{\Omega} \rho(\mathbf{x}, t)\dot{\mathbf{v}}(\mathbf{x}, t)\mathrm{d}v \quad .$$

  (b) With reference to eqs. $(4.35)_1$ and (4.40) show that these relations imply

$$\mathbf{L}(t) = m(\Omega)\dot{\mathbf{g}}(t) \quad , \qquad \dot{\mathbf{L}}(t) = m(\Omega)\ddot{\mathbf{g}}(t) = \mathbf{F}(t) \quad ,$$

which characterize the **motion of the mass center** and which we already know from Newtonian mechanics (they simple govern the Newtonian particle).

Note the following important property of the mass center: linear momentum of a given arbitrary region Ω of a body $\mathcal{B}$ at t is the same as that of a particle with mass $m(\Omega)$ concentrated at the mass center of that region. Additionally, the resultant force $\mathbf{F}$ of a given region Ω is equal to the mass of that region multiplied by the acceleration of its mass center.

2. A material vector field $\mathbf{U}$ is given by its Piola transformation $\mathbf{U} = J\mathbf{F}^{-1}\mathbf{u}$. Using the Piola identity (4.58) show that $\mathrm{Div}\mathbf{U} = J\mathrm{div}\mathbf{u}$, i.e. in index notation $\partial U_A/\partial X_A = J\partial u_a/\partial x_a$.

3. Consider the Cauchy stress distribution for a continuum in equilibrium given with reference to a rectangular x_1, x_2, x_3-coordinate system. The components of the Cauchy stress tensor $\boldsymbol{\sigma}$ are given in the form

$$[\boldsymbol{\sigma}] = \begin{bmatrix} x_1 x_2 & x_1^2 & -x_2 \\ x_1^2 & 0 & 0 \\ -x_2 & 0 & x_1^2 + x_2^2 \end{bmatrix} \quad .$$

Find the body force $\mathbf{b}$ that acts on this continuum.

4. The Cauchy stress distribution of a continuum is expressed with respect to a rectangular x_1, x_2, x_3-coordinate system. The matrix representation of the Cauchy

stress tensor $\boldsymbol{\sigma}$ is given in the form

$$[\boldsymbol{\sigma}] = \begin{bmatrix} \alpha & 0 & 0 \\ 0 & x_2 + \alpha x_3 & \Phi(x_2, x_3) \\ 0 & \Phi(x_2, x_3) & x_2 + \beta x_3 \end{bmatrix} \quad ,$$

where α and β are scalar constants.

(a) Find $\Phi(x_2, x_3)$ so that the given spatial stress field satisfies $\mathrm{div}\boldsymbol{\sigma} = \mathbf{o}$.

(b) Consider $\Phi(x_2, x_3)$ from (a) and find the Cauchy traction vector $\mathbf{t}$ on the plane $\Psi = x_1 + x_2 + x_3$.

5. A polar continuum in motion is acted on by resultant couples in addition to the linear momentum $\rho\mathbf{v}$, the surface traction $\mathbf{t}$ and the body force $\mathbf{b}$. Consider a spin angular momentum $\mathbf{p}$ and a body couple $\mathbf{c}$ acting over an arbitrary region Ω and a couple traction vector $\mathbf{m}$ on its boundary surface $\partial\Omega$.

(a) Show that the presence of couples does not affect the linear momentum principle, and hence the equations of motion (4.53) are still valid.

(b) Show that balance of angular momentum (4.46) requires that

$$\mathrm{div}\mathbf{M} + \mathbf{c} + \boldsymbol{\mathcal{E}} : \boldsymbol{\sigma}^{\mathrm{T}} = \dot{\mathbf{p}} \qquad \text{or} \qquad \frac{\partial M_{ad}}{\partial x_d} + c_a + \varepsilon_{abc}\sigma_{cb} = \dot{p}_a \quad , \quad (4.81)$$

where $\mathbf{M}$ is called the **couple stress tensor**. It is a *linear operator* that acts on the outward unit vector field $\mathbf{n}$ (perpendicular to $\partial\Omega$) generating the vector $\mathbf{m}$ according to

$$\mathbf{m}(\mathbf{x}, t, \mathbf{n}) = \mathbf{M}(\mathbf{x}, t)\mathbf{n} \qquad \text{or} \qquad m_a = M_{ab}n_b \quad ,$$

which is analogous to Cauchy's stress theorem $(3.3)_1$.

(c) Show that the three partial differential equations (4.81) imply that the stress tensor $\boldsymbol{\sigma}$ is not symmetric.

4.4   Balance of Mechanical Energy

In this section we consider only *mechanical* energy. Other forms of energy, such as thermal, electric, magnetic, chemical or nuclear, are neglected. For this consideration the balance of energy is *not* an additional statement to be satisfied, it is a consequence of Cauchy's first equation of motion (balance of linear momentum).

For subsequent studies we assume a dynamical process given by a symmetric Cauchy stress tensor $\boldsymbol{\sigma} = \boldsymbol{\sigma}(\mathbf{x}, t)$ (i.e. a smooth function of $\mathbf{x}$) and a motion $\mathbf{x} = \boldsymbol{\chi}(\mathbf{X}, t)$ which deforms an arbitrary region Ω_0 to Ω.

External mechanical power, stress power, kinetic energy. We consider a set of particles occupying a region Ω in space with boundary surface $\partial\Omega$ and define quantities first in terms of spatial coordinates.

The **external mechanical power** or the **rate of external mechanical work** $\mathcal{P}_{\text{ext}}$ is defined to be the power input on a region Ω at time t done by the system of forces $(\mathbf{t}, \mathbf{b})$, i.e.

$$\mathcal{P}_{\text{ext}}(t) = \int_{\partial\Omega} \mathbf{t} \cdot \mathbf{v} \mathrm{d}s + \int_{\Omega} \mathbf{b} \cdot \mathbf{v} \mathrm{d}v$$

$$\text{or} \quad \mathcal{P}_{\text{ext}}(t) = \int_{\partial\Omega} t_a v_a \mathrm{d}s + \int_{\Omega} b_a v_a \mathrm{d}v \quad . \tag{4.82}$$

The dimension of $\mathcal{P}_{\text{ext}}$ is work per time that is equal to power. As usual, the spatial velocity field is denoted by $\mathbf{v} = \dot{\mathbf{x}}$. The scalar quantities $\mathbf{t} \cdot \mathbf{v}$ and $\mathbf{b} \cdot \mathbf{v}$ give the *external* mechanical power per unit current surface s and current volume v, respectively.

The **kinetic energy** $\mathcal{K}$ of a continuum body occupying a region Ω at time t is basically a generalization of Newtonian mechanics to continuum mechanics. We have the definition

$$\mathcal{K}(t) = \int_{\Omega} \frac{1}{2}\rho \mathbf{v}^2 \mathrm{d}v = \int_{\Omega} \frac{1}{2}\rho \mathbf{v} \cdot \mathbf{v} \mathrm{d}v \quad \text{or} \quad \mathcal{K}(t) = \int_{\Omega} \frac{1}{2}\rho v_a v_a \mathrm{d}v \quad . \tag{4.83}$$

The **stress power** or the **rate of internal mechanical work** $\mathcal{P}_{\text{int}}$ which describes the response of a region Ω at time t, done by the stress field, is defined by the scalar

$$\mathcal{P}_{\text{int}}(t) = \int_{\Omega} \boldsymbol{\sigma} : \mathbf{d} \mathrm{d}v = \int_{\Omega} \mathrm{tr}(\boldsymbol{\sigma}^{\mathrm{T}}\mathbf{d}) \mathrm{d}v \quad \text{or} \quad \mathcal{P}_{\text{int}}(t) = \int_{\Omega} \sigma_{ab} d_{ab} \mathrm{d}v \quad . \tag{4.84}$$

For a rigid-body motion the stress power $\mathcal{P}_{\text{int}}$ is zero since the rate of deformation tensor $\mathbf{d}$, as characterized in eq. (2.148), vanishes (recall Example 2.12 on p. 99).

Balance of mechanical energy in spatial description. If only mechanical energy is considered, **balance of mechanical energy** (in the literature sometimes called the **theorem of power expended**) follows on using eqs. (4.82)–(4.84). Thus,

$$\frac{\mathrm{D}}{\mathrm{D}t}\mathcal{K}(t) + \mathcal{P}_{\text{int}}(t) = \mathcal{P}_{\text{ext}}(t) \quad , \tag{4.85}$$

$$\text{or} \quad \frac{\mathrm{D}}{\mathrm{D}t}\int_{\Omega} \frac{1}{2}\rho \mathbf{v}^2 \mathrm{d}v + \int_{\Omega} \boldsymbol{\sigma} : \mathbf{d} \mathrm{d}v = \int_{\partial\Omega} \mathbf{t} \cdot \mathbf{v} \mathrm{d}s + \int_{\Omega} \mathbf{b} \cdot \mathbf{v} \mathrm{d}v \quad . \tag{4.86}$$

It states that the rate of change of kinetic energy $\mathcal{K}$ of a mechanical system plus the rate of internal mechanical work (stress-power) $\mathcal{P}_{int}$ done by internal stresses equals the rate of external mechanical work (external mechanical power) $\mathcal{P}_{ext}$ done on that system by surface tractions $\mathbf{t}$ and body forces $\mathbf{b}$. Hence, the rate of change of kinetic energy $\mathcal{K}$ contains contributions from internal as well as external sources. Note that the kinetic energy $\mathcal{K}$ is not a conserved quantity, since the first term in eq. (4.86) does not vanish, in general.

If $\mathcal{P}_{ext}$ is zero, then we have a problem of **free vibration**, while if $\mathrm{D}\mathcal{K}/\mathrm{D}t$ is zero the problem is called **quasi-static** (the associated quantities can still depend on time).

Proof.   In order to prove (4.85) we look first at the term $\int_{\partial\Omega} \mathbf{t} \cdot \mathbf{v}\mathrm{d}s$ in eq. (4.86). With Cauchy's stress theorem $\mathbf{t} = \boldsymbol{\sigma}\mathbf{n}$, divergence theorem (1.301) and the product rule (1.290) we find that

$$\int_{\partial\Omega} \mathbf{t} \cdot \mathbf{v}\mathrm{d}s = \int_{\partial\Omega} (\boldsymbol{\sigma}\mathbf{n}) \cdot \mathbf{v}\mathrm{d}s = \int_{\Omega} \mathrm{div}(\boldsymbol{\sigma}^{\mathrm{T}}\mathbf{v})\mathrm{d}v$$

$$= \int_{\Omega} (\mathrm{div}\boldsymbol{\sigma} \cdot \mathbf{v} + \boldsymbol{\sigma} : \mathrm{grad}\mathbf{v})\mathrm{d}v \quad . \qquad (4.87)$$

After expansion with the scalar $\rho\dot{\mathbf{v}} \cdot \mathbf{v}$ (where $\dot{\mathbf{v}}$ denotes the spatial acceleration field) we may rewrite the external mechanical power $\mathcal{P}_{ext}$, i.e. eq. (4.82), as

$$\mathcal{P}_{ext}(t) = \int_{\Omega} (\rho\dot{\mathbf{v}} \cdot \mathbf{v} + \boldsymbol{\sigma} : \mathrm{grad}\mathbf{v})\mathrm{d}v + \int_{\Omega} \underbrace{(\mathrm{div}\boldsymbol{\sigma} + \mathbf{b} - \rho\dot{\mathbf{v}})}_{\mathbf{0}} \cdot \mathbf{v}\mathrm{d}v \quad . \qquad (4.88)$$

With (2.137) and decomposition (2.147) we find that $\boldsymbol{\sigma} : \mathrm{grad}\mathbf{v} = \boldsymbol{\sigma} : \mathbf{l} = \boldsymbol{\sigma} : (\mathbf{d} + \mathbf{w})$. Since the Cauchy stress tensor $\boldsymbol{\sigma}$ is symmetric and the spin tensor $\mathbf{w}$ is antisymmetric we find, using property (1.117), the result $\boldsymbol{\sigma} : \mathrm{grad}\mathbf{v} = \boldsymbol{\sigma} : \mathbf{d}$. Consequently, with the equation of motion (4.53) we deduce from (4.88) that

$$\mathcal{P}_{ext}(t) = \int_{\Omega} (\rho\dot{\mathbf{v}} \cdot \mathbf{v} + \boldsymbol{\sigma} : \mathbf{d})\mathrm{d}v \quad , \qquad (4.89)$$

which means that the skew part of the spatial velocity gradient $\mathbf{l}$, i.e. the spin tensor $\mathbf{w}$, does not contribute to the rate of work.

By $\dot{\mathbf{v}} \cdot \mathbf{v} = (\overline{\mathbf{v} \cdot \mathbf{v}})/2$ and relation (4.34), which allows $\mathrm{D}(\bullet)/\mathrm{D}t$ to be written in front of the integral, we conclude from (4.89) that

$$\mathcal{P}_{ext}(t) = \frac{\mathrm{D}}{\mathrm{D}t} \underbrace{\int_{\Omega} \frac{1}{2}\rho v^2 \mathrm{d}v}_{\mathcal{K}} + \underbrace{\int_{\Omega} \boldsymbol{\sigma} : \mathbf{d}\mathrm{d}v}_{\mathcal{P}_{int}} \quad , \qquad (4.90)$$

which, by means of eqs. (4.83) and (4.84), is the left-hand side of the balance of mechanical energy (4.85), i.e. the rate of change of kinetic energy $\mathcal{K}$ plus the stress power $\mathcal{P}_{\text{int}}$. ∎

Next, we introduce a fundamental quantity at current position $\mathbf{x} \in \Omega$ and time t which is the sum of all the microscopic forms of energy called the **internal energy**. It is a (thermodynamic) state variable denoted by $e_c = e_c(\mathbf{x}, t)$ (or in other texts frequently designated by u) and is defined per unit current volume. The internal energy possessed by a continuum body occupying a certain region Ω, denoted by $\mathcal{E}$, is expressible as

$$\mathcal{E}(t) = \int_{\Omega} e_c(\mathbf{x}, t) \mathrm{d}v \quad . \tag{4.91}$$

The term 'internal energy' was introduced by *Clausius* and *Rankine* in the second half of the nineteenth century.

Since only mechanical energy is considered, we state that the rate of work done on the continuum body by internal stresses, i.e. the stress power $\mathcal{P}_{\text{int}}$, equals the rate of internal energy $\mathcal{E}$. We write

$$\mathcal{P}_{\text{int}}(t) = \frac{\mathrm{D}}{\mathrm{D}t}\mathcal{E}(t) \quad . \tag{4.92}$$

Now, eq. (4.85) may be expressed in terms of the internal energy, i.e.

$$\frac{\mathrm{D}}{\mathrm{D}t}\mathcal{K}(t) + \frac{\mathrm{D}}{\mathrm{D}t}\mathcal{E}(t) = \mathcal{P}_{\text{ext}}(t) \quad , \tag{4.93}$$

or, when written in the explicit form,

$$\frac{\mathrm{D}}{\mathrm{D}t}\int_{\Omega}(\frac{1}{2}\rho v^2 + e_c)\mathrm{d}v = \int_{\partial\Omega} \mathbf{t} \cdot \mathbf{v}\mathrm{d}s + \int_{\Omega} \mathbf{b} \cdot \mathbf{v}\mathrm{d}v \quad . \tag{4.94}$$

The term on the left-hand side in the form

$$\int_{\Omega}(\frac{1}{2}\rho v^2 + e_c)\mathrm{d}v \tag{4.95}$$

characterizes the **total energy**. It is the sum of the kinetic and internal energies. Note that the contributions to the total energy are only due to external sources, i.e. the external mechanical power $\mathcal{P}_{\text{ext}}$.

Balance of mechanical energy in material description. In order to express the terms of the balance of mechanical energy (4.85) with respect to material coordinates at time t, we must establish the external mechanical power $\mathcal{P}_{\text{ext}}$, the kinetic energy $\mathcal{K}$ and the stress power $\mathcal{P}_{\text{int}}$ in the material description.

By means of relation (3.1) and identity (2.8), the first term on the right-hand side of eq. (4.82) transforms into

$$\int_{\partial\Omega} \mathbf{t} \cdot \mathbf{v} ds = \int_{\partial\Omega_0} \mathbf{T} \cdot \mathbf{V} dS \ , \tag{4.96}$$

and with the local form (4.48), $dv = JdV$ and (2.8) the remaining term in eq. (4.82) gives

$$\int_{\Omega} \mathbf{b} \cdot \mathbf{v} dv = \int_{\Omega_0} \mathbf{B} \cdot \mathbf{V} dV \ . \tag{4.97}$$

Hence, the external mechanical power $\mathcal{P}_{\text{ext}}$ can be obtained as

$$\mathcal{P}_{\text{ext}}(t) = \int_{\partial\Omega_0} \mathbf{T} \cdot \mathbf{V} dS + \int_{\Omega_0} \mathbf{B} \cdot \mathbf{V} dV$$

$$\text{or} \qquad \mathcal{P}_{\text{ext}}(t) = \int_{\partial\Omega_0} T_A V_A dS + \int_{\Omega_0} B_A V_A dV \ . \tag{4.98}$$

With identity (2.8) and conservation of mass (4.6) the kinetic energy $\mathcal{K}$ is, in view of (4.83), given as

$$\mathcal{K}(t) = \int_{\Omega_0} \frac{1}{2}\rho_0 \mathbf{V}^2 dV = \int_{\Omega_0} \frac{1}{2}\rho_0 \mathbf{V} \cdot \mathbf{V} dV$$

$$\text{or} \qquad \mathcal{K}(t) = \int_{\Omega_0} \frac{1}{2}\rho_0 V_A V_A dV \ . \tag{4.99}$$

Next, we write the rate of internal mechanical work (stress-power) $\mathcal{P}_{\text{int}}$ in terms of the first Piola-Kirchhoff stress tensor $\mathbf{P}$. For that, recall eq. $(4.84)_1$, the additive decomposition of the spatial velocity gradient, i.e. $\mathbf{l} = \mathbf{d} + \mathbf{w} = \dot{\mathbf{F}}\mathbf{F}^{-1}$ and conclude that the (skew) spin tensor $\mathbf{w}$ acting on the symmetric Cauchy stress tensor $\boldsymbol{\sigma}$ yields zero ($\boldsymbol{\sigma} : \mathbf{w} = 0$). Hence, by means of the property (1.95) and eq. (2.51) we find that

$$\mathcal{P}_{\text{int}}(t) = \int_{\Omega} \boldsymbol{\sigma} : \mathbf{d} dv = \int_{\Omega} \boldsymbol{\sigma} : (\dot{\mathbf{F}}\mathbf{F}^{-1}) dv$$

$$= \int_{\Omega} \boldsymbol{\sigma}\mathbf{F}^{-\text{T}} : \dot{\mathbf{F}} dv = \int_{\Omega_0} J\boldsymbol{\sigma}\mathbf{F}^{-\text{T}} : \dot{\mathbf{F}} dV \ . \tag{4.100}$$

With the Piola transformation (3.8) and manipulations according to $(1.93)_1$ we obtain from $(4.100)_4$ the equivalence for the rate of internal mechanical work, i.e.

$$\mathcal{P}_{\text{int}}(t) = \int_{\Omega_0} \mathbf{P} : \dot{\mathbf{F}} dV = \int_{\Omega_0} \text{tr}(\mathbf{P}^{\text{T}}\dot{\mathbf{F}}) dV$$

$$\text{or} \qquad \mathcal{P}_{\text{int}}(t) = \int_{\Omega_0} P_{aA}\dot{F}_{aA} dV \ . \tag{4.101}$$

Substituting (4.98), (4.99)$_1$, (4.101)$_1$ into (4.85), we have finally, equivalently to (4.86), the balance of mechanical energy in the material description, i.e.

$$\frac{D}{Dt} \int_{\Omega_0} \frac{1}{2}\rho_0 \mathbf{V}^2 dV + \int_{\Omega_0} \mathbf{P} : \dot{\mathbf{F}} dV = \int_{\partial\Omega_0} \mathbf{T} \cdot \mathbf{V} dS + \int_{\Omega_0} \mathbf{B} \cdot \mathbf{V} dV \quad . \tag{4.102}$$

By analogy with the relation (4.94) we now express the last equation in terms of the internal energy defined with respect to the reference volume denoted by $e = e(\mathbf{X}, t)$. With $dv = J dV$ and motion $\mathbf{x} = \chi(\mathbf{X}, t)$ we may write the internal energy (4.91) as

$$\mathcal{E}(t) = \int_{\Omega} e_c(\mathbf{x}, t) dv = \int_{\Omega_0} e_c(\chi(\mathbf{X}, t), t) J(\mathbf{X}, t) dV$$

$$= \int_{\Omega_0} e(\mathbf{X}, t) dV \quad , \tag{4.103}$$

with the important transformation

$$e(\mathbf{X}, t) = J(\mathbf{X}, t) e_c(\mathbf{x}, t) \quad . \tag{4.104}$$

Adopting the expression (4.99)$_1$ for the kinetic energy $\mathcal{K}$ and (4.103)$_3$ for the internal energy $\mathcal{E}$, we may write the total energy $\int_{\Omega}(\rho \mathbf{v}^2/2 + e_c) dv$ in the material description as

$$\int_{\Omega_0} (\frac{1}{2}\rho_0 \mathbf{V}^2 + e) dV \quad . \tag{4.105}$$

Using the equivalence of eqs. (4.82) and (4.98) and the total energy in the form of (4.105) we obtain finally from (4.94) the desired result

$$\frac{D}{Dt} \int_{\Omega_0} (\frac{1}{2}\rho_0 \mathbf{V}^2 + e) dV = \int_{\partial\Omega_0} \mathbf{T} \cdot \mathbf{V} dS + \int_{\Omega_0} \mathbf{B} \cdot \mathbf{V} dV \quad , \tag{4.106}$$

which may directly be derived from the balance of mechanical energy (4.102). This requires eq. (4.92) with (4.101)$_1$ and (4.103)$_3$, leading to the transformation $\int_{\Omega_0} \mathbf{P} : \dot{\mathbf{F}} dV = D/Dt \int_{\Omega_0} e dV$.

Alternative expressions for the stress power. In the following we write the stress power $\mathcal{P}_{\text{int}}$ in different equivalent versions.

We start from eq. (4.84)$_1$. By using (2.51), the pull-back of the rate of deformation tensor $\mathbf{d}$, i.e. the inverse of relation (2.165)$_4$, the rule (1.95) and the stress relation (3.65)$_1$ we find $\mathcal{P}_{\text{int}}$ for a region Ω_0 to be

$$\mathcal{P}_{\text{int}}(t) = \int_{\Omega_0} J\boldsymbol{\sigma} : \mathbf{d} dV = \int_{\Omega_0} J\boldsymbol{\sigma} : \mathbf{F}^{-T}\dot{\mathbf{E}}\mathbf{F}^{-1} dV$$

$$= \int_{\Omega_0} J\mathbf{F}^{-1}\boldsymbol{\sigma}\mathbf{F}^{-T} : \dot{\mathbf{E}} dV = \int_{\Omega_0} \mathbf{S} : \dot{\mathbf{E}} dV \quad , \tag{4.107}$$

with the second Piola-Kirchhoff stress tensor $\mathbf{S}$ and the material strain rate tensor $\dot{\mathbf{E}}$.

Another alternative form for the stress power $\mathcal{P}_{\text{int}}$ may be obtained by use of $\mathbf{C} = 2\dot{\mathbf{E}}$ (see eq. $(2.168)_1$) and the non-symmetric Mandel stress tensor $\boldsymbol{\Sigma} = \mathbf{CS}$ which is defined in an intermediate configuration. Starting from eq. $(4.107)_4$ we have, by means of (1.95),

$$\mathcal{P}_{\text{int}}(t) = \int_{\Omega_0} \mathbf{S} : \dot{\mathbf{E}} dV = \int_{\Omega_0} \mathbf{S} : \frac{1}{2}\dot{\mathbf{C}} dV = \int_{\Omega_0} \boldsymbol{\Sigma} : \frac{1}{2}\mathbf{C}^{-1}\dot{\mathbf{C}} dV \quad , \tag{4.108}$$

where the symmetry of the right Cauchy-Green tensor $\mathbf{C}$ is to be used.

Next, we recall the rotated Cauchy stress tensor $\boldsymbol{\sigma}_{\text{u}} = \mathbf{R}^{\text{T}}\boldsymbol{\sigma}\mathbf{R}$ (see eq. $(3.70)_3$), the rotated rate of deformation tensor $\mathbf{D}_{\text{R}} = \mathbf{R}^{\text{T}}\mathbf{dR}$ (see eq. $(2.169)_1$) and the relation $\mathbf{R}^{-1} = \mathbf{R}^{\text{T}}$. Then, from $(4.84)_1$ we find, using eq. (2.51), that

$$\mathcal{P}_{\text{int}}(t) = \int_{\Omega_0} J\boldsymbol{\sigma} : \mathbf{d} dV = \int_{\Omega_0} J\mathbf{R}^{\text{T}}\boldsymbol{\sigma}\mathbf{R} : \mathbf{R}^{-1}\mathbf{dR}^{-\text{T}} dV$$

$$= \int_{\Omega_0} J\boldsymbol{\sigma}_{\text{u}} : \mathbf{D}_{\text{R}} dV \quad , \tag{4.109}$$

where the integrand has been manipulated according to (1.95).

EXAMPLE 4.5   Express the stress power $\mathcal{P}_{\text{int}}$ in terms of the Biot stress tensor $\mathbf{T}_{\text{B}}$ and show that

$$\mathcal{P}_{\text{int}}(t) = \int_{\Omega_0} \mathbf{P} : \dot{\mathbf{F}} dV = \int_{\Omega_0} \text{sym}\mathbf{T}_{\text{B}} : \dot{\mathbf{U}} dV \quad , \tag{4.110}$$

where the notation $\text{sym}(\bullet)$ is used to indicate the symmetric part of $\mathbf{T}_{\text{B}}$, and $\dot{\mathbf{U}}$ is the material time derivative of the right stretch tensor $\mathbf{U}$, which is symmetric (compare with Section 2.6).

Solution.   Starting from $(4.101)_1$ and using the polar decomposition $\mathbf{F} = \mathbf{RU}$, the product rule of differentiation and the identity $\mathbf{R}^{\text{T}}\mathbf{R} = \mathbf{I}$, we have

$$\int_{\Omega_0} \mathbf{P} : \dot{\mathbf{F}} dV = \int_{\Omega_0} \mathbf{P} : [\dot{\mathbf{R}}(\mathbf{R}^{\text{T}}\mathbf{R})\mathbf{U} + \mathbf{R}\dot{\mathbf{U}}] dV$$

$$= \int_{\Omega_0} \mathbf{P} : \dot{\mathbf{R}}\mathbf{R}^{\text{T}}\mathbf{F} dV + \int_{\Omega_0} \mathbf{P} : \mathbf{R}\dot{\mathbf{U}} dV \quad . \tag{4.111}$$

With the definition of the symmetric Kirchhoff stress tensor $\boldsymbol{\tau} = \mathbf{P}\mathbf{F}^{\text{T}}$ (see Section 3.4) and the Biot stress tensor $\mathbf{T}_{\text{B}} = \mathbf{R}^{\text{T}}\mathbf{P}$ (see eq. (3.68)), we find from eq. $(4.111)_2$ using manipulations according to (1.95) that

$$\int_{\Omega_0} \mathbf{P} : \dot{\mathbf{F}} dV = \int_{\Omega_0} \mathbf{P}\mathbf{F}^T : \dot{\mathbf{R}}\mathbf{R}^T dV + \int_{\Omega_0} \mathbf{R}^T\mathbf{P} : \dot{\mathbf{U}} dV$$

$$= \int_{\Omega_0} \boldsymbol{\tau} : \dot{\mathbf{R}}\mathbf{R}^T dV + \int_{\Omega_0} \mathbf{T}_B : \dot{\mathbf{U}} dV \quad . \tag{4.112}$$

Knowing that $\boldsymbol{\tau}$ is a symmetric tensor and $\dot{\mathbf{R}}\mathbf{R}^T$ is a skew tensor according to (2.160), we conclude by analogy with property (1.117) that $\boldsymbol{\tau} : \dot{\mathbf{R}}\mathbf{R}^T = 0$, consequently the first term in $(4.112)_2$ vanishes. Since $\dot{\mathbf{U}}$ is a symmetric tensor the second term gives the desired result $(4.110)_2$. ■

In summary, the alternative expressions for the rate of internal mechanical work $\mathcal{P}_{int}$ are the relations $(4.107)_1$, $(4.101)_1$, $(4.107)_4$, $(4.108)_3$, $(4.109)_3$ and $(4.110)_2$, and we have finally the important identities

$$w_{int}(t) = J\boldsymbol{\sigma} : \mathbf{d} = \mathbf{P} : \dot{\mathbf{F}} = \mathbf{S} : \dot{\mathbf{E}} = \boldsymbol{\Sigma} : \frac{1}{2}\mathbf{C}^{-1}\dot{\mathbf{C}}$$

$$= J\boldsymbol{\sigma}_u : \mathbf{D}_R = \text{sym}\mathbf{T}_B : \dot{\mathbf{U}} \quad . \tag{4.113}$$

The double contraction of a stress tensor and the associated rate of deformation tensor (for example, $\mathbf{P} : \dot{\mathbf{F}}$) describes the real physical power during a dynamical process, i.e. the rate of internal mechanical work (or stress power) per unit reference volume, denoted by w_{int}. In this sense the stress fields $J\boldsymbol{\sigma}, \mathbf{P}, \mathbf{S}, \boldsymbol{\Sigma}, J\boldsymbol{\sigma}_u, \text{sym}\mathbf{T}_B$ are said to be **work conjugate** to the strain fields $\mathbf{d}, \dot{\mathbf{F}}, \dot{\mathbf{E}}, \mathbf{C}^{-1}\dot{\mathbf{C}}/2, \mathbf{D}_R, \dot{\mathbf{U}}$, respectively. Hence, for example, $\mathbf{P}$ and $\dot{\mathbf{F}}$ is said to be a **work conjugate pair** (for a more comprehensive survey of work conjugate pairs see, for example, ATLURI [1984]).

Conservative system.   Let the scalar-valued functions Π_{ext} and Π_{int} be called the **potential energy of the external loading** (or the **external potential energy**) and the **total strain energy** (or the **internal potential energy**) of the body, respectively.

The sum of Π_{ext} and Π_{int} we call the **total potential energy** (or frequently referred to as the **energy functional**) Π of the mechanical system, i.e.

$$\Pi(t) = \Pi_{ext}(t) + \Pi_{int}(t) \quad . \tag{4.114}$$

A mechanical system is known as **conservative** if the external mechanical power $\mathcal{P}_{ext}$ expended on a certain region of the body and the internal mechanical power $\mathcal{P}_{int}$ are expressible as

$$\mathcal{P}_{ext}(t) = -\frac{D\Pi_{ext}(t)}{Dt} = -\dot{\Pi}_{ext}(t) \quad , \tag{4.115}$$

$$\mathcal{P}_{\text{int}}(t) = \frac{\text{D}\Pi_{\text{int}}(t)}{\text{D}t} = \dot{\Pi}_{\text{int}}(t) \qquad \text{with} \qquad \Pi_{\text{int}}(t) = \int_{\Omega_0} \Psi \text{d}V \ , \quad (4.116)$$

where Ψ represents the **strain-energy function** (referred to briefly as the **strain energy** or **stored energy**) defined subsequently per unit reference volume rather than per unit mass and dealt with in Section 6.1 in more detail. In the literature the strain-energy function is sometimes denoted by W.

By means of the assumptions (4.115) and (4.116), the balance of mechanical energy (4.85) implies that the sum of the total potential energy Π and the kinetic energy $\mathcal{K}$ is **conserved** (constant) during a dynamical process (σ, χ). We write

$$\Pi_{\text{ext}}(t) + \Pi_{\text{int}}(t) + \mathcal{K}(t) = \text{const} \ . \tag{4.117}$$

It is important to note, that, for example, most surface tractions and all problems associated with dissipation of energy (due to external or internal friction, viscous or plastic effects) lead to non-mechanical energies which are non-conservative in the sense that they cannot be derived from a potential.

EXERCISES

1. If only mechanical energy is considered, show that the energy equation is merely Cauchy's first equation of motion. Thus, take the dot product of each term of $\text{div}\boldsymbol{\sigma} + \mathbf{b} = \rho\dot{\mathbf{v}}$ with the spatial velocity field $\mathbf{v}$ and integrate the result over the current region Ω with volume v.

2. A hydrostatic stress state at a certain point is given by the Cauchy stress tensor in the form of $\boldsymbol{\sigma} = -p\mathbf{I}$.

   (a) Show that the stress power w_{int} per unit referential volume is given by

   $$w_{\text{int}} = J\boldsymbol{\sigma} : \mathbf{d} = p\frac{J\,\text{D}\rho}{\rho\,\text{D}t} \ .$$

   (b) Formulate the balance of mechanical energy for this particular stress state and conclude that if the considered motion is isochoric and if there exists no external mechanical power, the kinetic energy is conserved ($\mathcal{K} = \text{const}$).

3. A rigid body is rotating about a fixed point O with angular velocity $\boldsymbol{\omega}$. Show that the kinetic energy may be expressed as

   $$\mathcal{K}(t) = \frac{1}{2}\boldsymbol{\omega} \cdot \mathbf{D}\boldsymbol{\omega} \ , \qquad \mathbf{D} = \int_{\Omega} \rho[(\mathbf{x} \cdot \mathbf{x})\mathbf{I} - \mathbf{x} \otimes \mathbf{x}]\text{d}v \ ,$$

   where the current position of an arbitrary point in the rigid body is characterized by $\mathbf{x}$. Note that $\mathbf{D}$ is called the **inertia tensor** of the body relative to O occupying a region Ω.

4.5 Balance of Energy in Continuum Thermodynamics

In this section we consider both *mechanical* and *thermal* energy which are essential in many problems of physics and engineering. The effect of other energies (such as electric, chemical ...) on the behavior of a continuum is a topic beyond the scope of this text.

In a thermodynamic context the conservation of mass and the momentum balance principles are supplemented by the balance of energy and the entropy inequality law. The development of the concepts of energy and entropy has been one of the most important achievements in the evolution of physics. For an extensive account of these concepts see, for example, TRUESDELL and TOUPIN [1960, Chapter E], the two-volume work of KESTIN [1979], or ŠILHAVÝ [1997] and WILMAŃSKI [1998].

In this section we discuss briefly the balance of energy equation in a thermodynamic context. To begin with, we introduce some important terminology common in thermodynamics.

General notation. A continuum which possesses both mechanical and thermal energy is called a **thermodynamic continuum**. We say that the **thermodynamic state** or the **condition** of a system is known if all quantities throughout the entire system are known. All quantities characterizing a system at a certain state are called **thermodynamic state variables**. They are macroscopic quantities and, in general, they depend on *position* and *time*. For example, the thermodynamic state of a thermoelastic solid can be represented in a seven-dimensional state space with six variables corresponding to the strains and one to temperature. The function that describes a certain state variable is called a **thermodynamic state function**. Any equation that interrelates state variables is called **equation of state** or **constitutive equation**.

Any change that a system undergoes from one thermodynamic state to another is called a **thermodynamic process**. The path that connects the two states is parameterized by time t. If a system returns to its initial state at the end of a thermodynamic process this system is said to have undergone a so-called **cycle** (initial and final states are identical).

A **non-equilibrium state** is a state of imbalance (there exists a gradient of temperature and velocity), while an **equilibrium state** is a state of balance (of uniform temperature and zero velocity). To study non-equilibrium or equilibrium states is a central goal of continuum thermodynamics.

A system within equilibrium has no tendency to change when it is isolated from its surroundings. If there is no change in the values of the state variables at any particle of the system with time we say that this system is in **thermodynamic equilibrium** or **thermal equilibrium**. The **transition** from one state of thermodynamic equilibrium to another is studied in **thermostatics**. A process in a system that remains close to a state of thermodynamic equilibrium at each time is referred to as a **quasi-static process** or **quasi-equilibrium process**. A quasi-static process is a sufficiently slow process whereby enough time remains for the system to adjust itself internally. It is basically a process during which the system is in equilibrium at all times; the contributions due to dynamical quantities are negligible.

Heat is the form of (thermal) energy that is transferred between a system and its surroundings (or between two systems) by virtue of a temperature gradient. The term *'heat'* is understood to mean **heat transfer** in thermodynamics.

Thermal power.   Let the **thermal (non-mechanical) power** or the **rate of thermal work** be denoted by $\mathcal{Q}$ and defined by

$$\mathcal{Q}(t) = \int_{\partial\Omega} q_n \mathrm{d}s + \int_{\Omega} r \mathrm{d}v = \int_{\partial\Omega_0} Q_N \mathrm{d}S + \int_{\Omega_0} R \mathrm{d}V \quad , \tag{4.118}$$

which is represented in the spatial and material descriptions, respectively.

The time-dependent scalar functions q_n and Q_N denote **heat fluxes**, determining heat per unit time and per unit current and reference surface area, respectively. The total heat fluxes $\int_{\partial\Omega} q_n \mathrm{d}s$ and $\int_{\partial\Omega_0} Q_N \mathrm{d}S$ measure the rate at which heat *enters* (inward normal flux) the body across the current and the reference boundary surfaces $\partial\Omega$ and $\partial\Omega_0$, respectively.

The time-dependent scalar fields $r = r(\mathbf{x}, t)$ and $R = R(\mathbf{X}, t)$ in eq. (4.118) denote **heat sources** per unit time and per unit current and reference volume, respectively (see Figure 4.3). A heat source is a reservoir that supplies energy in the form of heat. The total heat sources $\int_{\Omega} r \mathrm{d}v$ and $\int_{\Omega_0} R \mathrm{d}V$ measure the rate at which heat is generated (or destroyed) into a certain region of a body.

The counterpart of Cauchy's stress theorem (3.3) in continuum mechanics is the **Stokes' heat flux theorem** in thermodynamics. It postulates that the scalar functions q_n and Q_N are linear functions of the outward unit normals so that

$$\left.\begin{array}{llll} q_n(\mathbf{x}, t, \mathbf{n}) = -\mathbf{q}(\mathbf{x}, t) \cdot \mathbf{n} & \quad \text{or} & \quad q_n = -q_a n_a & , \\[2mm] Q_N(\mathbf{X}, t, \mathbf{N}) = -\mathbf{Q}(\mathbf{X}, t) \cdot \mathbf{N} & \quad \text{or} & \quad Q_N = -Q_A N_A & . \end{array}\right\} \tag{4.119}$$

The time-dependent spatial vector field $\mathbf{q} = \mathbf{q}(\mathbf{x}, t)$ is the so-called **Cauchy heat flux** (or **true heat flux**) defined per unit surface area in Ω, and $\mathbf{n}$ is the *outward* unit normal to an infinitesimal spatial surface element $\mathrm{d}s \in \partial\Omega$ at the current position $\mathbf{x}$. The **Piola-Kirchhoff heat flux** (or **nominal heat flux**) and the *outward* unit normal to an infinitesimal material surface element $\mathrm{d}S \in \partial\Omega_0$ at $\mathbf{X}$ are denoted by the vectors $\mathbf{Q} = \mathbf{Q}(\mathbf{X}, t)$ and $\mathbf{N}$, respectively. The time-dependent material vector field $\mathbf{Q}$ determines the heat flux per unit surface area in Ω_0.

The negative signs in eqs. (4.119) are needed (see Figure 4.3) because the unit vectors $\mathbf{n}$ and $\mathbf{N}$ are outward normals to $\partial\Omega$ and $\partial\Omega_0$, respectively. However, we claim in (4.118) that heat enters the body (inward normal flux).

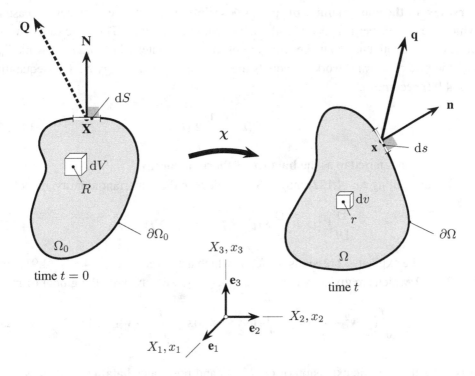

Figure 4.3 Heat flux vectors $\mathbf{q}$, $\mathbf{Q}$ and heat sources r, R.

In order to relate the contravariant vectors $\mathbf{q}$ and $\mathbf{Q}$ to one another, we equate the total heat fluxes $\int_{\partial\Omega_0} Q_N dS$ and $\int_{\partial\Omega} q_n ds$. Hence, with the fundamental Stokes' heat flux theorem (4.119), Nanson's formula (2.55) and manipulations according to (1.95) we obtain

$$\int_{\partial\Omega_0} \mathbf{Q}\cdot\mathbf{N} dS = \int_{\partial\Omega} \mathbf{q}\cdot\mathbf{n} ds = \int_{\partial\Omega_0} \mathbf{q}\cdot J\mathbf{F}^{-T}\mathbf{N} dS$$

$$= \int_{\partial\Omega_0} J\mathbf{F}^{-1}\mathbf{q}\cdot\mathbf{N} dS \ . \tag{4.120}$$

Comparing the left and right-hand sides of (4.120) we conclude that the Piola-Kirchhoff heat flux $\mathbf{Q}$ may be related to the Cauchy heat flux $\mathbf{q}$. We have, by analogy with (3.8),

$$\mathbf{Q} = J\mathbf{F}^{-1}\mathbf{q} = J\boldsymbol{\chi}_*^{-1}(\mathbf{q}^\sharp) \qquad \text{or} \qquad Q_A = JF_{Aa}^{-1}q_a \ , \tag{4.121}$$

which may be viewed as the pull-back operation on the contravariant vector $\mathbf{q}^\sharp$ by the inverse motion $\boldsymbol{\chi}^{-1}$ scaled by the volume ratio J. Such a transformation we called the *Piola transformation*, here relating the heat flux vectors $\mathbf{Q}$ and $\mathbf{q}$ (compare with the definition on p. 84).

First law of thermodynamics in spatial description. Here we consider the case in which thermal power is added to a thermodynamic continuum. Thus, the rate of work done on the continuum body, i.e. the sum of the rate of internal *mechanical* work $\mathcal{P}_{\mathrm{int}}$ and the rate of *thermal* work $\mathcal{Q}$, equals the rate of internal energy $\mathcal{E}$. Consequently, eq. (4.92) becomes

$$\mathcal{P}_{\mathrm{int}}(t) + \mathcal{Q}(t) = \frac{D}{Dt}\mathcal{E}(t) \quad, \tag{4.122}$$

which is often referred to as the **balance of thermal energy**.

By substituting eq. (4.122) into (4.85) we deduce the important identity, namely

$$\frac{D}{Dt}\mathcal{K}(t) + \frac{D}{Dt}\mathcal{E}(t) = \mathcal{P}_{\mathrm{ext}}(t) + \mathcal{Q}(t) \quad. \tag{4.123}$$

Using the explicit expressions for $\mathcal{K}$, $\mathcal{E}$, which are given by eqs. (4.83), (4.91), and for $\mathcal{P}_{\mathrm{ext}}$, $\mathcal{Q}$, which are given by eqs. (4.82), (4.118)$_1$, we may write the global form

$$\frac{D}{Dt}\int_{\Omega}(\frac{1}{2}\rho\mathbf{v}^2 + e_c)dv = \int_{\partial\Omega}(\mathbf{t}\cdot\mathbf{v} + q_{\mathrm{n}})ds + \int_{\Omega}(\mathbf{b}\cdot\mathbf{v} + r)dv \quad. \tag{4.124}$$

It is a thermodynamic extension of eq. (4.94) and postulates **balance of energy** (mechanical and thermal), a fundamental axiom in mechanics, known as the **first law of thermodynamics**. In particular, eq. (4.124) is the first law of thermodynamics in the spatial description.

It states that the rate of change of total energy (kinetic $\mathcal{K}$ and internal energy $\mathcal{E}$) of a thermodynamic system equals the rate at which external mechanical work (external mechanical power) $\mathcal{P}_{\mathrm{ext}}$ is done on that system by surface tractions and body forces plus the rate at which thermal work $\mathcal{Q}$ is done by heat fluxes and heat sources. The first law of thermodynamics governs the transformation from one type of energy involved in a thermodynamic process into another, but it *never* governs the direction of that energy transfer.

In order to express eq. (4.122) more explicitly we rewrite the thermal power $\mathcal{Q}$ first, i.e. (4.118)$_1$. By means of the Stokes' heat flux theorem (4.119)$_1$ and the divergence theorem according to eq. (1.295), we may deduce that $\mathcal{Q}(t) = \int_{\Omega}(-\mathrm{div}\mathbf{q} + r)dv$. By recalling the rate of internal mechanical work $\mathcal{P}_{\mathrm{int}}$ and the internal energy $\mathcal{E}$ from eqs. (4.84)$_1$ and (4.91), we obtain

$$\frac{D}{Dt}\int_{\Omega} e_c dv = \int_{\Omega}(\boldsymbol{\sigma}:\mathbf{d} - \mathrm{div}\mathbf{q} + r)dv \quad, \tag{4.125}$$

which is a reduced global form of balance of energy (4.124) in the spatial description.

First law of thermodynamics in material description.   In order to rewrite balance of energy (4.124) in terms of material coordinates, we recall the equivalent forms of the total energy, the external mechanical power $\mathcal{P}_{\text{ext}}$ and the thermal power $\mathcal{Q}$, that are given by eqs. (4.95), (4.105), and (4.82), (4.98) and (4.118), respectively.

Thus, the first law of thermodynamics in the material description reads

$$\frac{\text{D}}{\text{Dt}} \int_{\Omega_0} (\frac{1}{2}\rho_0 \mathbf{V}^2 + e)\text{d}V = \int_{\partial\Omega_0} (\mathbf{T} \cdot \mathbf{V} + Q_N)\text{d}S + \int_{\Omega_0} (\mathbf{B} \cdot \mathbf{V} + R)\text{d}V \quad . \quad (4.126)$$

Following arguments analogous to those which led to (4.125), we find the reduced global form of balance of energy in the material description, i.e.

$$\frac{\text{D}}{\text{Dt}} \int_{\Omega_0} e\,\text{d}V = \int_{\Omega_0} (\mathbf{P} : \dot{\mathbf{F}} - \text{Div}\mathbf{Q} + R)\text{d}V \quad . \quad (4.127)$$

In order to achieve the local form we must rewrite the term on the left-hand side of eq. (4.127). Since the reference volume V is independent of time we may write $\text{D}/\text{Dt} \int_{\Omega_0} e\,\text{d}V = \int_{\Omega_0} \dot{e}\,\text{d}V$. Note that the volume is arbitrary, leading to the local form of the balance of energy in the material description

$$\dot{e} = \mathbf{P} : \dot{\mathbf{F}} - \text{Div}\mathbf{Q} + R \qquad \text{or} \qquad \dot{e} = P_{aA}\dot{F}_{aA} - \frac{\partial Q_A}{\partial X_A} + R \quad , \quad (4.128)$$

presented in symbolic and index notation, respectively.

The first-order partial differential equation is due to *Kirchhoff* and holds at any particle of the body for all times.

EXERCISES

1. Assume a continuum body not subjected to body forces and heat sources. Establish the balance equations

   $$\frac{\text{D}}{\text{Dt}} \int_{\Omega} (\frac{1}{2}\rho\mathbf{v}^2 + e_c)\text{d}v = \int_{\partial\Omega} (\boldsymbol{\sigma}\mathbf{v} - \mathbf{q}) \cdot \mathbf{n}\text{d}s \quad ,$$

   $$\frac{\text{D}}{\text{Dt}} \int_{\Omega_0} (\frac{1}{2}\rho_0\mathbf{V}^2 + e)\text{d}V = \int_{\partial\Omega_0} (\mathbf{P}^{\text{T}}\mathbf{V} - \mathbf{Q}) \cdot \mathbf{N}\text{d}S \quad .$$

   These relations equate the rate of change of the total energy of a continuum body occupying a certain region with the total flux out of the boundary surfaces of that region. The vectors $\boldsymbol{\sigma}\mathbf{v} - \mathbf{q}$ and $\mathbf{P}^{\text{T}}\mathbf{V} - \mathbf{Q}$ are often referred to as the **energy flux vectors** with respect to the current and reference configurations, respectively.

2. Starting from (4.124), show that the local spatial form of the balance equation has the form

$$\frac{\partial}{\partial t}\left(\frac{1}{2}\rho\mathbf{v}^2 + e_c\right) = \operatorname{div}[\boldsymbol{\sigma}\mathbf{v} - \mathbf{q} - (\frac{1}{2}\rho\mathbf{v}^2 + e_c)\mathbf{v}] + \mathbf{b}\cdot\mathbf{v} + r \quad .$$

Hint: Use eqs. (2.51), (2.25), $(2.177)_6$ and the rule according to (1.288) for the left-hand side of (4.124).

3. Derive the local form of the balance equation in the spatial description corresponding to eq. (4.128) and note that we have used the internal energy per unit *volume* rather than per unit *mass* (see eq. (4.91)).

4.6 Entropy Inequality Principle

The first law of thermodynamics governs the energy transfer within a thermodynamic process, but is insensitive to the direction of the energy transfer. In the following we derive a much finer principle, that is the second law of thermodynamics, which is responsible for the direction of an energy transfer process. We introduce the concept of entropy and discuss some of the consequences of the second law of thermodynamics.

Second law of thermodynamics.   Physical observations show that heat always flows from the warmer to the colder region of a body (free from sources of heat), *not vice versa*; mechanical energy can be transformed into heat by friction, and this can never be converted back into mechanical energy.

We introduce a fundamental state variable, the **entropy** (coming from the Greek words ἐν and τρόπος meaning 'in' and 'turning, direction', respectively). It is an important thermodynamic property first described in the works of *Clausius* in the second half of the nineteenth century. The entropy can be viewed as the quantitative measure of microscopic randomness and disorder (see, for example, CALLEN [1985, Chapter 17]). A physical interpretation is provided by the subject of *statistical mechanics*. For a more detailed physical interpretation the reader is referred to Section 7.1 of this text. We introduce the notations $\eta_c = \eta_c(\mathbf{x}, t)$ and $\eta = \eta(\mathbf{X}, t)$ for the entropy per unit current and reference volume, respectively, at a certain point and time t. In other texts the entropy is frequently designated by s or S.

The entropy possessed by a continuum body occupying a certain region, denoted by $\mathcal{S}$, is defined to be

$$\mathcal{S}(t) = \int_{\Omega} \eta_c(\mathbf{x}, t)\mathrm{d}v = \int_{\Omega_0} \eta(\mathbf{X}, t)\mathrm{d}V \quad , \tag{4.129}$$

with $\eta(\mathbf{X}, t) = J(\mathbf{X}, t)\eta_c(\mathbf{x}, t)$, which should be compared with the analogues of (4.103) and (4.104), respectively. This definition is commonly introduced in continuum mechanics and differs from the definitions used in statistical mechanics.

Let the **rate of entropy input** into a certain region of a continuum body consist of the value of entropy transferred across its boundary surface and the entropy generated (or destroyed) inside that region. We denote it by $\tilde{\mathcal{Q}}$ and write

$$\tilde{\mathcal{Q}}(t) = -\int_{\partial\Omega} \mathbf{h} \cdot \mathbf{n} ds + \int_{\Omega} \tilde{r} dv = -\int_{\partial\Omega_0} \mathbf{H} \cdot \mathbf{N} dS + \int_{\Omega_0} \tilde{R} dV . \qquad (4.130)$$

The time-dependent scalar fields $\tilde{r} = \tilde{r}(\mathbf{x}, t)$ and $\tilde{R} = \tilde{R}(\mathbf{X}, t)$ denote **entropy sources** per unit time and per unit current and reference volume, respectively. The total entropy sources $\int_{\Omega} \tilde{r} dv$ and $\int_{\Omega_0} \tilde{R} dV$ measure the rate at which entropy is generated (or destroyed) into a region of a body. The time-dependent vector fields $\mathbf{h} = \mathbf{h}(\mathbf{x}, t)$ and $\mathbf{H} = \mathbf{H}(\mathbf{X}, t)$ determine the **Cauchy entropy flux** (or **true entropy flux**) defined per unit current surface area in Ω and the **Piola-Kirchhoff entropy flux** (or **nominal entropy flux**) per unit reference surface area in Ω_0, respectively.

As usual, the outward unit normals to the infinitesimal surface elements $ds \in \partial\Omega$ at $\mathbf{x}$ and $dS \in \partial\Omega_0$ at $\mathbf{X}$ are denoted by $\mathbf{n}$ and $\mathbf{N}$, respectively. Since we have defined the unit vectors $\mathbf{n}$ and $\mathbf{N}$ to be *outward* normals to $\partial\Omega$ and $\partial\Omega_0$ and since we compute the rate of entropy *input* (entropy entering the body), we need the negative signs in relation (4.130).

The difference between the rate of change of entropy $\dot{S}$ and the rate of entropy input $\tilde{\mathcal{Q}}$ into a body determines the **total production of entropy** per unit time, which we denote by Γ. We postulate that the total entropy production for all thermodynamic processes is never negative, following the mathematical expression

$$\Gamma(t) = \frac{D}{Dt}\mathcal{S}(t) - \tilde{\mathcal{Q}}(t) \geq 0 , \qquad (4.131)$$

which is known as the **second law of thermodynamics**. Unlike the considerations in previous sections, it is an inequality rather than an equation often referred to as the **entropy inequality principle**.

More explicitly, with eqs. (4.129)$_1$ and (4.130)$_1$ we find from (4.131) the global spatial form of the second law of thermodynamics in the notation of continuum mechanics, i.e.

$$\Gamma(t) = \frac{D}{Dt}\int_{\Omega} \eta_c(\mathbf{x}, t) dv + \int_{\partial\Omega} \mathbf{h} \cdot \mathbf{n} ds - \int_{\Omega} \tilde{r} dv \geq 0 . \qquad (4.132)$$

These relations assert clearly a *trend in time* by describing the direction of the energy transfer and postulating **irreversibility** of various thermodynamic processes.

The second law of thermodynamics is *not* a balance principle. It indicates a trend in both living and inanimate systems, where the situation $\Gamma < 0$ never occurs (it would mean that molecules organize themselves globally). Unlike the mass or the energy, in general, the entropy is *not* a conserved quantity, i.e. $\Gamma \geq 0$.

A thermodynamic process is called **reversible** if it is not accompanied by any entropy production, i.e. $\Gamma = 0$. For each cycle the material response returns to its initial state. A reversible process is a very useful idealized limit of a real process. The associated state of a reversible process is in equilibrium; thus, the equal signs hold in (4.131) and (4.132). Reversible processes belong to the realm of **equilibrium thermodynamics**, also known as **reversible thermodynamics** which is (classical) thermostatics.

A real process is **irreversible** and characterized by (4.131) and (4.132) in which the strict inequalities hold. This indicates that the rate of change of entropy $\dot{S}$ is always greater than the rate of entropy input $\tilde{Q}$. Irreversible processes are always associated with dissipation of energy studied in the realm of **non-equilibrium thermodynamics**, also known as **irreversible thermodynamics**. For the concept of different thermodynamic theories the reader is referred to the review article by HUTTER [1977].

Clausius-Duhem inequality.   The rate of entropy input is often closely related to the rate of thermal work (the total heat fluxes and sources). Very often the entropy fluxes $\mathbf{h}, \mathbf{H}$ and the entropy sources $\tilde{r}, \tilde{R}$ are assumed to be related to the heat fluxes $\mathbf{q}, \mathbf{Q}$ and heat sources r, R by the proportional factor $1/\Theta$. Thereby $\Theta = \Theta(\mathbf{x}, t) > 0$ denotes a time-dependent scalar field known as the **absolute temperature**. The corresponding unit of temperature is called a **Kelvin**, denoted by K, which always has positive values.

However, in practice Θ is often measured in *'Celsius temperature'* or *'Fahrenheit temperature'* for which the unit is called the **degree Celsius** $°C$ or the **degree Fahrenheit** $°F$, respectively. It is important to note that from the thermodynamic point of view the Celsius scale and the Fahrenheit scale are not true temperature scales at all. The temperatures on these scales can be negative, the zero is incorrect and temperature ratios are inconsistent with those required by the thermodynamic principles.

We postulate

$$\mathbf{h} = \frac{\mathbf{q}}{\Theta} \; , \qquad \tilde{r} = \frac{r}{\Theta} \qquad \text{and} \qquad \mathbf{H} = \frac{\mathbf{Q}}{\Theta} \; , \qquad \tilde{R} = \frac{R}{\Theta} \; , \qquad (4.133)$$

but note, however, that these relations are, for example, inconsistent with the consequences of the kinetic theory of ideal gases and therefore not valid for the thermodynamics of diffusion.

With relations (4.133) and eqs. (4.129) and (4.130) we find from the second law of thermodynamics in the form of (4.132) that

$$\Gamma(t) = \frac{\mathrm{D}}{\mathrm{D}t} \int\limits_{\Omega} \eta_c \mathrm{d}v + \int\limits_{\partial\Omega} \frac{\mathbf{q}}{\Theta} \cdot \mathbf{n} \mathrm{d}s - \int\limits_{\Omega} \frac{r}{\Theta} \mathrm{d}v \geq 0 \; , \qquad (4.134)$$

$$\Gamma(t) = \frac{D}{Dt} \int_{\Omega_0} \eta dV + \int_{\partial\Omega_0} \frac{\mathbf{Q}}{\Theta} \cdot \mathbf{N} dS - \int_{\Omega_0} \frac{R}{\Theta} dV \geq 0 \ , \qquad (4.135)$$

which is known as the **Clausius-Duhem inequality** here presented in the spatial and material descriptions, respectively. The Clausius-Duhem inequality is widely used in modern thermodynamic research initiated by TRUESDELL and TOUPIN [1960] and first clearly studied in the work of COLEMAN and NOLL [1963].

In order to derive the local form, which we only present in the material description, we convert the surface integral in eq. (4.135) to a volume integral according to the divergence theorem (1.299). By the product rule (1.288) we obtain

$$\int_{\partial\Omega_0} \frac{\mathbf{Q}}{\Theta} \cdot \mathbf{N} dS = \int_{\Omega_0} \mathrm{Div}\left(\frac{\mathbf{Q}}{\Theta}\right) dV = \int_{\Omega_0} \left(\frac{1}{\Theta}\mathrm{Div}\mathbf{Q} - \frac{1}{\Theta^2}\mathbf{Q}\cdot\mathrm{Grad}\Theta\right) dV \ , \quad (4.136)$$

where $\mathrm{Grad}\Theta$ denotes the material gradient of the smooth temperature field Θ.

By substituting eq. $(4.136)_2$ back into (4.135) and noting that the reference volume V is arbitrary and independent of time, the local form of the Clausius-Duhem inequality in the material description reads, in symbolic and index notation,

$$\dot{\eta} - \frac{R}{\Theta} + \frac{1}{\Theta}\mathrm{Div}\mathbf{Q} - \frac{1}{\Theta^2}\mathbf{Q}\cdot\mathrm{Grad}\Theta \geq 0 \ ,$$

$$(4.137)$$

$$\text{or} \qquad \dot{\eta} - \frac{R}{\Theta} + \frac{1}{\Theta}\frac{\partial Q_A}{\partial X_A} - \frac{1}{\Theta^2}Q_A\frac{\partial\Theta}{\partial X_A} \geq 0 \ .$$

An alternative local version results, for example, from (4.137) by elimination of the heat source R by means of eq. (4.128), leading to

$$\mathbf{P}:\dot{\mathbf{F}} - \dot{e} + \Theta\dot{\eta} - \frac{1}{\Theta}\mathbf{Q}\cdot\mathrm{Grad}\Theta \geq 0 \ ,$$

$$(4.138)$$

$$\text{or} \qquad P_{aA}\dot{F}_{aA} - \dot{e} + \Theta\dot{\eta} - \frac{1}{\Theta}Q_A\frac{\partial\Theta}{\partial X_A} \geq 0 \ .$$

The last term in eq. (4.138) determines **entropy production by conduction of heat**.

Clausius-Planck inequality and heat conduction.   Based on physical observations heat flows from the warmer to the colder region of a body (free from sources of heat), not vice versa. Hence, entropy production by conduction of heat must be *non-negative*, i.e. $-(1/\Theta)\mathbf{Q}\cdot\mathrm{Grad}\Theta \geq 0$ (see the last term in (4.138)). The spatial version of this condition reads $-(1/\Theta)\mathbf{q}\cdot\mathrm{grad}\Theta \geq 0$, where $\mathrm{grad}\Theta$ denotes the spatial gradient of the smooth temperature field Θ.

Knowing that Θ is non-negative, we have

$$\mathbf{q}\cdot\mathrm{grad}\Theta \leq 0 \qquad \text{or} \qquad q_a\frac{\partial\Theta}{\partial x_a} \leq 0 \ , \qquad (4.139)$$

$$\mathbf{Q}\cdot\mathrm{Grad}\Theta \leq 0 \qquad \text{or} \qquad Q_A\frac{\partial\Theta}{\partial X_A} \leq 0 \ , \qquad (4.140)$$

which is known as the classical **heat conduction inequality**, here presented in terms of spatial and material coordinates, respectively. The heat conduction inequality expresses that heat does flow against a temperature gradient. It imposes an essential restriction on the heat flux vector.

Obviously the Cauchy heat flux $\mathbf{q}$ depends on the spatial temperature gradient. With restriction (4.139) we may deduce that $\mathbf{q}(\mathrm{grad}\Theta) \cdot \mathrm{grad}\Theta \leq 0$ holds identically if $\mathrm{grad}\Theta = \mathbf{o}$. Therefore, the dot product takes on a local maximum at $\mathrm{grad}\Theta = \mathbf{o}$ and its derivative must vanish. Consequently, $\mathbf{q} = \mathbf{o}$ is implied, which means that there is no heat flux without a temperature gradient.

According to restriction (4.140), the Clausius-Duhem inequality (4.138) leads to an alternative *stronger form* of the second law of thermodynamics, often referred to as the **Clausius-Planck inequality**, i.e.

$$\mathcal{D}_{\mathrm{int}} = \mathbf{P} : \dot{\mathbf{F}} - \dot{e} + \Theta\dot{\eta} \geq 0 \qquad \text{or} \qquad \mathcal{D}_{\mathrm{int}} = P_{aA}\dot{F}_{aA} - \dot{e} + \Theta\dot{\eta} \geq 0 \ , \quad (4.141)$$

with the **internal dissipation** or **local production of entropy** $\mathcal{D}_{\mathrm{int}} \geq 0$, which is required to be non-negative at any particle of a body for all times.

The non-negative internal dissipation $\mathcal{D}_{\mathrm{int}}$ consists of three terms: the work conjugate pair $\mathbf{P} : \dot{\mathbf{F}}$, i.e. the rate of internal mechanical work (or stress-power) per unit reference volume, then the rate of internal energy, $\dot{e}$, and the absolute temperature multiplied by the rate of entropy, $\Theta\dot{\eta}$. The internal dissipation is zero for reversible processes while the inequality holds for irreversible processes.

By means of the Clausius-Planck inequality (4.141), the local form of the balance of energy (4.128) can be rephrased in the convenient **entropy form**

$$\Theta\dot{\eta} = -\mathrm{Div}\mathbf{Q} + \mathcal{D}_{\mathrm{int}} + R \qquad \text{or} \qquad \Theta\dot{\eta} = -\frac{\partial Q_A}{\partial X_A} + \mathcal{D}_{\mathrm{int}} + R \ , \quad (4.142)$$

in which the local evolution (change) of the entropy η appears explicitly.

A suitable constitutive assertion which relates the Cauchy heat flux $\mathbf{q} = \mathbf{q}(\mathbf{x}, t)$ in $\mathbf{x} = \chi(\mathbf{X}, t)$ to the spatial temperature gradient $\mathrm{grad}\Theta = \mathrm{grad}\Theta(\mathbf{x}, t)$ is furnished by

$$\mathbf{q} = -\kappa\,\mathrm{grad}\Theta \qquad \text{or} \qquad q_a = -\kappa_{ab}\frac{\partial\Theta}{\partial x_b} \ . \quad (4.143)$$

This classical (phenomenological) law is motivated by experimental observations and is known as **Duhamel's law of heat conduction**, here presented in the spatial description.

The symmetric second-order tensor κ denotes the **spatial thermal conductivity tensor**; its components $\kappa_{ab} = \kappa_{ba}$ are either constants or functions of deformation and temperature. Substituting Duhamel's law of heat conduction (4.143) into restriction (4.139) we obtain $\kappa\,\mathrm{grad}\Theta \cdot \mathrm{grad}\Theta \geq 0$. This implies that κ is a positive semi-definite tensor at each $\mathbf{x}$.

EXAMPLE 4.6   Derive Duhamel's law of heat conduction in terms of material co-ordinates,

$$\mathbf{Q} = -\mathbf{F}^{-1}\boldsymbol{\kappa}_0\mathbf{F}^{-T}\text{Grad}\Theta \qquad \text{or} \qquad Q_A = -F_{AB}^{-1}\kappa_{0\,BC}F_{DC}^{-1}\frac{\partial\Theta}{\partial X_D} \quad , \qquad (4.144)$$

where $\mathbf{F}^{-1}$ is the inverse of the deformation gradient $\mathbf{F}$ according to eq. (2.40) and $\boldsymbol{\kappa}_0 = J\boldsymbol{\kappa}$ denotes the positive semi-definite **material thermal conductivity tensor**. The volume ratio is $J = \det\mathbf{F} > 0$.

Solution.   In order to show eq. (4.144) we need an expression between the spatial and the material temperature gradients. By using the absolute temperature Θ instead of Φ in relation (2.47) we find that

$$\text{grad}\Theta = \mathbf{F}^{-T}\text{Grad}\Theta \quad . \qquad (4.145)$$

Subsequently, by recalling the Piola transformation (4.121) for the heat flux vector we obtain from eq. (4.143)

$$J^{-1}\mathbf{F}\mathbf{Q} - -\boldsymbol{\kappa}\mathbf{F}^{-T}\text{Grad}\Theta \quad , \qquad (4.146)$$

which after recasting gives the desired result (4.144). Since $\boldsymbol{\kappa}_0$ is a positive semi-definite tensor the imposed restriction on the inequality (4.140) is satisfied.   ∎

If $\boldsymbol{\kappa}$ is an isotropic tensor we say that the material is **thermally isotropic** (no pre-ferred direction for the heat conduction). For such a type of material the conductivity tensors become $\boldsymbol{\kappa} = k\mathbf{I}$ and $\boldsymbol{\kappa}_0 = k_0\mathbf{I}$. These relations substituted into (4.143) and (4.144) give

$$\mathbf{q} = -k\,\text{grad}\Theta \qquad \text{or} \qquad q_a = -k\frac{\partial\Theta}{\partial x_a} \quad , \qquad (4.147)$$

$$\mathbf{Q} = -k_0\mathbf{C}^{-1}\text{Grad}\Theta \qquad \text{or} \qquad Q_A = -k_0 C_{AB}^{-1}\frac{\partial\Theta}{\partial X_B} \quad , \qquad (4.148)$$

where $\mathbf{C}^{-1} = \mathbf{F}^{-1}\mathbf{F}^{-T}$ characterizes the inverse of the right Cauchy-Green tensor $\mathbf{C}$.

The scalars $k \geq 0$ and $k_0 = Jk \geq 0$ denote **coefficients of thermal conductivity** (constants or, in general, deformation and temperature-dependent) and are naturally associated with the current and reference configurations of a body, respectively. The conditions $k \geq 0$ and $k_0 \geq 0$ imply that heat is conducted in the direction of decreasing temperature.

The relations (4.147) and (4.148) are well-known as **Fourier's law of heat con-duction** which are basically constitutive relations.

Types of thermodynamic processes.   Finally, we formulate special cases of thermodynamic processes and introduce some common terminology.

If the heat flux $q_n = -\mathbf{q}\cdot\mathbf{n}$ across the surface $\partial\Omega$ of a certain region Ω ($Q_N = -\mathbf{Q}\cdot\mathbf{N}$ on $\partial\Omega_0$), and the heat source r in that region Ω (R in Ω_0) vanish for all points of a body at each time, then a thermodynamic process is said to be **adiabatic** (coming from the Greek word $\alpha\delta\iota\acute{\alpha}\beta\alpha\tau o\varsigma$, which means impassable). This definition is based on the work of TRUESDELL and TOUPIN [1960, Section 258]. However, in engineering thermodynamics an adiabatic process is frequently defined as a process which cannot involve any heat transfer, $q_n = 0$ (or $Q_N = 0$). This means that the heat source r (or R) need not be zero.

In our case an adiabatic process is based on the condition that thermal energy can neither cross (enter or leave) the boundary surface nor be generated or destroyed within the body, so that the thermal power $\mathcal{Q}$ is zero (compare with eq. (4.118)). Additionally, under the assumption of relations (4.133) the rate of entropy input $\tilde{\mathcal{Q}}$ is also zero (compare with eq. (4.130)). Consequently, the second law of thermodynamics in the form of (4.131) reads $\Gamma = \dot{\mathcal{S}} \geq 0$, which means that the total entropy $\mathcal{S}$ cannot decrease in an adiabatic process. Nevertheless, the entropy η at all points and times is not necessarily non-decreasing.

For an adiabatic process the energy balance equation, for example, in the entropy form (4.142), degenerates to

$$\Theta\dot{\eta} = \mathcal{D}_{\text{int}} \quad . \tag{4.149}$$

If, in addition, an adiabatic process is *reversible* (no entropy is produced ($\Gamma = 0$), entailing that $\dot{\mathcal{S}}$ is zero), the balance equation (4.149) reduces further to

$$\Theta\dot{\eta} = \mathcal{D}_{\text{int}} = 0 \quad , \tag{4.150}$$

which has some important applications in thermodynamics.

If the absolute temperature Θ during a thermodynamic process remains constant ($\dot{\Theta} = 0$), the process is said to be **isothermal** and if the entropy $\mathcal{S}$ possessed by a body remains constant ($\dot{\mathcal{S}} = 0$), the process is said to be **isentropic**.

For an isentropic process the second law of thermodynamics (4.131) takes on the form $\Gamma = -\tilde{\mathcal{Q}} \geq 0$. Hence, the local form of the Clausius-Duhem inequality in the material description (4.137) in symbolic and index notation degenerates to

$$-R + \text{Div}\mathbf{Q} - \frac{1}{\Theta}\mathbf{Q}\cdot\text{Grad}\Theta \geq 0 \quad ,$$
$$\text{or} \quad -R + \frac{\partial Q_A}{\partial X_A} - \frac{1}{\Theta}Q_A\frac{\partial\Theta}{\partial X_A} \geq 0 \quad . \tag{4.151}$$

If an isentropic process is *reversible*, the rate of entropy input $\tilde{\mathcal{Q}}$ (and the thermal power $\mathcal{Q}$ via assumption (4.133)) is zero since there is no entropy production Γ anymore. For this special case the equal signs in (4.151) hold. Consequently, it is important to note that *adiabatic processes* and *isentropic processes* are identical for the case in which both are *reversible*.

In discussing deformations of elastic materials (see Chapter 6) it is convenient to work with the strain-energy function Ψ introduced in eq. $(4.116)_2$. However, for the case in which Ψ is used within the thermodynamic regime incorporating thermal variables such as Θ or η, then Ψ is commonly referred to as the **Helmholtz free-energy function** or referred to briefly as the **free energy**.

Next, we express the Helmholtz free-energy function in terms of the internal energy e and the entropy η. Having in mind this aim, we apply the **Legendre transformation**, generally defining a procedure which replaces one or some variables with the conjugate variables, particularly used in analytical mechanics (see, for example, ABRAHAM and MARSDEN [1978]). We may write

$$\Psi = e - \Theta\eta \ . \tag{4.152}$$

Note that all three quantities Ψ, e, η are introduced so that they refer to a unit volume of the reference configuration.

By using the material time derivative of the free energy, i.e. $\dot{\Psi} = D\Psi/Dt$, we may write the Clausius-Planck inequality (4.141) in the convenient form

$$\mathcal{D}_{\text{int}} = \mathbf{P} : \dot{\mathbf{F}} - \dot{\Psi} - \eta\dot{\Theta} \geq 0 \ , \tag{4.153}$$

often employed when the absolute temperature is used as an independent variable.

Note that for the case of a purely mechanical theory, that is, if thermal effects are ignored (Θ and η are omitted), inequality (4.153) degenerates to

$$\mathcal{D}_{\text{int}} = w_{\text{int}} - \dot{\Psi} \geq 0 \ , \tag{4.154}$$

and the free energy Ψ coincides with the internal energy e (see the Legendre transformation (4.152)). For a reversible process for which the internal dissipation $\mathcal{D}_{\text{int}}$ is zero (no entropy production, $\Gamma = 0$), we conclude from (4.154) that the rate of internal mechanical work (or stress-power) $w_{\text{int}} = \mathbf{P} : \dot{\mathbf{F}}$ per unit reference volume (see eq. (4.113)) equals $\dot{\Psi}$.

EXERCISES

1. Recall relation (4.145) and Piola transformation (4.121) for the heat flux vector. Show the equivalence of relations (4.139) and (4.140) with manipulations according to identity (1.81).

2. Starting from (4.132), show that the local spatial form of the second law of thermodynamics is

$$\frac{\partial \eta_c}{\partial t} \geq -\text{div}(\mathbf{h} + \eta_c\mathbf{v}) + \tilde{r}$$

   (compare with Exercise 2 on p. 166).

3. Derive the local forms of the Clausius-Duhem inequality and the Clausius-Planck inequality in the spatial description associated with (4.137), (4.138) and (4.141), (4.142), respectively. Note the fact that the internal energy and the entropy possessed by a body were introduced per unit *volume* rather than per unit *mass* (see relations (4.91) and (4.129)).

   Hint: Use Reynold's transport theorem in the form of (4.26).

4. Show that if the absolute temperature Θ is merely a function of time, the Clausius-Duhem inequality and the Clausius-Planck inequality coincide, leading to the local material form

$$\Theta\dot{\eta} \geq R - \mathrm{Div}\mathbf{Q} \qquad \text{or} \qquad \Theta\dot{\eta} \geq R - \frac{\partial Q_A}{\partial X_A} \quad,$$

   also valid for isothermal processes.

4.7   Master Balance Principle

We recognize from previous sections that conservation of mass, the momentum balance principles and the two fundamental laws of thermodynamics are all of the same mathematical structure. To generalize these relations is the objective of the following section.

Master balance principle in the global form.   The present status of a set of particles occupying an arbitrary region Ω of a continuum body $\mathcal{B}$ with boundary surface $\partial\Omega$ at time t may be characterized by a tensor-valued function $I(t) = \int_{\Omega} f \mathrm{d}v$ (compare with eq. (4.22)). In the following let $f = f(\mathbf{x}, t)$ be a smooth spatial tensor field per unit current volume of order n. In particular, f may characterize some physical scalar, vector or tensor quantity such as *density, linear* and *angular momentum, total energy* and so forth.

A change of these quantities may now be expressed as the following **master balance principle** here presented in the global spatial form

$$\frac{\mathrm{D}}{\mathrm{D}t} \int_{\Omega} f(\mathbf{x}, t)\mathrm{d}v = \int_{\partial\Omega} \phi(\mathbf{x}, t, \mathbf{n})\mathrm{d}s + \int_{\Omega} \Sigma(\mathbf{x}, t)\mathrm{d}v \quad . \tag{4.155}$$

The first term on the right-hand side is the integral of the so-called **surface density** $\phi = \phi(\mathbf{x}, t, \mathbf{n})$, which is a tensor of order n. The surface density is defined per unit current area and distributed over the boundary surface $\partial\Omega$. Note that ϕ depends not only on position $\mathbf{x}$ and time t, but also on the orientation of the infinitesimal spatial

Balance principles	f	ϕ	Σ
Mass (4.8)$_2$	ρ	0	0
Linear momentum (4.44)	$\rho \mathbf{v}$	$\mathbf{t}$	$\mathbf{b}$
Angular momentum (4.45)	$\mathbf{r} \times \rho \mathbf{v}$	$\mathbf{r} \times \mathbf{t}$	$\mathbf{r} \times \mathbf{b}$
First law of thermodynamics (4.124)	$\rho \mathbf{v}^2/2 + e_c$	$\mathbf{t} \cdot \mathbf{v} + q_n$	$\mathbf{b} \cdot \mathbf{v} + r$

Table 4.1 Identified quantities for the master balance principle.

surface element $ds \in \partial\Omega$ characterized by the unit vector field $\mathbf{n}$ normal to $\partial\Omega$ at $\mathbf{x}$. Thus, ϕ is not a tensor field. The remaining term on the right-hand side is the integral of the so-called **volume density** $\Sigma = \Sigma(\mathbf{x}, t)$, which consists of external and internal sources. It is a spatial tensor field of order n defined per unit current volume and distributed over region Ω.

The spatial quantities ϕ and Σ describe the action of the surroundings (in the form of a *resultant force* and *moment, external mechanical power* and *thermal power*) on the set of particles within region Ω. Expression (4.155) represents the most general structure of tensor-valued balance principles in the global form.

Now the statements of balance of mass (4.8)$_2$, linear momentum (4.44), angular momentum (4.45) and the first law of thermodynamics (4.124) may be viewed within the context of the master balance principle (4.155) (see Table 4.1). The entries in the table summarize the spatial fields f, ϕ, Σ which are here identified as scalars $(n = 0)$ and vectors $(n = 1)$.

If equality (4.155) is replaced by an inequality in the form

$$\frac{\mathrm{D}}{\mathrm{D}t} \int_\Omega f(\mathbf{x}, t) \mathrm{d}v \geq \int_{\partial\Omega} \phi(\mathbf{x}, t, \mathbf{n}) \mathrm{d}s + \int_\Omega \Sigma(\mathbf{x}, t) \mathrm{d}v \quad , \tag{4.156}$$

we may refer to this as the **master inequality principle**, essential in thermodynamics. Hence, in regard to the second law of thermodynamics in the global spatial form of (4.132), the fields f, ϕ, Σ are identified as scalars $(n = 0)$ according to

$$f = \eta_c \quad , \qquad \phi = -\mathbf{h} \cdot \mathbf{n} \quad , \qquad \Sigma = \tilde{r} \quad . \tag{4.157}$$

We now introduce Cauchy's stress theorem, as presented by eq. (3.3), in a more general setting. There exists a unique tensor field $\boldsymbol{\Phi} = \boldsymbol{\Phi}(\mathbf{x}, t)$ (with respect to current volume) on Ω of order $n + 1$ so that

$$\phi(\mathbf{x}, t, \mathbf{n}) = \boldsymbol{\Phi}(\mathbf{x}, t)\mathbf{n} \tag{4.158}$$

(the proof is omitted). Thus, $\boldsymbol{\Phi}$ is a *linear operator* that acts on the vector $\mathbf{n}$ generating the surface density ϕ.

With the fundamental relation (4.158) and the divergence theorem by analogy with (1.296) (and (1.295)), the surface integral in the master balance principle (4.155) transforms into a volume integral according to

$$\int_{\partial\Omega} \phi(\mathbf{x}, t, \mathbf{n})ds = \int_{\partial\Omega} \boldsymbol{\Phi}(\mathbf{x}, t)\mathbf{n}ds = \int_{\Omega} \mathrm{div}\boldsymbol{\Phi}(\mathbf{x}, t)dv \quad . \tag{4.159}$$

The spatial field $\boldsymbol{\Phi}$, which we have assumed to be smooth, characterizes the Cauchy flux out of the boundary surface $\partial\Omega$. Hence, the integral over the surface density ϕ, i.e. $\int_{\partial\Omega} \phi ds$, describes the total flux of $\boldsymbol{\Phi}$. To be consistent with the notation, if $\boldsymbol{\Phi}$ is a vector field the term $\boldsymbol{\Phi}(\mathbf{x}, t)\mathbf{n}$ in eq. $(4.159)_1$ is replaced by $\boldsymbol{\Phi}(\mathbf{x}, t) \cdot \mathbf{n}$.

The corresponding relations of (4.155), (4.156) and (4.159) in terms of material coordinates may be derived in an analogous manner.

Master balance principle in the local form.   We are now interested in the equivalent local spatial forms of master balance principle (4.155) and master inequality principle (4.156).

By the material time derivative of function $I(t) = \int_{\Omega} f dv$, recall relation $(4.33)_2$, and by means of eq. $(4.159)_2$, the master balance principle (4.155) takes on the form

$$\int_{\Omega} [\frac{\partial f}{\partial t} + \mathrm{div}(f \otimes \mathbf{v})]dv = \int_{\Omega} \mathrm{div}\boldsymbol{\Phi}(\mathbf{x}, t)dv + \int_{\Omega} \Sigma(\mathbf{x}, t)dv \quad . \tag{4.160}$$

Since equation (4.160) holds for any current volume v, we conclude that

$$\frac{\partial f}{\partial t} = \mathrm{div}\boldsymbol{\Psi} + \Sigma \qquad \text{with} \qquad \boldsymbol{\Psi} = \boldsymbol{\Phi} - f \otimes \mathbf{v} \quad , \tag{4.161}$$

which is the master balance principle in the spatial description at a point $\mathbf{x}$ in region Ω and at time t, also called the **master field equation**. Here, $\boldsymbol{\Psi} = \boldsymbol{\Psi}(\mathbf{x}, t)$ is a tensor field of order $n + 1$. For the case that f is a scalar field, the notation $f \otimes \mathbf{v}$ in eqs. (4.160) and (4.161) should be understood as $f\mathbf{v}$.

For example, recall Cauchy's first equation of motion in the form of (4.55) which may be identified by setting

$$f = \rho\mathbf{v} \quad , \qquad \boldsymbol{\Psi} = \boldsymbol{\sigma} - \rho\mathbf{v} \otimes \mathbf{v} \quad , \qquad \Sigma = \mathbf{b} \quad . \tag{4.162}$$

Equation (4.161), which is replaced by an inequality of the form

$$\frac{\partial f}{\partial t} \geq \mathrm{div}(\boldsymbol{\Phi} - f \otimes \mathbf{v}) + \Sigma \quad , \tag{4.163}$$

gives the associated local spatial form of master inequality principle (4.156).

EXERCISES

1. Find the Cauchy flux $\boldsymbol{\Psi}$, as derived in (4.161), for each of the balance principles (see Table 4.1).

2. Define the Piola-Kirchhoff flux $\boldsymbol{\Phi}_R = \boldsymbol{\Phi}_R(\mathbf{X}, t)$ so that the corresponding surface density per unit reference area is $\boldsymbol{\Phi}_R\mathbf{N}$, where $\mathbf{N}$ is the unit outward normal to the boundary surface $\partial\Omega_0$. Find the master balance principle in the global and local material forms equivalent to eqs. (4.155) and (4.161), respectively.

5   Some Aspects of Objectivity

If laws of nature discovered at different places and times were not the same, scientific work would have to be redone at every new place and at each time. We know that the laws of nature we discover have to take the same form, however we are oriented or we set our clock; there is no difference whether we measure distances relative to, for example, east or west, or we date events from, for example, the birth of Christ or the death of Newton. Qualitative and quantitative descriptions of physical phenomena have to remain unchanged even if we make any changes in the point of view from which we observe them. Thus, physical processes do not depend on the change of observer. To ancient natural philosophers this was not so obvious. The mathematical representation of physical phenomena must reflect this invariance.

The following chapter has the task to express this fundamental finding with the concept of objectivity, or frame-indifference, which constitutes an essential part in non-linear continuum mechanics. We introduce the terminology of an observer, consider changes of observers and apply the concept of objectivity to tensor fields. Transformation rules for various kinematical, stress and stress rate quantities under changes of observers are also derived.

It is obvious to claim that material properties must be invariant under changes of observers. This fundamental requirement is expressed through the principle of material frame-indifference. To show how this objectivity requirement restricts elastic material response is the aim of the last section.

Additional information is found in the same monographs which have already been suggested for Chapter 2 (see the reference list on p. 56). Of course, the short list does not contain a comprehensive review of the large number of papers and books available on this subject.

5.1   Change of Observer, and Objective Tensor Fields

Before examining specific constitutive equations for some elastic materials it is first necessary to present a mathematical foundation for the change of an observer and to introduce the concept of objectivity for tensor fields. We study further how velocity and acceleration fields behave under changes of observers.

Observer and Euclidean transformation.   The description of a physical process is related directly to the choice of an **observer**, which we denote subsequently by O. An arbitrarily chosen observer in the three-dimensional Euclidean space and in time is equipped to measure

  (i) *relative positions* of points in space (with a ruler), and

  (ii) *instants of time* (with a clock).

An **event** is noticed by an observer in terms of *position (place* $\mathbf{x}$*)* and *time* t.

   Consider two arbitrary events in the Euclidean space characterized by the pairs $(\mathbf{x}_0, t_0)$ and $(\mathbf{x}, t)$ (compare with OGDEN [1997]). We assume that $(\mathbf{x}_0, t_0)$ is 'frozen' as long as the event $(\mathbf{x}, t)$ occurs. An observer records that the pair of points in space is separated by the *distance* $|\mathbf{x} - \mathbf{x}_0|$, and that the *time interval (lapse)* between the events under observation is $t - t_0$. In the following we let the pairs $(\mathbf{x}_0, t_0)$ and $(\mathbf{x}, t)$ map to $(\mathbf{x}_0^+, t_0^+)$ and $(\mathbf{x}^+, t^+)$ so that both the distance $|\mathbf{x} - \mathbf{x}_0|$ and the time interval $t - t_0$ are preserved (see Figure 5.1).

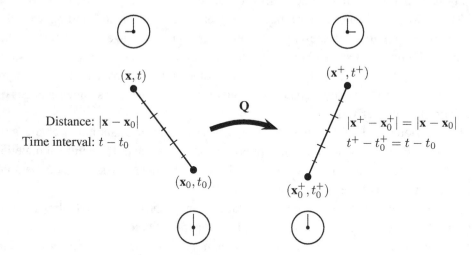

Figure 5.1 Map of two points preserving distance and time interval.

   A spatial mapping which satisfies the requirements above may be represented by the time-dependent transformation

$$\mathbf{x}^+ - \mathbf{x}_0^+ = \mathbf{Q}(t)(\mathbf{x} - \mathbf{x}_0) \quad . \tag{5.1}$$

The point differences $\mathbf{x}^+ - \mathbf{x}_0^+$ and $\mathbf{x} - \mathbf{x}_0$ can be interpreted as vectors which are related through the orthogonal tensor $\mathbf{Q}(t)$, with the well-known property $\mathbf{Q}^{-1}(t) = \mathbf{Q}(t)^{\mathrm{T}}$ (compare with Section 1.2, p. 16). In order to maintain orientations we admit only rotation, consequently, $\mathbf{Q}$ is assumed to be proper orthogonal ($\det\mathbf{Q} = +1$). Hence, with (5.1), we may write the following mathematical expression

$$\mathbf{x}^+ = \mathbf{c}(t) + \mathbf{Q}(t)\mathbf{x} \ , \qquad t^+ = t + \alpha \ , \tag{5.2}$$

where in regard to $\mathbf{x}^+$ and $\mathbf{x}$ we think of position vectors characterizing two points. For eq. (5.2) we have introduced a vector $\mathbf{c}(t)$, and a real number α denoting the *time-shift*, which are defined to be

$$\mathbf{c}(t) = \mathbf{x}_0^+ - \mathbf{Q}(t)\mathbf{x}_0 \ , \qquad \alpha = t_0^+ - t_0 \ . \tag{5.3}$$

Note that both $\mathbf{c}$ and $\mathbf{Q}$ are continuous functions of time which, for convenience, are assumed to be continuously differentiable. The one-to-one mapping of the form (5.2) connecting the pair $(\mathbf{x}, t)$ with its corresponding pair $(\mathbf{x}^+, t^+)$ is frequently referred to as a **Euclidean transformation**.

Change of observer.   We assert, for example, that macroscopic properties of materials are not affected by the choice of an observer, a fundamental principle of physics. However, the general aim is to ensure that the stress state in a body and any physical quantity with an intrinsic feature must be **invariant** relative to a particular **change of observer**. Therefore, we expect from a change of observer that *distances* between arbitrary pairs of points in space and *time intervals* between events are preserved. In other words, we require that a different observer O^+ monitors the same relative distances of points and the same time intervals between events under observation.

We can show that the spatial transformation (5.1) or (5.2)$_1$, combined with the time-shift (5.2)$_2$, denote the most general time-dependent change of observer from O to O^+. The event at place $\mathbf{x}$ and at time t recorded by observer O is the same event as that recorded by a different observer O^+ at $\mathbf{x}^+$ and t^+. Note that we are now considering **one** event recorded by **two** (different) observers O and O^+ who are moving relative to each other.

In order to describe a physical process in the (three-dimensional) Euclidean space and on the real time axis we assign to each of the observers a rectangular Cartesian coordinate system, which we characterize by a set of *fixed* basis vectors, i.e. $\{\mathbf{e}_a\}$ and $\{\mathbf{e}_a^+\}$ relative to O and O^+, respectively. We call them **reference frames of the observers**. Hence, any points $\mathbf{x}$ and $\mathbf{x}^+$ may be represented by the position vectors $\mathbf{x} = x_a\mathbf{e}_a$ and $\mathbf{x}^+ = x_a^+\mathbf{e}_a^+$, with x_a and x_a^+ denoting rectangular Cartesian coordinates, as usual. The shift in the time scale between the observer O and O^+ is $t^+ = t + \alpha$.

In order to formulate the change of observer in index notation, we identify $\mathbf{Q}(t)$ as the relative rotation of the reference frames of the observers. Hence, $\mathbf{e}_a^+ = \mathbf{Q}(t)\mathbf{e}_a$

(compare with eq. $(1.182)_1$). Multiplying eqs. $(5.2)_1$ and $(5.3)_1$ by $\mathbf{e}_a^+$ and using identities (1.81), $\mathbf{Q}^T\mathbf{Q} = \mathbf{I}$ and the relation according to $(1.23)_3$, we find that

$$x_a^+ = \mathbf{x}^+ \cdot \mathbf{e}_a^+ = c_a^+(t) + x_a , \qquad c_a^+(t) = \mathbf{c}(t) \cdot \mathbf{e}_a^+ = x_{0a}^+ - x_{0a} . \qquad (5.4)$$

The mathematical expression $(5.4)_1$ states that the observers O and O^+ assign, with the exception of the shift c_a^+, the same coordinates to the corresponding points $\mathbf{x}$ and $\mathbf{x}^+$.

The reader should be aware that the introduced mapping, i.e. a change of observer, affects the points in space-time and *not* the coordinates of points. However, a change of reference frame changes the coordinates of points (and not the points themselves) and is simply governed by a coordinate transformation, as introduced in Section 1.5.

EXAMPLE 5.1   Consider two arbitrary points of a continuum body identified by their position vectors $\mathbf{x}$ and $\mathbf{y}$ at time t. The events $(\mathbf{x}, t)$ and $(\mathbf{y}, t)$ are recorded by an observer O with the reference frame $\mathbf{e}_a$. A second observer O^+ with the reference frame $\mathbf{e}_a^+$ records the same events at the associated points $\mathbf{x}^+$ and $\mathbf{y}^+$ at time t^+.

Compute the transformation of the spatial vector field $\mathbf{u} = \mathbf{y} - \mathbf{x} = u_a \mathbf{e}_a$ into its counterpart $\mathbf{u}^+ = \mathbf{y}^+ - \mathbf{x}^+ = u_a^+ \mathbf{e}_a^+$ and determine the components u_a (of $\mathbf{u}$) and u_a^+ (of $\mathbf{u}^+$), recorded by the two (different) observers O and O^+, respectively.

Solution.   Using (5.2), $\mathbf{u}$ transforms according to

$$\mathbf{u}^+ = \mathbf{y}^+ - \mathbf{x}^+ = \mathbf{c}(t) + \mathbf{Q}(t)(\mathbf{y} - \mathbf{x}) - \mathbf{c}(t) = \mathbf{Q}(t)\mathbf{u} . \qquad (5.5)$$

With reference to eqs. $(1.23)_3$ and (1.21), the components of the spatial vector fields $\mathbf{u}^+$ and $\mathbf{u}$ relative to one and to the other reference frames of the observers read

$$u_a^+ = \mathbf{u}^+ \cdot \mathbf{e}_a^+ = (y_a^+ - x_a^+)\mathbf{e}_a^+ \cdot \mathbf{e}_a^+ = y_a^+ - x_a^+ , \qquad (5.6)$$

$$u_a = \mathbf{u} \cdot \mathbf{e}_a = (y_a - x_a)\mathbf{e}_a \cdot \mathbf{e}_a = y_a - x_a . \qquad (5.7)$$

By means of $(5.4)_1$ we deduce from $(5.6)_3$ that

$$u_a^+ = y_a^+ - x_a^+ = c_a^+ + y_a - c_a^+ - x_a = y_a - x_a . \qquad (5.8)$$

With $(5.7)_3$ this shows that $u_a^+ = u_a$, signifying that both observers measure the *same* distance.   ∎

Any spatial vector field $\mathbf{u}$ that transforms according to eq. $(5.5)_3$, i.e.

$$\mathbf{u}^+ = \mathbf{Q}(t)\mathbf{u} , \qquad (5.9)$$

is said to be **objective** or equivalently **frame-indifferent**. According to Example 5.1, transformation (5.9) implies that two observers O^+ and O, moving relative to each other, record the same coordinates, $u_a^+ = u_a$, which is the meaning of an objective spatial vector field. In general, if a physical quantity is objective then it is independent of an observer.

Consider now a motion $\mathbf{x} = \chi(\mathbf{X}, t)$ of a continuum body as seen by an arbitrary observer O in space. The motion specifies the place $\mathbf{x}$ at current time t of a certain material point initially at $\mathbf{X}$. A second observer O^+ monitors the *same* motion at the place $\mathbf{x}^+$ and at current time t^+, we write $\mathbf{x}^+ = \chi^+(\mathbf{X}, t^+)$. Note that the reference configuration (and any referential position $\mathbf{X}$) is fixed and therefore independent of the change of observer.

Hence, the motion $\mathbf{x}^+ = \chi^+(\mathbf{X}, t^+)$ is related to $\mathbf{x} = \chi(\mathbf{X}, t)$ by the Euclidean transformation (5.2), i.e.

$$\chi^+(\mathbf{X}, t^+) = \mathbf{c}(t) + \mathbf{Q}(t)\chi(\mathbf{X}, t) \ , \qquad t^+ = t + \alpha \ , \tag{5.10}$$

for each point $\mathbf{X}$ and time t.

Velocity and acceleration fields under changes of observers.   In general, observers are located at different places in space and move relative to each other, as implied by the time-dependence of $\mathbf{c}(t)$ and $\mathbf{Q}(t)$. Therefore, the descriptions of motions depend on the observers and, consequently, the velocity and acceleration of motion are, in general, *not* objective (frame-indifferent), as shown in the following.

To begin with, we introduce

$$\mathbf{v}(\mathbf{x}, t) = \dot{\chi}(\mathbf{X}, t) = \frac{\partial \chi(\mathbf{X}, t)}{\partial t} \ ,$$
$$\mathbf{v}^+(\mathbf{x}^+, t^+) = \dot{\chi}^+(\mathbf{X}, t^+) = \frac{\partial \chi^+(\mathbf{X}, t^+)}{\partial t^+} \ , \tag{5.11}$$

$$\mathbf{a}(\mathbf{x}, t) = \dot{\mathbf{v}}(\mathbf{x}, t) = \frac{\partial \mathbf{v}(\mathbf{x}, t)}{\partial t} \ ,$$
$$\mathbf{a}^+(\mathbf{x}^+, t^+) = \dot{\mathbf{v}}^+(\mathbf{x}^+, t^+) = \frac{\partial \mathbf{v}^+(\mathbf{x}^+, t^+)}{\partial t^+} \ , \tag{5.12}$$

which are the spatial velocities and accelerations of a certain point as observed by O and O^+, respectively. Before examining the transformation rules for the velocity and acceleration fields under changes of observers it is first necessary to formulate the inverse relation of $(5.2)_1$ and its material time derivative. Hence, with $\mathbf{Q}(t)^{\mathrm{T}}\mathbf{Q}(t) = \mathbf{I}$, we deduce from $(5.2)_1$

$$\mathbf{x} = \mathbf{Q}(t)^{\mathrm{T}}[\mathbf{x}^+ - \mathbf{c}(t)] \ , \tag{5.13}$$

and using $(2.28)_1$, the product rule of differentiation and $(5.11)_2$, we find that

$$\mathbf{v}(\mathbf{x}, t) = \overline{\dot{\mathbf{Q}(t)^{\mathrm{T}}}}[\mathbf{x}^+ - \mathbf{c}(t)] + \mathbf{Q}(t)^{\mathrm{T}}[\mathbf{v}^+ - \dot{\mathbf{c}}(t)] \ . \tag{5.14}$$

The overbar covers the quantity to which the time differentiation is applied. Further, we define the skew tensor

$$\mathbf{\Omega}(t) = \dot{\mathbf{Q}}(t)\mathbf{Q}(t)^{\mathrm{T}} = -\mathbf{\Omega}(t)^{\mathrm{T}} \tag{5.15}$$

with the property

$$\mathbf{\Omega}^2(t) = -\dot{\mathbf{Q}}(t)\mathbf{Q}(t)^{\mathrm{T}}[\dot{\mathbf{Q}}(t)\mathbf{Q}(t)^{\mathrm{T}}]^{\mathrm{T}} = -\dot{\mathbf{Q}}(t)\overline{\dot{\mathbf{Q}}(t)^{\mathrm{T}}} \quad . \tag{5.16}$$

The tensor $\mathbf{\Omega}$ represents the spin of the reference frame of observer O relative to the reference frame of observer O^+.

Hence, material time differentiation of the spatial part of (5.10) gives, using (5.11) and the product rule,

$$\mathbf{v}^+(\mathbf{x}^+, t^+) = \dot{\mathbf{c}}(t) + \dot{\mathbf{Q}}(t)\mathbf{x} + \mathbf{Q}(t)\mathbf{v}(\mathbf{x}, t) \quad . \tag{5.17}$$

From (5.17) we find, with the aid of (5.13) and (5.15)$_1$, the transformation law for the spatial velocity field, namely

$$\mathbf{v}^+ = \mathbf{Q}\mathbf{v} + \dot{\mathbf{c}} + \mathbf{\Omega}(\mathbf{x}^+ - \mathbf{c}) \tag{5.18}$$

(suppressing the arguments of functions).

We deduce from relation (5.18) that the spatial velocity field $\mathbf{v}$ is *not* objective under changes of observers following the arbitrary transformation $\mathbf{x}^+ = \mathbf{c}(t) + \mathbf{Q}(t)\mathbf{x}$. Since the extra terms $\dot{\mathbf{c}}$ and $\mathbf{\Omega}(\mathbf{x}^+ - \mathbf{c})$ are present, the requirement for objectivity, i.e. eq. (5.9), is not satisfied. Hence, the velocity field $\mathbf{v}$ is only objective if

$$\dot{\mathbf{c}} + \mathbf{\Omega}(\mathbf{x}^+ - \mathbf{c}) = \mathbf{o} \quad , \tag{5.19}$$

implying a change of observer according to the following time-independent transformation

$$\mathbf{x}^+ = \mathbf{c}_0 + \mathbf{Q}_0\mathbf{x} \qquad \text{with} \qquad \dot{\mathbf{c}}_0 = \mathbf{o} \quad , \qquad \dot{\mathbf{Q}}_0 = \mathbf{O} \quad , \tag{5.20}$$

referred to as a **time-independent rigid transformation**. Therein, the vector $\mathbf{c}_0$ and the orthogonal tensor $\mathbf{Q}_0$ are assumed to be time-independent (constant) quantities. For a time-independent rigid transformation the magnitudes of $\mathbf{v}$ and $\mathbf{v}^+$ are equal, so that $\mathbf{v}^+ = \mathbf{Q}\mathbf{v}$, as would be required for an objective vector field.

The material time differentiation of (5.18) gives, by means of the product rule and eqs. (5.12) and (5.11)$_2$,

$$\mathbf{a}^+ = \mathbf{Q}\mathbf{a} + \ddot{\mathbf{c}} + \dot{\mathbf{Q}}\mathbf{v} + \dot{\mathbf{\Omega}}(\mathbf{x}^+ - \mathbf{c}) + \mathbf{\Omega}(\mathbf{v}^+ - \dot{\mathbf{c}}) \tag{5.21}$$

(suppressing the arguments of functions).

Finally, by means of (5.14) and eqs. (5.15)$_1$, (5.16)$_2$, we obtain from (5.21) the transformation

$$\mathbf{a}^+ = \mathbf{Q}\mathbf{a} + \ddot{\mathbf{c}} + (\dot{\boldsymbol{\Omega}} - \boldsymbol{\Omega}^2)(\mathbf{x}^+ - \mathbf{c}) + 2\boldsymbol{\Omega}(\mathbf{v}^+ - \dot{\mathbf{c}}) \qquad (5.22)$$

for the spatial acceleration field $\mathbf{a}$. Like for the spatial velocity field, the acceleration field is *not* objective for a general change of observer. The terms $\dot{\boldsymbol{\Omega}}(\mathbf{x}^+ - \mathbf{c})$ and $-\boldsymbol{\Omega}^2(\mathbf{x}^+ - \mathbf{c})$ are called the **Euler acceleration** and **centrifugal acceleration**, while the last term in eq. (5.22), i.e. $2\boldsymbol{\Omega}(\mathbf{v}^+ - \dot{\mathbf{c}})$, represents the **Coriolis acceleration**.

An acceleration field is objective under all changes of observer *if and only if* the lengths of $\mathbf{a}^+$ and $\mathbf{a}$ are equal, i.e. $\mathbf{a}^+(\mathbf{x}^+, t^+) = \mathbf{Q}(t)\mathbf{a}(\mathbf{x}, t)$, requiring that

$$\ddot{\mathbf{c}} + (\dot{\boldsymbol{\Omega}} - \boldsymbol{\Omega}^2)(\mathbf{x}^+ - \mathbf{c}) + 2\boldsymbol{\Omega}(\mathbf{v}^+ - \dot{\mathbf{c}}) = \mathbf{o} \quad . \qquad (5.23)$$

Consequently, this implies that $\dot{\mathbf{c}}$ is constant and that the orthogonal tensor $\mathbf{Q}$ is also constant. A change of observer from O to O^+ of this type, for which the spatial acceleration is *objective*, is called a **Galilean transformation**, and is governed by

$$\mathbf{x}^+ = \mathbf{c}(t) + \mathbf{Q}_0\mathbf{x} \qquad \text{with} \qquad \ddot{\mathbf{c}}(t) = \mathbf{o} \quad , \qquad \dot{\mathbf{Q}}_0 = \mathbf{O} \qquad (5.24)$$

for all times t. Here, $\mathbf{Q}_0$ denotes the time-independent orthogonal tensor, and $\mathbf{c}(t) = \mathbf{v}_0 t + \mathbf{c}_0$, with the initial (constant) quantities for $\mathbf{c}_0$ and the velocity $\mathbf{v}_0$.

Objective higher-order tensor fields.   A spatial tensor field of order n, $n = 1, 2, \ldots$, i.e. $\mathbf{u}_1 \otimes \cdots \otimes \mathbf{u}_n$, is called **objective** or equivalently **frame-indifferent**, if, during any change of observer, $\mathbf{u}_1 \otimes \cdots \otimes \mathbf{u}_n$ transforms according to

$$(\mathbf{u}_1 \otimes \cdots \otimes \mathbf{u}_n)^+ = \mathbf{Q}\mathbf{u}_1 \otimes \cdots \otimes \mathbf{Q}\mathbf{u}_n \quad , \qquad (5.25)$$

which holds for every tensor $\mathbf{Q}$ and every vector $\mathbf{u}_n$.

By introducing a spatial second-order tensor field $\mathbf{A}(\mathbf{x}, t)$ to be $\mathbf{u}_1(\mathbf{x}, t) \otimes \mathbf{u}_2(\mathbf{x}, t)$ ($n = 2$), we find, using (5.9), the important relation

$$\mathbf{A}^+(\mathbf{x}^+, t^+) = [\mathbf{u}_1(\mathbf{x}, t) \otimes \mathbf{u}_2(\mathbf{x}, t)]^+ = \mathbf{Q}(t)\mathbf{u}_1(\mathbf{x}, t) \otimes \mathbf{Q}(t)\mathbf{u}_2(\mathbf{x}, t)$$

$$= \mathbf{Q}(t)[\mathbf{u}_1(\mathbf{x}, t) \otimes \mathbf{u}_2(\mathbf{x}, t)]\mathbf{Q}(t)^{\mathrm{T}} = \mathbf{Q}(t)\mathbf{A}(\mathbf{x}, t)\mathbf{Q}(t)^{\mathrm{T}} \quad . \qquad (5.26)$$

For any spatial vector field $\mathbf{u}$ ($n = 1$), eq. (5.25) reduces to $\mathbf{u}^+(\mathbf{x}^+, t^+) = \mathbf{Q}(t)\mathbf{u}(\mathbf{x}, t)$, which we found through eq. (5.9).

In particular, for $n = 0$ we have a scalar field. It is obvious that any spatial scalar field $\Phi(\mathbf{x}, t)$, recorded by O, is unaffected by a change of observer. Hence, a spatial scalar field Φ is objective if, under all Euclidean transformations (5.2), Φ transforms according to

$$\Phi^+(\mathbf{x}^+, t^+) = \Phi(\mathbf{x}, t) \quad , \qquad (5.27)$$

where Φ^+ is the corresponding scalar field recorded by observer O^+.

In summary: the *requirement of objectivity* means that tensor, vector and scalar fields transform under changes of observers according to the laws

$$\left.\begin{array}{rcl} \mathbf{A}^+(\mathbf{x}^+, t^+) & = & \mathbf{Q}(t)\mathbf{A}(\mathbf{x}, t)\mathbf{Q}(t)^{\mathrm{T}} \quad , \\[2mm] \mathbf{u}^+(\mathbf{x}^+, t^+) & = & \mathbf{Q}(t)\mathbf{u}(\mathbf{x}, t) \quad , \\[2mm] \Phi^+(\mathbf{x}^+, t^+) & = & \Phi(\mathbf{x}, t) \quad , \end{array}\right\} \tag{5.28}$$

where $\mathbf{x}^+$ and $\mathbf{x}$ are related by the Euclidean transformation (5.2).

EXAMPLE 5.2   Show that the spatial gradient of an objective vector field $\mathbf{u} = \mathbf{u}(\mathbf{x}, t)$ transforms according to

$$[\mathrm{grad}\mathbf{u}(\mathbf{x}, t)]^+ = \mathbf{Q}(t)\mathrm{grad}\mathbf{u}(\mathbf{x}, t)\mathbf{Q}(t)^{\mathrm{T}} \quad , \tag{5.29}$$

where $(\mathrm{grad}\mathbf{u})^+ = \partial\mathbf{u}^+/\partial\mathbf{x}^+$ denotes the spatial gradient of the vector $\mathbf{u}^+$ recorded by an observer O^+. Note that in view of $(5.28)_1$ the second-order tensor field $\mathrm{grad}\mathbf{u}$ retains the objectivity property.

Solution.   A vector field $\mathbf{u}$ remains unchanged during any change of observer if $\mathbf{u}^+(\mathbf{x}^+, t^+) = \mathbf{Q}(t)\mathbf{u}(\mathbf{x}, t)$. Hence, by the chain rule,

$$\frac{\partial\mathbf{u}^+}{\partial\mathbf{x}^+}\frac{\partial\boldsymbol{\chi}^+}{\partial\mathbf{x}} = \mathbf{Q}\frac{\partial\mathbf{u}}{\partial\mathbf{x}} \tag{5.30}$$

(the arguments of the functions have been omitted). With the aid of transformation $(5.10)_1$ we obtain the desired result. Note that, in general, the gradient of an objective tensor field of order n is also objective.   ∎

EXERCISES

1. Consider objective scalar, vector and tensor fields $\Phi(\mathbf{x}, t)$, $\mathbf{u}(\mathbf{x}, t)$ and $\mathbf{A}(\mathbf{x}, t)$, respectively.

   (a) Show that $\mathrm{grad}\Phi$, $\mathrm{div}\mathbf{u}$ and $\mathrm{div}\mathbf{A}$ are objective fields during any change of observer.

   (b) Show that the Lie time derivative of a contravariant spatial vector field $\mathbf{u}$, as determined in eq. (2.194), is objective.

2. Using eqs. (5.13)–(5.15) obtain the alternative relation for the spatial acceleration field (5.21) in terms of $\mathbf{x}$, $\mathbf{v}$ and $\mathbf{a}$ in the form

$$\mathbf{a}^+ = \mathbf{Q}\mathbf{a} + \ddot{\mathbf{c}} + \ddot{\mathbf{Q}}\mathbf{x} + 2\dot{\mathbf{Q}}\mathbf{v} \quad,$$

and show that its gradient is given by

$$(\mathrm{grad}\,\mathbf{a})^+ = (\ddot{\mathbf{Q}} + 2\dot{\mathbf{Q}}\mathbf{l} + \mathbf{Q}\mathrm{grad}\,\mathbf{a})\mathbf{Q}^{\mathrm{T}} \quad, \tag{5.31}$$

where $\mathbf{l} = \mathrm{grad}\,\mathbf{v}$ denotes the spatial velocity gradient.

3. Assume an objective transformation of the body force (per unit volume), i.e. $\mathbf{b}^+(\mathbf{x}^+, t^+) = \mathbf{Q}(t)\mathbf{b}(\mathbf{x}, t)$. Show that the local form of Cauchy's first equation of motion in the spatial description, i.e. (4.53), is only objective under a Galilean transformation.

5.2 Superimposed Rigid-body Motions

In the following section we show that a change of observer may equivalently be viewed as certain rigid-body motions superimposed on the current configuration. We apply this concept to various kinematical quantities and to some stress tensors of importance.

Rigid-body motion. As noted, the fundamental relationship (5.2) describes a change of observer, preserving both the distances between arbitrary pairs of points in space, and time intervals between events under observation.

It is essential to introduce an important equivalent mechanical statement of the specification (5.2): for this purpose we consider a motion $\mathbf{x}^+ = \boldsymbol{\chi}^+(\mathbf{X}, t^+)$ of a continuum body which differs from another motion $\mathbf{x} = \boldsymbol{\chi}(\mathbf{X}, t)$ of the same body by a *superimposed* (possibly time-dependent) *rigid-body motion* and by a *time-shift*, as depicted in Figure 5.2. We emphasize that, in contrast to the considerations of the last section, $\mathbf{x}^+ = \boldsymbol{\chi}^+(\mathbf{X}, t^+)$ and $\mathbf{x} = \boldsymbol{\chi}(\mathbf{X}, t)$ are motions of **two** events recorded by a **single** observer O. The rigid-body motion moves the region Ω in space occupied by the body at time t, defined by the motion $\mathbf{x} = \boldsymbol{\chi}(\mathbf{X}, t)$, to a new region Ω^+ occupied by the same body at t^+, which is given by $\mathbf{x}^+ = \boldsymbol{\chi}^+(\mathbf{X}, t^+)$. Here and elsewhere we will employ the symbol $(\bullet)^+$ to designate quantities associated with the new region Ω^+.

According to the principle of relativity, the description of a single motion monitored by two (different) observers, as described in the last section, is equivalent to the description of two (different) motions monitored by a single observer. Hence, the pairs $(\mathbf{x}, t)$ and $(\mathbf{x}^+, t^+)$, which are defined on regions Ω and Ω^+, are precisely related by the Euclidean transformation (5.2), i.e. $\mathbf{x}^+ = \mathbf{c}(t) + \mathbf{Q}(t)\mathbf{x}$ and $t^+ = t + \alpha$.

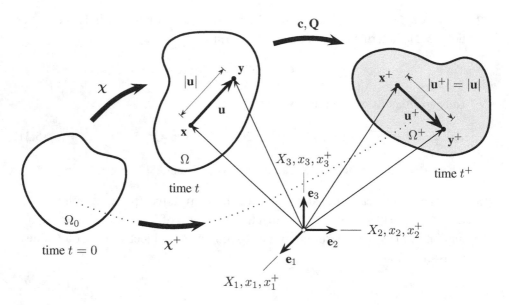

Figure 5.2 Two motions χ^+ and χ of a body (monitored by a single observer) which differ by a superimposed rigid-body motion and by a time-shift. A spatial vector field $\mathbf{u}$ transforms into $\mathbf{u}^+ = \mathbf{Q}\mathbf{u}$, with length $|\mathbf{u}^+| = |\mathbf{u}|$.

Within this context the vector $\mathbf{c}$ describes a superimposed (time-dependent, pure) *rigid-body translation* for which any material point moves an identical distance, with the same magnitude and direction at time t. Since $\mathbf{Q}$ is a *proper* orthogonal tensor ($\det\mathbf{Q} = +1$), the orientation is preserved and $\mathbf{Q}$ describes a superimposed (time-dependent, pure) *rigid-body rotation*. For a pure rigid-body rotation, transformation (5.2) reduces to $\mathbf{x}^+ = \mathbf{Q}(t)\mathbf{x}$.

Hence, at each instant of time a rigid-body motion is the composition of a rigid-body translation $\mathbf{c}$ and a rigid-body rotation $\mathbf{Q}$ about an axis of rotation, combined with a time-shift $\alpha = t^+ - t$. The material points occupy the same relative position in each motion (the *angle* between two arbitrary vectors and their *lengths* remain constant).

We now recall Example 5.1 and apply the described concept to the spatial vector field $\mathbf{u} = \mathbf{y} - \mathbf{x}$ located at region Ω (see Figure 5.2). Hence, a rigid-body motion maps the points $\mathbf{x}, \mathbf{y}$ to the associated points $\mathbf{x}^+, \mathbf{y}^+$ located in Ω^+ and the spatial vector $\mathbf{u} = \mathbf{y} - \mathbf{x}$ to $\mathbf{u}^+ = \mathbf{y}^+ - \mathbf{x}^+$.

With (5.2) we may conclude that the distance between the two points $\mathbf{y}^+$ and $\mathbf{x}^+$ remain unchanged. Namely, $\mathbf{y}^+ - \mathbf{x}^+ = \mathbf{Q}(t)(\mathbf{y} - \mathbf{x})$ (compare with eq. (5.1)), which immediately implies, on use of definition (1.15), identity (1.81) and the orthogonality condition $\mathbf{Q}(t)^{\mathrm{T}}\mathbf{Q}(t) = \mathbf{I}$, that $|\mathbf{y}^+ - \mathbf{x}^+| = |\mathbf{y} - \mathbf{x}|$. Consequently the lengths of the vectors $\mathbf{u}^+$ and $\mathbf{u}$ are equal, i.e. $|\mathbf{u}^+| = |\mathbf{u}|$. We say that the spatial vector field $\mathbf{u}$ is objective during the rigid-body motion.

It is trivial but worthy of mention that any material field $\mathcal{F}(\mathbf{X}, t)$ of some physical scalar, vector or tensor quantity, which is characterized as a function of the referential position $\mathbf{X}$ and time t, is unaffected by a rigid-body motion superimposed on Ω. Hence, $\mathcal{F}^+(\mathbf{X}, t^+) = \mathcal{F}(\mathbf{X}, t)$.

Euclidean transformation of various kinematical quantities. The following discussion is concerned with the behavior of various kinematical quantities during a superimposed rigid-body motion.

To begin with, we consider the deformation gradient at the point $\mathbf{x} \in \Omega$ and its associated point $\mathbf{x}^+ \in \Omega^+$, i.e.

$$\mathbf{F}(\mathbf{X}, t) = \frac{\partial \mathbf{x}(\mathbf{X}, t)}{\partial \mathbf{X}} \quad , \qquad \mathbf{F}^+(\mathbf{X}, t^+) = \frac{\partial \mathbf{x}^+(\mathbf{X}, t^+)}{\partial \mathbf{X}} \quad . \tag{5.32}$$

Differentiating (5.2) with respect to $\mathbf{X}$ gives the transformation rule

$$\mathbf{F}^+ = \frac{\partial \mathbf{x}^+}{\partial \mathbf{X}} = \mathbf{Q}\frac{\partial \mathbf{x}}{\partial \mathbf{X}} = \mathbf{Q}\mathbf{F} \qquad \text{or} \qquad F^+_{aA} = \frac{\partial x^+_a}{\partial X_A} = Q_{ab}\frac{\partial x_b}{\partial X_A} = Q_{ab}F_{bA} \tag{5.33}$$

for the deformation gradient (for convenience, we will not indicate subsequently the dependence for the above functions on position and time). Note that the second-order tensor $\mathbf{F}$ is *objective* even though (5.33) does not coincide with the fundamental (objectivity) requirement (5.28)$_1$. However, recall that the deformation gradient $\mathbf{F}$ is a two-point tensor field, in which one index describes material coordinates X_A which are intrinsically independent of the observer. That is why the deformation gradient transforms like a vector according to (5.28)$_2$ and why $\mathbf{F}$ is regarded as objective.

Moreover, let $J = \det\mathbf{F}$ and $J^+ = \det\mathbf{F}^+$. Since the tensor $\mathbf{Q}$ is proper orthogonal ($\det\mathbf{Q} = +1$), eq. (5.33)$_3$ implies, through the property (1.101), that

$$J^+ = J > 0 \quad . \tag{5.34}$$

Hence, the scalar field J remains unaltered by a superimposed rigid-body motion. Also the sign of the volume ratio J is preserved, since $\det\mathbf{Q} = +1$.

Next, we recall the unique polar decomposition of the deformation gradient at $\mathbf{x} \in \Omega$ and $\mathbf{x}^+ \in \Omega^+$, i.e.

$$\mathbf{F} = \mathbf{R}\mathbf{U} = \mathbf{v}\mathbf{R} \quad , \qquad \mathbf{F}^+ = \mathbf{R}^+\mathbf{U}^+ = \mathbf{v}^+\mathbf{R}^+ \quad . \tag{5.35}$$

Applying (5.33)$_3$ to (5.35), we arrive at the representations

$$\mathbf{R}^+\mathbf{U}^+ = \mathbf{Q}\mathbf{R}\,\mathbf{U} \quad , \qquad \mathbf{v}^+\mathbf{R}^+ = \mathbf{Q}\mathbf{v}\,\mathbf{R} \quad . \tag{5.36}$$

Since the tensor $\mathbf{Q}\mathbf{R}$ is orthogonal it follows from (5.36)$_1$ that the transformation rules for the rotation tensor $\mathbf{R}$ and the right stretch tensor $\mathbf{U}$ are

$$\mathbf{R}^+ = \mathbf{Q}\mathbf{R} \quad , \qquad \mathbf{U}^+ = \mathbf{U} \quad . \tag{5.37}$$

By analogy with the deformation gradient, $\mathbf{R}$ is an objective two-point tensor field. The right stretch tensor is defined with respect to the reference configuration. Hence, $\mathbf{U}$ remains unaltered by a superimposed rigid-body motion and $\mathbf{U}$ is therefore also objective.

From eq. $(5.36)_2$, we obtain, using result $(5.37)_1$ and the orthogonality condition $\mathbf{R}^{\mathrm{T}}\mathbf{R} = \mathbf{I}$, the transformation rule

$$\mathbf{v}^+ = \mathbf{Q}\mathbf{v}\mathbf{Q}^{\mathrm{T}} \ . \tag{5.38}$$

Clearly, the left stretch tensor $\mathbf{v}$ is objective.

Next, we discuss the spatial velocity gradient according to $(2.141)_4$, i.e. $\mathbf{l} = \dot{\mathbf{F}}\mathbf{F}^{-1}$. In an analogous manner, the spatial velocity gradient generated by the motion $\boldsymbol{\chi}^+$ reads

$$\mathbf{l}^+ = (\dot{\mathbf{F}}\mathbf{F}^{-1})^+ = \dot{\mathbf{F}}^+ (\mathbf{F}^+)^{-1} \ . \tag{5.39}$$

By deriving eq. $(5.33)_3$ with respect to time, using the product rule, i.e. $\dot{\mathbf{F}}^+ = \dot{\mathbf{Q}}\mathbf{F} + \mathbf{Q}\dot{\mathbf{F}}$, and the inverse relation of $(5.33)_3$, i.e. $(\mathbf{F}^+)^{-1} = \mathbf{F}^{-1}\mathbf{Q}^{\mathrm{T}}$, the spatial velocity gradient follows from $(5.39)_2$ as

$$\mathbf{l}^+ = (\dot{\mathbf{Q}}\mathbf{F} + \mathbf{Q}\dot{\mathbf{F}})(\mathbf{F}^{-1}\mathbf{Q}^{\mathrm{T}}) = \boldsymbol{\Omega} + \mathbf{Q}\mathbf{l}\mathbf{Q}^{\mathrm{T}} \ , \tag{5.40}$$

with the skew tensor $\boldsymbol{\Omega} = \dot{\mathbf{Q}}\mathbf{Q}^{\mathrm{T}}$. Since $\boldsymbol{\Omega}$ is present, the spatial velocity gradient $\mathbf{l}$ fails to satisfy the objectivity requirement $(\mathbf{l}^+ \neq \mathbf{Q}\mathbf{l}\mathbf{Q}^{\mathrm{T}})$. Hence, the kinematical quantity $\mathbf{l}$ is not a suitable candidate for formulating constitutive equations, which must be objective.

Euclidean transformation of stress tensors.   Let dynamical processes be given by the pairs $(\boldsymbol{\sigma}^+, \boldsymbol{\chi}^+)$ and $(\boldsymbol{\sigma}, \boldsymbol{\chi})$, where $\boldsymbol{\chi}^+$ and $\boldsymbol{\chi}$ are related through (5.10). We now want to show how the Cauchy stress tensors $\boldsymbol{\sigma}^+$ and $\boldsymbol{\sigma}$ are related.

We recall the Cauchy traction vector $\mathbf{t} = \boldsymbol{\sigma}\mathbf{n}$ with the unit vector $\mathbf{n}$ at point $\mathbf{x}$ directed along the outward normal to the boundary surface $\partial\Omega$ of an arbitrary region Ω at time t. A superimposed rigid-body motion transforms region Ω to a new region Ω^+ which is bounded by the associated boundary surface $\partial\Omega^+$ at a later time $t^+ = t + \alpha$. The Cauchy traction vector transforms to $\mathbf{t}^+ = \boldsymbol{\sigma}^+\mathbf{n}^+$ with the unit vector $\mathbf{n}^+$ at point $\mathbf{x}^+$ normal to $\partial\Omega^+$. By taking note that the vectors $\mathbf{t}$ and $\mathbf{n}$ transform according to the objectivity requirement $(5.28)_2$, we obtain that $\mathbf{Q}\mathbf{t} = \boldsymbol{\sigma}^+\mathbf{Q}\mathbf{n}$. A comparison with $\mathbf{t} = \boldsymbol{\sigma}\mathbf{n}$ gives the fundamental transformation rule

$$\boldsymbol{\sigma}^+ = \mathbf{Q}\boldsymbol{\sigma}\mathbf{Q}^{\mathrm{T}} \tag{5.41}$$

for the stress tensor. This means that the Cauchy stress tensor is objective.

In order to describe the first Piola-Kirchhoff stress tensor which is generated by the motion $\mathbf{x}^+ = \boldsymbol{\chi}^+(\mathbf{X}, t^+)$, we may write the Piola transformation (3.8) as $\mathbf{P}^+(\mathbf{F}^+)^{\mathrm{T}} =$

$J^+\sigma^+$. Knowing that the scalar J is objective according to eq. (5.34), and using (5.33)$_3$ and (5.41), we find, with the help of (3.8), that

$$\mathbf{P}^+(\mathbf{QF})^\mathrm{T} = J\mathbf{Q}\sigma\mathbf{Q}^\mathrm{T} \ ,$$
$$\mathbf{P}^+(\mathbf{F}^\mathrm{T}\mathbf{Q}^\mathrm{T}) = \mathbf{Q}\,J\sigma\mathbf{Q}^\mathrm{T} = \mathbf{QP}(\mathbf{F}^\mathrm{T}\mathbf{Q}^\mathrm{T}) \ ,$$
$$\mathbf{P}^+ = \mathbf{QP} \ . \tag{5.42}$$

Since the two-point stress tensor field $\mathbf{P}$ transforms like a vector field according to the objectivity requirement (5.28)$_2$, $\mathbf{P}$ is objective. The second Piola-Kirchhoff stress tensor $\mathbf{S}$ is parameterized by material coordinates only. Therefore, the material tensor field does not depend on any superimposed rigid-body motion, and hence $\mathbf{S} = \mathbf{S}^+$.

Note that all the stress tensors σ, $\mathbf{P}$ and $\mathbf{S}$ discussed are suitable candidates for the description of material response, which fundamentally is required to be independent of the observer.

EXERCISES

1. Recall the kinematic relations (2.65), (2.69) and (2.79), (2.83). Using the transformation rule $\mathbf{F}^+ = \mathbf{QF}$, show that the material strain tensors $\mathbf{C}$ and $\mathbf{E}$ are unaffected by any possible rigid-body motion, i.e.

$$\mathbf{C}^+ = \mathbf{C} \ , \qquad \mathbf{E}^+ = \mathbf{E} \ , \tag{5.43}$$

   and that the spatial strain tensors $\mathbf{b}$ and $\mathbf{e}$ transform according to the rules

$$\mathbf{b}^+ = \mathbf{Qb}\mathbf{Q}^\mathrm{T} \ , \qquad \mathbf{e}^+ = \mathbf{Qe}\mathbf{Q}^\mathrm{T} \ . \tag{5.44}$$

   Note that all these kinematical quantities are objective, since $\mathbf{C}$ and $\mathbf{E}$ are defined with respect to the reference configuration and the second-order tensor fields $\mathbf{b}$ and $\mathbf{e}$ conform with the requirement of objectivity given in eq. (5.28)$_1$.

2. By eqs. (2.148) and (2.149), the spatial velocity gradient $\mathbf{l} = \dot{\mathbf{F}}\mathbf{F}^{-1}$ is recalled to be the sum of the rate of deformation tensor $\mathbf{d} = \mathbf{d}^\mathrm{T}$ and the spin tensor $\mathbf{w} = -\mathbf{w}^\mathrm{T}$.

   Show that rigid-body motions involve the transformations

$$\mathbf{d}^+ = \mathbf{Qd}\mathbf{Q}^\mathrm{T} \ , \qquad \mathbf{w}^+ = \mathbf{\Omega} + \mathbf{Qw}\mathbf{Q}^\mathrm{T} \ , \tag{5.45}$$

   where $\mathbf{d}$ is objective. Note that $\mathbf{w}$, which is expressed through the skew tensor $\mathbf{\Omega}$, is affected by rigid-body motions, and hence $\mathbf{w}$ is not objective.

5.3 Objective Rates

One aim of this section is to perform objective time derivatives, which are essential in order to formulate constitutive equations in the rate form. We focus attention on some important objective stress rates associated with the names *Oldroyd, Green, Naghdi, Jaumann, Zaremba* or *Truesdell*.

Objective rates.   The material time derivatives of the objective vector field $\mathbf{u} = \mathbf{u}(\mathbf{x}, t)$ and the objective second-order tensor field $\mathbf{A} = \mathbf{A}(\mathbf{x}, t)$, which transform according to eqs. $(5.28)_1$ and $(5.28)_2$, are given by means of the product rule of differentiation as

$$\dot{\mathbf{u}}^+ = \mathbf{Q}\dot{\mathbf{u}} + \dot{\mathbf{Q}}\mathbf{u} \ , \qquad \dot{\mathbf{A}}^+ = \dot{\mathbf{Q}}\mathbf{A}\mathbf{Q}^{\mathrm{T}} + \mathbf{Q}\dot{\mathbf{A}}\mathbf{Q}^{\mathrm{T}} + \mathbf{Q}\mathbf{A}\dot{\overline{\mathbf{Q}^{\mathrm{T}}}} \ . \tag{5.46}$$

Clearly, neither $\dot{\mathbf{u}}$ nor $\dot{\mathbf{A}}$ retains the objectivity requirements (5.28) ($\dot{\mathbf{u}}^+ \neq \mathbf{Q}\dot{\mathbf{u}}$ and $\dot{\mathbf{A}}^+ \neq \mathbf{Q}\dot{\mathbf{A}}\mathbf{Q}^{\mathrm{T}}$). Note that material time derivatives of objective spatial tensor fields will not, in general, be objective and they are not, therefore, suitable quantities for formulating constitutive equations in the rate form.

 This motivates the introduction of objective time derivatives called **objective rates**, which are basically modified material time derivatives. Before proceeding to examine objective rate forms it is first necessary to express the material time derivatives of $\mathbf{Q}$ and $\mathbf{Q}^{\mathrm{T}}$ from relation $(5.45)_2$. With definition $(5.15)_1$ and property $\mathbf{w} = -\mathbf{w}^{\mathrm{T}}$ for the spin tensor we find that

$$\dot{\mathbf{Q}} = \mathbf{w}^+\mathbf{Q} - \mathbf{Q}\mathbf{w} \ , \qquad \dot{\overline{\mathbf{Q}^{\mathrm{T}}}} = -\mathbf{Q}^{\mathrm{T}}\mathbf{w}^+ + \mathbf{w}\mathbf{Q}^{\mathrm{T}} \ . \tag{5.47}$$

Hence, substituting $(5.47)_1$ into $(5.46)_1$, we find immediately, by analogy with the transformation rule $(5.28)_2$, that

$$(\dot{\mathbf{u}} - \mathbf{w}\mathbf{u})^+ = \mathbf{Q}(\dot{\mathbf{u}} - \mathbf{w}\mathbf{u}) \ , \tag{5.48}$$

$$(\mathring{\mathbf{u}})^+ = \mathbf{Q}\mathring{\mathbf{u}} \ , \tag{5.49}$$

where we have introduced the definition for the **co-rotational rate** of the objective vector field $\mathbf{u}$, i.e.

$$\mathring{\mathbf{u}} = \dot{\mathbf{u}} - \mathbf{w}\mathbf{u} \ . \tag{5.50}$$

In general, we denote co-rotational rates with the accent ($\mathring{\bullet}$).

 By analogy with the above we introduce the co-rotational rate of the objective second-order tensor field $\mathbf{A}$. With the help of eqs. (5.47) and $(5.28)_1$ we find from $(5.46)_2$ after some straightforward recasting that

$$(\dot{\mathbf{A}} - \mathbf{w}\mathbf{A} + \mathbf{A}\mathbf{w})^+ = \mathbf{Q}(\dot{\mathbf{A}} - \mathbf{w}\mathbf{A} + \mathbf{A}\mathbf{w})\mathbf{Q}^{\mathrm{T}} \ , \tag{5.51}$$

$$(\mathring{\mathbf{A}})^+ = \mathbf{Q}\mathring{\mathbf{A}}\mathbf{Q}^{\mathrm{T}} \ , \tag{5.52}$$

where we have introduced the definition

$$\mathring{\mathbf{A}} = \dot{\mathbf{A}} - \mathbf{w}\mathbf{A} + \mathbf{A}\mathbf{w} \ , \tag{5.53}$$

known as the **Jaumann-Zaremba rate**, which is often used in plasticity theory. Obviously, in regard to eqs. (5.49) and (5.52), the co-rotational rates of $\mathbf{u}$ and $\mathbf{A}$ are indeed objective.

If $\mathbf{A}$ is a symmetric tensor we can easily show an interesting property connecting the Jaumann-Zaremba rate and the material time derivative of $\mathbf{A}$. Using (5.53) and the property of double contraction according to (1.95), i.e. $\mathbf{A} : \mathbf{w}\mathbf{A} = \mathbf{A} : \mathbf{A}\mathbf{w}$, we obtain

$$2\mathbf{A} : \mathring{\mathbf{A}} = 2\mathbf{A} : \dot{\mathbf{A}} - 2\mathbf{A} : \mathbf{w}\mathbf{A} + 2\mathbf{A} : \mathbf{A}\mathbf{w} = \overline{\mathbf{A} : \mathbf{A}} \ . \tag{5.54}$$

Finally we define the **convected rates** of $\mathbf{u}$ and $\mathbf{A}$. These are the objective fields

$$\overset{\triangle}{\mathbf{u}} = \dot{\mathbf{u}} + \mathbf{l}^{\mathrm{T}}\mathbf{u} \ , \qquad \overset{\triangle}{\mathbf{A}} = \dot{\mathbf{A}} + \mathbf{l}^{\mathrm{T}}\mathbf{A} + \mathbf{A}\mathbf{l} \ , \tag{5.55}$$

where the accent $(\overset{\triangle}{\bullet})$ indicates convected rates. The rate $\overset{\triangle}{\mathbf{A}}$ is also called the **Cotter-Rivlin rate**.

Objective stress rates. We now focus attention on some of the infinitely many possible objective stress rates that may be defined. The choice of suitable, i.e. objective, stress rates is essential in the formulation of constitutive rate equations, which must be objective.

The **Oldroyd stress rate** of a spatial stress field is defined to be the Lie time derivative of that field. We shall indicate Oldroyd stress rates by the abbreviation Oldr. By recalling the concept of Lie time derivatives from Section 2.8, in particular, rule $(2.188)_1$, the Lie time derivative of the contravariant Cauchy stress tensor $\boldsymbol{\sigma}$ is given by

$$\pounds_{\mathbf{v}}(\boldsymbol{\sigma}^{\sharp}) = \mathbf{F}\left[\frac{D(\mathbf{F}^{-1}\boldsymbol{\sigma}\mathbf{F}^{-\mathrm{T}})}{Dt}\right]\mathbf{F}^{\mathrm{T}}$$

$$= \mathbf{F}(\overline{\mathbf{F}^{-1}}\boldsymbol{\sigma}\mathbf{F}^{-\mathrm{T}} + \mathbf{F}^{-1}\dot{\boldsymbol{\sigma}}\mathbf{F}^{-\mathrm{T}} + \mathbf{F}^{-1}\boldsymbol{\sigma}\overline{\mathbf{F}^{-\mathrm{T}}})\mathbf{F}^{\mathrm{T}} \ , \tag{5.56}$$

where the transformations (2.87) and the product rule of differentiation are to be used. Hence, using the identities $(2.145)_2$ and $(2.146)_2$, we conclude that the Oldroyd stress rate of the Cauchy stress $\boldsymbol{\sigma}$ is

$$\mathtt{Oldr}(\boldsymbol{\sigma}) = \dot{\boldsymbol{\sigma}} - \mathbf{l}\boldsymbol{\sigma} - \boldsymbol{\sigma}\mathbf{l}^{\mathrm{T}} \ , \tag{5.57}$$

where $\dot{\boldsymbol{\sigma}}$ denotes the material time derivative of the Cauchy stress tensor.

We now show how the Oldroyd stress rate, $\mathtt{Oldr}(\boldsymbol{\sigma})$, generated by the motion χ, is related to its counterpart $\mathtt{Oldr}(\boldsymbol{\sigma})^+$, generated by χ^+. Considering $\mathtt{Oldr}(\boldsymbol{\sigma})^+ =$

$\dot{\sigma}^+ - \mathbf{l}^+\sigma^+ - \sigma^+\mathbf{l}^{+\mathrm{T}}$ and using transformations (5.41) and (5.40)$_2$ in combination with eqs. (5.15) and (5.57), we obtain

$$\mathtt{Oldr}(\sigma)^+ = \overline{\mathbf{Q}\sigma\mathbf{Q}^{\mathrm{T}}} - (\mathbf{\Omega} + \mathbf{QlQ}^{\mathrm{T}})\mathbf{Q}\sigma\mathbf{Q}^{\mathrm{T}} - \mathbf{Q}\sigma\mathbf{Q}^{\mathrm{T}}(\mathbf{\Omega} + \mathbf{QlQ}^{\mathrm{T}})^{\mathrm{T}}$$
$$= \mathbf{Q}(\dot{\sigma} - \mathbf{l}\sigma - \sigma\mathbf{l}^{\mathrm{T}})\mathbf{Q}^{\mathrm{T}}$$
$$= \mathbf{Q}\,\mathtt{Oldr}(\sigma)\mathbf{Q}^{\mathrm{T}} \ . \tag{5.58}$$

Hence, the Oldroyd stress rate is objective. In general, we can prove that Lie time derivatives of objective spatial tensor fields yield objective spatial tensor fields.

By analogy with (5.57), the Oldroyd stress rate of the Kirchhoff stress τ is

$$\mathtt{Oldr}(\tau) = \dot{\tau} - \mathbf{l}\tau - \tau\mathbf{l}^{\mathrm{T}} \ . \tag{5.59}$$

Adopting rule (2.188) for the contravariant Kirchhoff stress tensor τ and using relation (3.63)$_1$, we obtain the important equation

$$\mathcal{L}_v(\tau^\sharp) = \chi_* \left(\frac{\mathrm{D}}{\mathrm{D}t} \chi_*^{-1}(\tau^\sharp) \right) = \mathbf{F}\dot{\mathbf{S}}\mathbf{F}^{\mathrm{T}} \ , \tag{5.60}$$

relating the Lie time derivative of τ, i.e. $\mathtt{Oldr}(\tau)$, and the material time derivative $\dot{\mathbf{S}}$ of the second Piola-Kirchhoff stress $\mathbf{S}$ according to the push-forward operation (2.87)$_1$.

EXAMPLE 5.3   Consider the definitions

$$\overset{\triangledown}{\sigma} = \dot{\sigma} - \dot{\mathbf{R}}\mathbf{R}^{\mathrm{T}}\sigma + \sigma\dot{\mathbf{R}}\mathbf{R}^{\mathrm{T}} \ , \qquad \overset{\circ}{\sigma} = \dot{\sigma} - \mathbf{w}\sigma + \sigma\mathbf{w} \tag{5.61}$$

of objective stress rates, where $\overset{\triangledown}{\sigma}$ is called the **Green-Naghdi stress rate** and $\overset{\circ}{\sigma}$ the **Jaumann-Zaremba stress rate** (compare with the Jaumann-Zaremba rate (5.53) for any objective second-order tensor field). The spin tensor $\mathbf{w}$ is given by definition (2.149).

Show that both $\overset{\triangledown}{\sigma}$ and $\overset{\circ}{\sigma}$ are special cases of the Oldroyd stress rate of σ in the sense that $\overset{\triangledown}{\sigma}$ corresponds to the Lie time derivative (5.56), with $\mathbf{F}$ replaced by the rotation tensor $\mathbf{R}$, and $\overset{\circ}{\sigma}$ is the Lie time derivative (5.56), with the rate of deformation tensor $\mathbf{d}$ set to zero. Discuss the case in which the Green-Naghdi stress rate and the Jaumann-Zaremba stress rate coincide.

Solution.   Setting $\mathbf{F} = \mathbf{R}$ in relation (5.56) and employing identities (2.145)$_1$ and (2.146)$_1$ (change $\mathbf{F}$ to $\mathbf{R}$) with the orthogonality condition $\mathbf{R}^{\mathrm{T}}\mathbf{R} = \mathbf{I}$, we obtain

$$\mathcal{L}_v(\sigma^\sharp)|_{\mathbf{F}=\mathbf{R}} = \dot{\sigma} - \dot{\mathbf{R}}\mathbf{R}^{-1}\sigma - \sigma\mathbf{R}^{-\mathrm{T}}\dot{\mathbf{R}^{\mathrm{T}}} = \dot{\sigma} - \dot{\mathbf{R}}\mathbf{R}^{\mathrm{T}}\sigma + \sigma\dot{\mathbf{R}}\mathbf{R}^{\mathrm{T}} \ , \tag{5.62}$$

where $\dot{\mathbf{R}}\mathbf{R}^{\mathrm{T}}$ is a skew tensor according to (2.160).

However, setting $\mathbf{d} = \mathbf{O}$ in the Lie time derivative (5.56), which equivalently means that $\mathbf{l} = \mathbf{w}$, we deduce from (5.57) that

$$\pounds_{\mathbf{v}}(\sigma^{\sharp})|_{\mathbf{d}=\mathbf{O}} = \dot{\sigma} - \mathbf{w}\sigma + \sigma\mathbf{w} \quad , \tag{5.63}$$

with the skew tensor $\mathbf{w} = -\mathbf{w}^{\mathrm{T}}$.

The Green-Naghdi stress rate and the Jaumann-Zaremba stress rate clearly coincide for $\mathbf{w} = \dot{\mathbf{R}}\mathbf{R}^{\mathrm{T}}$. This is the case for a rigid-body rotation (recall Example 2.12, eq. (2.162)). Note, however, that both stress rates reveal the occurrence of undesirable stress oscillations in simple shear. ∎

The **Truesdell stress rate** of the Cauchy stress, denoted by $\mathtt{Trues}(\sigma)$, is defined as the Piola transformation of $\dot{\mathbf{S}}$. Thus,

$$\mathtt{Trues}(\sigma) = J^{-1}\mathbf{F}\dot{\mathbf{S}}\mathbf{F}^{\mathrm{T}} \quad , \tag{5.64}$$

that is the push-forward of $\dot{\mathbf{S}}$ scaled by the inverse of the volume ratio, $J^{-1} = (\det\mathbf{F})^{-1}$. Hence, using the Piola transformation $(3.65)_1$ and the product rule of differentiation we find from (5.64) that

$$\mathtt{Trues}(\sigma) = J^{-1}\mathbf{F}\left[\frac{\mathrm{D}(J\mathbf{F}^{-1}\sigma\mathbf{F}^{-\mathrm{T}})}{\mathrm{D}t}\right]\mathbf{F}^{\mathrm{T}} = J^{-1}\mathbf{F}(\dot{J}\mathbf{F}^{-1}\sigma\mathbf{F}^{-\mathrm{T}}$$

$$+ J\dot{\overline{\mathbf{F}^{-1}}}\sigma\mathbf{F}^{-\mathrm{T}} + J\mathbf{F}^{-1}\dot{\sigma}\mathbf{F}^{-\mathrm{T}} + J\mathbf{F}^{-1}\sigma\dot{\overline{\mathbf{F}^{-\mathrm{T}}}})\mathbf{F}^{\mathrm{T}} \quad . \tag{5.65}$$

Using relations $(2.178)_2$ and $(2.145)_2$, $(2.146)_2$ we deduce from $(5.65)_2$ that

$$\mathtt{Trues}(\sigma) = \dot{\sigma} - \mathbf{l}\sigma - \sigma\mathbf{l}^{\mathrm{T}} + \sigma\mathrm{tr}\mathbf{d} \quad . \tag{5.66}$$

By comparing eqs. (5.57) and (5.59) with (5.66) we may easily deduce the relationships

$$\mathtt{Oldr}(\sigma) = \mathtt{Trues}(\sigma) - \sigma\mathrm{tr}\mathbf{d} \quad , \tag{5.67}$$

$$\mathtt{Oldr}(\tau) = J\,\mathtt{Trues}(\sigma) \tag{5.68}$$

between the Oldroyd stress rate and the Truesdell stress rate.

EXAMPLE 5.4   Suppose that the transformation (5.68) is given. Express σ as $J^{-1}\tau$ and derive the relation (5.67) by simply applying the product rule of differentiation to the Oldroyd stress rate.

Solution.   With $\sigma = J^{-1}\tau$ and the fact that the directional derivative (in our case the Lie time derivative) satisfies the common rules of differentiation, for example, the product rule, we may write

$$\text{Oldr}(\boldsymbol{\sigma}) = \pounds_v(J^{-1}\boldsymbol{\tau}) = J^{-1}\pounds_v(\boldsymbol{\tau}) + \pounds_v(J^{-1})\boldsymbol{\tau} \ . \tag{5.69}$$

According to considerations of Section 2.8, the Lie time derivative of a scalar field is equal to the material time derivative of that scalar field, and hence $\pounds_v(J^{-1}) = \overline{J^{-1}}$. Therefore, with the chain rule and eqs. $(2.178)_2$ and (5.68), relation $(5.69)_2$ leads to the desired result,

$$\text{Oldr}(\boldsymbol{\sigma}) = J^{-1}\,\text{Oldr}(\boldsymbol{\tau}) + \overline{J^{-1}}\boldsymbol{\tau} = J^{-1}\text{Oldr}(\boldsymbol{\tau}) - J^{-1}\boldsymbol{\tau}\text{trd}$$
$$= \text{Trues}(\boldsymbol{\sigma}) - \boldsymbol{\sigma}\text{trd} \ , \tag{5.70}$$

where the relation $\boldsymbol{\sigma} = J^{-1}\boldsymbol{\tau}$ is to be used again.   ■

EXERCISES

1. Consider the Cotter-Rivlin rate $\overset{\triangle}{\mathbf{A}}$ defined by $(5.55)_2$.

   (a) Show that $\overset{\triangle}{\mathbf{A}}$ is the Lie time derivative of a covariant tensor field $\mathbf{A}$.

   (b) With (5.53) and $\mathbf{l} = \mathbf{d} + \mathbf{w}$ show that the connection between the Cotter-Rivlin and Jaumann-Zaremba rates is

   $$\overset{\triangle}{\mathbf{A}} = \overset{\circ}{\mathbf{A}} + \mathbf{dA} + \mathbf{Ad} \ .$$

2. Recall the convected rates $\overset{\triangle}{\mathbf{u}}$ and $\overset{\triangle}{\mathbf{A}}$, as defined in eq. (5.55). Using eqs. $(5.40)_2$, (5.15) and (5.46), (5.28), show that they are objective according to

   $$(\overset{\triangle}{\mathbf{u}})^+ = \mathbf{Q}\overset{\triangle}{\mathbf{u}} \ , \qquad (\overset{\triangle}{\mathbf{A}})^+ = \mathbf{Q}\overset{\triangle}{\mathbf{A}}\mathbf{Q}^{\text{T}} \ .$$

5.4   Invariance of Elastic Material Response

In this section we introduce the principle of material frame-indifference which states basically that material properties do not depend on the change of observer. In particular, we show how it restricts the response of elastic materials and derive objective constitutive equations which are defined to be invariant for all changes of observer. This principle is crucial when constitutive theories such as the theory of elasticity or plasticity are considered.

In the following we consider only the isothermal case for which the absolute temperature Θ remains constant during the process.

Cauchy-elastic materials.   A material is called **Cauchy-elastic** or **elastic** if the stress field at time t depends only on the state of deformation (and the state of temperature) at this time t and not on the deformation history (and temperature history). Hence, the stress field of a Cauchy-elastic material is independent of the deformation path (independent of the time). However, note that the actual work done by the stress field on a Cauchy-elastic material does, in general, depend on the deformation path.

A **constitutive equation** (or **equation of state**) represents the intrinsic physical properties of a continuum body. It determines generally the state of stress at any point of that body to any arbitrary motion at time t. A constitutive equation is either regarded as mathematically generalized (axiomatic) or is based upon experimental data (empirical).

The constitutive equation of an isothermal elastic body relates the Cauchy stress tensor $\boldsymbol{\sigma} = \boldsymbol{\sigma}(\mathbf{x}, t)$ at each place $\mathbf{x} = \boldsymbol{\chi}(\mathbf{X}, t)$ with the deformation gradient $\mathbf{F} = \mathbf{F}(\mathbf{X}, t)$. We may express the constitutive equation in the general form

$$\boldsymbol{\sigma}(\mathbf{x}, t) = \mathbf{g}(\mathbf{F}(\mathbf{X}, t), \mathbf{X}) \quad , \tag{5.71}$$

where $\mathbf{g}$ is referred to as the **response function** associated with the Cauchy stress tensor $\boldsymbol{\sigma}$.

In equation (5.71), $\boldsymbol{\sigma}$ was allowed to depend upon the referential position $\mathbf{X} \in \Omega_0$ in addition to $\mathbf{F}$. Hence, the stress response varies from one particle to the other. However, for subsequent introductory treatments, it is convenient to restrict our attention on continuum bodies, in which both the Cauchy stress tensor $\boldsymbol{\sigma}$ and the reference mass density ρ_0 are independent of the position $\mathbf{X}$; such bodies are called **homogeneous**.

Hence, instead of (5.71), we write the constitutive equation in the form

$$\boldsymbol{\sigma} = \mathbf{g}(\mathbf{F}) \quad , \tag{5.72}$$

which determines the stresses $\boldsymbol{\sigma}$ from the given deformation gradient $\mathbf{F}$. From the *mechanical* point of view $\mathbf{g}$ characterizes the material properties of a (isothermal) Cauchy-elastic material, while from the *mathematical* point of view $\mathbf{g}$ is a tensor-valued function of one tensor variable $\mathbf{F}$. The concept of tensor functions, as we will use it here, is explained in Section 1.7. A constitutive equation of the type of (5.72) is often referred to as a **stress relation**.

Note that for homogeneous deformations the corresponding stresses are constant (since $\mathbf{F}$ has the same value at every point of the body) and, interestingly enough, Cauchy's equation of equilibrium (4.54) trivially reduces to $\mathrm{div}\boldsymbol{\sigma} = \mathbf{o}$. For this case the body force $\mathbf{b}$ is zero, which means that a homogeneous deformation of a continuum body occurs without body force.

Principle of material frame-indifference.   As already mentioned in previous sections, constitutive equations must be objective (frame-indifferent) with respect to the Euclidean transformation (5.2). In other words, if a constitutive equation is satisfied for a dynamical process (σ, χ) then it must also be satisfied for any associated (equivalent) dynamical process (σ^+, χ^+) which is generated by the transformations (5.41) and (5.10). This is a fundamental axiom of mechanics which is known as the **principle of material frame-indifference** or the **principle of material objectivity** or simply as **objectivity** (see TRUESDELL and NOLL [2004, Sections 19, 19A]). If this principle is violated, the constitutive equations are affected by rigid-body motions and meaningless results are obtained.

To begin with, the material frame-indifference of the stress relation (5.72) imposes certain restrictions on the response function $\mathbf{g}$. We consider a motion χ^+ which differs from χ by a rigid-body motion superimposed on the current configuration (compare with Figure 5.2). The rigid-body motion maps the region Ω to a new region Ω^+ and the stress relation (5.72) to $\sigma^+ = \mathbf{g}(\mathbf{F}^+)$. We demand that both regions, namely Ω and Ω^+, are associated with the same function $\mathbf{g}$ because it is for the *same* elastic material. Hence, using $(5.33)_3$ on the one hand and (5.41), (5.72) on the other hand, we arrive at

$$\sigma^+ = \mathbf{g}(\mathbf{F}^+) = \mathbf{g}(\mathbf{QF}) \;, \qquad \sigma^+ = \mathbf{Q}\sigma\mathbf{Q}^{\mathrm{T}} = \mathbf{Qg}(\mathbf{F})\mathbf{Q}^{\mathrm{T}} \;. \tag{5.73}$$

Combining $(5.73)_1$ and $(5.73)_2$, we find the restriction

$$\mathbf{Qg}(\mathbf{F})\mathbf{Q}^{\mathrm{T}} = \mathbf{g}(\mathbf{QF}) \tag{5.74}$$

on $\mathbf{g}$ for every nonsingular $\mathbf{F}$ and orthogonal $\mathbf{Q}$. In other words, constitutive equation (5.72) is independent of the observer if the response function $\mathbf{g}$ satisfies the invariance relation (5.74).

Employing the right polar decomposition $\mathbf{F} = \mathbf{RU}$ on the right-hand side of (5.74), we may write $\mathbf{Qg}(\mathbf{F})\mathbf{Q}^{\mathrm{T}} = \mathbf{g}(\mathbf{QRU})$, where $\mathbf{R}$ is the orthogonal rotation tensor and $\mathbf{U}$ the right stretch tensor. Since the latter relation holds for all proper orthogonal tensors $\mathbf{Q}$, it also holds for the special choice $\mathbf{Q} = \mathbf{R}^{\mathrm{T}}$. Hence, using the orthogonality condition $\mathbf{R}^{\mathrm{T}}\mathbf{R} = \mathbf{I}$, we obtain a corresponding *reduced* form of eq. (5.74), i.e.

$$\mathbf{g}(\mathbf{F}) = \mathbf{Rg}(\mathbf{U})\mathbf{R}^{\mathrm{T}} \;, \tag{5.75}$$

for the function $\mathbf{g}$ and for every $\mathbf{F}$ and $\mathbf{R}$. Therefore, the associated stress relation reads

$$\sigma = \mathbf{Rg}(\mathbf{U})\mathbf{R}^{\mathrm{T}} \;, \tag{5.76}$$

which shows that the properties of an elastic material are independent of the rotational part of $\mathbf{F} = \mathbf{RU}$, characterized by $\mathbf{R}$. Note that the reduced constitutive equation (5.76) is compatible with the principle of material frame-indifference which can be shown as follows. By analogy with (5.76), let $\boldsymbol{\sigma}^+ = \mathbf{R}^+ \mathbf{g}(\mathbf{U}^+)(\mathbf{R}^+)^{\mathrm{T}}$ and use eq. (5.37) in order to obtain $\boldsymbol{\sigma}^+ = \mathbf{QRg}(\mathbf{U})\mathbf{R}^{\mathrm{T}}\mathbf{Q}^{\mathrm{T}}$. Hence, by use of (5.76), we obtain once more $\boldsymbol{\sigma}^+ = \mathbf{Q}\boldsymbol{\sigma}\mathbf{Q}^{\mathrm{T}}$ (compare with (5.41)).

An alternative form of constitutive equation (5.72) follows from Piola transformation (3.8). With (5.72) and the volume ratio $J = \det\mathbf{F}$, we obtain

$$\mathbf{P} = J\boldsymbol{\sigma}\mathbf{F}^{-\mathrm{T}} = \det\mathbf{F}\mathbf{g}(\mathbf{F})\mathbf{F}^{-\mathrm{T}} = \mathfrak{G}(\mathbf{F}) \ , \tag{5.77}$$

where we have defined the tensor-valued tensor function $\mathfrak{G}$ associated with the first Piola-Kirchhoff stress tensor $\mathbf{P}$.

By analogy with the above we may now show the material frame-indifference of the stress relation $(5.77)_3$. Considering $\mathbf{P}^+ = \mathfrak{G}(\mathbf{F}^+)$ and, using $(5.33)_3$ on the one hand and (5.42) and $(5.77)_3$ on the other hand, we find that

$$\mathbf{P}^+ = \mathfrak{G}(\mathbf{F}^+) = \mathfrak{G}(\mathbf{QF}) \ , \qquad \mathbf{P}^+ = \mathbf{QP} = \mathbf{Q}\mathfrak{G}(\mathbf{F}) \ . \tag{5.78}$$

Equating $(5.78)_1$ and $(5.78)_2$, we find the invariant relation

$$\mathbf{Q}\mathfrak{G}(\mathbf{F}) = \mathfrak{G}(\mathbf{QF}) \tag{5.79}$$

for the function $\mathfrak{G}$ and for every $\mathbf{F}$ and $\mathbf{Q}$. Relations (5.74) and (5.79) are necessary and sufficient conditions for the constitutive equations (5.72) and $(5.77)_3$ to satisfy the principle of material frame-indifference.

A reduced form of constitutive equation $(5.77)_3$ is obtained from restriction (5.79). Setting $\mathbf{Q} = \mathbf{R}^{\mathrm{T}}$ and replacing $\mathbf{F}$ by its right polar decomposition $\mathbf{RU}$ on the right-hand side of (5.79), we obtain, using $\mathbf{R}^{\mathrm{T}}\mathbf{R} = \mathbf{I}$,

$$\mathfrak{G}(\mathbf{F}) = \mathbf{R}\mathfrak{G}(\mathbf{U}) \ . \tag{5.80}$$

The restriction on $\mathfrak{G}$ expressed through eqs. (5.79) and (5.80) are equivalent to the restriction on $\mathbf{g}$, as given in eqs. (5.74) and (5.75).

Another alternative form of the constitutive equation which turns out to be very useful in the theory of elasticity follows from relation $(3.65)_1$. With the volume ratio $J = \det\mathbf{U}$, i.e. $(2.96)_2$, the polar decomposition $\mathbf{F} = \mathbf{RU}$, the stress relation (5.76), and the fact that the right stretch tensor $\mathbf{U}$ is symmetric, we obtain

$$\mathbf{S} = J\mathbf{F}^{-1}\boldsymbol{\sigma}\mathbf{F}^{-\mathrm{T}} = \det\mathbf{U}\mathbf{U}^{-1}\mathbf{g}(\mathbf{U})\mathbf{U}^{-1} \ . \tag{5.81}$$

By recalling that $\mathbf{U}$ is the unique square root of the right Cauchy-Green tensor $\mathbf{C}$, we may write $\mathbf{C}^{1/2}$ in the place of $\mathbf{U}$. Defining a tensor-valued tensor function $\mathfrak{H}$ we may introduce finally the second Piola-Kirchhoff stress tensor $\mathbf{S}$ in the form

$$\mathbf{S} = \mathfrak{H}(\mathbf{C}) \ . \tag{5.82}$$

Since the reference configuration is unaffected by superimposed rigid-body motions, we know that the second Piola-Kirchhoff stress tensor and the right Cauchy-Green tensor simply transform according to $\mathbf{S}^+ = \mathbf{S}$ and $\mathbf{C}^+ = \mathbf{C}$. We conclude that the stress relation (5.82) is independent of the observer.

EXAMPLE 5.5   Investigate if the elastic material given by

$$\sigma = \mathfrak{J}(\mathbf{E}) \quad , \tag{5.83}$$

associated with motion χ, satisfies the principle of material frame-indifference.

Solution.   For another motion χ^+ (recorded by a single observer O), assumption (5.83) implies $\sigma^+ = \mathfrak{J}(\mathbf{E}^+)$. By recalling the transformations for the Cauchy stress tensor σ and the Green-Lagrange strain tensor $\mathbf{E}$, which satisfy eqs. (5.41) and (5.43)$_2$, and knowing that $\mathfrak{J}$ is the same function for the two (different) motions χ and χ^+ (it is for the *same* elastic material), we conclude that

$$\mathbf{Q}\sigma\mathbf{Q}^{\mathrm{T}} = \mathfrak{J}(\mathbf{E}) \quad . \tag{5.84}$$

Note that this relation is only true for $\mathbf{Q} = \mathbf{I}$; thus, constitutive equation (5.83) does *not* satisfy the principle of material frame-indifference.   ∎

Isotropic Cauchy-elastic materials.   We assume that the Cauchy stress tensor σ depends on the left Cauchy-Green tensor $\mathbf{b} = \mathbf{F}\mathbf{F}^{\mathrm{T}}$. The constitutive equation (5.72) may then be written in the alternative form

$$\sigma = \mathfrak{h}(\mathbf{b}) \quad , \tag{5.85}$$

where $\mathfrak{h}$ is a tensor-valued function of the symmetric second-order tensor $\mathbf{b}$ associated with the Cauchy stress tensor σ.

In order to find the restriction imposed on the response function $\mathfrak{h}$ by the assumption of material frame-indifference, let $\sigma^+ = \mathfrak{h}(\mathbf{b}^+)$, where the response function $\mathfrak{h}$ is the same for the two motions χ and χ^+. We now use (5.44)$_1$ on the one hand and (5.41), (5.85) on the other hand in order to obtain

$$\sigma^+ = \mathfrak{h}(\mathbf{b}^+) = \mathfrak{h}(\mathbf{Q}\mathbf{b}\mathbf{Q}^{\mathrm{T}}) \quad , \qquad \sigma^+ = \mathbf{Q}\sigma\mathbf{Q}^{\mathrm{T}} = \mathbf{Q}\mathfrak{h}(\mathbf{b})\mathbf{Q}^{\mathrm{T}} \quad . \tag{5.86}$$

Combining eqs. (5.86)$_1$ and (5.86)$_2$ we find the fundamental invariance relation

$$\mathbf{Q}\mathfrak{h}(\mathbf{b})\mathbf{Q}^{\mathrm{T}} = \mathfrak{h}(\mathbf{Q}\mathbf{b}\mathbf{Q}^{\mathrm{T}}) \tag{5.87}$$

for the function $\mathfrak{h}$ and for every tensor $\mathbf{b}$ and orthogonal tensor $\mathbf{Q}$. Hence, constitutive equation (5.85) is independent of the observer if $\mathfrak{h}$ satisfies restriction (5.87).

A specific elastic material which may be described by the constitutive equation in the form (5.85), with property (5.87), is said to be **isotropic**. A tensor-valued function such as $\mathfrak{h}(\mathbf{b})$ is said to be **isotropic** if it satisfies relations of type (5.87). Hence, we refer to $\mathfrak{h}(\mathbf{b})$ as a **tensor-valued isotropic tensor function** of one variable $\mathbf{b}$.

From the physical point of view the condition of isotropy is expressed by the property that the material exhibits no preferred directions. In fact, the stress response of an isotropic elastic material is not affected by the choice of the reference configuration. For a piece of wood, for example, which is of cellular structure, the properties in the direction of the grain differ from those in other directions, so the material certainly is *not* isotropic.

The isotropic tensor function $\mathfrak{h}(\mathbf{b})$, which satisfies (5.87), may be represented in the explicit form

$$\boldsymbol{\sigma} = \mathfrak{h}(\mathbf{b}) = \alpha_0 \mathbf{I} + \alpha_1 \mathbf{b} + \alpha_2 \mathbf{b}^2 \ , \qquad \alpha_a = \alpha_a[I_1(\mathbf{b}), I_2(\mathbf{b}), I_3(\mathbf{b})] \ , \quad (5.88)$$

for each $\mathbf{b}$, which is known as the **Rivlin-Ericksen representation theorem**. For a proof of this crucial relation see RIVLIN and ERICKSEN [1955], SPENCER [1980, Appendix], GURTIN [1981a, pp. 233-235] and TRUESDELL and NOLL [2004, Section 12]. Note that the representation theorem (5.88) represents a fundamental requirement for the mathematical form of the stress relation.

Here, α_a, $a = 0, 1, 2$, are three scalar functions called **response coefficients** or **material functions**. Hence, in general, for an *isotropic* material only three parameters are needed in order to describe the stress state. The scalar functions α_a depend on the three invariants of tensor $\mathbf{b}$ and therefore on the current deformation state. The invariants are defined with respect to eqs. (1.170)–(1.172) as

$$I_1(\mathbf{b}) = \mathrm{tr}\mathbf{b} = \lambda_1^2 + \lambda_2^2 + \lambda_3^2 \ , \tag{5.89}$$

$$I_2(\mathbf{b}) = \frac{1}{2}\left[(\mathrm{tr}\mathbf{b})^2 - \mathrm{tr}(\mathbf{b}^2)\right] = \mathrm{tr}\mathbf{b}^{-1}\mathrm{det}\mathbf{b} = \lambda_1^2\lambda_2^2 + \lambda_1^2\lambda_3^2 + \lambda_2^2\lambda_3^2 \ , \tag{5.90}$$

$$I_3(\mathbf{b}) = \mathrm{det}\mathbf{b} = J^2 = \lambda_1^2\lambda_2^2\lambda_3^2 \ , \tag{5.91}$$

where λ_a^2 are the three eigenvalues of the symmetric spatial tensor $\mathbf{b}$, see eq. (2.119). In eq. (5.91), relation $(2.81)_2$ was used.

Representation (5.88) is the *most general form* of a constitutive equation for isotropic elastic materials also known as the **first representation theorem for isotropic tensor functions**. From the constitutive equation (5.88) we deduce that the principal directions of the Cauchy stress tensor $\boldsymbol{\sigma}$ and the left Cauchy-Green tensor $\mathbf{b}$ coincide. Hence, for isotropic elastic materials the two symmetric tensors $\boldsymbol{\sigma}$ and $\mathbf{b}$ are said to be **coaxial** in every configuration.

By analogy with (5.88), the constitutive equation

$$\sigma = \overline{\alpha}_0 \mathbf{I} + \overline{\alpha}_1 \mathbf{d} + \overline{\alpha}_2 \mathbf{d}^2 \ , \qquad \overline{\alpha}_a = \overline{\alpha}_a[\rho, I_1(\mathbf{d}), I_2(\mathbf{d}), I_3(\mathbf{d})] \ , \qquad (5.92)$$

characterizes the behavior of a viscous fluid, in particular, of a so-called **Reiner-Rivlin fluid**. Therein, ρ and $\mathbf{d}$ are the spatial mass density and the rate of deformation tensor, while $\overline{\alpha}_a$, $a = 0, 1, 2$, are scalar functions of the invariants I_a, $a = 1, 2, 3$, given in (5.89)–(5.91) with $\mathbf{d}$ replacing $\mathbf{b}$.

In order to find an alternative explicit representation for (5.88), we recall the *Cayley-Hamilton* equation (1.174). Since any tensor satisfies its own characteristic equation, we may write (1.174) as $\mathbf{b}^3 - I_1 \mathbf{b}^2 + I_2 \mathbf{b} - I_3 \mathbf{I} = \mathbf{O}$ and find, by multiplying this equation with $\mathbf{b}^{-1}$, that

$$\mathbf{b}^2 = I_1 \mathbf{b} - I_2 \mathbf{I} + I_3 \mathbf{b}^{-1} \ . \qquad (5.93)$$

Eliminating $\mathbf{b}^2$ from (5.88) in favor of $\mathbf{b}^{-1}$ we obtain an alternative representation of a constitutive equation for isotropic elastic materials, i.e.

$$\sigma = \mathfrak{h}(\mathbf{b}) = \beta_0 \mathbf{I} + \beta_1 \mathbf{b} + \beta_{-1} \mathbf{b}^{-1} \ , \qquad \beta_a = \beta_a[I_1(\mathbf{b}), I_2(\mathbf{b}), I_3(\mathbf{b})] \ , \qquad (5.94)$$

where β_a, $a = 0, 1, -1$, are three scalar functions (response coefficients) which, in terms of the three invariants of $\mathbf{b}$, are expressed as

$$\beta_0 = \alpha_0 - I_2 \alpha_2 \ , \qquad \beta_1 = \alpha_1 + I_1 \alpha_2 \ , \qquad \beta_{-1} = I_3 \alpha_2 \ . \qquad (5.95)$$

Representation (5.94) is also known as the **second representation theorem for isotropic tensor functions**, see, for example, GURTIN [1981a, p. 235].

Incompressible Cauchy-elastic materials.   If the Cauchy-elastic material is **incompressible**, then the stress relation is determined only up to an arbitrary scalar p which can be identified as a pressure-like quantity. The constitutive equations (5.72), (5.76) and (5.82) are then replaced by

$$\sigma = -p\mathbf{I} + \mathbf{g}(\mathbf{F}) = -p\mathbf{I} + \mathbf{R}\mathbf{g}(\mathbf{U})\mathbf{R}^{\mathrm{T}} \ , \qquad (5.96)$$

$$\mathbf{S} = -p\mathbf{C}^{-1} + \mathfrak{H}(\mathbf{C}) \ , \qquad (5.97)$$

where the tensor-valued tensor functions $\mathbf{g}(\mathbf{F})$, $\mathbf{g}(\mathbf{U})$ and $\mathfrak{H}(\mathbf{C})$ need only be defined for the kinematic constraints $\det \mathbf{F} = 1$, $\det \mathbf{U} = 1$ and $\det \mathbf{C} = 1$, respectively. The indeterminate terms $-p\mathbf{I}$ and $-p\mathbf{C}^{-1}$ are known as **reaction stresses**, which do no work in any motion compatible with above constraints. For incompressible materials the (indeterminate) scalar p required to maintain incompressibility may only be found by means of the equilibrium conditions and the boundary conditions and is *not* specified by a constitutive equation. Note that the scalar p must always be included in a stress relation of an incompressible material.

In an incompressible and isotropic Cauchy-elastic material we replace constitutive equations $(5.88)_1$ and $(5.94)_1$ by

$$\boldsymbol{\sigma} = -p\mathbf{I} + \alpha_1\mathbf{b} + \alpha_2\mathbf{b}^2 \ , \qquad \boldsymbol{\sigma} = -p\mathbf{I} + \beta_1\mathbf{b} + \beta_{-1}\mathbf{b}^{-1} \ , \qquad (5.98)$$

respectively, where the response coefficients α_1, α_2 and β_1, β_{-1} depend now only on the two scalar invariants I_1 and I_2 (since $I_3 = \det\mathbf{b} = 1$). Note that the scalars p in eqs. (5.98) differ by the term $\alpha_2 I_2$. For an incompressible and isotropic Cauchy-elastic material the stresses given in (5.98) are determined only up to p. The scalar functions α_0 and β_0 multiplying $\mathbf{I}$ in $(5.88)_1$ and $(5.94)_1$ are absorbed into the reaction stresses. The response coefficients in (5.98) are related by

$$\alpha_1 = \beta_1 - I_1\beta_{-1} \ , \qquad \alpha_2 = \beta_{-1} \ . \qquad (5.99)$$

Two special cases result directly from $(5.98)_2$, i.e. the so-called **Mooney-Rivlin model** for incompressible materials, for which β_1 and β_{-1} are constants, and the **neo-Hookean model** for incompressible materials, for which β_1 is constant and $\beta_{-1} = 0$. For a study of these types of material see more in Section 6.5.

EXERCISES

1. Consider the classical **Newtonian fluid**, for which the viscous stress depends linearly on the rate of deformation tensor $\mathbf{d}$. It is the simplest model for a viscous fluid and is given by the constitutive equation

$$\boldsymbol{\sigma} = [-p(\rho) + \lambda(\rho)\mathrm{tr}\mathbf{d}]\mathbf{I} + 2\eta(\rho)\mathbf{d} \ ,$$

characterizing (low molecular weight) liquids and gases such as water, oil or air. Therein, the function $p(\rho)$ depends on the spatial mass density ρ, and λ and η are two parameters characterizing the **viscosity** of the Newtonian fluid.

(a) Show that the response of this type of fluid conforms to the principle of material frame-indifference.

(b) Apply the Newtonian constitutive equation to a motion which causes simple shear deformation (see eq. (2.3)). Take a constant viscosity η and show that the only non-vanishing shear stress σ_{12} is

$$\sigma_{12} = \sigma_{21} = \eta\dot{c} \ , \qquad (5.100)$$

with the shear rate $\dot{c}$ (compare with Exercise 4(a),(b) on p. 105).

Note that the Newtonian viscous fluid is simply a special case of the Reiner-Rivlin fluid (5.92) obtained by choosing the response coefficients $\overline{\alpha}_a$, $a = 0, 1, 2$, as $\overline{\alpha}_0 = -p(\rho) + \lambda(\rho)\mathrm{tr}\mathbf{d}$, $\overline{\alpha}_1 = 2\eta(\rho)$ and $\overline{\alpha}_2 = 0$, respectively. However, by setting $\overline{\alpha}_0 = -p(\rho)$ and $\overline{\alpha}_1 = \overline{\alpha}_2 = 0$ we obtain the constitutive equation $\boldsymbol{\sigma} = -p(\rho)\mathbf{I}$ characterizing the material properties of an **elastic fluid**.

2. By recalling the transformation rules (5.41), (5.31) and (5.40)$_2$, (5.45)$_1$ (from Sections 5.1 and 5.2), show that the constitutive equations

$$\boldsymbol{\sigma} = \mathrm{grad}\,\mathbf{a} + \mathrm{grad}^{\mathrm{T}}\mathbf{a} + 2\mathbf{l}^{\mathrm{T}}\mathbf{l} \quad , \qquad \dot{\boldsymbol{\sigma}} = \mathbf{l}\boldsymbol{\sigma} + \boldsymbol{\sigma}\mathbf{l}^{\mathrm{T}} + \alpha\mathbf{d}$$

are acceptable forms which conform to the principle of material frame-indifference, where α is a material parameter and $\mathbf{a}, \mathbf{l}$ and $\mathbf{d}$ denote the spatial acceleration field, the spatial velocity gradient and the rate of deformation tensor, respectively.

3. Consider a material which is isotropic and Cauchy-elastic. Let a uniform extension (or compression) in all three directions according to relation (2.131), which corresponds to a triaxial stress state, be given.

By determining the deformation gradient $\mathbf{F}$ and the left Cauchy-Green tensor $\mathbf{b}$ (with respect to a set of some orthonormal basis vectors $\mathbf{e}_a$) find the most general representation for the Cauchy stress tensor $\boldsymbol{\sigma}$.

4. By means of a pull-back of (5.88) to the reference configuration and a scaling with the inverse volume ratio J^{-1}, show that for isotropic elastic materials the second Piola-Kirchhoff stress tensor $\mathbf{S}$ is coaxial with the right Cauchy-Green tensor $\mathbf{C}$ *if and only if* $\boldsymbol{\sigma}$ is coaxial with $\mathbf{b}$.

6 Hyperelastic Materials

The fundamental equations introduced in Chapters 2-4 are essential to characterize kinematics, stresses and balance principles, and hold for any continuum body for all times. However, they do not distinguish one material from another and remain valid in all branches of continuum mechanics.

For the case of deformable bodies the equations mentioned are certainly not sufficient on their own to determine the material response. Hence, we must establish additional equations in the form of appropriate *constitutive laws* which are furnished to specify the *ideal material* in question. A constitutive law should approximate the observed physical behavior of a *real material* under specific conditions of interest.

Generally we use a functional relationship as a *constitutive equation* and this enables us to specify the stress components in terms of other field functions such as strain and temperature. A constitutive equation determines the state of stress at any point **x** of a continuum body at time t and is necessarily different for different types of continuous bodies.

Each field of continuum mechanics deals with certain continuous media including **fluids,** which are liquids or gases (such as water, oil, air etc.) and **solids** (such as rubber, metal, ceramics, wood, living tissue etc.). If the constitutive equations are valid for physical objects such as fluids we call the field of continuum mechanics **fluid mechanics**. Another important field in which constitutive equations are valid for solids is known as **solid mechanics**. Note that fluid and solid mechanics differ only with respect to constitutive equations, but they share the same set of field equations.

The main goal of the next two chapters is to study various constitutive equations within the field of solid mechanics appropriate for approximation techniques such as the finite element method. For the most part we follow the so-called **phenomeno-logical approach**, describing the macroscopic nature of materials as continua. The phenomenological approach is mainly concerned with fitting mathematical equations to experimental data and is particularly successful in solid mechanics (such as classical elastoplasticity). However, phenomenological modeling is not capable of relating the mechanism of deformation to the underlying physical (microscopic) structure of the material.

In Chapter 6 we discuss phenomenological constitutive equations which interrelate the stress components and the strain components within a nonlinear regime. Since we are studying so-called purely *mechanical* theories, thermodynamic variables such as the entropy and the temperature are ignored. However, they are taken into account and elaborated on within Chapter 7. No attempt is made to present a comprehensive list of the various important contributions on constitutive modeling to date. This text discusses some selected material models essential in science, industrial engineering practice and in the field of biomechanics, where structures exhibit large strain behavior, very often within the coupled thermodynamic regime.

Sections 6.1-6.8 consider hyperelastic materials in general, and cover a wide range of important types of material such as isotropic and transversely isotropic materials, incompressible and compressible hyperelastic materials and composite materials, in particular. Some important specifications for rubber-like (or other) materials are also presented. The remaining Sections 6.9-6.11 focus attention on inelastic materials. Based on the concept of internal variables, viscoelastic materials and isotropic hyperelastic materials with damage at finite strains are introduced. Plastic and viscoplastic materials which have the ability to undergo irreversible or permanent deformations are not considerd in this text.

6.1 General Remarks on Constitutive Equations

It is the aim of **constitutive theories** to develop mathematical models for representing the real behavior of matter. Constitutive theories of materials are very important but they are a difficult subject in modern nonlinear continuum mechanics. We make no attempt to conduct a comprehensive review of the large number of constitutive theories. For more on formulating nonlinear constitutive theories see, for example, TRUESDELL and NOLL [2004] and the excellent contributions by *Rivlin* in the 1940s and 1950s collected by BARENBLATT and JOSEPH [1997].

In particular, we present a nonlinear constitutive theory suitable to describe a wide variety of physical phenomena in which the strains may be large, i.e. finite. For the case of a (hyper)elastic material the resulting theory is called **finite (hyper)elasticity theory** or just **finite (hyper)elasticity** for which nonlinear continuum mechanics is the fundamental basis (see GREEN and ADKINS [1970] for an analytical treatment and inter alia LE TALLEC [1994] for numerical solution techniques).

Constitutive equations for hyperelastic materials.   A so-called **hyperelastic material** (or in the literature often called a **Green-elastic material**) postulates the existence of a **Helmholtz free-energy function** Ψ, which is defined per unit reference *volume* rather than per unit *mass*.

For the case in which $\Psi = \Psi(\mathbf{F})$ is solely a function of $\mathbf{F}$ or some strain tensor, as introduced in Section 2.4, the Helmholtz free-energy function is referred to as the **strain-energy function** or **stored-energy function** (see Section 4.4, p. 159). Subsequently, we often use the common terminology **strain energy** or **stored energy**. The strain-energy function $\Psi = \Psi(\mathbf{F})$ is a typical example of a scalar-valued function of one tensor variable $\mathbf{F}$, which we assume to be continuous.

A concept of importance in elasticity is **polyconvexity** of strain-energy functions. The global existence theory of solutions, for example, is based on the condition of polyconvexity of strain-energy functions. For an extensive discussion on the underlying issue, see BALL [1977], CIARLET [1988, Chapters 4, 7], MARSDEN and HUGHES [1994, Section 6.4] and ŠILHAVÝ [1997, Sections 17.5, 18.5].

We now restrict attention to **homogeneous materials** in which the distributions of the internal constituents are assumed to be uniform on the continuum scale. For this type of ideal material the strain-energy function Ψ depends only upon the deformation gradient $\mathbf{F}$. Of course, for so-called **heterogeneous materials** (a material that is not homogeneous) Ψ will depend additionally upon the position of a point in the medium.

A hyperelastic material is defined as a subclass of an elastic material, as given in eqs. $(5.77)_3$ and (5.72), whose response functions $\mathfrak{G}$ and $\mathfrak{g}$ have physical expressions of the form

$$\mathbf{P} = \mathfrak{G}(\mathbf{F}) = \frac{\partial \Psi(\mathbf{F})}{\partial \mathbf{F}} \qquad \text{or} \qquad P_{aA} = \frac{\partial \Psi}{\partial F_{aA}} \;, \qquad (6.1)$$

and by use of relation (3.9) for the symmetric Cauchy stress tensor, i.e. $\boldsymbol{\sigma} = J^{-1}\mathbf{P}\mathbf{F}^{\mathrm{T}} = \boldsymbol{\sigma}^{\mathrm{T}}$,

$$\boldsymbol{\sigma} = \mathfrak{g}(\mathbf{F}) = J^{-1}\frac{\partial \Psi(\mathbf{F})}{\partial \mathbf{F}}\mathbf{F}^{\mathrm{T}} = J^{-1}\mathbf{F}\left(\frac{\partial \Psi(\mathbf{F})}{\partial \mathbf{F}}\right)^{\mathrm{T}}$$

$$\text{or} \qquad \sigma_{ab} = J^{-1}F_{bA}\frac{\partial \Psi}{\partial F_{aA}} = J^{-1}F_{aA}\frac{\partial \Psi}{\partial F_{bA}} \;. \qquad (6.2)$$

These types of equation we already know as (purely mechanical) constitutive equations (or equations of state). They establish an axiomatic or empirical model as the basis for approximating the behavior of a real material. Such a model we call a **material model** or a **constitutive model**. As is clear from the constitutive equations (6.1) and (6.2) the stress response of hyperelastic materials is derived from a given scalar-valued energy function, which implies that hyperelasticity has a conservative structure.

The derivative of the scalar-valued function Ψ with respect to the tensor variable $\mathbf{F}$ determines the gradient of Ψ and is understood according to the definition introduced in (1.239). It is a second-order tensor which we know as the first Piola-Kirchhoff stress tensor $\mathbf{P}$. The derivation requires that the component function $\Psi(F_{aA})$ is differentiable with respect to all components F_{aA}.

A so-called **perfectly elastic material** is by definition a material which produces locally no entropy (see TRUESDELL and NOLL [2004, Section 80]). In other words, we use subsequently the term 'perfectly' for a certain class of materials which has the special merits that for every admissible process the internal dissipation $\mathcal{D}_{int}$ is zero (naturally, damage, viscous mechanisms and plastic deformations are excluded). We will consider perfectly elastic materials up to Section 6.9.

We may derive the constitutive equation (6.1) directly from the Clausius-Planck form of the second law of thermodynamics (4.154) which degenerates to an equality for the class of perfectly elastic materials. With the expression (4.154) for $\mathcal{D}_{int}$ the time differentiation of the strain-energy function, i.e. $\dot{\Psi}(\mathbf{F}) = \partial\Psi(\mathbf{F})/\partial\mathbf{F} : \dot{\mathbf{F}}$, gives

$$\mathcal{D}_{int} = \mathbf{P} : \dot{\mathbf{F}} - \dot{\Psi} = \left(\mathbf{P} - \frac{\partial\Psi(\mathbf{F})}{\partial\mathbf{F}}\right) : \dot{\mathbf{F}} = 0 \ , \tag{6.3}$$

at every point of the continuum body and for all times during the process.

As $\mathbf{F}$ and hence $\dot{\mathbf{F}}$ can be chosen arbitrarily, the expressions in parentheses must be zero. Therefore, as a consequence of the second law of thermodynamics, the physical expression (6.1) holds. We often say that $\mathbf{P}$ is the **thermodynamic force** *work conjugate* to $\mathbf{F}$. This procedure goes back to COLEMAN and NOLL [1963] and COLEMAN and GURTIN [1967], and in the literature is sometimes referred to as the **Coleman-Noll procedure**.

For convenience, throughout this text we require that the strain-energy function vanishes in the reference configuration, i.e. where $\mathbf{F} = \mathbf{I}$. We express this assumption by the **normalization condition**

$$\Psi = \Psi(\mathbf{I}) = 0 \ . \tag{6.4}$$

From the physical observation we know that the strain-energy function Ψ increases with deformation. In addition to (6.4), we therefore require that

$$\Psi = \Psi(\mathbf{F}) \geq 0 \ , \tag{6.5}$$

which restricts the ranges of admissible functions occurring in expressions for the strain energy.

The strain-energy function Ψ attains its *global* minimum for $\mathbf{F} = \mathbf{I}$ at thermodynamic equilibrium (in fact, from (6.4), $\Psi(\mathbf{I})$ is zero). We assume that Ψ has no other stationary points in the strain space. Relations (6.4) and (6.5) ensure that the stress in the reference configuration, which we call the **residual stress**, is zero. We say that the reference configuration is **stress-free**.

For the behavior at finite strains we require additionally that the scalar-valued function Ψ must satisfy so-called **growth conditions**. This implies that Ψ tends to $+\infty$ if either $J = \det\mathbf{F}$ approaches $+\infty$ or 0^+, i.e.

$$\left.\begin{array}{lll}\Psi(\mathbf{F}) \rightarrow +\infty & \text{as} & \det\mathbf{F} \rightarrow +\infty \quad, \\[2mm] \Psi(\mathbf{F}) \rightarrow +\infty & \text{as} & \det\mathbf{F} \rightarrow 0^+ \quad.\end{array}\right\} \tag{6.6}$$

Physically, that means that we would require an infinite amount of strain energy in order to expand a continuum body to the infinite range or to compress it to a point with vanishing volume.

For further discussions see the books by, for example, CIARLET [1988] and OGDEN [1997].

Equivalent forms of the strain-energy function.   In order to illustrate Ψ we imagine a stretched (rubber) band with a certain amount of energy stored. The strain energy $\Psi(\mathbf{F})$ generated by the motion $\mathbf{x} = \boldsymbol{\chi}(\mathbf{X}, t)$ is assumed to be *objective*. This means, after a (possibly time-dependent) *translation* and *rotation* of the stretched (rubber) band in space, that the amount of energy stored is unchanged.

Hence, the strain energy $\Psi(\mathbf{F})$ must be equal to the strain energy $\Psi(\mathbf{F}^+)$ generated by a second motion $\mathbf{x}^+ = \boldsymbol{\chi}^+(\mathbf{X}, t^+)$ which differs from $\boldsymbol{\chi}$ by a superimposed rigid-body motion (recall Section 5.2). Employing the transformation rule for the deformation gradient $(5.33)_3$, we see that Ψ cannot be an arbitrary function of $\mathbf{F}$. In particular, it must obey the restriction

$$\Psi(\mathbf{F}) = \Psi(\mathbf{F}^+) = \Psi(\mathbf{Q}\mathbf{F}) \tag{6.7}$$

for all tensors $\mathbf{F}$, with $\det\mathbf{F} > 0$, and for all orthogonal tensors $\mathbf{Q}$, since $\mathbf{F}$ transforms into $\mathbf{Q}\mathbf{F}$, i.e. $F_{aA}^+ = Q_{ab}F_{bA}$.

In order to obtain equivalent formulations of (6.7) we take a special choice for $\mathbf{Q}$, namely the transpose of the proper orthogonal rotation tensor, $\mathbf{R}^{\mathrm{T}}$, and use the right polar decomposition $(2.93)_1$. Then, from (6.7), we find that $\Psi(\mathbf{F}) = \Psi(\mathbf{R}^{\mathrm{T}}\mathbf{F}) = \Psi(\mathbf{R}^{\mathrm{T}}\mathbf{R}\mathbf{U})$, and finally,

$$\Psi(\mathbf{F}) = \Psi(\mathbf{U}) \quad, \tag{6.8}$$

which holds for arbitrary $\mathbf{F}$.

From (6.8) we learn that Ψ is independent of the rotational part of $\mathbf{F} = \mathbf{R}\mathbf{U}$. We conclude that a hyperelastic material depends only on the stretching part of $\mathbf{F}$, i.e. the symmetric right stretch tensor $\mathbf{U}$. It is important to note that the relation $\Psi(\mathbf{F}) = \Psi(\mathbf{U})$ specifies the necessary and sufficient condition for the strain energy to be objective during superimposed rigid-body motions.

Since the right Cauchy-Green tensor and the Green-Lagrange strain tensor are given by $\mathbf{C} = \mathbf{U}^2$ and $\mathbf{E} = (\mathbf{U}^2 - \mathbf{I})/2$, we may express Ψ as a function of the six components C_{AB}, E_{AB} of the symmetric material tensors $\mathbf{C}, \mathbf{E}$, respectively. Hence, we may write

$$\Psi(\mathbf{F}) = \Psi(\mathbf{C}) = \Psi(\mathbf{E}) \quad. \tag{6.9}$$

For notational simplicity, here and elsewhere we will use the same Greek letter Ψ for different strain-energy functions.

Reduced forms of constitutive equations.   In the following we present some reduced forms of constitutive equations for hyperelastic materials at finite strains.

Consider the derivative of the strain-energy function $\Psi(\mathbf{F}) = \Psi(\mathbf{C})$ with respect to time t. By means of the chain rule of differentiation, property $(1.93)_1$ and the combination of $(2.168)_1$ and $(2.165)_1$, we obtain the expressions

$$\dot{\Psi} = \mathrm{tr}\left[\left(\frac{\partial\Psi(\mathbf{F})}{\partial\mathbf{F}}\right)^{\mathrm{T}}\dot{\mathbf{F}}\right] = \mathrm{tr}\left[\left(\frac{\partial\Psi(\mathbf{C})}{\partial\mathbf{C}}\right)\dot{\mathbf{C}}\right]$$

$$= \mathrm{tr}\left[\frac{\partial\Psi(\mathbf{C})}{\partial\mathbf{C}}\left(\dot{\overline{\mathbf{F}^{\mathrm{T}}\mathbf{F}}} + \mathbf{F}^{\mathrm{T}}\dot{\mathbf{F}}\right)\right] = 2\mathrm{tr}\left(\frac{\partial\Psi(\mathbf{C})}{\partial\mathbf{C}}\mathbf{F}^{\mathrm{T}}\dot{\mathbf{F}}\right) \quad , \qquad (6.10)$$

which must be valid for arbitrary tensors $\dot{\mathbf{F}}$. Since $\mathbf{C}$ is a symmetric second-order tensor, the gradient of the scalar-valued tensor function $\Psi(\mathbf{C})$, used in (6.10), is also symmetric. From (6.10) we deduce immediately that

$$\left(\frac{\partial\Psi(\mathbf{F})}{\partial\mathbf{F}}\right)^{\mathrm{T}} = 2\frac{\partial\Psi(\mathbf{C})}{\partial\mathbf{C}}\mathbf{F}^{\mathrm{T}} \quad , \qquad (6.11)$$

which, when substituted back into $(6.2)_3$, gives an important reduced form of the constitutive equation for hyperelastic materials, namely

$$\boldsymbol{\sigma} = J^{-1}\mathbf{F}\left(\frac{\partial\Psi(\mathbf{F})}{\partial\mathbf{F}}\right)^{\mathrm{T}} = 2J^{-1}\mathbf{F}\frac{\partial\Psi(\mathbf{C})}{\partial\mathbf{C}}\mathbf{F}^{\mathrm{T}}$$

$$(6.12)$$

$$\text{or} \qquad \sigma_{ab} = J^{-1}F_{aA}\frac{\partial\Psi}{\partial F_{bA}} = 2J^{-1}F_{aA}F_{bB}\frac{\partial\Psi}{\partial C_{AB}} \quad .$$

Alternative expressions may be obtained for the Piola-Kirchhoff stress tensors $\mathbf{P}$ (which is non-symmetric) and $\mathbf{S}$ (which is symmetric). From (3.8) and $(3.65)_1$ we find, by means of the stress relation $(6.12)_2$, the chain rule and $2\mathbf{E} = \mathbf{C} - \mathbf{I}$, that

$$\mathbf{P} = 2\mathbf{F}\frac{\partial\Psi(\mathbf{C})}{\partial\mathbf{C}} \quad , \qquad \mathbf{S} = 2\frac{\partial\Psi(\mathbf{C})}{\partial\mathbf{C}} = \frac{\partial\Psi(\mathbf{E})}{\partial\mathbf{E}}$$

$$(6.13)$$

$$\text{or} \qquad P_{aA} = 2F_{aB}\frac{\partial\Psi}{\partial C_{AB}} \quad , \qquad S_{AB} = 2\frac{\partial\Psi}{\partial C_{AB}} = \frac{\partial\Psi}{\partial E_{AB}} \quad .$$

Note that in the expression for Ψ we cannot use the symmetry condition $C_{AB} = C_{BA}$ before carrying out the differentiation $\partial\Psi/\partial C_{AB}$. Hence, the components in $\Psi(\mathbf{C})$ must be treated independently in evaluating (6.13). The same holds for Ψ in terms of the strain components E_{AB}. The response function occurring in the general constitutive equation (5.82) is with reference to $(6.13)_2$ determined by $\mathfrak{H}(\mathbf{C}) = 2\partial\Psi(\mathbf{C})/\partial\mathbf{C}$.

The Eshelby tensor and the tensor of chemical potential.   The **Eshelby tensor** (or the **(elastic) energy-momentum tensor**) is a crucial quantity in *fracture mechanics* and the *continuum theory of dislocations* (which are not discussed in this text). However, for completeness we present the **(isothermal) referential Eshelby tensor G**, which is, in general, non-symmetric and is defined as

$$
\mathbf{G} = -J\mathbf{F}^{\mathrm{T}}\frac{\partial}{\partial \mathbf{F}}\left(\frac{\Psi(\mathbf{F})}{J}\right) \qquad \text{or} \qquad G_{AB} = -JF_{aA}\frac{\partial}{\partial F_{aB}}\left(\frac{\Psi}{J}\right) \qquad (6.14)
$$

(see ESHELBY [1975] and CHADWICK [1975]), where $J = \det\mathbf{F}$ denotes the volume ratio. The physical dimension of the Eshelby tensor is the same as that of the strain energy.

The symmetric **tensor of chemical potential k** is related to **G** according to

$$
\mathbf{k} = \mathbf{F}^{-\mathrm{T}}\mathbf{G}\mathbf{F}^{\mathrm{T}} \qquad \text{or} \qquad k_{ab} = F_{Aa}^{-1}G_{AB}F_{bB} \qquad (6.15)
$$

see, for example, BOWEN [1976a].

The spatial tensor of chemical potential is required in the theory of *diffusing mixtures* (see TRUESDELL [1984] which contains more details).

Work done on hyperelastic materials.   We consider a dynamical process within some closed **time interval** denoted by $[t_1, t_2]$ in which the two arbitrary instants t_1 and t_2 are elements of the interval. The dynamical process is given by a motion $\mathbf{x} = \chi(\mathbf{X}, t)$ and the stress σ with the corresponding Cauchy traction vector $\mathbf{t}$ and the body force $\mathbf{b}$. During the process the body deforms according to the deformation gradient $\mathbf{F} = \mathbf{F}(t)$, with $t \in [t_1, t_2]$.

We say that a dynamical process is **closed** if $\mathbf{F}_1 = \mathbf{F}_2$. We introduce the definitions $\mathbf{F}_1 = \mathbf{F}(t_1)$ and $\mathbf{F}_2 = \mathbf{F}(t_2)$ of the deformation gradients in the **initial configuration** and the **final configuration** of the dynamical process, respectively.

Our next step is to determine the work done by the stress field on a continuum body of unit volume during a certain time interval $[t_1, t_2]$. Consider a body whose material properties are hyperelastic according to the general constitutive equation given in (6.1). Hence, from $(4.101)_1$ and the above definitions we find by means of the chain rule that

$$
\int_{t_1}^{t_2} \mathbf{P} : \dot{\mathbf{F}}\, dt = \int_{t_1}^{t_2} \frac{\partial\Psi}{\partial\mathbf{F}} : \dot{\mathbf{F}}\, dt = \int_{t_1}^{t_2} \frac{\mathrm{D}\Psi(\mathbf{F})}{\mathrm{D}t}\, dt = \Psi(\mathbf{F}_2) - \Psi(\mathbf{F}_1) \ , \qquad (6.16)
$$

which, for a closed dynamical process with $\mathbf{F}_1 = \mathbf{F}_2$, reduces to

$$
\int_{t_1}^{t_2} \mathbf{P} : \dot{\mathbf{F}}\, dt = \Psi(\mathbf{F}_2) - \Psi(\mathbf{F}_1) = 0 \ . \qquad (6.17)
$$

Thus, as distinct from Cauchy-elastic materials (see Section 5.4), the actual work done by the stress field on a hyperelastic material during a certain (closed) time interval depends only on the *initial* and *final* configurations (path independent). In fact, the work is *zero* in closed dynamical processes. This important result also holds for continuum bodies which may undergo inhomogeneous deformations, in which $\Psi = \Psi(\mathbf{F}, \mathbf{X})$ and $\rho_0 = \rho_0(\mathbf{X})$, with $\mathbf{F} = \mathbf{F}(\mathbf{X}, t)$.

EXERCISES

1. Expand the strain-energy function $\Psi(\mathbf{F})$ in the form of tensorial polynomials in the vicinity of the reference configuration, i.e. for $\mathbf{F} = \mathbf{I}$,

$$\Psi(\mathbf{F}) = \Psi(\mathbf{I}) + (\mathbf{F} - \mathbf{I}) : \frac{\partial \Psi(\mathbf{F})}{\partial \mathbf{F}} + \cdots \geq 0$$

   for all $\mathbf{F} - \mathbf{I}$. Using relations (6.4) and (6.5) show that the stress in the reference configuration is zero.

2. Recall the definitions of the referential Eshelby tensor $\mathbf{G}$ and the tensor of chemical potential $\mathbf{k}$.

   (a) Show that the forms

$$\mathbf{G} = \Psi(\mathbf{F})\mathbf{I} - J\mathbf{F}^{\mathrm{T}}\boldsymbol{\sigma}\mathbf{F}^{-\mathrm{T}} \quad, \qquad \mathbf{k} = \Psi(\mathbf{F})\mathbf{I} - \boldsymbol{\tau} \qquad (6.18)$$

   are equivalent to those given in (6.14) and (6.15), where $\boldsymbol{\tau}$ is the Kirchhoff stress tensor, as defined in eq. (3.62).

   (b) By applying requirement (3.10), show with $(6.18)_1$ that

$$\mathbf{G}\mathbf{C} = \mathbf{C}\mathbf{G}^{\mathrm{T}} \quad,$$

   with the right Cauchy-Green tensor $\mathbf{C} = \mathbf{F}^{\mathrm{T}}\mathbf{F}$, as defined in eq. (2.66).

6.2 Isotropic Hyperelastic Materials

We now restrict the strain-energy function by a particular property that the material may possess, namely *isotropy*. This property is based on the physical idea that the response of the material, when studied in a stress-strain experiment, is the same in all directions. One example of an (approximately) isotropic material with a wide range of applications is rubber.

In this section we are concerned with the mathematical formulation of isotropy within the context of hyperelasticity.

Scalar-valued isotropic tensor function. We consider an arbitrary point $\mathbf{X}$ of an elastic continuum body occupying the region Ω_0 (reference configuration) at time $t = 0$. A motion χ may carry this point $\mathbf{X} \in \Omega_0$ to a place $\mathbf{x} = \chi(\mathbf{X}, t)$ specifying a location in the region Ω (current configuration) at time t.

We now study the effect of a rigid-body motion superimposed on the *reference configuration*. We postulate that the body occupying the region Ω_0 is translated by the vector $\mathbf{c}$ and rotated by the orthogonal tensor $\mathbf{Q}$ according to

$$\mathbf{X}^\star = \mathbf{c} + \mathbf{Q}\mathbf{X} \ , \tag{6.19}$$

which moves Ω_0 to a new region $\Omega_0^\star$ (new reference configuration), and the arbitrary point with position vector $\mathbf{X}$ to a new location identified by the position vector $\mathbf{X}^\star \in \Omega_0^\star$ (see Figure 6.1).

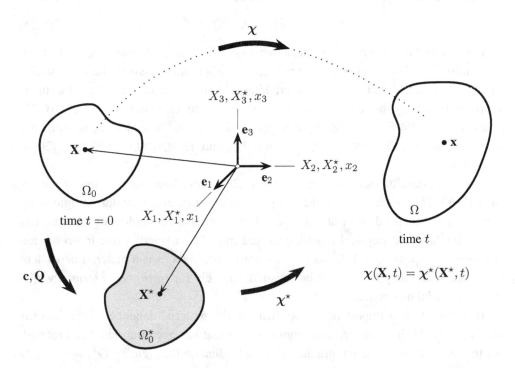

Figure 6.1 Rigid-body motion superimposed on the reference configuration.

We now demand that a different motion $\mathbf{x} = \chi^\star(\mathbf{X}^\star, t)$ moves $\Omega_0^\star$ to the current configuration Ω so that

$$\mathbf{x} = \chi(\mathbf{X}, t) = \chi^\star(\mathbf{X}^\star, t) \ , \tag{6.20}$$

mapping $\mathbf{X}^\star$ to place $\mathbf{x}$. By the chain rule and relation (6.19) the deformation gradient

$\mathbf{F}$ may be expressed as

$$\mathbf{F} = \frac{\partial \mathbf{x}}{\partial \mathbf{X}} = \frac{\partial \mathbf{x}}{\partial \mathbf{X}^\star}\mathbf{Q} = \mathbf{F}^\star \mathbf{Q}$$

$$\text{or} \qquad F_{aA} = \frac{\partial x_a}{\partial X_A} = \frac{\partial x_a}{\partial X_B^\star}Q_{BA} = F_{aB}^\star Q_{BA} \quad , \tag{6.21}$$

where $\mathbf{F}^\star = \partial \mathbf{x}/\partial \mathbf{X}^\star$ is defined to be the deformation gradient relative to the region $\Omega_0^\star$. From $(6.21)_3$ we find the important transformation, namely

$$\mathbf{F}^\star = \mathbf{F}\mathbf{Q}^\mathrm{T} \qquad \text{or} \qquad F_{aA}^\star = F_{aB}Q_{AB} \quad . \tag{6.22}$$

Hence, we say that a hyperelastic material is **isotropic** relative to the reference configuration Ω_0 if the values of the strain energy $\Psi(\mathbf{F})$ and $\Psi(\mathbf{F}^\star)$ are the same for all orthogonal tensors $\mathbf{Q}$. With (6.22) we may write

$$\Psi(\mathbf{F}) = \Psi(\mathbf{F}^\star) = \Psi(\mathbf{F}\mathbf{Q}^\mathrm{T}) \quad . \tag{6.23}$$

In other words, if we can show that a motion of an elastic body superimposed on any particularly translated and/or rotated reference configuration leads to the same strain-energy function at time t, then the material is said to be isotropic. However, if a superimposed rigid-body motion changes the strain-energy function in the sense that (6.23) is not satisfied ($\Psi(\mathbf{F}) \neq \Psi(\mathbf{F}^\star)$) the hyperelastic material is said to be **anisotropic** (see, for example, OGDEN [1997, Section 4.2.5] and TRUESDELL and NOLL [2004, Section 33]).

It is important to mention that relation (6.23) is *fundamentally distinct* from requirement (6.7), which says that the strain energy must be objective during rigid-body motions, i.e. independent of an observer. The later condition holds for **all** materials (it is a fundamental physical requirement and must be satisfied) while the condition of isotropic response (6.23) holds only for **some** materials (it is a material-dependent requirement and may or may not be satisfied), namely for *isotropic materials*, which makes a crucial difference.

In addition, it is important to note that for the material-dependent requirements (6.22) and (6.23) it is the *reference* configuration that has been translated and rotated. For that case the deformation gradient $\mathbf{F}$ is multiplied on the *right* by $\mathbf{Q}^\mathrm{T}$, whereby $\mathbf{Q}$ acts with material coordinates, i.e. Q_{AB}. However, for the objectivity requirements $(5.33)_3$ and (6.7), it is the *current* configuration that has been translated and rotated, and $\mathbf{F}$ is multiplied on the *left* by $\mathbf{Q}$, acting with spatial coordinates, i.e. Q_{ab}.

We now suppose that during motion $\mathbf{x} = \chi(\mathbf{X}, t)$ the strain-energy function may adopt the form $\Psi(\mathbf{F}) = \Psi(\mathbf{C})$; recall eq. $(6.9)_1$. If we restrict the hyperelastic material to isotropic hyperelastic response we require that $\Psi(\mathbf{C}) = \Psi(\mathbf{C}^\star)$, with $\mathbf{C}^\star = \mathbf{F}^{\star\mathrm{T}}\mathbf{F}^\star$. With reference to (6.22) we conclude that

$$\Psi(\mathbf{C}) = \Psi(\mathbf{F}^{\star\mathrm{T}}\mathbf{F}^\star) = \Psi(\mathbf{Q}\mathbf{F}^\mathrm{T}\mathbf{F}\mathbf{Q}^\mathrm{T}) \quad , \tag{6.24}$$

implying with the right Cauchy-Green tensor $\mathbf{C} = \mathbf{F}^T\mathbf{F}$, that

$$\Psi(\mathbf{C}) = \Psi(\mathbf{QCQ}^T) \ . \tag{6.25}$$

If the requirement for isotropy (6.25) holds for all *symmetric* tensors $\mathbf{C}$ and or-thogonal tensors $\mathbf{Q}$, we say that the strain-energy function $\Psi(\mathbf{C})$ is a **scalar-valued isotropic tensor function** of one variable $\mathbf{C}$ or simply an **invariant** of the symmetric tensor $\mathbf{C}$. We can show, if the strain-energy function is an invariant, then its gradient is a tensor-valued isotropic tensor function.

EXAMPLE 6.1 Assume that the hyperelastic material is restricted to isotropic re-sponse. Show that the strain energy may be expressed by the identity

$$\Psi(\mathbf{C}) = \Psi(\mathbf{b}) \ , \tag{6.26}$$

where $\Psi(\mathbf{b})$ characterizes an isotropic function of the (spatial) left Cauchy-Green ten-sor $\mathbf{b} = \mathbf{FF}^T$.

Solution. We substitute for $\mathbf{Q}$ in condition (6.25) the proper orthogonal rotation ten-sor $\mathbf{R}$. Then, kinematic relation (2.109)$_3$ implies (6.26), which holds for any isotropic deformation. ∎

Constitutive equations in terms of invariants. If a scalar-valued tensor function is an invariant under a rotation, according to (6.25), it may be expressed in terms of the principal invariants of its argument (for example, $\mathbf{C}$ or $\mathbf{b}$), which is a fundamental result for isotropic scalar functions, known as the **representation theorem for invariants** (for a proof see, for example, GURTIN [1981a, p. 231] or TRUESDELL and NOLL [2004, Section 10]).

Having this in mind, the strain energies, as established in eq. (6.26), may be ex-pressed as a set of independent strain invariants of the symmetric Cauchy-Green ten-sors $\mathbf{C}$ and $\mathbf{b}$, namely, through $I_a = I_a(\mathbf{C})$ and $I_a = I_a(\mathbf{b})$, $a = 1, 2, 3$, respectively. With reference to (6.26), we may write equivalently

$$\Psi = \Psi[I_1(\mathbf{C}), I_2(\mathbf{C}), I_3(\mathbf{C})] = \Psi[I_1(\mathbf{b}), I_2(\mathbf{b}), I_3(\mathbf{b})] \ . \tag{6.27}$$

Again, (6.27) is exclusively valid for *isotropic* hyperelastic materials satisfying con-dition (6.25) for all orthogonal tensors $\mathbf{Q}$. Since $\mathbf{C}$ and $\mathbf{b}$ have the same eigenvalues, which are the squares of the principal stretches λ_a^2, $a = 1, 2, 3$, we conclude that

$$I_1(\mathbf{C}) = I_1(\mathbf{b}) \ , \qquad I_2(\mathbf{C}) = I_2(\mathbf{b}) \ , \qquad I_3(\mathbf{C}) = I_3(\mathbf{b}) \ , \tag{6.28}$$

where the three principal invariants are explicitly given in accordance with eqs. (5.89)–

(5.91). Note that for the stress-free reference configuration, the strain-energy functions (6.27), with (5.89)–(5.91), must satisfy the normalization condition (6.4), i.e. $\Psi = 0$, for $I_1 = I_2 = 3$ and $I_3 = 1$. The representation in the form of invariants was established in the classical work of RIVLIN [1948].

In order to determine constitutive equations for isotropic hyperelastic materials in terms of strain invariants, consider a differentiation of $\Psi(\mathbf{C}) = \Psi(I_1, I_2, I_3)$ with respect to tensor $\mathbf{C}$. We assume that $\Psi(\mathbf{C})$ has continuous derivatives with respect to the principal invariants I_a, $a = 1, 2, 3$. By means of the chain rule of differentiation we find

$$\frac{\partial \Psi(\mathbf{C})}{\partial \mathbf{C}} = \frac{\partial \Psi}{\partial I_1}\frac{\partial I_1}{\partial \mathbf{C}} + \frac{\partial \Psi}{\partial I_2}\frac{\partial I_2}{\partial \mathbf{C}} + \frac{\partial \Psi}{\partial I_3}\frac{\partial I_3}{\partial \mathbf{C}} = \sum_{a=1}^{3} \frac{\partial \Psi}{\partial I_a}\frac{\partial I_a}{\partial \mathbf{C}} \quad . \tag{6.29}$$

The derivative of the first invariant I_1 with respect to $\mathbf{C}$, as needed for (6.29), gives with $(6.28)_1$, $(5.89)_1$ and the property (1.94) of double contraction

$$\frac{\partial I_1}{\partial \mathbf{C}} = \frac{\partial \mathrm{tr}\mathbf{C}}{\partial \mathbf{C}} = \frac{\partial (\mathbf{I}:\mathbf{C})}{\partial \mathbf{C}} = \mathbf{I} \quad \text{or} \quad \frac{\partial I_1}{\partial C_{AB}} = \delta_{AB} \quad . \tag{6.30}$$

The derivatives of the remaining two invariants with respect to $\mathbf{C}$ follow from eqs. $(5.90)_1$ and $(5.91)_1$, by means of (6.28), $(6.30)_3$, the chain rule and relations $(1.252)_2$, (1.241) (use the *symmetric* tensor $\mathbf{C}$ in the place of $\mathbf{A}$), and have the forms

$$\frac{\partial I_2}{\partial \mathbf{C}} = \frac{1}{2}\left(2\mathrm{tr}\mathbf{C}\,\mathbf{I} - \frac{\partial \mathrm{tr}(\mathbf{C}^2)}{\partial \mathbf{C}} \right) = I_1\mathbf{I} - \mathbf{C} \quad , \qquad \frac{\partial I_3}{\partial \mathbf{C}} = I_3\mathbf{C}^{-1}$$

$$\text{or} \quad \frac{\partial I_2}{\partial C_{AB}} = I_1\delta_{AB} - C_{AB} \quad , \qquad \frac{\partial I_3}{\partial C_{AB}} = I_3 C_{AB}^{-1} \quad . \tag{6.31}$$

Substituting (6.29)–(6.31) into constitutive equation $(6.13)_2$ gives the *most general form* of a stress relation in terms of the three strain invariants, which characterizes isotropic hyperelastic materials at finite strains, i.e.

$$\mathbf{S} = 2\frac{\partial \Psi(\mathbf{C})}{\partial \mathbf{C}} = 2\left[\left(\frac{\partial \Psi}{\partial I_1} + I_1\frac{\partial \Psi}{\partial I_2} \right)\mathbf{I} - \frac{\partial \Psi}{\partial I_2}\mathbf{C} + I_3\frac{\partial \Psi}{\partial I_3}\mathbf{C}^{-1} \right] \quad . \tag{6.32}$$

The gradient of the invariant $\Psi(\mathbf{C}) = \Psi(I_1, I_2, I_3)$ has the simple representation (6.32), which is a fundamental relationship in the theory of finite hyperelasticity. Note that (6.32) is a general representation for three dimensions, in which Ψ may adopt any scalar-valued isotropic function of one symmetric second-order tensor variable.

Multiplication of $(6.32)_2$ by tensor $\mathbf{C}$ from the right-hand side or from the left-hand side leads to the same result. We say that $\partial \Psi(\mathbf{C})/\partial \mathbf{C}$ **commutes** (or, is **coaxial**) with $\mathbf{C}$ in the sense that

$$\frac{\partial \Psi(\mathbf{C})}{\partial \mathbf{C}}\mathbf{C} = \mathbf{C}\frac{\partial \Psi(\mathbf{C})}{\partial \mathbf{C}} \quad , \tag{6.33}$$

which is an essential consequence of isotropy.

Next, we present the spatial counterpart of constitutive equation (6.32). According to relation (3.66), the Cauchy stress $\boldsymbol{\sigma}$ follows from the second Piola-Kirchhoff stress $\mathbf{S}$ by the Piola transformation $\boldsymbol{\sigma} = J^{-1}\mathbf{F}\mathbf{S}\mathbf{F}^{\mathrm{T}}$. By multiplying the tensor variables $\mathbf{I}, \mathbf{C}, \mathbf{C}^{-1}$ with $\mathbf{F}$ from the left-hand side and with $\mathbf{F}^{\mathrm{T}}$ from the right-hand side, we may write by means of the left Cauchy-Green tensor $\mathbf{b} = \mathbf{F}\mathbf{F}^{\mathrm{T}}$, that $\mathbf{F}\mathbf{I}\mathbf{F}^{\mathrm{T}} = \mathbf{F}\mathbf{F}^{\mathrm{T}} = \mathbf{b}$, $\mathbf{F}\mathbf{C}\mathbf{F}^{\mathrm{T}} = (\mathbf{F}\mathbf{F}^{\mathrm{T}})^2 = \mathbf{b}^2$, $\mathbf{F}\mathbf{C}^{-1}\mathbf{F}^{\mathrm{T}} = (\mathbf{F}\mathbf{F}^{-1})(\mathbf{F}^{-\mathrm{T}}\mathbf{F}^{\mathrm{T}}) = \mathbf{I}$. With (6.32) we deduce from $\boldsymbol{\sigma} = J^{-1}\mathbf{F}\mathbf{S}\mathbf{F}^{\mathrm{T}}$ that

$$\boldsymbol{\sigma} = 2J^{-1}\left[I_3\frac{\partial\Psi}{\partial I_3}\mathbf{I} + \left(\frac{\partial\Psi}{\partial I_1} + I_1\frac{\partial\Psi}{\partial I_2}\right)\mathbf{b} - \frac{\partial\Psi}{\partial I_2}\mathbf{b}^2 \right] \ . \tag{6.34}$$

Following arguments analogous to those which led from (5.88) to (5.94), we find an alternative form to (6.34), namely

$$\boldsymbol{\sigma} = 2J^{-1}\left[\left(I_2\frac{\partial\Psi}{\partial I_2} + I_3\frac{\partial\Psi}{\partial I_3}\right)\mathbf{I} + \frac{\partial\Psi}{\partial I_1}\mathbf{b} - I_3\frac{\partial\Psi}{\partial I_2}\mathbf{b}^{-1} \right] \ . \tag{6.35}$$

By comparing (6.34) and (6.35) with (5.88) and (5.94) we may derive the response coefficients α_a, $a = 0, 1, 2$, and β_a, $a = 0, 1, -1$, in terms of the strain-energy function (6.27). We obtain the forms

$$\alpha_0 = 2I_3^{1/2}\frac{\partial\Psi}{\partial I_3} \ , \qquad \alpha_1 = 2I_3^{-1/2}\left(\frac{\partial\Psi}{\partial I_1} + I_1\frac{\partial\Psi}{\partial I_2}\right) \ ,$$

$$\alpha_2 = -2I_3^{-1/2}\frac{\partial\Psi}{\partial I_2} \ , \tag{6.36}$$

$$\beta_0 = 2I_3^{-1/2}\left(I_2\frac{\partial\Psi}{\partial I_2} + I_3\frac{\partial\Psi}{\partial I_3}\right) \ , \qquad \beta_1 = 2I_3^{-1/2}\frac{\partial\Psi}{\partial I_1} \ ,$$

$$\beta_{-1} = -2I_3^{1/2}\frac{\partial\Psi}{\partial I_2} \ , \tag{6.37}$$

which specify the first and the second representation theorem for isotropic tensor functions, i.e. (5.88) and (5.94), respectively.

Note that in order to formulate constitutive equations which are not restricted to isotropic response and which satisfy the objectivity requirement, it is appropriate to use quantities which are referred to the reference configuration. Within the context of hyperelasticity it is obvious to use the right Cauchy-Green tensor $\mathbf{C}$ and its work conjugate stress field, i.e. the second Piola-Kirchhoff stress tensor $\mathbf{S}$.

Constitutive equations derived from $\Psi(\mathbf{b})$ and $\Psi(\mathbf{v})$. If the strain-energy function depends on the (symmetric) left Cauchy-Green tensor $\mathbf{b}$, then the isotropic hyperelastic response is

$$\boldsymbol{\sigma} = 2J^{-1}\frac{\partial\Psi(\mathbf{b})}{\partial\mathbf{b}}\mathbf{b} = 2J^{-1}\mathbf{b}\frac{\partial\Psi(\mathbf{b})}{\partial\mathbf{b}} \qquad \text{or} \qquad \sigma_{ab} = 2J^{-1}b_{ac}\frac{\partial\Psi}{\partial b_{cb}} \tag{6.38}$$

(see also TRUESDELL and NOLL [2004, Section 85, p. 313]), where $\Psi(\mathbf{b})$ is a scalar-valued isotropic function of the tensor variable $\mathbf{b} = \mathbf{FF}^{\mathrm{T}}$. This constitutive equation plays an essential part in isotropic finite hyperelasticity. On comparison with eq. (5.85) we deduce that the right-hand side of eq. (6.38) corresponds to the response function $\mathfrak{h}(\mathbf{b})$.

Proof.   In order to obtain the constitutive equation (6.38), we start by differentiating the postulated strain-energy function $\Psi(\mathbf{b})$ with respect to time t. Considering symmetries, a straightforward algebraic manipulation gives, by means of the chain rule, relation (2.171) and the property (1.95) of double contraction,

$$\dot{\Psi} = \frac{\partial \Psi(\mathbf{b})}{\partial \mathbf{b}} : \dot{\mathbf{b}} = \frac{\partial \Psi(\mathbf{b})}{\partial \mathbf{b}} : (\mathbf{lb} + \mathbf{bl}^{\mathrm{T}})$$

$$= 2\frac{\partial \Psi(\mathbf{b})}{\partial \mathbf{b}} : \mathbf{lb} = 2\frac{\partial \Psi(\mathbf{b})}{\partial \mathbf{b}}\mathbf{b} : \mathbf{l} \quad , \tag{6.39}$$

where $\mathbf{l}$ is the spatial velocity gradient, in general a non-symmetric tensor.

With respect to eqs. (6.32) and (6.33), taking $\mathbf{b}$ in place of $\mathbf{C}$, we deduce that $\partial \Psi(\mathbf{b})/\partial \mathbf{b}$ commutes with the symmetric second-order tensor $\mathbf{b}$, in the sense that

$$\frac{\partial \Psi(\mathbf{b})}{\partial \mathbf{b}}\mathbf{b} = \mathbf{b}\frac{\partial \Psi(\mathbf{b})}{\partial \mathbf{b}} \quad , \tag{6.40}$$

implying the symmetry of tensor $(\partial \Psi(\mathbf{b})/\partial \mathbf{b})\mathbf{b}$.

However, from (4.154), we know that for perfectly elastic materials (for which $\mathcal{D}_{\mathrm{int}} = 0$) the stress power w_{int} per unit reference volume equals $\dot{\Psi}$. The combination of identity (4.113)$_1$, i.e. $w_{\mathrm{int}} = J\boldsymbol{\sigma} : \mathbf{d}$, and eq. (6.39)$_4$ with the requirement (6.40) for isotropy implies

$$J\boldsymbol{\sigma} : \mathbf{d} = 2\frac{\partial \Psi(\mathbf{b})}{\partial \mathbf{b}}\mathbf{b} : \mathbf{d} = 2\mathbf{b}\frac{\partial \Psi(\mathbf{b})}{\partial \mathbf{b}} : \mathbf{d} \quad , \tag{6.41}$$

where the rate of deformation tensor $\mathbf{d}$ is the symmetric part of $\mathbf{l}$. For this relation we used the fact that the double contraction of a symmetric tensor and a skew tensor is zero.

Simple arguments reduce relation (6.41) to the desired fundamental constitutive equation (6.38) for isotropic response.   ∎

Since the left stretch tensor $\mathbf{v} = \mathbf{b}^{1/2}$ is the unique square root of $\mathbf{b}$, the strain energy may also be expressed as an isotropic function of $\mathbf{v}$; thus, relation (6.26) may be extended to $\Psi(\mathbf{C}) = \Psi(\mathbf{b}) = \Psi(\mathbf{v})$. Note that for notational convenience, we do not distinguish between different strain-energy functions Ψ.

With respect to constitutive equation (6.38) we derive an equivalent form in terms of the left stretch tensor $\mathbf{v}$. Analogous to the procedure which was used to establish eq. (6.11) we find that

$$2\frac{\partial \Psi(\mathbf{b})}{\partial \mathbf{b}} = \frac{\partial \Psi(\mathbf{v})}{\partial \mathbf{v}}\mathbf{v}^{-1} \quad , \tag{6.42}$$

so that (6.38) reads

$$\boldsymbol{\sigma} = J^{-1}\frac{\partial \Psi(\mathbf{v})}{\partial \mathbf{v}}\mathbf{v} = J^{-1}\mathbf{v}\frac{\partial \Psi(\mathbf{v})}{\partial \mathbf{v}} \quad , \tag{6.43}$$

which is another important stress relation characterizing the behavior of isotropic hyperelastic materials at finite strains. Note that, since $\mathbf{v}$ is the unique square root of $\mathbf{b}$, $\partial \Psi/\partial \mathbf{b}$ also commutes with $\mathbf{v}$.

Constitutive equations in terms of principal stretches. If the strain-energy function Ψ is an invariant, we may regard Ψ as a function of the principal stretches λ_a, $a = 1, 2, 3$. In the place of (6.27), we may represent Ψ in the form

$$\Psi = \Psi(\mathbf{C}) = \Psi(\lambda_1, \lambda_2, \lambda_3) \quad . \tag{6.44}$$

For the stress-free reference configuration the normalization condition (6.4) takes on the form $\Psi(1, 1, 1) = 0$.

Consider the left stretch tensor $\mathbf{v} = \mathbf{b}^{1/2}$ describing the deformed state of an isotropic hyperelastic material. From the eigenvalue problem $(2.118)_3$ we know that λ_a denote the three principal stretches (the real eigenvalues) of $\mathbf{v}$. Since the principal directions of $\mathbf{v}$ coincide with those of $\mathbf{b}$ (compare with eqs. $(2.118)_3$ and $(2.119)_2$) they also coincide with the principal directions of the Cauchy stress tensor $\boldsymbol{\sigma}$ (recall representation (5.88)).

Consequently, with respect to (6.43) the principal Cauchy stresses σ_a, $a = 1, 2, 3$, simply result in

$$\sigma_a = J^{-1}\lambda_a\frac{\partial \Psi}{\partial \lambda_a} \quad , \qquad a = 1, 2, 3 \quad , \tag{6.45}$$

with the volume ratio

$$J = \lambda_1\lambda_2\lambda_3 \quad , \tag{6.46}$$

according to $(5.91)_3$.

In addition to (6.45), we introduce equivalent relations for the three principal Piola-Kirchhoff stresses P_a and S_a, namely

$$P_a = \frac{\partial \Psi}{\partial \lambda_a} \quad , \qquad S_a = \frac{1}{\lambda_a}\frac{\partial \Psi}{\partial \lambda_a} \quad , \qquad a = 1, 2, 3 \tag{6.47}$$

(compare with the following Example 6.2), which may be expressed in terms of the Cauchy stresses (6.45) as

$$P_a = J\lambda_a^{-1}\sigma_a \quad , \qquad S_a = J\lambda_a^{-2}\sigma_a \quad , \qquad a = 1, 2, 3 \quad . \tag{6.48}$$

Constitutive relations (6.45) and (6.47) show that principal stresses in an isotropic hyperelastic material depend only upon the principal stretches. They are simply obtained by differentiating the strain-energy function with respect to the corresponding principal stretches.

EXAMPLE 6.2   Consider the strain energy $\Psi(\mathbf{C}) = \Psi(\lambda_1, \lambda_2, \lambda_3)$. Obtain the constitutive equations in the spectral forms

$$\boldsymbol{\sigma} = \sum_{a=1}^{3} \sigma_a \hat{\mathbf{n}}_a \otimes \hat{\mathbf{n}}_a \ , \tag{6.49}$$

$$\mathbf{P} = \sum_{a=1}^{3} P_a \hat{\mathbf{n}}_a \otimes \hat{\mathbf{N}}_a \ , \qquad \mathbf{S} = \sum_{a=1}^{3} S_a \hat{\mathbf{N}}_a \otimes \hat{\mathbf{N}}_a \ , \tag{6.50}$$

where σ_a and P_a, S_a, $a = 1, 2, 3$, are the principal values of the Cauchy stress tensor $\boldsymbol{\sigma}$ and the two Piola-Kirchhoff stress tensors $\mathbf{P}$, $\mathbf{S}$ according to the expressions (6.45) and (6.47), respectively. The orthonormal vectors $\hat{\mathbf{N}}_a$ and $\hat{\mathbf{n}}_a = \mathbf{R}\hat{\mathbf{N}}_a$, $a = 1, 2, 3$, denote the principal referential and spatial directions (axes of stretch), respectively.

These constitutive equations describe isotropic response of hyperelastic materials and hold *if and only if* $\lambda_1 \neq \lambda_2 \neq \lambda_3 \neq \lambda_1$.

Solution.   We start with constitutive equation (6.50)$_2$ which is expressed in terms of second Piola-Kirchhoff stresses S_a, $a = 1, 2, 3$. We compute the derivative of the isotropic function $\Psi(\mathbf{C})$ with respect to the symmetric tensor $\mathbf{C}$. By means of the chain rule and kinematic relation (2.125), we obtain for the general case $\lambda_1 \neq \lambda_2 \neq \lambda_3 \neq \lambda_1$,

$$\frac{\partial \Psi(\mathbf{C})}{\partial \mathbf{C}} = \sum_{a=1}^{3} \frac{\partial \Psi}{\partial \lambda_a^2} \frac{\partial \lambda_a^2}{\partial \mathbf{C}} = \sum_{a=1}^{3} \frac{\partial \Psi}{\partial \lambda_a^2} \hat{\mathbf{N}}_a \otimes \hat{\mathbf{N}}_a \ . \tag{6.51}$$

In (6.51)$_2$, λ_a^2 are the eigenvalues (the squares of the principal stretches) and $\hat{\mathbf{N}}_a$ the corresponding eigenvectors (principal referential directions) of $\mathbf{C}$ (compare with the eigenvalue problem (2.117)). With (6.51)$_2$ and the chain rule we find from (6.13)$_2$ that

$$\mathbf{S} = 2\frac{\partial \Psi(\mathbf{C})}{\partial \mathbf{C}} = \sum_{a=1}^{3} \underbrace{\frac{1}{\lambda_a} \frac{\partial \Psi}{\partial \lambda_a}}_{S_a} \hat{\mathbf{N}}_a \otimes \hat{\mathbf{N}}_a \ , \tag{6.52}$$

which gives the desired results (6.47)$_2$ and (6.50)$_2$.

By the use of (6.52)$_2$, the relation according to (1.58) and eq. (2.134)$_1$, i.e. $\mathbf{F}\hat{\mathbf{N}}_a = \lambda_a \hat{\mathbf{n}}_a$, $a = 1, 2, 3$, the spectral form of the first Piola-Kirchhoff stress tensor $\mathbf{P}$ may be found from transformation (3.67) as

$$\mathbf{P} = \mathbf{F}\mathbf{S} = \mathbf{F} \left(\sum_{a=1}^{3} \frac{1}{\lambda_a} \frac{\partial \Psi}{\partial \lambda_a} \hat{\mathbf{N}}_a \otimes \hat{\mathbf{N}}_a \right)$$

$$= \sum_{a=1}^{3} \frac{1}{\lambda_a} \frac{\partial \Psi}{\partial \lambda_a} (\mathbf{F}\hat{\mathbf{N}}_a) \otimes \hat{\mathbf{N}}_a = \sum_{a=1}^{3} \underbrace{\frac{\partial \Psi}{\partial \lambda_a}}_{P_a} \hat{\mathbf{n}}_a \otimes \hat{\mathbf{N}}_a \ . \tag{6.53}$$

Similarly, having in mind the results $(6.53)_4$ and $(2.134)_1$, transformation (3.9) gives, using (3.10), the spectral form

$$\boldsymbol{\sigma} = J^{-1}\mathbf{F}\mathbf{P}^{\mathrm{T}} = J^{-1}\mathbf{F} \left(\sum_{a=1}^{3} \frac{\partial \Psi}{\partial \lambda_a} (\hat{\mathbf{n}}_a \otimes \hat{\mathbf{N}}_a)^{\mathrm{T}} \right) = \sum_{a=1}^{3} \underbrace{J^{-1} \lambda_a \frac{\partial \Psi}{\partial \lambda_a}}_{\sigma_a} \hat{\mathbf{n}}_a \otimes \hat{\mathbf{n}}_a \tag{6.54}$$

of the Cauchy stress tensor, where the property $\mathbf{F}(\hat{\mathbf{n}}_a \otimes \hat{\mathbf{N}}_a)^{\mathrm{T}} = (\mathbf{F}\hat{\mathbf{N}}_a) \otimes \hat{\mathbf{n}}_a$, $a = 1, 2, 3$, was used (compare with relations which are analogous to (1.85) and (1.58)). ■

EXERCISES

1. By analogy with the procedure which led to (6.11), obtain the eq. (6.42) and relation

$$\left(\frac{\partial \Psi(\mathbf{F})}{\partial \mathbf{F}} \right)^{\mathrm{T}} = 2\mathbf{F}^{\mathrm{T}} \frac{\partial \Psi(\mathbf{b})}{\partial \mathbf{b}} \ . \tag{6.55}$$

2. Rewrite the spectral representations of constitutive equations (6.49) and (6.50) for a given strain-energy function of the particular form $\Psi = \Psi(\ln\lambda_1, \ln\lambda_2, \ln\lambda_3)$. Consider the general case $\lambda_1 \neq \lambda_2 \neq \lambda_3 \neq \lambda_1$.

3. Take the strain energy Ψ as a function of the principal stretches characterizing the behavior of isotropic hyperelastic materials. Let at least one principal stretch be equal to the other.

    (a) For the case in which we have two equal principal stretches, namely $\lambda_1 = \lambda_2 \neq \lambda_3$, obtain the constitutive equations

$$\boldsymbol{\sigma} = J^{-1}\lambda_1 \frac{\partial \Psi}{\partial \lambda_1}(\hat{\mathbf{n}}_1 \otimes \hat{\mathbf{n}}_1 + \hat{\mathbf{n}}_2 \otimes \hat{\mathbf{n}}_2) + J^{-1}\lambda_3 \frac{\partial \Psi}{\partial \lambda_3}\hat{\mathbf{n}}_3 \otimes \hat{\mathbf{n}}_3 \ ,$$

$$\mathbf{P} = \frac{\partial \Psi}{\partial \lambda_1}(\hat{\mathbf{n}}_1 \otimes \hat{\mathbf{N}}_1 + \hat{\mathbf{n}}_2 \otimes \hat{\mathbf{N}}_2) + \frac{\partial \Psi}{\partial \lambda_3}\hat{\mathbf{n}}_3 \otimes \hat{\mathbf{N}}_3 \ ,$$

$$\mathbf{S} = \frac{1}{\lambda_1} \frac{\partial \Psi}{\partial \lambda_1}(\hat{\mathbf{N}}_1 \otimes \hat{\mathbf{N}}_1 + \hat{\mathbf{N}}_2 \otimes \hat{\mathbf{N}}_2) + \frac{1}{\lambda_3} \frac{\partial \Psi}{\partial \lambda_3}\hat{\mathbf{N}}_3 \otimes \hat{\mathbf{N}}_3 \ .$$

(b) Using the property $(1.65)_2$ for the second-order unit tensor $\mathbf{I}$ and relation $(2.124)_3$ for the rotation tensor $\mathbf{R}$, show that for $\lambda_1 = \lambda_2 = \lambda_3 = \lambda$

$$\boldsymbol{\sigma} = \sigma \sum_{a=1}^{3} \hat{\mathbf{n}}_a \otimes \hat{\mathbf{n}}_a = \sigma \mathbf{I} \quad ,$$

$$\mathbf{P} = P \sum_{a=1}^{3} \hat{\mathbf{n}}_a \otimes \hat{\mathbf{N}}_a = P\mathbf{R} \quad , \qquad \mathbf{S} = S \sum_{a=1}^{3} \hat{\mathbf{N}}_a \otimes \hat{\mathbf{N}}_a = S\mathbf{I} \quad ,$$

with the scalar-valued scalar functions $\sigma = J^{-1}\lambda \partial \Psi / \partial \lambda$ and with $P = \partial \Psi / \partial \lambda$, $S = \lambda^{-1} \partial \Psi / \partial \lambda$.

6.3　Incompressible Hyperelastic Materials

Numerous polymeric materials can sustain finite strains without noticeable volume changes. Such types of material may be regarded as incompressible so that only isochoric motions are possible. For many cases, this is a common idealization and accepted assumption often invoked in continuum and computational mechanics. In this section we present the constitutive foundation of incompressible hyperelastic materials.

Incompressible hyperelasticity.　Materials which keep the volume constant throughout a motion are characterized by the incompressibility constraint

$$J = 1 \quad , \tag{6.56}$$

or by some other equivalent expressions according to (2.179) (recall the expression (2.52) for the volume ratio J). In general, a material which is subjected to an internal constraint, of which incompressibility is the most common, is referred to as a **constrained material**.

In order to derive general constitutive equations for incompressible hyperelastic materials, we may postulate the strain-energy function

$$\Psi = \Psi(\mathbf{F}) - p(J - 1) \quad , \tag{6.57}$$

where the strain energy Ψ is defined for $J = \det\mathbf{F} = 1$. The scalar p introduced in (6.57) serves as an indeterminate *Lagrange multiplier*, which can be identified as a **hydrostatic pressure**. Note that the scalar p may only be determined from the equilibrium equations and the boundary conditions. It represents a workless reaction to the kinematic constraint on the deformation field.

Differentiating eq. (6.57) with respect to the deformation gradient $\mathbf{F}$ and using identity (2.176), we arrive at a general constitutive equation for the first Piola-Kirchhoff stress tensor $\mathbf{P}$. Hence, eq. (6.1) may be adopted in the form

$$\mathbf{P} = -p\mathbf{F}^{-T} + \frac{\partial \Psi(\mathbf{F})}{\partial \mathbf{F}} \quad . \tag{6.58}$$

An alternative derivation of (6.58) is obtained by reference to the expression $(6.3)_2$. For incompressible hyperelasticity, $\dot{\mathbf{F}}$ is not arbitrary anymore and the expressions in parentheses of $(6.3)_2$ need not be zero. However, $(6.3)_2$ must be satisfied for every $\dot{\mathbf{F}}$ which is governed by the incompressibility constraint in the form of $\dot{J} = \mathbf{F}^{-T} : \dot{\mathbf{F}} = 0$ (recall $(2.179)_4$). Consequently, adding the zero term to $(6.3)_2$, we find that

$$\left(\mathbf{P} - \frac{\partial \Psi(\mathbf{F})}{\partial \mathbf{F}} + p\mathbf{F}^{-T} \right) : \dot{\mathbf{F}} = 0 \quad . \tag{6.59}$$

With standard arguments, the Coleman-Noll procedure implies physical expression (6.58).

Multiplying eq. (6.58) by $\mathbf{F}^{-1}$ from the left-hand side, we conclude from $(3.65)_2$ that the second Piola-Kirchhoff stress tensor $\mathbf{S}$ takes on the form

$$\mathbf{S} = -p\mathbf{F}^{-1}\mathbf{F}^{-T} + \mathbf{F}^{-1}\frac{\partial \Psi(\mathbf{F})}{\partial \mathbf{F}} = -p\mathbf{C}^{-1} + 2\frac{\partial \Psi(\mathbf{C})}{\partial \mathbf{C}} \quad , \tag{6.60}$$

where the inverse of relation (2.65), i.e. $\mathbf{C}^{-1} = \mathbf{F}^{-1}\mathbf{F}^{-T}$, and identity (6.11) are to be used.

However, multiplying eq. (6.58) by $\mathbf{F}^{T}$ from the right-hand side, we conclude from (3.9) that the symmetric Cauchy stress tensor $\boldsymbol{\sigma}$ may be expressed as

$$\boldsymbol{\sigma} = -p\mathbf{I} + \frac{\partial \Psi(\mathbf{F})}{\partial \mathbf{F}}\mathbf{F}^{T} = -p\mathbf{I} + \mathbf{F}\left(\frac{\partial \Psi(\mathbf{F})}{\partial \mathbf{F}} \right)^{T} \quad . \tag{6.61}$$

The fundamental constitutive equations (6.58), (6.60) and (6.61) are the *most general forms* used to define incompressible hyperelastic materials at finite strains. Equations $(6.60)_2$ and (6.61) are associated with (5.97) and (5.96). Note that the response functions $\mathfrak{H}(\mathbf{C})$ and $\mathbf{g}(\mathbf{F})$ occurring in (5.97) and $(5.96)_1$ are identified by $\mathfrak{H}(\mathbf{C}) = 2\partial \Psi(\mathbf{C})/\partial \mathbf{C}$ and $\mathbf{g}(\mathbf{F}) = (\partial \Psi(\mathbf{F})/\partial \mathbf{F})\mathbf{F}^{T} = \mathbf{F}(\partial \Psi(\mathbf{F})/\partial \mathbf{F})^{T}$, i.e. constitutive equation $(6.2)_2$ for $J = 1$.

Incompressible isotropic hyperelasticity. For the case of isotropy we have already pointed out that the dependence of Ψ on the Cauchy-Green tensors $\mathbf{C}$ or $\mathbf{b}$ may be expressed by their three strain invariants (see eq. (6.27)). However, for the incompressible case we consider the kinematic constraint, namely $I_3 = \det\mathbf{C} = \det\mathbf{b} = 1$. Therefore, the two principal invariants I_1 and I_2 are the only independent deformation variables.

For a review on the theory of incompressible isotropic hyperelasticity see, for example, OGDEN [1982, 1986].

A suitable strain-energy function for incompressible isotropic hyperelastic materials is, in view of (6.27), given by

$$\Psi = \Psi[I_1(\mathbf{C}), I_2(\mathbf{C})] - \frac{1}{2}p(I_3 - 1) = \Psi[I_1(\mathbf{b}), I_2(\mathbf{b})] - \frac{1}{2}p(I_3 - 1) \quad , \qquad (6.62)$$

where $p/2$ serves as an indeterminate *Lagrange multiplier*.

In order to examine the associated constitutive equation in terms of the two principal strain invariants I_1, I_2, we proceed by deriving $(6.62)_1$ with respect to tensor $\mathbf{C}$. Analogous to the procedure which led to (6.32) we find, using the chain rule, eqs. $(6.30)_3$, (6.31) and the constraint $I_3 = 1$, that

$$\mathbf{S} = 2\frac{\partial \Psi(I_1, I_2)}{\partial \mathbf{C}} - \frac{\partial[p(I_3 - 1)]}{\partial \mathbf{C}} = -p\mathbf{C}^{-1} + 2\left(\frac{\partial \Psi}{\partial I_1} + I_1\frac{\partial \Psi}{\partial I_2}\right)\mathbf{I} - 2\frac{\partial \Psi}{\partial I_2}\mathbf{C} \quad , \quad (6.63)$$

which is basically constitutive equation (6.32), in which the term $I_3(\partial\Psi/\partial I_3)$ is substituted by $-p/2$.

A push-forward operation of $(6.63)_2$ and an elimination of $\mathbf{b}^2$ in favor of $\mathbf{b}^{-1}$ (see relation (5.93)) yields two alternative forms of $\boldsymbol{\sigma}$, corresponding to eqs. (6.34) and (6.35), namely

$$\boldsymbol{\sigma} = -p\mathbf{I} + 2\left(\frac{\partial \Psi}{\partial I_1} + I_1\frac{\partial \Psi}{\partial I_2}\right)\mathbf{b} - 2\frac{\partial \Psi}{\partial I_2}\mathbf{b}^2 \quad , \qquad (6.64)$$

$$\boldsymbol{\sigma} = -p\mathbf{I} + 2\frac{\partial \Psi}{\partial I_1}\mathbf{b} - 2\frac{\partial \Psi}{\partial I_2}\mathbf{b}^{-1} \quad . \qquad (6.65)$$

Note that the scalars p in eqs. (6.64) and (6.65) differ by the term $2I_2(\partial\Psi/\partial I_2)$. By comparing (6.64) and (6.65) with (5.98) we obtain explicitly the response coefficients α_1, α_2 and β_1, β_{-1} as

$$\alpha_1 = 2\left(\frac{\partial \Psi}{\partial I_1} + I_1\frac{\partial \Psi}{\partial I_2}\right) \quad , \qquad \alpha_2 = -2\frac{\partial \Psi}{\partial I_2} \quad , \qquad (6.66)$$

$$\beta_1 = 2\frac{\partial \Psi}{\partial I_1} \quad , \qquad \beta_{-1} = -2\frac{\partial \Psi}{\partial I_2} \quad . \qquad (6.67)$$

In order to find a constitutive equation for incompressible materials which is associated with (6.38), we recall the transformation (6.55). Then, (6.61) gives the constitutive equation

$$\boldsymbol{\sigma} = -p\mathbf{I} + 2\frac{\partial \Psi(\mathbf{b})}{\partial \mathbf{b}}\mathbf{b} = -p\mathbf{I} + 2\mathbf{b}\frac{\partial \Psi(\mathbf{b})}{\partial \mathbf{b}}$$

$$\text{or} \qquad \sigma_{ab} = -p\delta_{ab} + 2b_{ac}\frac{\partial \Psi}{\partial b_{cb}} \qquad (6.68)$$

in terms of the spatial strain variable **b**. This is only valid for incompressible isotropic hyperelastic materials.

If we express Ψ as a function of the three principal stretches λ_a we write $\Psi = \Psi(\lambda_1, \lambda_2, \lambda_3) - p(J-1)$ in the place of (6.57), with the indeterminate *Lagrange multiplier* p. Using $\partial J / \partial \lambda_a = J \lambda_a^{-1}$, $a = 1, 2, 3$, which is relation (2.176) expressed in principal stretches, eqs. (6.45) and (6.47) are then replaced by

$$\sigma_a = -p + \lambda_a \frac{\partial \Psi}{\partial \lambda_a} \quad , \qquad a = 1, 2, 3 \quad ; \tag{6.69}$$

$$P_a = -\frac{1}{\lambda_a} p + \frac{\partial \Psi}{\partial \lambda_a} \quad , \qquad S_a = -\frac{1}{\lambda_a^2} p + \frac{1}{\lambda_a} \frac{\partial \Psi}{\partial \lambda_a} \quad , \qquad a = 1, 2, 3 \quad , \tag{6.70}$$

with the three principal Cauchy stresses σ_a and the Piola-Kirchhoff stresses P_a, S_a. These stress relations incorporate the unknown scalar p, which must be determined from the equilibrium equations and the boundary conditions. The incompressibility constraint $J = 1$ takes on the form

$$\lambda_1 \lambda_2 \lambda_3 = 1 \quad , \tag{6.71}$$

leaving two independent stretches as the deformation measures. Expressing the first and second Piola-Kirchhoff stresses in terms of the Cauchy stresses (6.69), we obtain, by analogy with (6.48), $P_a = \lambda_a^{-1} \sigma_a$ and $S_a = \lambda_a^{-2} \sigma_a$, $a = 1, 2, 3$.

EXAMPLE 6.3 Consider a thin sheet of incompressible hyperelastic material which is embedded in a reference frame of (right-handed) coordinate axes with a fixed set of orthonormal basis vectors $\mathbf{e}_a$, $a = 1, 2, 3$. Suppose that the axes are aligned with the major faces of the sheet.

A deformation created by the stretch ratios λ_1, λ_2 along the directions $\mathbf{e}_1$, $\mathbf{e}_2$ results in a *(homogeneous) biaxial deformation* with the kinematic relation (2.132). The associated stress state is assumed to be *plane* throughout the sheet so that the Cauchy stress components $\sigma_{13}, \sigma_{23}, \sigma_{33}$ are equal to zero which is in accordance with (3.59).

Show that the biaxial stress state of the homogeneous problem is of the form

$$\sigma_1 = 2(\lambda_1^2 - \lambda_1^{-2}\lambda_2^{-2}) \left(\frac{\partial \Psi}{\partial I_1} + \lambda_2^2 \frac{\partial \Psi}{\partial I_2} \right) \quad , \tag{6.72}$$

$$\sigma_2 = 2(\lambda_2^2 - \lambda_1^{-2}\lambda_2^{-2}) \left(\frac{\partial \Psi}{\partial I_1} + \lambda_1^2 \frac{\partial \Psi}{\partial I_2} \right) \tag{6.73}$$

(see RIVLIN [1948, eq. (6.5)]), with the principal invariants $I_1 = \lambda_1^2 + \lambda_2^2 + \lambda_1^{-2}\lambda_2^{-2}$, $I_2 = \lambda_1^2 \lambda_2^2 + \lambda_1^{-2} + \lambda_2^{-2}$ and $I_3 = 1$.

Solution.   Since the tensors σ and $\mathbf{b}$ are coaxial for isotropic elastic materials (recall p. 201), the principal stresses follow from (6.65),

$$\sigma_a = -p + 2 \left(\lambda_a^2 \frac{\partial \Psi}{\partial I_1} - \frac{1}{\lambda_a^2} \frac{\partial \Psi}{\partial I_2} \right) , \qquad a = 1, 2, 3 , \qquad (6.74)$$

where λ_a^2 are the three eigenvalues of the left Cauchy-Green tensor $\mathbf{b}$ (see the eigen-value problem $(2.119)_2$). This relation was first presented by RIVLIN [1948].

With the condition of incompressibility (6.71) in the form of $\lambda_3 = (\lambda_1 \lambda_2)^{-1}$ and the boundary condition $\sigma_3 = 0$ we may determine p explicitly. For $a = 3$, we deduce from (6.74) that

$$\sigma_3 = 0 \qquad \longrightarrow \qquad p = 2 \left(\frac{1}{\lambda_1^2 \lambda_2^2} \frac{\partial \Psi}{\partial I_1} - \lambda_1^2 \lambda_2^2 \frac{\partial \Psi}{\partial I_2} \right) . \qquad (6.75)$$

This result substituted back into eq. (6.74) leads to the nonzero stress components σ_1 and σ_2.   ∎

EXERCISES

1. Consider a thin sheet of incompressible hyperelastic material $(I_3 = 1)$ with the same setting as formulated in Example 6.3.

   (a) Consider a **simple tension** for which $\lambda_1 = \lambda$. Then, obeying incompress-ibility constraint $\lambda_1 \lambda_2 \lambda_3 = 1$, the equal stretch ratios in the transverse di-rections are, by symmetry, $\lambda_2 = \lambda_3 = \lambda^{-1/2}$. Show that for this mode of deformation the homogeneous stress state reduces to $\sigma_1 = \sigma, \sigma_2 = \sigma_3 = 0$, with

$$\sigma = 2 \left(\lambda^2 - \frac{1}{\lambda} \right) \left(\frac{\partial \Psi}{\partial I_1} + \frac{1}{\lambda} \frac{\partial \Psi}{\partial I_2} \right) ,$$

   where the invariants are $I_1 = 2\lambda^{-1} + \lambda^2, I_2 = \lambda^{-2} + 2\lambda$.

   As a special case of the biaxial deformation, as discussed in Example 6.3, consider an *equibiaxial deformation* for which $\lambda_1 = \lambda_2 = \lambda, \lambda_3 = \lambda^{-2}$ and $\sigma_1 = \sigma_2 = \sigma, \sigma_3 = 0$. Show that

$$\sigma = 2 \left(\lambda^2 - \frac{1}{\lambda^4} \right) \left(\frac{\partial \Psi}{\partial I_1} + \lambda^2 \frac{\partial \Psi}{\partial I_2} \right) ,$$

   with $I_1 = 2\lambda^2 + \lambda^{-4}, I_2 = \lambda^4 + 2\lambda^{-2}$.

   (b) Consider a homogeneous *pure shear deformation* with the kinematic rela-tion $\lambda_1 = \lambda, \lambda_2 = 1, \lambda_3 = 1/\lambda$ (compare with eq. (2.133)). Show that the nonzero Cauchy stress components are

$$\sigma_1 = 2\left(\lambda^2 - \frac{1}{\lambda^2}\right)\left(\frac{\partial\Psi}{\partial I_1} + \frac{\partial\Psi}{\partial I_2}\right) \quad, \tag{6.76}$$

$$\sigma_2 = 2\left(1 - \frac{1}{\lambda^2}\right)\left(\frac{\partial\Psi}{\partial I_1} + \lambda^2\frac{\partial\Psi}{\partial I_2}\right) \quad, \tag{6.77}$$

evaluated for $I_1 = I_2 = \lambda^2 + \lambda^{-2} + 1$.

2. Consider a thin sheet of incompressible hyperelastic material with the same set-ting as formulated in Example 6.3 but subjected to a homogeneous *simple shear deformation* which is caused by a motion in the form of (2.3) (compare also with Exercise 2 on p. 93).

   (a) Show that the associated stress state is completely defined by

$$\sigma_{11} = -p + 2(1 + c^2)\frac{\partial\Psi}{\partial I_1} - 2\frac{\partial\Psi}{\partial I_2} \quad,$$

$$\sigma_{22} = -p + 2\frac{\partial\Psi}{\partial I_1} - 2(1 + c^2)\frac{\partial\Psi}{\partial I_2} \quad,$$

$$\sigma_{33} = -p + 2\frac{\partial\Psi}{\partial I_1} - 2\frac{\partial\Psi}{\partial I_2} \quad,$$

$$\sigma_{12} = \mu c \quad, \qquad \mu = 2\left(\frac{\partial\Psi}{\partial I_1} + \frac{\partial\Psi}{\partial I_2}\right) \tag{6.78}$$

   (with $I_1 = I_2 = 3 + c^2$), where $\mu > 0$, called the **shear modulus**, is a measure of resistance to distortion and p is a scalar to be determined from the boundary conditions.

   In addition, show that the angle θ of the two principal directions acting in the plane normal to $\hat{\mathbf{n}}_3$ is given by $\tan 2\theta = 2/c$.

   (b) Consider a *plane stress state* throughout the sheet in the sense that the face of the body normal to the direction $\mathbf{e}_3$ is free of surface tractions, i.e. $\sigma_{13} = \sigma_{23} = \sigma_{33} \equiv 0$. Show that the nonzero Cauchy stress components are

$$\sigma_{11} = 2c^2\frac{\partial\Psi}{\partial I_1} \quad, \qquad \sigma_{22} = -2c^2\frac{\partial\Psi}{\partial I_2} \quad, \qquad \sigma_{12} = \mu c \quad.$$

6.4 Compressible Hyperelastic Materials

A material which can undergo changes of volume is said to be **compressible**. Foamed elastomers, for example, are able to sustain finite strains with volume changes. The only restriction on this class of materials is that the volume ratio J must be positive.

In this section we introduce suitable constitutive equations in order to characterize compressible hyperelastic materials, and we discuss isotropy as a special case.

Compressible hyperelasticity.   Since some materials behave quite differently in bulk and shear it is most beneficial to split the deformation locally into a so-called *volumetric part* and an *isochoric part*, originally proposed by FLORY [1961] and successfully applied within the context of isothermal finite strain elasticity by, for example, LUBLINER [1985], SIMO and TAYLOR [1991a], OGDEN [1997] and within the context of finite strain elastoplasticity by, for example, SIMO et al. [1985] among many others.

In particular, we consider the deformation gradient $\mathbf{F}$ and the corresponding strain measure $\mathbf{C} = \mathbf{F}^T\mathbf{F}$. Rather than dealing directly with $\mathbf{F}$ and $\mathbf{C}$ we perform a multiplicative decomposition of $\mathbf{F}$ into *volume-changing (dilational)* and *volume-preserving (distortional)* parts, often used in elastoplasticity (see, for example, LEE [1969]). We write

$$\mathbf{F} = (J^{1/3}\mathbf{I})\overline{\mathbf{F}} = J^{1/3}\overline{\mathbf{F}} \ , \qquad \mathbf{C} = (J^{2/3}\mathbf{I})\overline{\mathbf{C}} = J^{2/3}\overline{\mathbf{C}} \ . \tag{6.79}$$

The terms $J^{1/3}\mathbf{I}$ and $J^{2/3}\mathbf{I}$ are associated with volume-changing deformations, while $\overline{\mathbf{F}}$ and $\overline{\mathbf{C}} = \overline{\mathbf{F}}^T\overline{\mathbf{F}}$ are associated with volume-preserving deformations of the material, with

$$\det\overline{\mathbf{F}} = \overline{\lambda}_1\overline{\lambda}_2\overline{\lambda}_3 = 1 \qquad \text{and} \qquad \det\overline{\mathbf{C}} = (\det\overline{\mathbf{F}})^2 = 1 \ , \tag{6.80}$$

where

$$\overline{\lambda}_a = J^{-1/3}\lambda_a \ , \qquad a = 1, 2, 3 \tag{6.81}$$

characterize the so-called **modified principal stretches**. We call $\overline{\mathbf{F}}$ and $\overline{\mathbf{C}}$ the **modified deformation gradient** and the **modified right Cauchy-Green tensor**, respectively. A material for which dilational changes require a much higher exterior work than volume-preserving changes is called a **nearly incompressible** (or **slightly compressible**) material, for which the compressibility effects are small.

The concept of the multiplicative decomposition of $\mathbf{F}$ is supported additionally by the field of computational mechanics. For example, to avoid numerical complications in the finite element analysis of slightly compressible materials it is often advantageous to separate numerical treatments of the volumetric and isochoric parts of the deformation gradient $\mathbf{F}$; this will be discussed in Sections 8.5 and 8.6.

Before proceeding to examine constitutive equations for compressible hyperelastic materials it is first necessary to stick to kinematics and to compute the derivative of the modified right Cauchy-Green tensor $\overline{\mathbf{C}}$ relative to the symmetric tensor $\mathbf{C}$. By means of $(5.91)_2$, we obtain from $(6.31)_2$, $\partial J^2/\partial \mathbf{C} = J^2\mathbf{C}^{-1}$. Using the chain rule we arrive at

$$\frac{\partial J}{\partial \mathbf{C}} = \frac{J}{2}\mathbf{C}^{-1} \qquad \text{and} \qquad \frac{\partial J^{-2/3}}{\partial \mathbf{C}} = -\frac{1}{3}J^{-2/3}\mathbf{C}^{-1} \ . \tag{6.82}$$

Finally, according to the inverse of relation (6.79)$_2$, property (1.256) and relation (6.82)$_2$, we obtain the fourth-order tensor

$$\frac{\partial \overline{\mathbf{C}}}{\partial \mathbf{C}} = \frac{\partial (J^{-2/3}\mathbf{C})}{\partial \mathbf{C}} = J^{-2/3} \left(\mathbb{I} + J^{2/3}\mathbf{C} \otimes \frac{\partial J^{-2/3}}{\partial \mathbf{C}} \right)$$

$$= J^{-2/3} \underbrace{(\mathbb{I} - \frac{1}{3}\mathbf{C} \otimes \mathbf{C}^{-1})}_{\mathbb{P}^{\mathrm{T}}} = J^{-2/3}\mathbb{P}^{\mathrm{T}} \quad , \tag{6.83}$$

in which $\mathbb{P}^{\mathrm{T}}$ defines the transpose of the fourth-order tensor $\mathbb{P}$ governed by the identity (1.157). We call $\mathbb{P}$ the **projection tensor** with respect to the reference configuration, therefore expressed through $\mathbf{C}$. With the associated property (1.159), the relation for the projection tensor $\mathbb{P}$ reads, with reference to (6.83)$_3$, as

$$\mathbb{P} = \mathbb{I} - \frac{1}{3}\mathbf{C}^{-1} \otimes \mathbf{C} \quad , \tag{6.84}$$

where $\mathbb{I}$ denotes the fourth-order unit tensor with the components $(\mathbb{I})_{ABCD} = (\delta_{AC}\delta_{BD} + \delta_{AD}\delta_{BC})/2$.

Earlier we agreed to study the purely mechanical theory. To characterize processes within an isothermal situation at constant temperature, we postulate a unique *decoupled* representation of the strain-energy function $\Psi = \Psi(\mathbf{C})$ (per unit reference volume). It is based on kinematic assumption (6.79)$_2$ and of the specific form

$$\Psi(\mathbf{C}) = \Psi_{\mathrm{vol}}(J) + \Psi_{\mathrm{iso}}(\overline{\mathbf{C}}) \quad , \tag{6.85}$$

where $\Psi_{\mathrm{vol}}(J)$ and $\Psi_{\mathrm{iso}}(\overline{\mathbf{C}})$ are given scalar-valued functions of J and $\overline{\mathbf{C}}$ which are assumed to be objective. They describe the so-called **volumetric** (or **dilational**) **elastic response** and the **isochoric** (or **distortional**) **elastic response** of the material, respectively.

Additionally, we require that Ψ_{vol} is a *strictly convex* function taking on its unique minimum at $J = 1$ (for formal definitions of strictly convex functions see, for example, OGDEN [1997, Appendix 1]). With reference to normalization condition (6.4) we claim that $\Psi_{\mathrm{vol}}(J) = 0$ and $\Psi_{\mathrm{iso}}(\overline{\mathbf{C}}) = 0$ hold *if and only if* $J = 1$ and $\overline{\mathbf{C}} = \mathbf{I}$, respectively.

We now determine constitutive equations for compressible hyperelastic materials. In order to particularize the second law of thermodynamics through the Clausius-Planck inequality (4.154) to the specific strain energy (6.85) at hand we determine the derivative of Ψ with respect to time t first. By means of the chain rule we obtain from (6.85)

$$\dot{\Psi} = \frac{\mathrm{d}\Psi_{\mathrm{vol}}(J)}{\mathrm{d}J}\dot{J} + \frac{\partial \Psi_{\mathrm{iso}}(\overline{\mathbf{C}})}{\partial \overline{\mathbf{C}}} : \dot{\overline{\mathbf{C}}} \quad . \tag{6.86}$$

Hence, we need to compute $\dot{J}$ and $\dot{\overline{\mathbf{C}}}$, which, with eqs.(6.82)$_1$ and (6.83)$_4$, simply results in $\dot{J} = \partial J/\partial \mathbf{C} : \dot{\mathbf{C}} = J\mathbf{C}^{-1} : \dot{\mathbf{C}}/2$ and $\dot{\overline{\mathbf{C}}} = 2(\partial \overline{\mathbf{C}}/\partial \mathbf{C}) : \dot{\mathbf{C}}/2 = 2J^{-2/3}\mathbb{P}^{\mathrm{T}} :$ $\dot{\mathbf{C}}/2$. Having this in mind, with the stress power $w_{\mathrm{int}} = \mathbf{S} : \dot{\mathbf{C}}/2$ per unit reference volume and relation (6.86), we may deduce from (4.154) that

$$\mathcal{D}_{\mathrm{int}} = \left(\mathbf{S} - J\frac{\mathrm{d}\Psi_{\mathrm{vol}}(J)}{\mathrm{d}J}\mathbf{C}^{-1} - J^{-2/3}\mathbb{P} : 2\frac{\partial\Psi_{\mathrm{iso}}(\overline{\mathbf{C}})}{\partial\overline{\mathbf{C}}} \right) : \frac{\dot{\mathbf{C}}}{2} = 0 \ , \qquad (6.87)$$

where the identity (1.157) is to be used. Since we consider perfectly elastic materials the internal dissipation $\mathcal{D}_{\mathrm{int}}$ must vanish.

The standard Coleman-Noll procedure leads to constitutive equations for compressible hyperelastic materials, in which the stress response constitutes an additive split of $(6.13)_2$, i.e. $\mathbf{S} = 2\partial\Psi(\mathbf{C})/\partial\mathbf{C}$. In particular, the second Piola-Kirchhoff stress $\mathbf{S}$ consists of a purely **volumetric contribution** and a purely **isochoric contribution**, i.e. $\mathbf{S}_{\mathrm{vol}}$ and $\mathbf{S}_{\mathrm{iso}}$, respectively. We write

$$\mathbf{S} = 2\frac{\partial\Psi(\mathbf{C})}{\partial\mathbf{C}} = \mathbf{S}_{\mathrm{vol}} + \mathbf{S}_{\mathrm{iso}} \ . \qquad (6.88)$$

This split is based on the definitions

$$\mathbf{S}_{\mathrm{vol}} = 2\frac{\partial\Psi_{\mathrm{vol}}(J)}{\partial\mathbf{C}} = Jp\mathbf{C}^{-1} \ , \qquad (6.89)$$

$$\mathbf{S}_{\mathrm{iso}} = 2\frac{\partial\Psi_{\mathrm{iso}}(\overline{\mathbf{C}})}{\partial\mathbf{C}} = J^{-2/3}(\mathbb{I} - \frac{1}{3}\mathbf{C}^{-1} \otimes \mathbf{C}) : \overline{\mathbf{S}}$$
$$= J^{-2/3}\mathrm{Dev}\overline{\mathbf{S}} = J^{-2/3}\mathbb{P} : \overline{\mathbf{S}} \ , \qquad (6.90)$$

with the constitutive equations for the **hydrostatic pressure** p and the **fictitious second Piola-Kirchhoff stress** $\overline{\mathbf{S}}$ defined by

$$p = \frac{\mathrm{d}\Psi_{\mathrm{vol}}(J)}{\mathrm{d}J} \qquad \text{and} \qquad \overline{\mathbf{S}} = 2\frac{\partial\Psi_{\mathrm{iso}}(\overline{\mathbf{C}})}{\partial\overline{\mathbf{C}}} \ . \qquad (6.91)$$

It is important to note that in contrast to incompressible materials, the scalar function p is specified by a constitutive equation. The projection tensor $\mathbb{P} = \mathbb{I} - \frac{1}{3}\mathbf{C}^{-1} \otimes \mathbf{C}$ in (6.90) furnishes the physically correct deviatoric operator in the *Lagrangian description*, i.e. $\mathrm{Dev}(\bullet) = (\bullet) - (1/3)[(\bullet) : \mathbf{C}]\mathbf{C}^{-1}$, so that

$$\mathrm{Dev}\overline{\mathbf{S}} : \mathbf{C} = 0 \ . \qquad (6.92)$$

The characterization of the stress response in the material description in terms of the projection tensor $\mathbb{P}$ leads to a convenient short-hand notation (see also, for example, HOLZAPFEL [1996a]).

EXAMPLE 6.4 Consider the decoupled strain-energy function (6.85) with the associated stress relation (6.88). Perform a Piola transformation according to (3.66) and obtain the additive decomposition

$$\boldsymbol{\sigma} = \boldsymbol{\sigma}_{\text{vol}} + \boldsymbol{\sigma}_{\text{iso}} \tag{6.93}$$

of the Cauchy stress tensor $\boldsymbol{\sigma}$, where the purely volumetric and isochoric stress contributions are defined by

$$\boldsymbol{\sigma}_{\text{vol}} = p\mathbf{I} \;, \qquad \boldsymbol{\sigma}_{\text{iso}} = J^{-1}\overline{\mathbf{F}}(\mathbb{P} : \overline{\mathbf{S}})\overline{\mathbf{F}}^{\text{T}} \;, \tag{6.94}$$

$\overline{\mathbf{F}}$ being the modified deformation gradient. The constitutive equations for the hydrostatic pressure p and the fictitious second Piola-Kirchhoff stress $\overline{\mathbf{S}}$ are given by (6.91).

Solution. A push-forward operation on the second Piola-Kirchhoff stress tensor $\mathbf{S}$ to the current configuration and a scaling with the inverse of the volume ratio transforms (6.88)$_1$ to

$$\boldsymbol{\sigma} = J^{-1}\chi_*(\mathbf{S}^{\sharp}) = 2J^{-1}\mathbf{F}\left(\frac{\partial\Psi_{\text{vol}}(J)}{\partial\mathbf{C}} + \frac{\partial\Psi_{\text{iso}}(\overline{\mathbf{C}})}{\partial\mathbf{C}}\right)\mathbf{F}^{\text{T}} \;, \tag{6.95}$$

where the decoupled form (6.85) is to be used. Hence, considering the first term on the right-hand side, we obtain, using (6.89)$_2$ and $\mathbf{C}^{-1} = \mathbf{F}^{-1}\mathbf{F}^{-\text{T}}$,

$$2J^{-1}\mathbf{F}\frac{\partial\Psi_{\text{vol}}(J)}{\partial\mathbf{C}}\mathbf{F}^{\text{T}} = J^{-1}\mathbf{F}\left(Jp\mathbf{C}^{-1}\right)\mathbf{F}^{\text{T}} = p\mathbf{I} \;, \tag{6.96}$$

which is the volumetric Cauchy stress contribution $\boldsymbol{\sigma}_{\text{vol}}$ defining a hydrostatic stress state, as discussed on p. 125.

Considering the second term on the right-hand side of eq. (6.95)$_2$, we obtain, using (6.90)$_4$ and the kinematic assumption (6.79)$_1$,

$$2J^{-1}\mathbf{F}\frac{\partial\Psi_{\text{iso}}(\overline{\mathbf{C}})}{\partial\mathbf{C}}\mathbf{F}^{\text{T}} = J^{-5/3}\mathbf{F}(\mathbb{P} : \overline{\mathbf{S}})\mathbf{F}^{\text{T}} = J^{-1}\overline{\mathbf{F}}(\mathbb{P} : \overline{\mathbf{S}})\overline{\mathbf{F}}^{\text{T}} \;, \tag{6.97}$$

which is the isochoric Cauchy stress contribution $\boldsymbol{\sigma}_{\text{iso}}$. ∎

Compressible isotropic hyperelasticity. A suitable *decoupled* representation of the strain-energy function for compressible *isotropic* hyperelastic materials is, by analogy with assumption (6.85), given by

$$\Psi(\mathbf{b}) = \Psi_{\text{vol}}(J) + \Psi_{\text{iso}}(\overline{\mathbf{b}}) \;, \tag{6.98}$$

with the multiplicative split of the left Cauchy-Green tensor $\mathbf{b} = \mathbf{F}\mathbf{F}^{\mathrm{T}}$ in the form

$$\mathbf{b} = (J^{2/3}\mathbf{I})\overline{\mathbf{b}} = J^{2/3}\overline{\mathbf{b}} \tag{6.99}$$

(compare with eq. $(6.79)_2$). The terms $J^{2/3}\mathbf{I}$ and $\overline{\mathbf{b}} = \overline{\mathbf{F}}\,\overline{\mathbf{F}}^{\mathrm{T}}$ represent the volume-changing (dilational) and volume-preserving contributions to the deformation. We call $\overline{\mathbf{b}}$ the **modified left Cauchy-Green tensor**, with $\det\overline{\mathbf{b}} = 1$.

The derivative of the volume ratio $J = (\det\mathbf{b})^{1/2}$ and the modified left Cauchy-Green tensor $\overline{\mathbf{b}}$ relative to $\mathbf{b}$ is given by $(6.82)_1$ and $(6.83)_3$ (with $\mathbf{C}$ replaced by $\mathbf{b}$). We obtain

$$\frac{\partial J}{\partial \mathbf{b}} = \frac{J}{2}\mathbf{b}^{-1} \quad , \qquad \frac{\partial \overline{\mathbf{b}}}{\partial \mathbf{b}} = J^{-2/3}(\mathbb{I} - \frac{1}{3}\mathbf{b}\otimes\mathbf{b}^{-1}) \quad . \tag{6.100}$$

Following arguments analogous to those which led to eqs. (6.88)–(6.91), we obtain from (6.98) the spatial version of constitutive equations which are expressed in terms of J and $\overline{\mathbf{b}}$. Given entirely in the spatial description and characterizing the isotropic behavior of compressible hyperelastic materials, we have

$$\boldsymbol{\sigma} = 2J^{-1}\mathbf{b}\frac{\partial \Psi(\mathbf{b})}{\partial \mathbf{b}} = 2J^{-1}\frac{\partial \Psi(\mathbf{b})}{\partial \mathbf{b}}\mathbf{b} = \boldsymbol{\sigma}_{\mathrm{vol}} + \boldsymbol{\sigma}_{\mathrm{iso}} \quad , \tag{6.101}$$

where the stress contributions are defined by

$$\boldsymbol{\sigma}_{\mathrm{vol}} = 2J^{-1}\mathbf{b}\frac{\partial \Psi_{\mathrm{vol}}(J)}{\partial \mathbf{b}} = p\mathbf{I} \quad , \tag{6.102}$$

$$\boldsymbol{\sigma}_{\mathrm{iso}} = 2J^{-1}\mathbf{b}\frac{\partial \Psi_{\mathrm{iso}}(\overline{\mathbf{b}})}{\partial \mathbf{b}} = 2J^{-1}\mathbf{b}J^{-2/3}(\mathbb{I} - \frac{1}{3}\mathbf{b}^{-1}\otimes\mathbf{b}) : \frac{\partial \Psi_{\mathrm{iso}}(\overline{\mathbf{b}})}{\partial \overline{\mathbf{b}}}$$

$$= \mathbf{b}(J^{-2/3}\mathbb{I} - \frac{1}{3}\mathbf{b}^{-1}\otimes\overline{\mathbf{b}})\overline{\mathbf{b}}^{-1} : 2J^{-1}\frac{\partial \Psi_{\mathrm{iso}}(\overline{\mathbf{b}})}{\partial \overline{\mathbf{b}}}\overline{\mathbf{b}}$$

$$= \underbrace{(\mathbb{I} - \frac{1}{3}\mathbf{I}\otimes\mathbf{I})}_{\mathbb{P}} : \overline{\boldsymbol{\sigma}} = \mathrm{dev}\overline{\boldsymbol{\sigma}}$$

$$= \mathbb{P} : \overline{\boldsymbol{\sigma}} \quad . \tag{6.103}$$

Use has been made of eqs. (6.99) and (6.100) and properties (1.95) and (1.155).

The constitutive equation for the hydrostatic pressure p is given in $(6.91)_1$ and the **fictitious Cauchy stress tensor** $\overline{\boldsymbol{\sigma}}$ is defined to be

$$\overline{\boldsymbol{\sigma}} = 2J^{-1}\overline{\mathbf{b}}\frac{\partial \Psi_{\mathrm{iso}}(\overline{\mathbf{b}})}{\partial \overline{\mathbf{b}}} = 2J^{-1}\frac{\partial \Psi_{\mathrm{iso}}(\overline{\mathbf{b}})}{\partial \overline{\mathbf{b}}}\overline{\mathbf{b}} \quad . \tag{6.104}$$

In $(6.103)_4$ we have introduced, additionally, the projection tensor

$$\mathbb{P} = \mathbb{I} - \frac{1}{3}\mathbf{I}\otimes\mathbf{I} \tag{6.105}$$

which furnishes the physically correct deviatoric operator in the *Eulerian description*, i.e. $\text{dev}(\bullet) = (\bullet) - (1/3)[(\bullet) : \mathbf{I}]\mathbf{I}$, so that

$$\text{dev}\overline{\boldsymbol{\sigma}} : \mathbf{I} = 0 \ . \tag{6.106}$$

For the characterization of the stress response in the spatial description in terms of the projection tensor $\mathbb{P}$ see also, for example, the work of MIEHE [1994].

Note the similar structure of the stress relations in the Lagrangian description (6.88)–(6.91) to those presented in (6.101)–(6.104).

Compressible isotropic hyperelasticity in terms of invariants.   We now introduce a strain-energy function for compressible *isotropic* hyperelastic materials in terms of strain invariants. By analogy with the decoupled representation (6.85), or (6.98), we write

$$\Psi = \Psi_{\text{vol}}(J) + \Psi_{\text{iso}}[\bar{I}_1(\overline{\mathbf{C}}), \bar{I}_2(\overline{\mathbf{C}})] = \Psi_{\text{vol}}(J) + \Psi_{\text{iso}}[\bar{I}_1(\overline{\mathbf{b}}), \bar{I}_2(\overline{\mathbf{b}})] \ , \tag{6.107}$$

with the first two strain invariants $\bar{I}_1$ and $\bar{I}_2$ of the symmetric modified Cauchy-Green tensors. Since $\overline{\mathbf{C}}$ and $\overline{\mathbf{b}}$ have the same eigenvalues, we deduce that

$$\bar{I}_1 = \bar{I}_1(\overline{\mathbf{C}}) = \bar{I}_1(\overline{\mathbf{b}}) \ , \qquad \bar{I}_2 = \bar{I}_2(\overline{\mathbf{C}}) = \bar{I}_2(\overline{\mathbf{b}}) \ . \tag{6.108}$$

The strain invariants $\bar{I}_a$, $a = 1, 2, 3$, are referred to as the **modified invariants** and are defined by

$$\bar{I}_1 = \text{tr}\overline{\mathbf{C}} = \text{tr}\overline{\mathbf{b}} \ , \tag{6.109}$$

$$\bar{I}_2 = \frac{1}{2}\left[(\text{tr}\overline{\mathbf{C}})^2 - \text{tr}(\overline{\mathbf{C}}^2)\right] = \frac{1}{2}\left[(\text{tr}\overline{\mathbf{b}})^2 - \text{tr}(\overline{\mathbf{b}}^2)\right] \ , \tag{6.110}$$

$$\bar{I}_3 = \det\overline{\mathbf{C}} = \det\overline{\mathbf{b}} \ . \tag{6.111}$$

With the kinematic assumption (6.79)$_2$, or (6.99), and properties (1.92) and (1.101), we conclude from (6.109)–(6.111) with reference to (5.89)–(5.91) that

$$\bar{I}_1 = J^{-2/3}I_1 \ , \qquad \bar{I}_2 = J^{-4/3}I_2 \ , \qquad \bar{I}_3 = 1 \ . \tag{6.112}$$

Finally, we formulate the associated constitutive equation in terms of the volume ratio J and the modified invariants $\bar{I}_1, \bar{I}_2$, which reads in the material description as

$$\mathbf{S} = 2\frac{\partial\Psi(\mathbf{C})}{\partial\mathbf{C}} = \mathbf{S}_{\text{vol}} + \mathbf{S}_{\text{iso}} \ , \tag{6.113}$$

with the volumetric contribution $\mathbf{S}_{\text{vol}}$ to the second Piola-Kirchhoff stress, i.e. (6.89), and the isochoric contribution $\mathbf{S}_{iso}$, defined by

$$\mathbf{S}_{iso} = 2\frac{\partial\Psi_{iso}(\bar{I}_1,\bar{I}_2)}{\partial\mathbf{C}} = J^{-2/3}\mathbb{P} : \bar{\mathbf{S}} \quad . \tag{6.114}$$

The isochoric second Piola-Kirchhoff stress tensor $\mathbf{S}_{iso}$ is $J^{-2/3}$ multiplied by the double contraction of the fourth-order projection tensor $\mathbb{P}$, see eq.(6.84), with the fictitious second Piola-Kirchhoff stress tensor $\bar{\mathbf{S}}$, which is here defined as

$$\bar{\mathbf{S}} = 2\frac{\partial\Psi_{iso}(\bar{I}_1,\bar{I}_2)}{\partial\overline{\mathbf{C}}} = \bar{\gamma}_1\mathbf{I} + \bar{\gamma}_2\overline{\mathbf{C}} \quad , \tag{6.115}$$

with the two response coefficients given by

$$\bar{\gamma}_1 = 2\left(\frac{\partial\Psi_{iso}(\bar{I}_1,\bar{I}_2)}{\partial\bar{I}_1} + \bar{I}_1\frac{\partial\Psi_{iso}(\bar{I}_1,\bar{I}_2)}{\partial\bar{I}_2}\right) \quad , \qquad \bar{\gamma}_2 = -2\frac{\partial\Psi_{iso}(\bar{I}_1,\bar{I}_2)}{\partial\bar{I}_2} \quad . \tag{6.116}$$

The details are left to be supplied as an exercise by the reader.

For the stress response of compressible isotropic hyperelastic materials in terms of the volume ratio J and two of the modified principal stretches $\bar{\lambda}_a$ as independent variables, see the study by OGDEN [1997, Section 7.2.3].

EXERCISES

1. Consider the modified right Cauchy-Green tensor $\overline{\mathbf{C}}$ according to $(6.79)_2$ and properties (1.159), (1.134). For the fourth-order projection tensor $\mathbb{P}$ obtain the identities

$$\mathbb{P} = \mathbb{I} - \frac{1}{3}\mathbf{C}^{-1}\otimes\mathbf{C} = \mathbb{I} - \frac{1}{3}\overline{\mathbf{C}}^{-1}\otimes\overline{\mathbf{C}} = \mathbb{I} - \frac{1}{3}(\overline{\mathbf{C}}\otimes\overline{\mathbf{C}}^{-1})^{\mathrm{T}} \quad , \qquad \mathbb{P}^n = \mathbb{P} \quad ,$$

   where n is a positive integer.

2. Show that the properties (6.92) and (6.106) hold.

3. Consider the strain-energy function $\Psi_{iso}[\bar{I}_1(\overline{\mathbf{C}}), \bar{I}_2(\overline{\mathbf{C}})]$ in terms of the modified invariants $\bar{I}_1 = J^{-2/3}I_1$ and $\bar{I}_2 = J^{-4/3}I_2$.

   (a) Show that the derivatives of $\bar{I}_1$ and $\bar{I}_2$ with respect to tensor $\overline{\mathbf{C}}$ are

$$\frac{\partial\bar{I}_1}{\partial\overline{\mathbf{C}}} = \mathbf{I} \quad , \qquad \frac{\partial\bar{I}_2}{\partial\overline{\mathbf{C}}} = \bar{I}_1\mathbf{I} - \overline{\mathbf{C}} \quad . \tag{6.117}$$

   (b) Use the chain rule and the results (6.117) and $(6.83)_4$ in order to obtain the constitutive equation $(6.114)_2$ (with eqs. (6.115) and (6.116)) in the material description.

4. Consider the strain-energy function $\Psi = \Psi_{\mathrm{vol}}(J) + \Psi_{\mathrm{iso}}[\bar{I}_1(\overline{\mathbf{b}}), \bar{I}_2(\overline{\mathbf{b}})]$ with the associated constitutive equation for compressible isotropic hyperelastic materials in the spatial description, i.e. $\boldsymbol{\sigma} = \boldsymbol{\sigma}_{\mathrm{vol}} + \boldsymbol{\sigma}_{\mathrm{iso}}$.

   (a) By analogy with the above Exercise 3(b), obtain the constitutive equation for the isochoric (Cauchy) stress contribution $\boldsymbol{\sigma}_{\mathrm{iso}} = \mathbb{P} : \overline{\boldsymbol{\sigma}}$, where $\mathbb{P}$ is the fourth-order projection tensor, i.e. eq.(6.105), and $\overline{\boldsymbol{\sigma}}$ the fictitious Cauchy stress tensor, defined as

   $$\overline{\boldsymbol{\sigma}} = J^{-1}(\overline{\gamma}_1\overline{\mathbf{b}} + \overline{\gamma}_2\overline{\mathbf{b}}^2) \ . \tag{6.118}$$

   The response coefficients $\overline{\gamma}_1, \overline{\gamma}_2$ are equivalent to those given in (6.116).

   (b) Eliminate $\overline{\mathbf{b}}^2$ from (6.118) in favor of $\overline{\mathbf{b}}^{-1}$ and derive the equivalent form

   $$\overline{\boldsymbol{\sigma}} = J^{-1}(\overline{\gamma}_0\overline{\mathbf{b}} + \overline{\gamma}_2)\overline{\mathbf{b}}^{-1} \qquad \text{with} \qquad \overline{\gamma}_0 = 2\frac{\partial\Psi_{\mathrm{iso}}(\bar{I}_1, \bar{I}_2)}{\partial\bar{I}_1}$$

   for the fictitious Cauchy stress tensor, which is responsible for volume-preserving deformations. The response coefficient $\overline{\gamma}_2$ is given by $(6.116)_2$.

   Hint: Recall identity (5.93).

6.5   Some Forms of Strain-energy Functions

From previous sections we have learnt that the stress response of hyperelastic materials is derived from the given strain-energy function Ψ. Numerous *specific* forms of strain-energy functions to describe the elastic properties of incompressible as well compressible materials have been proposed in the literature and more or less efficient new specific forms are published on a daily basis.

The aim of this section is to specify some forms of strain-energy functions which are well tried within the constitutive theory of finite elasticity and frequently employed in the literature. In particular, we present a selection of representative examples of Ψ known from rubber elasticity describing *isotropic* hyperelastic materials within the isothermal regime (for a collection of constitutive models for rubber see the book edited by DORFMANN and MUHR [1999]). We start by presenting suitable strain-energy functions for *incompressible* materials and continue with some particular forms which are able to describe *compressibility*.

Ogden model for incompressible (rubber-like) materials.   The only materials undergoing finite strains relative to an equilibrium state are biomaterials such as biological soft tissues and solid polymers such as **rubber-like materials**. On the latter we

will focus subsequently. If we subject vulcanized rubber to very high hydrostatic pressures, we observe that it undergoes very small volume changes. To change the shape
of a piece of rubber is very much easier than to change its volume. For the purpose of
computational analyses, rubber is often regarded as incompressible with the constraint
condition $J = \lambda_1 \lambda_2 \lambda_3 = 1$.

A very sophisticated development for simulating incompressible (rubber-like) materials in the phenomenological context is due to OGDEN [1972a, 1982] and [1997,
Chapter 7]. The postulated strain energy is a function of the principal stretches λ_a,
$a = 1, 2, 3$, is computationally simple, and plays a crucial role in the theory of finite
elasticity. It describes the changes of the principal stretches from the reference to the
current configuration and has the form

$$\Psi = \Psi(\lambda_1, \lambda_2, \lambda_3) = \sum_{p=1}^{N} \frac{\mu_p}{\alpha_p} (\lambda_1^{\alpha_p} + \lambda_2^{\alpha_p} + \lambda_3^{\alpha_p} - 3) \ . \tag{6.119}$$

On comparison with the linear theory we obtain the (consistency) condition

$$2\mu = \sum_{p=1}^{N} \mu_p \alpha_p \qquad \text{with} \qquad \mu_p \alpha_p > 0 \ , \qquad p = 1, \ldots, N \ , \tag{6.120}$$

where the parameter μ denotes the classical **shear modulus** in the reference configuration, known from the linear theory.

In equation (6.119), N is a positive integer which determines the number of terms
in the strain-energy function, μ_p are (constant) shear moduli and α_p are dimensionless
constants, $p = 1, \ldots, N$. It emerges that only three pairs of constants ($N = 3$) are
required to give an excellent correlation with experimental stress-deformation data (see
TRELOAR [1944] and TRELOAR [2005]) for simple tension, equibiaxial tension and
pure shear of vulcanized rubber over a very large strain range. Many scientists consider
the experimental data of TRELOAR [1944] to be the essential rubber data. For a more
detailed discussion of the correlation with the experimental data and for additional
sources, see the works by, for example, OGDEN [1972a, 1986, 1987, 1992a, 1997],
TWIZELL and OGDEN [1983], BEATTY [1987, Sections 8-11] and TRELOAR [2005,
Section 11.2].

Typical values of the constants $\alpha_p, \mu_p, p = 1, 2, 3$, are

$$\left. \begin{array}{ll} \alpha_1 = 1.3 & \mu_1 = 6.3 \cdot 10^5 \text{N/m}^2 \\ \alpha_2 = 5.0 & \mu_2 = 0.012 \cdot 10^5 \text{N/m}^2 \\ \alpha_3 = -2.0 & \mu_3 = -0.1 \cdot 10^5 \text{N/m}^2 \end{array} \right\} \tag{6.121}$$

which determine the shear modulus $\mu = 4.225 \cdot 10^5 \text{N/m}^2$ according to $(6.120)_1$.

VALANIS and LANDEL [1967] have postulated the hypothesis that the strain energy
$\Psi = \Psi(\lambda_1, \lambda_2, \lambda_3)$ may be written as the sum of *three separate* functions $\omega(\lambda_a)$, $a =$

$1, 2, 3$, which depend on the principal stretches, we write $\Psi = \omega(\lambda_1) + \omega(\lambda_2) + \omega(\lambda_3)$. This additive decomposition of the strain energy is known as the **Valanis-Landel hypothesis**.

Hence, in view of the Valanis-Landel hypothesis the strain-energy function due to Ogden may be written in the equivalent form

$$\Psi(\lambda_1, \lambda_2, \lambda_3) = \sum_{a=1}^{3} \omega(\lambda_a) \qquad \text{with} \qquad \omega(\lambda_a) = \sum_{p=1}^{N} \frac{\mu_p}{\alpha_p}(\lambda_a^{\alpha_p} - 1) \ . \tag{6.122}$$

According to OGDEN [1986, 1997], separation (6.122) may also be motivated by data obtained from biaxial experiments of JONES and TRELOAR [2005].

EXAMPLE 6.5   Consider an incompressible hyperelastic membrane under *biaxial* deformation with kinematic assumption (2.132). In particular, the two principal stretches λ_1 and λ_2 are given. According to the membrane theory assume a plane stress state and specify the Cauchy stresses in the plane of the membrane by applying Ogden's strain-energy function.

Solution.   The three principal values σ_a of the Cauchy stresses are given according to relation (6.69). Using (6.119) we find, after differentiation, that

$$\sigma_a = -p + \sum_{p=1}^{N} \mu_p \lambda_a^{\alpha_p} \ , \qquad a = 1, 2, 3 \ , \tag{6.123}$$

where p is a scalar not specified by a constitutive equation. It is determined from a boundary condition, namely by the requirement that $\sigma_3 = 0$ which allows p to be expressed explicitly from (6.123), setting $a = 3$, as

$$p = \sum_{p=1}^{N} \mu_p \lambda_3^{\alpha_p} \ . \tag{6.124}$$

Combining (6.124) and (6.123) we obtain the two nonzero stress components

$$\sigma_1 = \sum_{p=1}^{N} \mu_p [\lambda_1^{\alpha_p} - (\lambda_1 \lambda_2)^{-\alpha_p}] \ , \tag{6.125}$$

$$\sigma_2 = \sum_{p=1}^{N} \mu_p [\lambda_2^{\alpha_p} - (\lambda_1 \lambda_2)^{-\alpha_p}] \ , \tag{6.126}$$

where the incompressibility constraint $\lambda_3 = (\lambda_1 \lambda_2)^{-1}$ has been used.   ∎

Mooney-Rivlin, neo-Hookean, Varga model for incompressible (rubber-like) materials. As a special case we obtain from eq. (6.119) the **Mooney-Rivlin model**, the **neo-Hookean model** and the **Varga model** (see MOONEY [1940], RIVLIN [1948, 1949a, b], TRELOAR [1943a, b] and VARGA [1966], respectively).

For example, the very useful *Mooney-Rivlin model* results from (6.119) by setting $N = 2$, $\alpha_1 = 2$, $\alpha_2 = -2$. Using the strain invariants I_1, I_2 as presented by $(5.89)_2$ and $(5.90)_3$, with the constraint condition $I_3 = \lambda_1^2 \lambda_2^2 \lambda_3^2 = 1$, we find from (6.119) that

$$\Psi = c_1(\lambda_1^2 + \lambda_2^2 + \lambda_3^2 - 3) + c_2(\lambda_1^{-2} + \lambda_2^{-2} + \lambda_3^{-2} - 3)$$
$$= c_1(I_1 - 3) + c_2(I_2 - 3) \quad, \tag{6.127}$$

with the constants $c_1 = \mu_1/2$ and $c_2 = -\mu_2/2$. Adopting $(6.120)_1$ the shear modulus μ has the value $\mu_1 - \mu_2$.

The classical strain energy $\Psi = \Psi(I_1, I_2)$ of the Mooney-Rivlin form is often employed in the description of the behavior of isotropic rubber-like materials. *Mooney* derived it on the basis of mathematical arguments employing considerations of symmetry.

The *neo-Hookean model* results from (6.119) by setting $N = 1$, $\alpha_1 = 2$. Using the first principal invariant I_1, see eq. $(5.89)_2$, we find from (6.119) that

$$\Psi = c_1(\lambda_1^2 + \lambda_2^2 + \lambda_3^2 - 3) = c_1(I_1 - 3) \quad, \tag{6.128}$$

with the constant $c_1 = \mu_1/2$ and the shear modulus $\mu = \mu_1$ according to $(6.120)_1$.

This strain-energy function involves a single parameter only and provides a mathematically simple model for the nonlinear deformation behavior of rubber-like materials. It relies on phenomenological considerations and includes typical effects known from nonlinear elasticity within the small strain domain. The function (6.128) may also be motivated from the **statistical theory** in which vulcanized rubber is regarded as a three-dimensional network of long-chain molecules that are connected at a few points. A discussion is given in Section 7.2 on p. 318. Note, however, that experimental data of several isotropic, incompressible elastic materials cannot be reproduced by (6.128), and for these material it is worthwhile considering also the dependence on I_2.

Constitutive relations for the Mooney-Rivlin and the neo-Hookean model follow from (6.65) by means of $(6.127)_2$ and $(6.128)_2$. Derivatives of Ψ with respect to the strain invariants I_1 and I_2 give the simple associated stress relations $\sigma = -p\mathbf{I} + 2c_1\mathbf{b} - 2c_2\mathbf{b}^{-1}$ and $\sigma = -p\mathbf{I} + 2c_1\mathbf{b}$, respectively. Compare also the considerations on p. 203.

As the last special case of Ogden's model we introduce the model by *Varga*. It results from (6.119) by setting $N = 1$, $\alpha_1 = 1$, i.e.

$$\Psi = c_1(\lambda_1 + \lambda_2 + \lambda_3 - 3) \quad, \tag{6.129}$$

with the constant $c_1 = \mu_1$ and the shear modulus $\mu = \mu_1/2$ according to $(6.120)_1$.

Note that of all constitutive approaches given, the Ogden model with $N = 3$ excellently replicates the finite strain behavior of rubber-like materials; see, for example, OGDEN [1972a] for an analytical treatment or DUFFETT and REDDY [1983], SUSSMAN and BATHE [1987], SIMO and TAYLOR [1991a] and MIEHE [1994] for a numerical simulation, among many others. The assumptions made in the Varga model, the neo-Hookean model (obeying *(Gaussian) statistical theory*) or in the Mooney-Rivlin model are rather simple. Consequently, these types of constitutive model are not able to capture the finite extensibility domain of polymer chains (see TRELOAR [1976]).

EXAMPLE 6.6 This example has the aim of investigating the inflation of a spherical *(incompressible rubber) balloon* with different material models. Analyses of balloon inflations have some applications in producing, for example, meteorological balloons for high-altitude measurements or balloon-tipped catheters for clinical treatments. Inflation experiments of spherical neoprene balloons were carried out by ALEXANDER [1971].

In particular, compute the inflation pressure p_i, i.e. the internal pressure in the balloon, and the circumferential Cauchy stress σ as a function of the circumferential stretch λ of the balloon. Let the initial (zero-pressure) radius of the rubber balloon be $R = 10.0$ and the initial thickness of the wall be $H = 0.1$. For the geometrical situation of the spherical balloon in the reference and the current configurations see Figure 6.2.

On the basis of the described four prototypes of constitutive models, that are the *Ogden, Mooney-Rivlin, neo-Hookean* and *Varga models*, study the different mechanical behavior and compare the solutions, drawn in a diagram. Do not consider aspherical modes which clearly develop during the inflation process.

The material properties for the Ogden model are given according to (6.121), with the shear modulus $\mu = 4.225 \cdot 10^5 \mathrm{N/m^2}$ in the reference configuration. For the Mooney-Rivlin model take $c_1 = 0.4375\mu, c_2 = 0.0625\mu$ ($c_1/c_2 = 7$), as suggested by ANAND [1986], for the neo-Hookean model $c_1 = \mu/2$, and for the Varga model take $c_1 = 2\mu$.

Solution. We know from a perfect sphere under inflation pressure p_i that every direction in the plane of the sphere is a principal direction. Hence, the stretch ratio is $\lambda = \lambda_1 = \lambda_2$ which characterizes an equibiaxial deformation. The associated circumferential Cauchy stress is $\sigma = \sigma_1 = \sigma_2$ (while $\sigma_3 = 0$ by the assumption of plane stress). Hence, constitutive equations (6.125) and (6.126) reduce to a single relation, namely

$$\sigma = \sum_{p=1}^{N} \mu_p(\lambda^{\alpha_p} - \lambda^{-2\alpha_p}) \ , \tag{6.130}$$

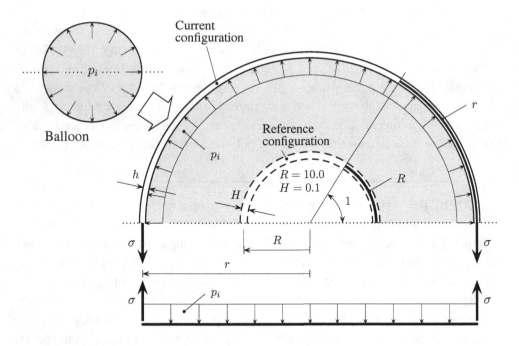

Figure 6.2 Geometry of a spherical balloon in the reference and the current configuration, showing only one hemisphere.

which is plotted in Figure 6.3 for different material parameters. The figure illustrates the relationships between the Cauchy stress σ and the circumferential stretch λ of any point of the rubber balloon for various constitutive models.

By equilibrium we find from Figure 6.2 (free-body diagram) that $r^2\pi p_i = 2r\pi h\sigma$, where r and h denote the radius and the wall thickness of the rubber balloon in the current configuration. According to this condition we find that

$$p_i = 2\frac{h}{r}\sigma \ . \tag{6.131}$$

In view of the kinematical situation of the inflated balloon (see Figure 6.2) the stretch λ at a certain point of the balloon is r/R. Incompressibility requires that the wall volume is conserved, which means that $4\pi r^2 h = 4\pi R^2 H$. With this condition we find that $\lambda_3 = h/H = 1/\lambda^2$ which denotes the stretch in the direction perpendicular to the surface of the sphere, indeed $\lambda_1\lambda_2\lambda_3 = \lambda^2\lambda_3 = 1$.

Using these relations and constitutive equation (6.130) we may find from (6.131) the analytical expression

$$p_i = 2\frac{H}{R}\sum_{p=1}^{N}\mu_p(\lambda^{\alpha_p-3} - \lambda^{-2\alpha_p-3}) \tag{6.132}$$

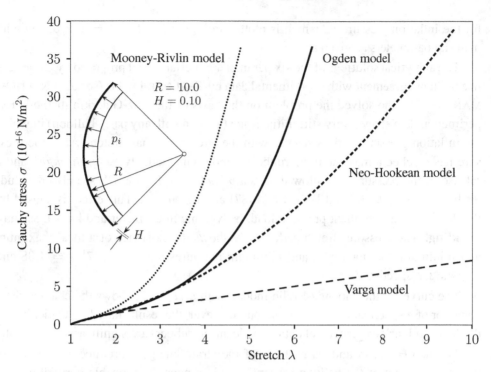

Figure 6.3 Geometry and Cauchy stress σ versus stretch λ of any point of the balloon.

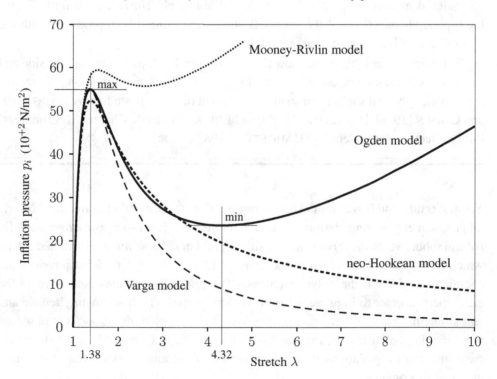

Figure 6.4 Inflation pressure p_i versus stretch λ of any point of the balloon.

for the inflation pressure p_i, which is plotted in Figure 6.4, for $N = 1, 2, 3$, and with material parameters as given above.

The analytical solutions of the six parameter material model proposed by Ogden are in excellent agreement with experimental data by TRELOAR [1944] (see also NEEDLE-MAN [1977]) who solved the problem on the basis of the Ritz-Galerkin method. Experimental data show a very stiff initial stage (as seen with any party balloon) in which the inflation pressure p_i rises steeply with the circumferential stretch. After the pressure has reached a maximum the rubber balloon will suddenly 'snap through', and a release of the pressure will allow it to 'snap back' (see Figure 6.4, see also the studies by OGDEN [1972a] and BEATTY [1987] among others). This effect is caused by the deformation dependent pressure load, is dynamic in character and known as **snap buckling**. The pressure-stretch path clearly shows the existence of a local maximum and minimum, the maximum and minimum pressures are at $\lambda = 7^{1/6} = 1.38$ and $\lambda = 4.32$, respectively.

The curve for the Mooney-Rivlin model, with $c_1/c_2 = 7$, shows the characteristic behavior of a spherical rubber balloon, but, however, the results based on Ogden's model (and Treloar's experimental data on vulcanized rubber) are significantly different.

The neo-Hookean and Varga form of the strain energy reproduces more or less the real behavior of the balloon for small strain ranges. Reasonable correlations for all material models are obtained at the low strain level. However, for finite strains the typical characteristic of the load-deflection curve cannot be reproduced with the simplified neo-Hookean and Varga model.

Experimental investigations show that the balloon develops a bulge on one side and becomes aspherical (compare with NEEDLEMAN [1977]). The bifurcations of pressurized elastic spherical shells from an analytical point of view are studied by HAUGHTON and OGDEN [1978], HAUGHTON [1980] and ERICKSEN [1998, Chapter 5]. Compressibility effects are considered by HAUGHTON [1987]. ■

Yeoh, Arruda and Boyce model for incompressible (rubber-like) materials. Nearly all practical engineering elastomers contain reinforcing fillers such as *carbon black* (in natural rubber vulcanizate) or *silica* (in silicone rubber). These finely distributed fillers, which have typical dimensions of the order of $1.0 - 2.0 \cdot 10^{-12}$m, form physical and chemical bonds with the polymer chains. The fine filler particles are added to the elastomers in order to improve their physical properties which are mainly tensile and tear strength, or abrasion resistance. The associated stress-strain behavior is observed to be highly nonlinear (see, for example, GENT [1962]). Carbon-black filled rubbers have important applications in the manufacture of automotive tyres and many other engineering components.

It turns out that for carbon-black filled rubbers the strain-energy functions described hitherto in this section are not adequate to approximate the observed physical behavior. For example, consider a simple shear deformation of a filler-loaded rubber. Physical observations show that the shear modulus μ of the material varies with deformation in a significant way. To be more specific, μ decreases with increasing deformation initially and then rises again at large deformations (see YEOH [1990, Figure 2]). The associated relation for the shear stress is clearly nonlinear. Now, taking, for example, the Mooney-Rivlin model according to strain-energy function $(6.127)_2$, then, from the explicit expression $(6.78)_2$ we may specify a shear modulus

$$\mu = 2 \left(\frac{\partial \Psi}{\partial I_1} + \frac{\partial \Psi}{\partial I_2} \right) = 2(c_1 + c_2) > 0 \quad . \tag{6.133}$$

The relation for the shear stress is, however, linear with the constant slope $2(c_1 + c_2)$, i.e. the shear modulus. Apparently the Mooney-Rivlin (and its neo-Hookean specialization) model is too simple for the characterization of the elastic properties of carbon-black filled rubber vulcanizates.

The phenomenological material model by YEOH [1990] is motivated in order to simulate the mechanical behavior of carbon-black filled rubber vulcanizates with the typical stiffening effect in the large strain domain. Published data for filled rubbers (see KAWABATA and KAWAI [1977] and SEKI et al. [1987]) suggest that $\partial \Psi / \partial I_2$ is numerically close to zero. *Yeoh* made a simplifying assumption that $\partial \Psi / \partial I_2$ is equal to zero and proposed a three-term strain-energy function where the second strain invariant does not appear. It has the specific form

$$\Psi = c_1(I_1 - 3) + c_2(I_1 - 3)^2 + c_3(I_1 - 3)^3 \quad , \tag{6.134}$$

where c_1, c_2, c_3 are material constants which must satisfy certain restrictions.

Since by (6.5) the strain-energy function Ψ is either *zero* (in which case Ψ has only one real root, corresponding to $I_1 = 3$) or *positive*, we must have $I_1 > 3$ (note that for an incompressible material $I_1 \geq 3$ with the equality only in the reference configuration). Hence, the (convex) strain-energy function increases monotonically with I_1 and $\partial \Psi / \partial I_1 = 0$ has no real roots. From the discriminants of the respective cubic and quadratic equations in $(I_1 - 3)$ the appropriate restrictions on the values for c_1, c_2, c_3 may be determined.

Recall the simple shear deformation example of a filled rubber from above once more. With the strain energy (6.134) we now conclude from eq. $(6.133)_1$ that

$$\mu = 2c_1 + 4c_2(I_1 - 3) + 6c_3(I_1 - 3)^2 > 0 \quad . \tag{6.135}$$

The shear modulus μ involves first-order and second-order terms in $(I_1 - 3)$ and approximates the observed nonlinear physical behavior with satisfying accuracy (provided $c_2 < 0$ and $c_1 > 0, c_3 > 0$).

Another material model for the response of rubber which has a similar structure to (6.134) is due to ARRUDA and BOYCE [1993]. It is, however, a statistical model where the parameters are physically linked to the chain orientations involved in the deformation of the three-dimensional network structure of the rubber. The molecular network structure is represented by an eight-chain model which replaces classical three and four-chain models. The individual polymer chains in the network are described by the non-Gaussian statistical theory and are able to capture the finite extensibility domain. The physically based constitutive model possesses symmetry with respect to the principal stretch space.

The strain-energy function is derived from the *inverse Langevin function* (see, for example, TRELOAR [2005, Chapter 6]) by means of Taylor's expansion (compare with TRELOAR [1954]). Here we present the first three terms for the strain energy, i.e.

$$\Psi = \mu \left[\frac{1}{2}(I_1 - 3) + \frac{1}{20n}(I_1^2 - 9) + \frac{11}{1050n^2}(I_1^3 - 27) + \ldots \right] \ , \qquad (6.136)$$

where μ denotes the shear modulus and n is the number of segments (each of the same length) in a chain, freely jointed together at chemical cross-links. For a more detailed explanation of the underlying concept of statistical thermodynamics the reader is referred to Section 7.2 of this text.

In this two parameter model the first strain invariant I_1 may be linked to the stretch in a chain, λ_{chain}, by the expression $\sqrt{I_1} = \sqrt{3}\lambda_{\text{chain}}$. The chain stretch λ_{chain} is defined to be the current chain length divided by the initial chain length. An advantageous feature of the eight-chain model (6.136) is that all chains stretch equally under uniform extension and biaxial extension.

For further network models which consider chain interactions see, for example, the book by TRELOAR [2005, Chapter 6] and the articles by FLORY and ERMAN [1982] and ANAND [1996].

Ogden model for compressible (rubber-like) materials.   Rubber-like materials in the 'rubbery' state used in engineering are often slightly compressible and associated with minor dilatational deformations. Compressibility is accounted for by the addition of a strain energy Ψ_{vol}, describing the purely *volumetric elastic response* (see the framework of compressible hyperelasticity, Section 6.4). For our considerations, in particular, we use the decoupled representation of the strain-energy function $\Psi(\lambda_1, \lambda_2, \lambda_3) = \Psi_{\text{vol}}(J) + \Psi_{\text{iso}}(\overline{\lambda}_1, \overline{\lambda}_2, \overline{\lambda}_3)$ expressed in terms of principal stretches.

For rubber-like materials, OGDEN [1972b] proposed a volumetric response function in terms of the volume ratio J of the following form

$$\Psi_{\text{vol}}(J) = \kappa \mathcal{G}(J) \qquad \text{with} \qquad \mathcal{G} = \beta^{-2}(\beta \ln J + J^{-\beta} - 1) \ , \qquad (6.137)$$

for $\beta > 0$. The scalar-valued scalar function $\mathcal{G}$ characterizes a strictly convex function, and κ and β denote the constant **bulk modulus** in the reference configuration and an **(empirical) coefficient**, respectively. The strain energy (6.137) satisfies the normalization condition, $\Psi_{\mathrm{vol}}(1) = 0$. Note that this empirical function meets experimental results with excellent accuracy (see OGDEN [1972b, Figure 1]), indicating that rubberlike materials are (slightly) compressible. In particular, for $\beta = 9$, the distribution of the hydrostatic pressure is in good agreement with experimental data of ADAMS and GIBSON [1930] and BRIDGMAN [1945].

An alternative version of $(6.137)_2$, due to SIMO and MIEHE [1992], is obtained by setting $\beta = -2$ to give

$$\mathcal{G} = \frac{1}{4}(J^2 - 1 - 2\ln J) \ . \tag{6.138}$$

The second part of the decoupled strain energy, i.e. $\Psi_{\mathrm{iso}}(\overline{\lambda}_1, \overline{\lambda}_2, \overline{\lambda}_3)$, describes the purely *isochoric elastic response* in terms of modified principal stretches $\overline{\lambda}_a = J^{-1/3}\lambda_a$, $a = 1, 2, 3$. We have

$$\Psi_{\mathrm{iso}}(\overline{\lambda}_1, \overline{\lambda}_2, \overline{\lambda}_3) = \sum_{a=1}^{3} \overline{\omega}(\overline{\lambda}_a) \qquad \text{with} \qquad \overline{\omega}(\overline{\lambda}_a) = \sum_{p=1}^{N} \frac{\mu_p}{\alpha_p}(\overline{\lambda}_a^{\alpha_p} - 1) \ , \tag{6.139}$$

and with the condition (6.120).

EXAMPLE 6.7 Consider the decomposed structure (6.137) and (6.139) of the strain energy and the additive split of the second Piola-Kirchhoff stress tensor $(6.88)_2$. With specifications $(6.137)_2$ and (6.138) find the purely *volumetric contribution* $\mathbf{S}_{\mathrm{vol}}$ of the stress response in the explicit form. In addition, with $(6.139)_2$, find the spectral decomposition of $\mathbf{S}_{\mathrm{iso}}$, i.e. the purely *isochoric contribution*.

Solution. In order to particularize the volumetric stress $\mathbf{S}_{\mathrm{vol}}$ it is only necessary to derive the term $\mathrm{d}\Psi_{\mathrm{vol}}(J)/\mathrm{d}J$ (see eq. $(6.91)_1$). From eq. (6.89) we find, using $(6.91)_1$ and the relation for the purely volumetric elastic response in the form of $(6.137)_1$, that

$$\mathbf{S}_{\mathrm{vol}} = 2\frac{\partial \Psi_{\mathrm{vol}}(J)}{\partial \mathbf{C}} = Jp\mathbf{C}^{-1} \ , \qquad p = \frac{\mathrm{d}\Psi_{\mathrm{vol}}(J)}{\mathrm{d}J} = \kappa\frac{\mathrm{d}\mathcal{G}(J)}{\mathrm{d}J} \ , \tag{6.140}$$

in which, with the strain-energy functions $(6.137)_2$ and (6.138), we obtain the specification

$$\frac{\mathrm{d}\mathcal{G}(J)}{\mathrm{d}J} = \frac{1}{\beta J}\left(1 - \frac{1}{J^\beta}\right) \ , \qquad \frac{\mathrm{d}\mathcal{G}(J)}{\mathrm{d}J} = \frac{1}{2J}\left(J^2 - 1\right) \ . \tag{6.141}$$

As a second step we particularize the isochoric stress $\mathbf{S}_{\mathrm{iso}}$ in respect of the strain energy (6.139). Before proceeding it is first necessary to provide the relation $\partial\overline{\lambda}_a/\partial\lambda_b$.

Recall (6.81), i.e. $\overline{\lambda}_a = J^{-1/3}\lambda_a$, $a = 1, 2, 3$, and relation $(6.82)_1$ which, when formulated in principal stretches, reads $\partial J/\partial \lambda_a = J\lambda_a^{-1}$. Thus, we have

$$
\frac{\partial \overline{\lambda}_a}{\partial \lambda_b} = \frac{\partial (J^{-1/3}\lambda_a)}{\partial \lambda_b} = J^{-1/3}(\delta_{ab} - \frac{1}{3}J^{-1}\frac{\partial J}{\partial \lambda_b}\lambda_a)
$$

$$
= J^{-1/3}(\delta_{ab} - \frac{1}{3}\overline{\lambda}_a\overline{\lambda}_b^{-1}) , \qquad a, b = 1, 2, 3 , \qquad (6.142)
$$

which is relation (6.83) expressed through the modified principal stretches $\overline{\lambda}_a$.

Hence, by analogy with (6.52), we obtain the isochoric stress response in terms of principal values for the general case $\lambda_1 \neq \lambda_2 \neq \lambda_3 \neq \lambda_1$, namely

$$
\mathbf{S}_{\text{iso}} = 2\frac{\partial \Psi_{\text{iso}}(\overline{\lambda}_1, \overline{\lambda}_2, \overline{\lambda}_3)}{\partial \mathbf{C}} = \sum_{a=1}^{3} \underbrace{\frac{1}{\lambda_a}\frac{\partial \Psi_{\text{iso}}}{\partial \lambda_a}}_{S_{\text{iso}\,a}} \hat{\mathbf{N}}_a \otimes \hat{\mathbf{N}}_a , \qquad (6.143)
$$

where $S_{\text{iso}\,a}$, $a = 1, 2, 3$, are the principal values of the second Piola-Kirchhoff stress tensor $\mathbf{S}_{\text{iso}}$ and $\hat{\mathbf{N}}_a$, $a = 1, 2, 3$, denote the principal referential directions.

By use of the chain rule and relation $(6.142)_3$, a straightforward computation from $(6.143)_2$ gives the explicit expressions

$$
S_{\text{iso}\,a} = \frac{1}{\lambda_a}\frac{\partial \Psi_{\text{iso}}}{\partial \lambda_a} = \frac{1}{\lambda_a^2}\left(\overline{\lambda}_a\frac{\partial \Psi_{\text{iso}}}{\partial \overline{\lambda}_a} - \frac{1}{3}\sum_{b=1}^{3}\overline{\lambda}_b\frac{\partial \Psi_{\text{iso}}}{\partial \overline{\lambda}_b}\right) , \qquad a = 1, 2, 3 \quad (6.144)
$$

for the principal isochoric stress values (compare with OGDEN [1997, Section 7.2.3]). The summation symbol (which could be omitted) emphasizes that the index b is repeated, meaning summation over $1, 2, 3$. However, there is no summation over the index a. Using the relation for the purely isochoric elastic response in the form of (6.139), we achieve finally the term $\partial \Psi_{\text{iso}}/\partial \overline{\lambda}_a$, $a = 1, 2, 3$, in the specific form

$$
\frac{\partial \Psi_{\text{iso}}}{\partial \overline{\lambda}_a} = \frac{\partial \overline{\omega}(\overline{\lambda}_a)}{\partial \overline{\lambda}_a} = \sum_{p=1}^{N} \mu_p \overline{\lambda}_a^{\alpha_p - 1} , \qquad a = 1, 2, 3 \qquad (6.145)
$$

(see also the derivation by SIMO and TAYLOR [1991a]).

The complete stress response, as given through (6.140), (6.141) and the spectral decomposition (6.143)–(6.145), serves as a meaningful basis for finite element analyses of constitutive models for isotropic hyperelastic materials at finite strains.   ∎

Similarly to the compressible version of Ogden's model we can reformulate the *Mooney-Rivlin, neo-Hookean, Varga, Yeoh, Arruda and Boyce* models, i.e. (6.127)–(6.129), (6.134), (6.136), as *decoupled* representations. We have just to replace λ_a, I_a by the *modified* quantities $\overline{\lambda}_a, \overline{I}_a$, as defined in (6.81), (6.109)–(6.111) and to add a suitable volumetric response function Ψ_{vol}, for example, $(6.137)_2$ or (6.138).

For example, the decoupled strain-energy function for the Mooney-Rivlin model has the form

$$\Psi(J, \bar{I}_1, \bar{I}_2) = \Psi_{\mathrm{vol}}(J) + c_1(\bar{I}_1 - 3) + c_2(\bar{I}_2 - 3) \ . \tag{6.146}$$

However, material models are often presented in a *coupled* form. The compressible Mooney-Rivlin model, for example, may be given as

$$\Psi(J, I_1, I_2) = c(J - 1)^2 - d \ln J + c_1(I_1 - 3) + c_2(I_2 - 3) \ , \tag{6.147}$$

where c is a material constant and d defines a (dependent) parameter with certain restrictions. By recalling the assumption that the reference configuration is stress-free we may deduce from (6.147) that $d = 2(c_1 + 2c_2)$. The first two terms in (6.147) were proposed by CIARLET and GEYMONAT [1982] in a slightly different form (see also CIARLET [1988, Section 4.10]).

Another example is the *coupled* form of the compressible neo-Hookean model given by the strain-energy function

$$\Psi(I_1, J) = \frac{c_1}{\beta}(J^{-2\beta} - 1) + c_1(I_1 - 3) \ , \qquad \beta = \frac{\nu}{1 - 2\nu} \tag{6.148}$$

(see, for example, BLATZ [1971]), with the constants $c_1 = \mu/2$ and β. The material parameters μ and ν denote the **shear modulus** and **Poisson's ratio**, respectively.

Blatz and Ko model.   For foamed elastomers which cannot be regarded as incompressible BLATZ and KO [1962] and OGDEN [1972b] proposed a strain-energy function which combines theoretical arguments and experimental data (performed on certain solid polyurethane rubbers and foamed polyurethane elastomers). It is based on a *coupled* function of volumetric and isochoric parts according to

$$\Psi(I_1, I_2, I_3) = f\frac{\mu}{2}\left[(I_1 - 3) + \frac{1}{\beta}(I_3^{-\beta} - 1)\right]$$
$$+ (1 - f)\frac{\mu}{2}\left[\left(\frac{I_2}{I_3} - 3\right) + \frac{1}{\beta}(I_3^{\beta} - 1)\right] \ , \tag{6.149}$$

in which μ and ν denote the shear modulus and Poisson's ratio, and $f \in [0, 1]$ is an interpolation parameter. By means of the incompressibility constraint $I_3 = 1$, eq. (6.149) reduces to the Mooney-Rivlin form introduced in eq. (6.127) (with the constants $c_1 = f\mu/2$ and $c_2 = (1 - f)\mu/2$).

Another special case of the strain energy (6.149) may be found by taking $f = 1$, leading to the compressible neo-Hookean model introduced in eq. (6.148) (with $I_3 = J^2$ and the constant $\mu/2 = c_1$). An interesting description of the Blatz and Ko model was presented by BEATTY and STALNAKER [1986] and BEATTY [1987].

EXERCISES

1. For the description of isotropic hyperelastic materials at finite strains we recall
   the important class of strain-energy functions Ψ in terms of principal invariants.
   Study some models suitable to describe compressible materials and particularize
   the associated stress relations.

   (a) Firstly, we consider the *coupled* form of the compressible *Mooney-Rivlin,
       neo-Hookean, Blatz and Ko* models according to the given strain energies
       (6.147)–(6.149), respectively. By means of (6.32) deduce the stress relation

$$S = 2\frac{\partial\Psi(I_1, I_2, I_3)}{\partial C} = \gamma_1 I + \gamma_2 C + \gamma_3 C^{-1} \ ,$$

   with the three response coefficients $\gamma_1 = 2(\partial\Psi/\partial I_1 + I_1\partial\Psi/\partial I_2)$, $\gamma_2 = -2\partial\Psi/\partial I_2$, $\gamma_3 = 2I_3\partial\Psi/\partial I_3$ for the second Piola-Kirchhoff stress tensor
   S as specified in Table 6.1.

	Mooney-Rivlin model (6.147)	neo-Hookean model (6.148)	Blatz and Ko model (6.149)
γ_1	$2(c_1 + c_2 I_1)$	$2c_1$	$\mu f + \xi I_1/2$
γ_2	$-2c_2$	0	$-\xi/2$
γ_3	$2cJ(J - 1) - d$	$-2c_1 I_3^{-\beta}$	$-\mu f I_3^{-\beta}$ $-\xi(I_2 - I_3^{\beta+1})/2$

Table 6.1   Specified coefficients for the constitutive equations of some materials in the coupled form.

For notational simplicity we have introduced the non-negative parameter
$\xi = 2\mu(1 - f)/I_3$. Note that the response coefficients $\gamma_1, \gamma_2, \gamma_3$ for the *neo-
Hookean* model may be found as a special case of the *Blatz and Ko* model
by taking $\xi = 0$ (with the constant $\mu f = 2c_1$).

   (b) Secondly, we consider the *decoupled* form of the compressible *Mooney-
       Rivlin* model (6.146) and the compressible *neo-Hookean* model (obtained
       by setting $c_2 = 0$ in the Mooney-Rivlin model). In addition, we consider
       the decoupled versions of the *Yeoh* model and the *Arruda and Boyce* model,
       i.e.

$$\Psi = \Psi_{\text{vol}} + c_1(\bar{I}_1 - 3) + c_2(\bar{I}_1 - 3)^2 + c_3(\bar{I}_1 - 3)^3 \ ,$$

$$\Psi = \Psi_{\text{vol}} + \mu\left[\frac{1}{2}(\bar{I}_1 - 3) + \frac{1}{20n}(\bar{I}_1^2 - 9) + \frac{11}{1050n^2}(\bar{I}_1^3 - 27) + \ldots\right]$$

(just replace I_1 in eqs. (6.134), (6.136) by the modified first invariant $\bar{I}_1$).

Derive the associated stress relations

$$\overline{\mathbf{S}} = 2\frac{\partial\Psi_{\text{iso}}(\bar{I}_1, \bar{I}_2)}{\partial\overline{\mathbf{C}}} = \overline{\gamma}_1\mathbf{I} + \overline{\gamma}_2\overline{\mathbf{C}}$$

for the fictitious second Piola-Kirchhoff stress tensor, with the specified response coefficients $\overline{\gamma}_1, \overline{\gamma}_2$ (see eq. (6.116)) according to Table 6.2.

	Mooney-Rivlin model (6.146)	neo-Hookean model	Yeoh model (6.134)	Arruda and Boyce model (6.136)
$\overline{\gamma}_1$	$2(c_1 + c_2\bar{I}_1)$	$2c_1$	$2c_1 + 4c_2(\bar{I}_1 - 3)$ $+6c_3(\bar{I}_1 - 3)^2$	$\mu[1 + (1/5n)\bar{I}_1$ $+(11/175n^2)\bar{I}_1^2 + \ldots]$
$\overline{\gamma}_2$	$-2c_2$	0	0	0

Table 6.2 Specified coefficients for the constitutive equations of some materials in the decoupled form.

2. Consider a spherical balloon of incompressible hyperelastic material ($\lambda_1\lambda_2\lambda_3 = 1$). The material is characterized by a strain-energy function in terms of principal stretches according to

$$\Psi = \frac{\mu}{\alpha}(\lambda_1^\alpha + \lambda_2^\alpha + \lambda_3^\alpha - 3) \ , \tag{6.150}$$

with material constants μ and α. For $\alpha = 2$ we obtain the classical neo-Hookean model.

(a) Determine the inflation pressure p_i as a function of the circumferential stretch λ in the form

$$p_i = 2\mu\frac{H}{R}(\lambda^{\alpha-3} - \lambda^{-2\alpha-3}) \ ,$$

where H and R are the initial (zero-pressure) thickness and radius of the spherical balloon, respectively.

(b) Show that the function $p_i = p_i(\lambda)$ has a relative maximum if $0 < \alpha < 3$ and a relative minimum if $-3/2 < \alpha < 0$ (see OGDEN [1972a]).

Consequently, balloons made of materials which are described by the strain-energy function (6.150) will not 'snap through' for $\alpha \gg 3$. Typical examples for this type of (stable) material behavior are biomaterials such as an *artery* (for the mechanics of the arterial wall see the excellent survey text by HUMPHREY [1995], and the papers by HOLZAPFEL et al. [2000] and HOLZAPFEL [2001]) or a *ventricle* (see NEEDLEMAN et al. [1983]). Specific results on membrane biomechanics including illustrative examples from the literature are reviewed in the article by HUMPHREY [1998].

3. Consider a thin sheet of incompressible hyperelastic material with the same setting as formulated in Example 6.3. The homogeneous stress response of the
material is assumed to be isotropic and based on Ogden's model. Discuss the
stress states (which are plane throughout the sheet) for the following two modes
of deformations (for the associated kinematic relations, compare with Exercise 1
on p. 226):

(a) Consider a *simple tension* for which $\lambda_1 = \lambda$ ($\lambda_2 = \lambda_3$). Show that the only
nonzero Cauchy stress σ, in the direction of the applied stretch λ, is

$$\sigma = \sum_{p=1}^{N} \mu_p(\lambda^{\alpha_p} - \lambda^{-\alpha_p/2}) \ .$$

In addition, find the stress required to produce a final extended length of
$\lambda = 2$ for each of the Mooney-Rivlin and the neo-Hookean models.

(b) Consider a homogeneous *pure shear deformation* and show that the biaxial
stress state ($\sigma_3 = 0$) of the problem is of the form

$$\sigma_1 = \sum_{p=1}^{N} \mu_p(\lambda^{\alpha_p} - \lambda^{-\alpha_p}) \ , \qquad \sigma_2 = \sum_{p=1}^{N} \mu_p(1 - \lambda^{-\alpha_p}) \ .$$

(c) Compute the associated constitutive equations given in (a) and (b) for the
Mooney-Rivlin, neo-Hookean and Varga models and compare the results
with Ogden's model (plot the relation between the Cauchy stress and the
associated stretch ratio for each material model).

4. The so-called **Saint-Venant Kirchhoff model** is characterized by the strain-
energy function

$$\Psi(\mathbf{E}) = \frac{\gamma}{2}(\mathrm{tr}\mathbf{E})^2 + \mu\mathrm{tr}\mathbf{E}^2 \tag{6.151}$$

(see, for example, CIARLET [1988, p. 155]), in which $\gamma > 0$ and $\mu > 0$ are the
two constants of **Lamé**. The Lamé constant γ is usually denoted in the literature
by the symbol λ. However, in order to avoid confusion with the stretch ratio λ
we use a different symbol for it. The Saint-Venant Kirchhoff model is a classical
nonlinear model for compressible hyperelastic materials often used for metals.
Note that the volume ratio J does not appear explicitly in this material model.

(a) From the given strain energy $\Psi(\mathbf{E})$ derive the second Piola-Kirchhoff stress
S, which linearly depends on the Green-Lagrange strain **E**.

(b) Consider the one-dimensional case of the constitutive equation derived in (a). For a *uniform deformation* of a rod (with uniform cross-section), i.e. $x = \lambda X$, derive the relation between the nominal stress P and the associated stretch ratio λ (which is a cubic equation in λ) and plot the function $P = P(\lambda)$.

Show that $P(\lambda)$ is not monotonic in compression and derive the critical stretch value $\lambda_{\text{crit}} = (1/3)^{1/2}$ at which the Saint-Venant Kirchhoff model fails (zero stiffness, the tangent of $P(\lambda)$ at λ_{crit} is horizontal). This failure is not influenced by the material constants γ and μ.

In addition show that the material model does not satisfy the growth condition $(6.6)_2$ (in fact for $\lambda \to 0^+$ the stress tends to zero which is physically unrealistic). CIARLET [1988] showed, with the proof by RAOULT [1986], that the Saint-Venant Kirchhoff model does not satisfy the requirement of polyconvexity either.

Note that this material model is suitable for large displacements but it is *not* recommended to use it for large compressive strains.

5. In the first term in eq. (6.151) replace $\text{tr}\mathbf{E}$ by $\ln J$ and γ by $\kappa > 0$ in order to obtain a **modified Saint-Venant Kirchhoff model** of the form

$$\Psi(\mathbf{E}) = \frac{\kappa}{2}(\ln J)^2 + \mu\text{tr}\mathbf{E}^2 \ , \tag{6.152}$$

where $J = \det\mathbf{F}$ denotes the volume ratio. The proposed material model (6.152) circumvents the serious drawbacks of the classical Saint-Venant Kirchhoff model (see Exercise 4) when used for large compressive strains.

(a) From the strain energy (6.152) derive the second Piola-Kirchhoff stress $\mathbf{S} = \mathbf{S}(\mathbf{C})$ as a function of the right Cauchy-Green tensor $\mathbf{C}$. The result is similar to a stress relation proposed by CURNIER [1994, eq. (6.113)] which also has the aim of avoiding the defects of the classical Saint-Venant Kirchhoff model occurring at large compressive strains.

(b) Consider a one-dimensional problem as described in Exercise 4(b) and obtain the nominal stress P as a function of λ. Discuss the function $P(\lambda)$ for the two regions $\lambda > 1$, $\lambda < 1$ and show that the modified Saint-Venant Kirchhoff model satisfies the growth condition in the sense that the stress tends to (minus) infinity for $\lambda \to 0^+$.

6.6   Elasticity Tensors

In order to obtain solutions of nonlinear (initial boundary-value) problems in computational finite elasticity and inelasticity so-called **incremental/iterative solution techniques** of *Newton's type* are frequently applied to solve a sequence of linearized problems.

   This strategy requires knowledge of the linearized constitutive equation, here presented in both the material and spatial descriptions. The underlying technique was first
introduced in the mechanics of solids and structures by HUGHES and PISTER [1978].
The process of linearizing constitutive equations is a very important task in computational mechanics and the main objective of this section. For more on the concept of
linearization, which is basically differentiation, see Section 8.4.

Material and spatial representations of the elasticity tensor.   Consider the nonlinear second Piola-Kirchhoff stress tensor $\mathbf{S}$ of a point at a certain time t. We look at
$\mathbf{S}$ as a nonlinear tensor-valued tensor function of one variable. We assume this variable
to be the right Cauchy-Green tensor $\mathbf{C}$.

   First of all we do not assume that the stress tensor is derived from a strain-energy
function Ψ. According to considerations (1.247) and (1.248) we are now in a position
to determine the total differential

$$\mathrm{d}\mathbf{S} = \mathbb{C} : \frac{1}{2}\mathrm{d}\mathbf{C} \quad , \tag{6.153}$$

in which we have introduced the definition

$$\mathbb{C} = 2\frac{\partial\mathbf{S}(\mathbf{C})}{\partial\mathbf{C}} \qquad \text{or} \qquad C_{ABCD} = 2\frac{\partial S_{AB}}{\partial C_{CD}} \quad , \tag{6.154}$$

which, by means of the chain rule, reads, in terms of the Green-Lagrange strain tensor
$\mathbf{E} = (\mathbf{C} - \mathbf{I})/2$,

$$\mathbb{C} = \frac{\partial\mathbf{S}(\mathbf{E})}{\partial\mathbf{E}} \qquad \text{or} \qquad C_{ABCD} = \frac{\partial S_{AB}}{\partial E_{CD}} \quad . \tag{6.155}$$

The quantity $\mathbb{C}$ characterizes the gradient of function $\mathbf{S}$ and relates the work conjugate
pairs of stress and strain tensors. It measures the change in stress which results from a
change in strain and is referred to as the **elasticity tensor** in the *material description*
or the **referential tensor of elasticities**. It is a tensor of rank four with the four indices
A, B, C, D.

   The elasticity tensor $\mathbb{C}$ is always symmetric in its first and second slots, i.e. AB,
and in its third and fourth slots, i.e. CD,

$$C_{ABCD} = C_{BACD} = C_{ABDC} \tag{6.156}$$

(we have, in general, 36 independent components at each strain state).

We say $\mathbb{C}$ possesses the **minor symmetries**. The symmetry condition (6.156) is independent of the existence of a strain-energy function Ψ and holds for *all* elastic materials. Note that the minor symmetry of $\mathbb{C}$ follows from the symmetries of the right Cauchy-Green tensor $\mathbf{C}$ (or equivalently from the Green-Lagrange strain tensor $\mathbf{E}$) and the second Piola-Kirchhoff stress tensor $\mathbf{S}$.

If we assume the existence of a scalar-valued energy function Ψ (hyperelasticity), then $\mathbf{S}$ may be derived from Ψ according to $\mathbf{S} = 2\partial\Psi(\mathbf{C})/\partial\mathbf{C}$ (see $(6.13)_2$). Hence, using (6.154), we arrive at the crucial relation

$$\mathbb{C} = 4\frac{\partial^2\Psi(\mathbf{C})}{\partial\mathbf{C}\partial\mathbf{C}} \qquad \text{or} \qquad C_{ABCD} = 4\frac{\partial^2\Psi}{\partial C_{AB}\partial C_{CD}} \tag{6.157}$$

for the elasticities in the material description, with the symmetries

$$\mathbb{C} = \mathbb{C}^\mathrm{T} \qquad \text{or} \qquad C_{ABCD} = C_{CDAB} \ . \tag{6.158}$$

We say $\mathbb{C}$ possesses the **major symmetries**. Thus, tensor $\mathbb{C}$ has only 21 independent components at each strain state. The condition (6.158) is a necessary and sufficient condition for a material to be *hyperelastic*. The symmetry condition $C_{ABCD} = C_{CDAB}$ is often referred to as the definition of hyperelasticity. Hence, the major symmetry of $\mathbb{C}$ is basically *equivalent* to the existence of a strain-energy function. Note that the major symmetry of the elasticity tensor is associated with the symmetry of the **(tangent) stiffness matrix** arising in a finite element discretization procedure.

The **elasticity tensor** in the *spatial description* or the **spatial tensor of elasticities**, denoted by $\mathfrak{c}$, is defined as the push-forward operation of $\mathbb{C}$ times a factor of J^{-1} (see MARSDEN and HUGHES [1994, Section 3.4]), in other texts the definition of $\mathfrak{c}$ frequently excludes the factor J^{-1}. It is the *Piola transformation* of $\mathbb{C}$ on each large index so that

$$\mathfrak{c} = J^{-1}\chi_*(\mathbb{C}) \ , \qquad c_{abcd} = J^{-1}F_{aA}F_{bB}F_{cC}F_{dD}\,C_{ABCD} \ , \tag{6.159}$$

with the minor symmetries

$$c_{abcd} = c_{bacd} = c_{abdc} \ , \tag{6.160}$$

and additionally for hyperelasticity we have the major symmetries $\mathfrak{c} = \mathfrak{c}^\mathrm{T}$ or $c_{abcd} = c_{cdab}$. The fourth-order tensors $\mathbb{C}$ and $\mathfrak{c}$ are crucial within the *concept of linearization*, as will become apparent in Chapter 8, particularly in Section 8.4.

The spatial representation of eq. (6.153) can be shown to be

$$\mathcal{L}_v(\tau^\sharp) = J\mathfrak{c} : \mathbf{d} \tag{6.161}$$

(for a proof see Section 8.4, p. 398), in which $\mathcal{L}_v(\tau^\sharp), \mathbf{d}$ and $\mathfrak{c}$ denote the objective Oldroyd stress rate (5.59) of the contravariant Kirchhoff stress tensor τ, the rate of deformation tensor (2.148), and the spatial elasticity tensor, as defined in eq. (6.159), respectively. A material is said to be **hypoelastic** if the associated rate equations of the form (6.161) are not obtained from a (scalar-valued) energy function. For more on hypoelastic materials see the classical and detailed account by TRUESDELL and NOLL [2004, Sections 99-103].

Systematic treatments of the elasticity tensors have been given by, for example, TRUESDELL and TOUPIN [1960, Sections 246-249], CHADWICK and OGDEN [1971a, b], HILL [1981], OGDEN [1997, Chapter 6] and TRUESDELL and NOLL [2004, Sections 45, 82].

Decoupled representation of the elasticity tensor.   Based on the kinematic assumption $(6.79)_2$ and the decoupled structure of the strain-energy function (6.85) we derive the associated elasticity tensor. We focus attention solely on the material description of the elasticity tensor.

The elasticity tensor (6.154) may be written in the *decoupled* form

$$\mathbb{C} = 2\frac{\partial \mathbf{S}(\mathbf{C})}{\partial \mathbf{C}} = \mathbb{C}_{\mathrm{vol}} + \mathbb{C}_{\mathrm{iso}} , \tag{6.162}$$

which represents the completion of the additive split of the stress response (6.88).

In relation (6.162) we introduced the definitions

$$\mathbb{C}_{\mathrm{vol}} = 2\frac{\partial \mathbf{S}_{\mathrm{vol}}}{\partial \mathbf{C}} , \qquad \mathbb{C}_{\mathrm{iso}} = 2\frac{\partial \mathbf{S}_{\mathrm{iso}}}{\partial \mathbf{C}} \tag{6.163}$$

of the purely *volumetric contribution* $\mathbb{C}_{\mathrm{vol}}$ and the purely *isochoric contribution* $\mathbb{C}_{\mathrm{iso}}$.

By analogy with eq. (6.157) we express the two contributions $\mathbb{C}_{\mathrm{vol}}$ and $\mathbb{C}_{\mathrm{iso}}$ in terms of the strain-energy function Ψ. Before this exploitation we introduce the definition

$$\frac{\partial \mathbf{C}^{-1}}{\partial \mathbf{C}} = -\mathbf{C}^{-1} \odot \mathbf{C}^{-1} \tag{6.164}$$

of the fourth-order tensor $\partial \mathbf{C}^{-1}/\partial \mathbf{C}$, for convenience (recall Example 1.11, p. 43, and take $\mathbf{A} = \mathbf{C}$ in relation (1.249)), where the symbol $\odot$ has been introduced to denote the tensor product according to the rule

$$-(\mathbf{C}^{-1} \odot \mathbf{C}^{-1})_{ABCD} = -\frac{1}{2}(C_{AC}^{-1}C_{BD}^{-1} + C_{AD}^{-1}C_{BC}^{-1}) = \frac{\partial C_{AB}^{-1}}{\partial C_{CD}} . \tag{6.165}$$

Starting with $(6.163)_1$, a straightforward computation yields, with definition $(6.89)_2$, property (1.256), the derivative of J and $\mathbf{C}^{-1}$ with respect to $\mathbf{C}$, i.e. eqs. $(6.82)_1$ and (6.164), and the product rule of differentiation,

$$\mathbb{C}_{\mathrm{vol}} = 2\frac{\partial \mathbf{S}_{\mathrm{vol}}}{\partial \mathbf{C}} = 2\frac{\partial(Jp\mathbf{C}^{-1})}{\partial \mathbf{C}}$$

$$= 2\mathbf{C}^{-1} \otimes \left(p\frac{\partial J}{\partial \mathbf{C}} + J\frac{\partial p}{\partial \mathbf{C}} \right) + 2Jp\frac{\partial \mathbf{C}^{-1}}{\partial \mathbf{C}}$$

$$= J\tilde{p}\mathbf{C}^{-1} \otimes \mathbf{C}^{-1} - 2Jp\mathbf{C}^{-1} \odot \mathbf{C}^{-1} . \tag{6.166}$$

For convenience, we have introduced the scalar function $\tilde{p}$, defined by

$$\tilde{p} = p + J\frac{\mathrm{d}p}{\mathrm{d}J} \quad , \tag{6.167}$$

with the constitutive equation for p given in (6.91)$_1$. Note that the only values which must be specified for a given material are p and $\tilde{p}$.

The following example shows a lengthy but *representative* derivation of an elasticity tensor.

EXAMPLE 6.8   Show the following explicit expression for the second contribution to the elasticity tensor, i.e. the isochoric part $\mathbb{C}_{\mathrm{iso}}$, as defined in eq. (6.163)$_2$,

$$\mathbb{C}_{\mathrm{iso}} = \mathbb{P} : \overline{\mathbb{C}} : \mathbb{P}^{\mathrm{T}} + \frac{2}{3}\mathrm{Tr}(J^{-2/3}\overline{\mathbf{S}})\tilde{\mathbb{P}}$$

$$-\frac{2}{3}(\mathbf{C}^{-1} \otimes \mathbf{S}_{\mathrm{iso}} + \mathbf{S}_{\mathrm{iso}} \otimes \mathbf{C}^{-1}) \tag{6.168}$$

(compare also with HOLZAPFEL [1996a]), which is based on the definitions

$$\overline{\mathbb{C}} = 2J^{-4/3}\frac{\partial\overline{\mathbf{S}}}{\partial\overline{\mathbf{C}}} = 4J^{-4/3}\frac{\partial^2\Psi_{\mathrm{iso}}(\overline{\mathbf{C}})}{\partial\overline{\mathbf{C}}\partial\overline{\mathbf{C}}} \quad , \qquad \mathrm{Tr}(\bullet) = (\bullet) : \mathbf{C} \quad , \tag{6.169}$$

$$\tilde{\mathbb{P}} = \mathbf{C}^{-1} \odot \mathbf{C}^{-1} - \frac{1}{3}\mathbf{C}^{-1} \otimes \mathbf{C}^{-1} \tag{6.170}$$

of the fourth-order **fictitious elasticity tensor** $\overline{\mathbb{C}}$ in the *material description*, the trace $\mathrm{Tr}(\bullet)$ and the **modified projection tensor** $\tilde{\mathbb{P}}$ of fourth-order.

Solution.   Starting from the definition of $\mathbb{C}_{\mathrm{iso}}$, i.e. (6.163)$_2$, we find, using (6.90)$_4$ and property (1.256), that

$$\mathbb{C}_{\mathrm{iso}} = 2\frac{\partial\mathbf{S}_{\mathrm{iso}}}{\partial\mathbf{C}} = 2\frac{\partial(J^{-2/3}\mathbb{P} : \overline{\mathbf{S}})}{\partial\mathbf{C}}$$

$$= 2(\mathbb{P} : \overline{\mathbf{S}}) \otimes \frac{\partial J^{-2/3}}{\partial\mathbf{C}} + 2J^{-2/3}\frac{\partial(\mathbb{P} : \overline{\mathbf{S}})}{\partial\mathbf{C}} \quad . \tag{6.171}$$

The first term in this equation yields, through property (6.82)$_2$ and definition (6.90)$_4$,

$$2(\mathbb{P} : \overline{\mathbf{S}}) \otimes \frac{\partial J^{-2/3}}{\partial\mathbf{C}} = -\frac{2}{3}(J^{-2/3}\mathbb{P} : \overline{\mathbf{S}}) \otimes \mathbf{C}^{-1} = -\frac{2}{3}\mathbf{S}_{\mathrm{iso}} \otimes \mathbf{C}^{-1} \quad , \tag{6.172}$$

which gives the last expression of (6.168) we show.

Hence, in the following we analyze exclusively the second term in (6.171). With the definition of the projection tensor (6.84), identities (1.160)$_1$, (1.152) and the chain rule, we have

$$2J^{-2/3}\frac{\partial(\mathbb{P}:\bar{\mathbf{S}})}{\partial\mathbf{C}} = 2J^{-2/3}\frac{\partial}{\partial\mathbf{C}}\left(\bar{\mathbf{S}} - \frac{1}{3}(\mathbf{C}^{-1}\otimes\mathbf{C}):\bar{\mathbf{S}}\right)$$

$$= 2J^{-2/3}\left(\frac{\partial\bar{\mathbf{S}}}{\partial\bar{\mathbf{C}}} - \frac{1}{3}\frac{\partial(\bar{\mathbf{S}}:\mathbf{C})\mathbf{C}^{-1}}{\partial\bar{\mathbf{C}}}\right):\frac{\partial\bar{\mathbf{C}}}{\partial\mathbf{C}} \quad , \tag{6.173}$$

and finally, using the definition of the fourth-order tensor $\partial\bar{\mathbf{C}}/\partial\mathbf{C}$, i.e. $(6.83)_4$, definition $(6.169)_1$ and property (1.256),

$$2J^{-2/3}\frac{\partial(\mathbb{P}:\bar{\mathbf{S}})}{\partial\mathbf{C}} = [\bar{\mathbb{C}} - \frac{1}{3}(\mathbb{A}_1 + \mathbb{A}_2)]:\mathbb{P}^{\mathrm{T}} \quad , \tag{6.174}$$

in which the definitions

$$\mathbb{A}_1 = 2J^{-4/3}\mathbf{C}^{-1}\otimes\frac{\partial(\bar{\mathbf{S}}:\mathbf{C})}{\partial\bar{\mathbf{C}}} \quad , \qquad \mathbb{A}_2 = 2J^{-4/3}(\bar{\mathbf{S}}:\mathbf{C})\frac{\partial\mathbf{C}^{-1}}{\partial\bar{\mathbf{C}}} \tag{6.175}$$

are to be used.

In order to study (6.175) in more detail we apply property (1.255), the chain rule, identity $(1.160)_1$, definition $(6.169)_1$ of $(6.175)_1$ and definition (6.164) of $(6.175)_2$ to give

$$\mathbb{A}_1 = \mathbf{C}^{-1}\otimes(\mathbf{C}:\bar{\mathbb{C}} + 2J^{-2/3}\bar{\mathbf{S}}) = \mathbf{C}^{-1}\otimes\mathbf{C}:\bar{\mathbb{C}} + 2\mathbf{C}^{-1}\otimes J^{-2/3}\bar{\mathbf{S}} \quad , \tag{6.176}$$

$$\mathbb{A}_2 = -2J^{-2/3}(\bar{\mathbf{S}}:\mathbf{C})\mathbf{C}^{-1}\odot\mathbf{C}^{-1} \quad . \tag{6.177}$$

Eqs. (6.176) and (6.177) substituted back into (6.174) yield, using identity $(1.160)_1$ and definition $(6.169)_2$,

$$2J^{-2/3}\frac{\partial(\mathbb{P}:\bar{\mathbf{S}})}{\partial\mathbf{C}} = (\mathbb{I} - \frac{1}{3}\mathbf{C}^{-1}\otimes\mathbf{C}):\bar{\mathbb{C}}:\mathbb{P}^{\mathrm{T}} - \frac{2}{3}\mathbf{C}^{-1}\otimes J^{-2/3}\bar{\mathbf{S}}:\mathbb{P}^{\mathrm{T}}$$

$$+ \frac{2}{3}\mathrm{Tr}(J^{-2/3}\bar{\mathbf{S}})\mathbf{C}^{-1}\odot\mathbf{C}^{-1}:\mathbb{P}^{\mathrm{T}} \quad . \tag{6.178}$$

With the definition of the projection tensor (6.84), then with definition $(6.90)_4$, according to identity (1.157), and by means of (6.164) with rule (1.254) and (6.170) we find finally

$$2J^{-2/3}\frac{\partial(\mathbb{P}:\bar{\mathbf{S}})}{\partial\mathbf{C}} = \mathbb{P}:\bar{\mathbb{C}}:\mathbb{P}^{\mathrm{T}} - \frac{2}{3}\mathbf{C}^{-1}\otimes\mathbf{S}_{\mathrm{iso}}$$

$$+ \frac{2}{3}\mathrm{Tr}(J^{-2/3}\bar{\mathbf{S}})\underbrace{(\mathbf{C}^{-1}\odot\mathbf{C}^{-1} - \frac{1}{3}\mathbf{C}^{-1}\otimes\mathbf{C}^{-1})}_{\tilde{\mathbb{P}}} \quad , \tag{6.179}$$

which is identical to the remaining terms in (6.168). ∎

The two tensor expressions $(6.166)_4$ and (6.168) in the material description represent explicit forms which are generally applicable to any compressible hyperelastic material of interest. Since we have already computed the stress relation $\mathbf{S} = \mathbf{S}_{\mathrm{vol}} + \mathbf{S}_{\mathrm{iso}}$, with the terms $\mathbf{S}_{\mathrm{vol}} = Jp\mathbf{C}^{-1}$ and $\mathbf{S}_{\mathrm{iso}} = J^{-2/3}\mathbb{P} : \overline{\mathbf{S}}$, it is a straightforward task to set up the associated elasticity tensor $\mathbb{C} = \mathbb{C}_{\mathrm{vol}} + \mathbb{C}_{\mathrm{iso}}$. All that remains is to determine $\tilde{p}$, $\overline{\mathbb{C}}$ and $\tilde{\mathbb{P}}$ from relations (6.167), $(6.169)_1$ and (6.170), respectively. Since the nonlinear functions p and $\overline{\mathbf{S}}$, $\mathbf{S}_{\mathrm{iso}}$, occurring in $(6.166)_4$ and (6.168), are already known from the stress relation, a different material model only affects the elasticity tensor $\mathbb{C}$ through the scalar function $\tilde{p} = \tilde{p}(\Psi_{\mathrm{vol}})$ and the fourth-order fictitious elasticity tensor $\overline{\mathbb{C}} = \overline{\mathbb{C}}(\Psi_{\mathrm{iso}})$.

Expressions $(6.166)_4$ and (6.168) exhibit a clear structure and are fundamental within the finite element method in preserving quadratic rate of convergence near the solution point, when **Newton's method** is used as the associated solution technique.

The two explicit relations $(6.166)_4$ and (6.168) were specified within a so-called **consistent linearization process** of the associated stress tensor. The notion 'consistent linearization' means a linearization of *all* quantities which are related to the nonlinear problem. In the community of computational mechanics the elasticity tensors $(6.166)_4$ and (6.168) are frequently referred to as **algorithmic** or **consistent linearized tangent moduli** in the material and spatial descriptions.

For some complex problems the set up of the analytical tangent moduli is a difficult and time-consuming task. This is why the tangent moduli are also computed on a numerical basis, which turns out to be a straightforward and convenient technique in order to linearize sophisticated stress relations such as, for example, the stress response of materials associated with anisotropic damage at finite strains. For a numerical computation of consistent tangent moduli in large-strain inelasticity see, for example, MIEHE [1996], which contains more details and references on the underlying concept of approximation. However, manipulations which have led to $(6.166)_4$ and (6.168) may also be carried out with some of the commercially available mathematical software-packages having the feature of symbolic computation.

Elasticity tensor in terms of principal stretches. Consider an isotropic hyperelastic material characterized by the strain-energy function $\Psi = \Psi(\lambda_1, \lambda_2, \lambda_3)$, with the principal stretches $\lambda_1, \lambda_2, \lambda_3$.

The aim is now to derive the spectral form of the elasticity tensor $\mathbb{C}$ in the *material description*, namely

$$\mathbb{C} = \sum_{a,b=1}^{3} \frac{1}{\lambda_b} \frac{\partial S_a}{\partial \lambda_b} \hat{\mathbf{N}}_a \otimes \hat{\mathbf{N}}_a \otimes \hat{\mathbf{N}}_b \otimes \hat{\mathbf{N}}_b$$

$$+ \sum_{\substack{a,b=1 \\ a \neq b}}^{3} \frac{S_b - S_a}{\lambda_b^2 - \lambda_a^2} (\hat{\mathbf{N}}_a \otimes \hat{\mathbf{N}}_b \otimes \hat{\mathbf{N}}_a \otimes \hat{\mathbf{N}}_b + \hat{\mathbf{N}}_a \otimes \hat{\mathbf{N}}_b \otimes \hat{\mathbf{N}}_b \otimes \hat{\mathbf{N}}_a) \ , \qquad (6.180)$$

with the principal second Piola-Kirchhoff stresses

$$S_a = 2\frac{\partial\Psi}{\partial\lambda_a^2} = \frac{1}{\lambda_a}\frac{\partial\Psi}{\partial\lambda_a} \quad , \qquad a = 1,2,3 \quad , \tag{6.181}$$

and the set $\{\hat{\mathbf{N}}_a\}$, $a = 1,2,3$, of orthonormal eigenvectors of the right Cauchy-Green tensor $\mathbf{C}$. They define principal referential directions at a point $\mathbf{X}$, with the conditions $|\hat{\mathbf{N}}_a| = 1$ and $\hat{\mathbf{N}}_a \cdot \hat{\mathbf{N}}_b = \delta_{ab}$.

The important fourth-order tensor $\mathbb{C}$ in a more general setting was given by OGDEN [1997, Section 6.1.4]. Compare also the work of CHADWICK and OGDEN [1971a, b] with some differences in notation.

The proof of representation (6.180) is as follows:

Proof.   In order to prove relation (6.180) we follow an approach which takes advantage of isotropy. We know from Chapter 5 that for isotropic elastic materials the second Piola-Kirchhoff stress tensor $\mathbf{S}$ is coaxial with the right Cauchy-Green tensor $\mathbf{C}$, so that $\mathbf{S}$ has the same principal directions as $\mathbf{C}$.

For notational convenience we use henceforth a rate formulation rather than an infinitesimal formulation. In particular, we now compute the material time derivatives of the stress and strain tensors $\mathbf{S}$ and $\mathbf{C}$, and compare them with the rate form of relation (6.153), i.e. $\dot{\mathbf{S}} = \mathbb{C} : \dot{\mathbf{C}}/2$, in order to obtain the elasticity tensor $\mathbb{C}$, as defined in (6.154) and specified in (6.180).

To begin with, consider a set of orthonormal basis vectors $\mathbf{e}_a$, $a = 1,2,3$, fixed in space. Consequently, the set $\{\hat{\mathbf{N}}_a\}$, $a = 1,2,3$, of orthonormal eigenvectors may be governed by the transformation law

$$\hat{\mathbf{N}}_a = \mathbf{Q}\mathbf{e}_a \quad , \qquad a = 1,2,3 \tag{6.182}$$

(compare with eq. (1.182), Section 1.5), where $\mathbf{Q}$ denotes an orthogonal tensor with components $Q_{ab} = \mathbf{e}_a \cdot \hat{\mathbf{N}}_b = \cos\theta(\mathbf{e}_a, \hat{\mathbf{N}}_b)$, representing the cosine of the angle between the fixed basis vectors $\mathbf{e}_a$ and the orthonormal eigenvectors $\hat{\mathbf{N}}_b$ (principal referential directions). Tensor $\mathbf{Q}$ is characterized by the *orthogonality condition*, i.e. $\mathbf{Q}^{\mathrm{T}}\mathbf{Q} = \mathbf{Q}\mathbf{Q}^{\mathrm{T}} = \mathbf{I}$.

Next, we compute the material time derivative of the principal referential directions $\hat{\mathbf{N}}_a$. Since the basis vectors are assumed to be fixed in space ($\dot{\mathbf{e}}_a = \mathbf{o}$), we may write $\dot{\hat{\mathbf{N}}}_a = \dot{\mathbf{Q}}\mathbf{e}_a$, $a = 1,2,3$. By expanding this equation with the orthogonality condition and by means of the skew tensor $\boldsymbol{\Omega}$, as introduced in (5.15), and transformation (6.182), we may eliminate the basis $\{\mathbf{e}_a\}$ and find that

$$\dot{\hat{\mathbf{N}}}_a = (\dot{\mathbf{Q}}\mathbf{Q}^{\mathrm{T}})\mathbf{Q}\mathbf{e}_a = \boldsymbol{\Omega}\hat{\mathbf{N}}_a \quad , \qquad a = 1,2,3 \quad . \tag{6.183}$$

Note that the components of the skew tensor $\boldsymbol{\Omega} = -\boldsymbol{\Omega}^{\mathrm{T}}$ with respect to the basis $\{\mathbf{e}_a\}$ are obtained from (6.183)$_2$ in the form

$$\Omega_{ab} = \hat{\mathbf{N}}_a \cdot \boldsymbol{\Omega}\hat{\mathbf{N}}_b = \hat{\mathbf{N}}_a \cdot \dot{\hat{\mathbf{N}}}_b = -\Omega_{ba} \quad , \tag{6.184}$$

with $\Omega_{aa} = 0$. By means of identity $(1.65)_2$, we may deduce from $(6.183)_2$ the representation $\mathbf{\Omega} = \sum_{a=1}^{3} \dot{\hat{\mathbf{N}}}_a \otimes \hat{\mathbf{N}}_a$.

Knowing $\dot{\hat{\mathbf{N}}}_a$, we now can determine the material time derivative of the spectral representation of $\mathbf{C}$. Converting eq. (2.128) to the rate form, we obtain, by means of eq. $(6.183)_2$ and the spectral decomposition of the right Cauchy-Green tensor, i.e. $\mathbf{C} = \sum_{a=1}^{3} \lambda_a^2 \hat{\mathbf{N}}_a \otimes \hat{\mathbf{N}}_a$,

$$\dot{\mathbf{C}} - \sum_{a=1}^{3} 2\lambda_a \dot{\lambda}_a \hat{\mathbf{N}}_a \otimes \hat{\mathbf{N}}_a = \sum_{a=1}^{3} \lambda_a^2 (\dot{\hat{\mathbf{N}}}_a \otimes \hat{\mathbf{N}}_a + \hat{\mathbf{N}}_a \otimes \dot{\hat{\mathbf{N}}}_a)$$

$$= \sum_{a=1}^{3} \lambda_a^2 (\mathbf{\Omega}\hat{\mathbf{N}}_a \otimes \hat{\mathbf{N}}_a + \hat{\mathbf{N}}_a \otimes \mathbf{\Omega}\hat{\mathbf{N}}_a)$$

$$= \mathbf{\Omega}\mathbf{C} - \mathbf{C}\mathbf{\Omega} \ . \tag{6.185}$$

From $(6.185)_2$ using (6.184) we deduce that

$$\dot{\mathbf{C}} = \sum_{a=1}^{3} 2\lambda_a \dot{\lambda}_a \hat{\mathbf{N}}_a \otimes \hat{\mathbf{N}}_a + \sum_{\substack{a,b=1 \\ a \neq b}}^{3} \Omega_{ab}(\lambda_b^2 - \lambda_a^2)\hat{\mathbf{N}}_a \otimes \hat{\mathbf{N}}_b \ , \tag{6.186}$$

where $2\lambda_a \dot{\lambda}_a = \hat{\mathbf{N}}_a \cdot \dot{\mathbf{C}}\hat{\mathbf{N}}_a = \dot{C}_{aa}$, $a = 1, 2, 3$ (compare with eq. (2.129)), denote *normal* components (diagonal elements) and $\Omega_{ab}(\lambda_b^2 - \lambda_a^2) = \hat{\mathbf{N}}_a \cdot \dot{\mathbf{C}}\hat{\mathbf{N}}_b = \dot{C}_{ab}$, $a \neq b$, denote *shear* components of $\dot{\mathbf{C}}$ (off-diagonal elements) with respect to the basis $\{\hat{\mathbf{N}}_a\}$.

By isotropy, $\mathbf{S}$ has the same principal directions as $\mathbf{C}$. Hence, recall $(6.52)_2$, i.e. $\mathbf{S} = \sum_{a=1}^{3} S_a \hat{\mathbf{N}}_a \otimes \hat{\mathbf{N}}_a$, with $S_a = 1/\lambda_a (\partial\Psi/\partial\lambda_a)$, $a = 1, 2, 3$, we obtain, by analogy with (6.186)

$$\dot{\mathbf{S}} = \sum_{a=1}^{3} \dot{S}_a \hat{\mathbf{N}}_a \otimes \hat{\mathbf{N}}_a + \sum_{\substack{a,b=1 \\ a \neq b}}^{3} \Omega_{ab}(S_b - S_a)\hat{\mathbf{N}}_a \otimes \hat{\mathbf{N}}_b \ , \tag{6.187}$$

in which the material time derivative of the principal second Piola-Kirchhoff stresses is defined to be

$$\dot{S}_a = \sum_{b=1}^{3} \frac{\partial S_a}{\partial \lambda_b^2} \dot{\lambda}_b^2 = \sum_{b=1}^{3} \frac{\partial S_a}{\partial \lambda_b} \dot{\lambda}_b \ . \tag{6.188}$$

By expanding the numerator and denominator of the second term in (6.187) with $\lambda_b^2 - \lambda_a^2$ and by means of $(6.188)_2$, eq. (6.187) can be rephrased as

$$\dot{\mathbf{S}} = \sum_{a,b=1}^{3} \frac{\partial S_a}{\partial \lambda_b} \dot{\lambda}_b \hat{\mathbf{N}}_a \otimes \hat{\mathbf{N}}_a + \sum_{\substack{a,b=1 \\ a \neq b}}^{3} \Omega_{ab}(\lambda_b^2 - \lambda_a^2) \frac{S_b - S_a}{\lambda_b^2 - \lambda_a^2} \hat{\mathbf{N}}_a \otimes \hat{\mathbf{N}}_b \ . \tag{6.189}$$

On comparing the derived eqs. (6.189) and (6.186) with (6.153) in the rate form, i.e. $\dot{\mathbf{S}} = \mathbb{C} : \dot{\mathbf{C}}/2$, we find by inspection that the elasticity tensor $\mathbb{C}$ emerges, as given by eq. (6.180). ∎

For the case in which two or even all three eigenvalues λ_a^2 of $\mathbf{C}$ (and also of $\mathbf{b}$) are equal, the associated two or three stresses S_a are also equal, by isotropy. Hence, the divided difference $(S_b - S_a)/(\lambda_b^2 - \lambda_a^2)$ in expression (6.180) represents an indeterminate form of type $\frac{0}{0}$. However, it can be shown that the divided difference is well-defined as λ_b approaches λ_a. Namely, applying *l'Hôpital's rule*, we see simply that

$$\lim_{\lambda_b \to \lambda_a} \frac{S_b - S_a}{\lambda_b^2 - \lambda_a^2} = \frac{\partial S_b}{\partial \lambda_b^2} - \frac{\partial S_a}{\partial \lambda_b^2} \tag{6.190}$$

(compare also with the work of CHADWICK and OGDEN [1971b]). Consequently, the elasticity tensor, as defined in eq. (6.180), is valid for the three cases: $\lambda_1 \neq \lambda_2 \neq \lambda_3 \neq \lambda_1$, $\lambda_1 = \lambda_2 \neq \lambda_3$ and $\lambda_1 = \lambda_2 = \lambda_3$.

Finally, in order to set up the spectral form of the elasticity tensor $\mathbb{c}$ in the *spatial description* we use the Piola transformation of $\mathbb{C}$ for principal values. According to (6.159), this gives $c_{abcd} = J^{-1}\lambda_a\lambda_b\lambda_c\lambda_d\,C_{abcd}$, with the principal stretches $\lambda_1, \lambda_2, \lambda_3$ and the volume ratio $J = \lambda_1\lambda_2\lambda_3$. A straightforward computation leads to

$$\mathbb{c} = \sum_{a,b=1}^{3} J^{-1}\lambda_a^2\lambda_b \frac{\partial S_a}{\partial \lambda_b} \hat{\mathbf{n}}_a \otimes \hat{\mathbf{n}}_a \otimes \hat{\mathbf{n}}_b \otimes \hat{\mathbf{n}}_b$$

$$+ \sum_{\substack{a,b=1 \\ a \neq b}}^{3} \frac{\sigma_b\lambda_a^2 - \sigma_a\lambda_b^2}{\lambda_b^2 - \lambda_a^2} (\hat{\mathbf{n}}_a \otimes \hat{\mathbf{n}}_b \otimes \hat{\mathbf{n}}_a \otimes \hat{\mathbf{n}}_b + \hat{\mathbf{n}}_a \otimes \hat{\mathbf{n}}_b \otimes \hat{\mathbf{n}}_b \otimes \hat{\mathbf{n}}_a) \,, \tag{6.191}$$

with the principal Cauchy stresses $\sigma_a = J^{-1}\lambda_a^2 S_a$, $a = 1,2,3$ (see the inverse of eq. $(6.48)_2$), and the principal spatial directions $\hat{\mathbf{n}}_a$, $a = 1,2,3$, which are the orthonormal eigenvectors of $\mathbf{v}$ (and also of $\mathbf{b}$), with $|\hat{\mathbf{n}}_a| = 1$ and $\hat{\mathbf{n}}_a \cdot \hat{\mathbf{n}}_b = \delta_{ab}$. From the property (2.120) we know that the two-point tensor $\mathbf{R}$ rotates the principal referential directions $\hat{\mathbf{N}}_a$ into the principal spatial directions $\hat{\mathbf{n}}_a$.

If $\lambda_a = \lambda_b$ we may conclude that $\sigma_a = \sigma_b$, by isotropy. Hence, the divided difference $(\sigma_b\lambda_a^2 - \sigma_a\lambda_b^2)/(\lambda_b^2 - \lambda_a^2)$ in expression (6.191) gives us $\frac{0}{0}$ and must therefore be determined applying *l'Hôpital's rule*. Differentiating the numerator and denominator by λ_b and taking the limits $\lambda_b \to \lambda_a$, the divided difference becomes

$$\lim_{\lambda_b \to \lambda_a} \frac{\sigma_b\lambda_a^2 - \sigma_a\lambda_b^2}{\lambda_b^2 - \lambda_a^2} = \left(\frac{\partial \sigma_b}{\partial \lambda_b}\lambda_a^2 - \frac{\partial \sigma_a}{\partial \lambda_b}\lambda_b^2 - 2\lambda_b\sigma_a \right) \frac{1}{2\lambda_b}$$

$$= \frac{1}{2}\lambda_a \left(\frac{\partial \sigma_b}{\partial \lambda_b} - \frac{\partial \sigma_a}{\partial \lambda_b} \right) - \sigma_a \,. \tag{6.192}$$

An alternative version of solution $(6.192)_2$ in terms of the principal second Piola-Kirchhoff stresses, frequently used in other texts, is left as an exercise.

EXERCISES

1. For the description of isotropic hyperelastic materials at finite strains consider the strain-energy function $\Psi = \Psi(I_1, I_2, I_3)$ in the *coupled* form, with the principal invariants I_a, $a = 1, 2, 3$.

(a) Use the stress relation $(6.32)_2$, the chain rule and the derivatives of the invariants with respect to $\mathbf{C}$, i.e. eqs. $(6.30)_3$ and (6.31), in order to obtain the *most general form* of the elasticity tensor $\mathbb{C}$ in terms of the three principal invariants

$$\mathbb{C} = 2\frac{\partial \mathbf{S}}{\partial \mathbf{C}} = 4\frac{\partial^2 \Psi(I_1, I_2, I_3)}{\partial \mathbf{C} \partial \mathbf{C}}$$

$$= \delta_1 \mathbf{I} \otimes \mathbf{I} + \delta_2(\mathbf{I} \otimes \mathbf{C} + \mathbf{C} \otimes \mathbf{I}) + \delta_3(\mathbf{I} \otimes \mathbf{C}^{-1} + \mathbf{C}^{-1} \otimes \mathbf{I})$$

$$+ \delta_4 \mathbf{C} \otimes \mathbf{C} + \delta_5(\mathbf{C} \otimes \mathbf{C}^{-1} + \mathbf{C}^{-1} \otimes \mathbf{C})$$

$$+ \delta_6 \mathbf{C}^{-1} \otimes \mathbf{C}^{-1} + \delta_7 \mathbf{C}^{-1} \odot \mathbf{C}^{-1} + \delta_8 \mathbb{S} \quad , \tag{6.193}$$

with the coefficients $\delta_1, \ldots, \delta_8$ defined by

$$\left.\begin{aligned}
\delta_1 &= 4\left(\frac{\partial^2 \Psi}{\partial I_1 \partial I_1} + 2I_1\frac{\partial^2 \Psi}{\partial I_1 \partial I_2} + \frac{\partial \Psi}{\partial I_2} + I_1^2\frac{\partial^2 \Psi}{\partial I_2 \partial I_2}\right) \quad , \\[2mm]
\delta_2 &= -4\left(\frac{\partial^2 \Psi}{\partial I_1 \partial I_2} + I_1\frac{\partial^2 \Psi}{\partial I_2 \partial I_2}\right) \quad , \\[2mm]
\delta_3 &= 4\left(I_3\frac{\partial^2 \Psi}{\partial I_1 \partial I_3} + I_1 I_3\frac{\partial^2 \Psi}{\partial I_2 \partial I_3}\right) \quad , \qquad \delta_4 = 4\frac{\partial^2 \Psi}{\partial I_2 \partial I_2} \quad , \\[2mm]
\delta_5 &= -4I_3\frac{\partial^2 \Psi}{\partial I_2 \partial I_3} \quad , \qquad \delta_6 = 4\left(I_3\frac{\partial \Psi}{\partial I_3} + I_3^2\frac{\partial^2 \Psi}{\partial I_3 \partial I_3}\right) \quad , \\[2mm]
\delta_7 &= -4I_3\frac{\partial \Psi}{\partial I_3} \quad , \qquad \delta_8 = -4\frac{\partial \Psi}{\partial I_2} \quad .
\end{aligned}\right\} \tag{6.194}$$

The fourth-order tensors $\mathbf{C}^{-1} \odot \mathbf{C}^{-1}$ and $\mathbb{S} = (\mathbb{I} + \bar{\mathbb{I}})/2$ in eq. $(6.193)_3$ are defined according to (6.164) and (1.161), (1.162), respectively.

(b) Particularize the coefficients δ_a, $a = 1, \ldots, 8$, (6.194), for the compressible *Mooney-Rivlin, neo-Hookean, Blatz and Ko* models, i.e. eqs. (6.147)–(6.149). For convenience, summarize the (nonzero) coefficients for the three material models to form the entries of Table 6.3.

For notational simplicity we have introduced the abbreviation $\xi = 2\mu(1 - f)/I_3$ with $\xi \geq 0$. Note that the coefficients δ_a, $a = 1, \ldots, 8$, for the *neo-Hookean* model are simple those of the *Blatz and Ko* model obtained by setting $\xi = 0$ and $\mu f = 2c_1$.

Compare the corresponding constitutive equations of the three material models of Exercise 1(a) on p. 248, with the specified coefficients summarized in Table 6.1.

	Mooney-Rivlin model (6.147)	neo-Hookean model (6.148)	Blatz and Ko model (6.149)
δ_1	$4c_2$		ξ
δ_3			$-\xi I_1$
δ_5			ξ
δ_6	$2cJ(2J-1)$	$4\beta c_1 I_3^{-\beta}$	$2\mu f \beta I_3^{-\beta} + \xi(I_2 + \beta I_3^{\beta+1})$
δ_7	$-2[2cJ(J-1)-d]$	$4c_1 I_3^{-\beta}$	$2\mu f I_3^{-\beta} + \xi(I_2 - I_3^{\beta+1})$
δ_8	$-4c_2$		$-\xi$

Table 6.3   Specified coefficients for the elasticity tensors of some materials in the coupled form.

2. The strain-energy function $\Psi(J, \bar{I}_1, \bar{I}_2) = \Psi_{\text{vol}}(J) + \Psi_{\text{iso}}(\bar{I}_1, \bar{I}_2)$ is given in terms of the volume ratio J and the two modified principal invariants $\bar{I}_1, \bar{I}_2$. This type of strain energy in *decoupled* representation is suitable for the characterization of compressible isotropic materials at finite strains.

   (a) The associated decoupled elasticity tensor $\mathbb{C}$ is given by (6.162), with the volumetric and isochoric parts (6.166)$_4$ and (6.168), respectively. Particularize the fictitious elasticity tensor $\overline{\mathbb{C}}$ in the *material description*, i.e. eq. (6.169)$_1$, to the specific strain energy at hand.

   Start with the constitutive equation for the fictitious second Piola-Kirchhoff stress $\overline{\mathbf{S}}$, as defined in eq. (6.115), and use the derivatives of $\bar{I}_1, \bar{I}_2$ with respect to tensor $\overline{\mathbf{C}}$, i.e. (6.117), in order to obtain the *most general form* of $\overline{\mathbb{C}}$ in terms of $\bar{I}_1$ and $\bar{I}_2$ in the form

$$J^{4/3}\overline{\mathbb{C}} = 2\frac{\partial \overline{\mathbf{S}}}{\partial \overline{\mathbf{C}}} = 4\frac{\partial^2 \Psi_{\text{iso}}(\bar{I}_1, \bar{I}_2)}{\partial \overline{\mathbf{C}} \partial \overline{\mathbf{C}}}$$
$$= \bar{\delta}_1 \mathbf{I} \otimes \mathbf{I} + \bar{\delta}_2 (\mathbf{I} \otimes \overline{\mathbf{C}} + \overline{\mathbf{C}} \otimes \mathbf{I}) + \bar{\delta}_3 \overline{\mathbf{C}} \otimes \overline{\mathbf{C}} + \bar{\delta}_4 \mathbb{S} \quad,$$

   with the fourth-order tensor $\mathbb{S} = (\mathbb{I} + \bar{\mathbb{I}})/2$, where $\mathbb{I}$ and $\bar{\mathbb{I}}$ are according to (1.161) and (1.162), and the coefficients $\bar{\delta}_a, a = 1, \ldots, 4$, defined by

$$\left. \begin{aligned} \bar{\delta}_1 &= 4\left(\frac{\partial^2 \Psi_{\text{iso}}}{\partial \bar{I}_1 \partial \bar{I}_1} + 2\bar{I}_1 \frac{\partial^2 \Psi_{\text{iso}}}{\partial \bar{I}_1 \partial \bar{I}_2} + \frac{\partial \Psi_{\text{iso}}}{\partial \bar{I}_2} + \bar{I}_1^2 \frac{\partial^2 \Psi_{\text{iso}}}{\partial \bar{I}_2 \partial \bar{I}_2}\right) \quad, \\ \bar{\delta}_2 &= -4\left(\frac{\partial^2 \Psi_{\text{iso}}}{\partial \bar{I}_1 \partial \bar{I}_2} + \bar{I}_1 \frac{\partial^2 \Psi_{\text{iso}}}{\partial \bar{I}_2 \partial \bar{I}_2}\right) \quad, \\ \bar{\delta}_3 &= 4\frac{\partial^2 \Psi_{\text{iso}}}{\partial \bar{I}_2 \partial \bar{I}_2} \quad, \qquad \bar{\delta}_4 = -4\frac{\partial \Psi_{\text{iso}}}{\partial \bar{I}_2} \quad. \end{aligned} \right\} \quad (6.195)$$

(b) To be specific, take the decoupled form of the compressible *Mooney-Rivlin* model (6.146) and the compressible *neo-Hookean* model (set $c_2 = 0$). Additionally, take the *Yeoh* model and the *Arruda and Boyce* model of the forms

$$\Psi = \Psi_{\mathrm{vol}} + c_1(\bar{I}_1 - 3) + c_2(\bar{I}_1 - 3)^2 + c_3(\bar{I}_1 - 3)^3 \quad,$$

$$\Psi = \Psi_{\mathrm{vol}} + \mu\left[\frac{1}{2}(\bar{I}_1 - 3) + \frac{1}{20n}(\bar{I}_1^2 - 9) + \frac{11}{1050n^2}(\bar{I}_1^3 - 27) + \ldots\right] \quad,$$

corresponding to eqs. (6.134) and (6.136), respectively.

Specify the coefficients $\bar{\delta}_a$, $a = 1,\ldots,4$, (6.195), for the four material models in question and summarize the result in the form of a table.

	Mooney-Rivlin model (6.146)	neo-Hookean model	Yeoh model (6.134)	Arruda and Boyce model (6.136)
$\bar{\delta}_1$	$4c_2$	0	$8[c_2 + 3c_3(\bar{I}_1 - 3)]$	$\mu[2/5n$ $+(44/175n^2)\bar{I}_1 + \ldots]$
$\bar{\delta}_4$	$-4c_2$	0	0	0

Table 6.4 Specified coefficients for the elasticity tensors of some materials in the decoupled form.

Note that the associated constitutive equations, with the specified coefficients as summarized in Table 6.2, are presented in Exercise 1(b) on p. 248.

3. Consider a compressible isotropic material characterized by the strain-energy function in the *decoupled* form of $\Psi(\lambda_1, \lambda_2, \lambda_3) = \Psi_{\mathrm{vol}}(J) + \Psi_{\mathrm{iso}}(\bar{\lambda}_1, \bar{\lambda}_2, \bar{\lambda}_3)$, with the volume ratio $J = \lambda_1\lambda_2\lambda_3$ and the modified principal stretches $\bar{\lambda}_a = J^{-1/3}\lambda_a$, $a = 1, 2, 3$. The associated decoupled structure of the elasticity tensor $\mathbb{C}$ in the material description is given as $\mathbb{C}(\lambda_1, \lambda_2, \lambda_3) = \mathbb{C}_{\mathrm{vol}} + \mathbb{C}_{\mathrm{iso}}$, with the *volumetric* contribution $\mathbb{C}_{\mathrm{vol}}$ specified by the expression (6.166)$_4$.

(a) Show that the spectral form of the *isochoric* contribution $\mathbb{C}_{\mathrm{iso}}$ may be given by

$$\mathbb{C}_{\mathrm{iso}} = \sum_{a,b=1}^{3} \frac{1}{\lambda_b}\frac{\partial S_{\mathrm{iso}\,a}}{\partial \lambda_b}\hat{\mathbf{N}}_a \otimes \hat{\mathbf{N}}_a \otimes \hat{\mathbf{N}}_b \otimes \hat{\mathbf{N}}_b$$

$$+ \sum_{\substack{a,b=1 \\ a\neq b}}^{3} \frac{S_{\mathrm{iso}\,b} - S_{\mathrm{iso}\,a}}{\lambda_b^2 - \lambda_a^2}(\hat{\mathbf{N}}_a \otimes \hat{\mathbf{N}}_b \otimes \hat{\mathbf{N}}_a \otimes \hat{\mathbf{N}}_b$$

$$+\hat{\mathbf{N}}_a \otimes \hat{\mathbf{N}}_b \otimes \hat{\mathbf{N}}_b \otimes \hat{\mathbf{N}}_a) \quad, \qquad (6.196)$$

where $S_{\mathrm{iso}\,a} = (\partial\Psi_{\mathrm{iso}}/\partial\lambda_a)/\lambda_a$, $a = 1, 2, 3$, denote the principal values of the second Piola-Kirchhoff stress tensor $\mathbf{S}_{\mathrm{iso}}$.

(b) In order to specify the elasticity tensor (6.196), take Ogden's model (6.139) and recall the isochoric hyperelastic stress response in terms of the three principal stresses $S_{\mathrm{iso}\,a}$, $a = 1, 2, 3$, determined by the closed form expression (6.144)$_2$ using (6.145) (compare with Example 6.7).

For this important class of material models particularize the coefficients of the first part of the elasticity tensor (6.196) by means of the given strain-energy function (6.139), i.e. $\Psi_{\mathrm{iso}}(\overline{\lambda}_1, \overline{\lambda}_2, \overline{\lambda}_3) = \sum_{a=1}^{3} \sum_{p=1}^{N} (\mu_p/\alpha_p)(\overline{\lambda}_a^{\alpha_p} - 1)$. Show that

$$\frac{1}{\lambda_b} \frac{\partial S_{\mathrm{iso}\,a}}{\partial \lambda_b} = \frac{J^{-1/3}}{\lambda_b} \sum_{c=1}^{3} \frac{\partial S_{\mathrm{iso}\,a}}{\partial \overline{\lambda}_c} \left(\delta_{cb} - \frac{1}{3} \overline{\lambda}_c \overline{\lambda}_b^{-1} \right) \tag{6.197}$$

$$= \begin{cases} \lambda_a^{-2} \lambda_b^{-2} \sum_{p=1}^{N} \mu_p \alpha_p \left(\frac{1}{3} \overline{\lambda}_a^{\alpha_p} + \frac{1}{9} \sum_{c=1}^{3} \overline{\lambda}_c^{\alpha_p} \right) & \text{for } a = b \ , \\[2ex] \lambda_a^{-2} \lambda_b^{-2} \sum_{p=1}^{N} \mu_p \alpha_p \left(-\frac{1}{3} \overline{\lambda}_a^{\alpha_p} - \frac{1}{3} \overline{\lambda}_b^{\alpha_p} + \frac{1}{9} \sum_{c=1}^{3} \overline{\lambda}_c^{\alpha_p} \right) & \text{for } a \neq b \ . \end{cases}$$

This representation was given by SIMO and TAYLOR [1991a, Example 2.2], with some differences in notation.

4. Consider a strain-energy function in terms of principal stretches λ_a, $a = 1, 2, 3$, and principal invariants I_a, $a = 1, 2, 3$, according to $\Psi = \Psi(\lambda_1, \lambda_2, \lambda_3) = \Psi(I_1, I_2, I_3)$.

Write down the associated constitutive equation for the principal second Piola-Kirchhoff stresses S_a, $a = 1, 2, 3$, in terms of the three principal invariants (compare with eq. (6.32)$_2$). By use of this result show that the divided difference $(S_b - S_a)/(\lambda_b^2 - \lambda_a^2)$ is well-defined as λ_b approaches λ_a and yields

$$\lim_{\lambda_b \to \lambda_a} \frac{S_b - S_a}{\lambda_b^2 - \lambda_a^2} = -2\frac{\partial \Psi}{\partial I_2} - 2I_3 \frac{\partial \Psi}{\partial I_3} \lambda_a^{-2} \lambda_b^{-2} \ .$$

5. By means of *l'Hôpital's rule* derive the alternative version of (6.192) in terms of the principal second Piola-Kirchhoff stresses S_a, i.e.

$$\lim_{\lambda_b \to \lambda_a} \frac{\sigma_b \lambda_a^2 - \sigma_a \lambda_b^2}{\lambda_b^2 - \lambda_a^2} = J^{-1} \lambda_a^2 \lambda_b^2 \left(\frac{\partial S_b}{\partial \lambda_b^2} - \frac{\partial S_a}{\partial \lambda_b^2} \right) \ .$$

6. Consider the constitutive equation (6.38) in terms of the (spatial) left Cauchy-Green tensor **b** representing isotropic hyperelastic response. Deduce from the constitutive rate equation (6.161) that the associated elasticity tensor $\mathbb{c}$ in the spatial description has the explicit form

$$J\mathbb{c} = 4\mathbf{b}\frac{\partial^2 \Psi(\mathbf{b})}{\partial \mathbf{b} \partial \mathbf{b}}\mathbf{b} \ . \tag{6.198}$$

Hint: Recall relation (3.62), the Oldroyd stress rate (5.59) of the Kirchhoff stress, kinematic relation (2.171) and use the chain rule.

Representation (6.198) was given by MIEHE and STEIN [1992]. In their work the definition of the elasticity tensor $\mathbb{c} = \chi_*(\mathbb{C})$ in the spatial description excludes the factor J^{-1}.

7. Consider the additive split of the Cauchy stress (6.101)–(6.104) in terms of J and $\overline{\mathbf{b}}$, based on the strain-energy function of the form (6.98). Show that the associated elasticity tensor $\mathbb{c}$ in the spatial description may be written in the *decoupled* form

$$\mathbb{c} = \mathbb{c}_{\text{vol}} + \mathbb{c}_{\text{iso}} \quad ,$$

with the definitions

$$J\mathbb{c}_{\text{vol}} = 4\mathbf{b}\frac{\partial^2 \Psi_{\text{vol}}(J)}{\partial\mathbf{b}\partial\mathbf{b}}\mathbf{b} = J(\tilde{p}\mathbf{I}\otimes\mathbf{I} - 2p\mathbb{I}) \qquad \text{with} \qquad \tilde{p} = p + J\frac{\mathrm{d}p}{\mathrm{d}J} \quad ,$$

$$J\mathbb{c}_{\text{iso}} = 4\mathbf{b}\frac{\partial^2 \Psi_{\text{iso}}(\overline{\mathbf{b}})}{\partial\mathbf{b}\partial\mathbf{b}}\mathbf{b}$$

$$= \mathbb{P} : \overline{\mathbb{c}} : \mathbb{P} + \frac{2}{3}\text{tr}(\overline{\boldsymbol{\tau}})\mathbb{P} - \frac{2}{3}(\mathbf{I}\otimes\boldsymbol{\tau}_{\text{iso}} + \boldsymbol{\tau}_{\text{iso}}\otimes\mathbf{I})$$

of the purely volumetric contribution $\mathbb{c}_{\text{vol}}$ and the purely isochoric contribution $\mathbb{c}_{\text{iso}}$, the latter being based on the spatial projection tensor $\mathbb{P} = \mathbb{I} - \frac{1}{3}\mathbf{I}\otimes\mathbf{I}$ and on the definitions $\overline{\boldsymbol{\tau}} = J\overline{\boldsymbol{\sigma}}$, $\boldsymbol{\tau}_{\text{iso}} = J\boldsymbol{\sigma}_{\text{iso}}$, with $\boldsymbol{\sigma}_{\text{iso}} = \mathbb{P} : \overline{\boldsymbol{\sigma}}$, as given in (6.103) and (6.104).

In addition, we introduced the definitions of the fourth-order **fictitious elasticity tensor** $\overline{\mathbb{c}}$ in the spatial description and the trace $\text{tr}(\bullet)$ according to

$$\overline{\mathbb{c}} = 4\overline{\mathbf{b}}\frac{\partial^2 \Psi_{\text{iso}}(\overline{\mathbf{b}})}{\partial\overline{\mathbf{b}}\partial\overline{\mathbf{b}}}\overline{\mathbf{b}} \quad , \qquad \text{tr}(\bullet) = (\bullet) : \mathbf{I} \quad .$$

For an explicit derivation, see MIEHE [1994, Appendix A].

6.7   Transversely Isotropic Materials

Numerous materials are composed of a **matrix material** (or in the literature often called **ground substance**) and one or more **families of fibers**. This type of material, which we call a **composite material** or **fiber-reinforced composite**, is heterogeneous in the sense that it has different compositions throughout the body. We consider only composite materials in which the fibers are continuously arranged in the matrix material. These types of composites have strong directional properties and their mechanical responses are regarded as **anisotropic**.

The challenge in the design of fiber-reinforced composites is to combine the matrix material and the fibers in such a way that the resulting material is most efficient for the desired application. For engineering applications composite materials provide many advantages over monolithic materials such as high stiffness and strength, low weight and thermal expansion and corrosion resistance. However, the drawbacks in using composite materials seem to be the high costs when compared with those of monolithic (more classical) materials and, from the practical point of view, limited knowledge of how to combine these types of material.

A material which is reinforced by only *one* family of fibers has a *single* preferred direction. The stiffness of this type of composite material in the fiber direction is typically much greater than in the directions orthogonal to the fibers. It is the simplest representation of material anisotropy, which we call **transversely isotropic** with respect to this preferred direction. The material response along directions orthogonal to this preferred direction is isotropic. These composite materials are employed in a variety of applications in industrial engineering and medicine. For manufacturing and fabrication processes and for typical features and properties of transversely isotropic materials the reader should consult, for example, the textbook by HERAKOVICH [1998].

The aim of the following section is to investigate transversely isotropic materials capable of supporting *finite elastic* strains. As mentioned above, all fibers have a single preferred direction. However, the fibers are assumed to be continuously distributed throughout the material. We derive appropriate constitutive equations which are based solely on a continuum approach (excluding micromechanical considerations). Constitutive equations which model transversely isotropic materials in the *small elastic* strain regime are well established and may be found, for example, in the textbooks by TSAI and HAHN [1980], DANIEL and ISHAI [1994], HERAKOVICH [1998] and JONES [1999].

Kinematic relation and structure of the free energy.   We consider a continuum body $\mathcal{B}$ which initially occupies a typical region Ω_0 at a fixed *reference time* $t = 0$. The region is known as a fixed *reference configuration* of that body $\mathcal{B}$. A point in Ω_0 may be characterized by the position vector $\mathbf{X}$ (with material coordinates X_A, $A = 1, 2, 3$) related to a fixed set of axes. At a subsequent time $t > 0$ the continuum body is in a *deformed configuration* occupying a region Ω. The associated point in Ω is characterized by the position vector $\mathbf{x}$ (with spatial coordinates x_a, $a = 1, 2, 3$) related to the same fixed set of axes. For more details about the relevant notation recall Section 2.1.

We suppose that the only anisotropic property of the solid comes from the presence of the fibers. To start with, for a material which is reinforced by only one family of fibers, the stress at a material point depends not only on the deformation gradient $\mathbf{F}$ but also on that single preferred direction, which we call the **fiber direction**. The direction of a fiber at point $\mathbf{X} \in \Omega_0$ is defined by a unit vector field $\mathbf{a}_0(\mathbf{X})$, $|\mathbf{a}_0| = 1$, with material coordinates $a_{0\,A}$. The fiber under a deformation moves with the material points of the continuum body and arrives at the deformed configuration Ω. Hence, the new fiber direction at the associated point $\mathbf{x} \in \Omega$ is defined by a unit vector field $\mathbf{a}(\mathbf{x}, t)$, $|\mathbf{a}| = 1$, with spatial coordinates a_a. For subsequent use it is beneficial to review the section on material and spatial strain tensors introduced on pp. 76-81.

Allowing length changes of the fibers, we must determine the *stretch* λ of the fiber along its direction $\mathbf{a}_0$. It is defined as the ratio between the length of a fiber element in the deformed and reference configuration. By combining eq. (2.62) with (2.73) we find that

$$\lambda \mathbf{a}(\mathbf{x}, t) = \mathbf{F}(\mathbf{X}, t)\mathbf{a}_0(\mathbf{X}) \quad , \tag{6.199}$$

which relates the fiber directions in the reference and the deformed configurations.

Consequently, since $|\mathbf{a}| = 1$, we find the square of stretch λ following the symmetry

$$\lambda^2 = \mathbf{a}_0 \cdot \mathbf{F}^{\mathrm{T}}\mathbf{F}\mathbf{a}_0 = \mathbf{a}_0 \cdot \mathbf{C}\mathbf{a}_0 \quad , \tag{6.200}$$

which we already have introduced in relation (2.64). This means, that the fiber stretch depends on the fiber direction of the undeformed configuration, i.e. the unit vector field $\mathbf{a}_0$, and the strain measure, i.e. the right Cauchy-Green tensor $\mathbf{C}$.

We now assume the transversely isotropic material to be hyperelastic, characterized by a Helmholtz free-energy function Ψ per unit reference volume. Because of the directional dependence on the deformation, expressed by the unit vector field $\mathbf{a}_0$, we require that the free energy depends explicitly on both the right Cauchy-Green tensor $\mathbf{C}$ and the fiber direction $\mathbf{a}_0$ in the reference configuration.

Since the sense of $\mathbf{a}_0$ is immaterial, Ψ is taken as an *even* function of $\mathbf{a}_0$. Hence, by introducing the tensor product $\mathbf{a}_0 \otimes \mathbf{a}_0$, Ψ may be expressed as a function of the two arguments $\mathbf{C}$ and $\mathbf{a}_0 \otimes \mathbf{a}_0$. The tensor $\mathbf{a}_0 \otimes \mathbf{a}_0$ (with Cartesian components $a_{0\,A}a_{0\,B}$) is of order two. For the Helmholtz free-energy function we may therefore write

$$\Psi = \Psi(\mathbf{C}, \mathbf{a}_0 \otimes \mathbf{a}_0) \quad . \tag{6.201}$$

From previous sections we know that the free energy must be independent of the coordinate system; hence $\Psi(\mathbf{C}, \mathbf{a}_0 \otimes \mathbf{a}_0)$ must be objective. Since $\mathbf{C}$ and $\mathbf{a}_0 \otimes \mathbf{a}_0$ are defined with respect to the reference configuration (which is fixed), they are unaffected by a rigid-body motion superimposed on the current configuration. Consequently, the principle of material frame-indifference of the postulated free energy $\Psi(\mathbf{C}, \mathbf{a}_0 \otimes \mathbf{a}_0)$ is satisfied trivially.

EXAMPLE 6.9 The free energy $\Psi(\mathbf{C}, \mathbf{a}_0 \otimes \mathbf{a}_0)$ must be unchanged if both the *matrix material* and the *fibers* in the reference configuration undergo a rotation around a certain axis described by the proper orthogonal tensor $\mathbf{Q}$.

Show that the requirement for transversely isotropic hyperelastic materials formally reads

$$\Psi(\mathbf{C}, \mathbf{a}_0 \otimes \mathbf{a}_0) = \Psi(\mathbf{QCQ}^\mathrm{T}, \mathbf{Qa}_0 \otimes \mathbf{a}_0\mathbf{Q}^\mathrm{T}) \quad , \tag{6.202}$$

which holds for all proper orthogonal tensors $\mathbf{Q}$.

Solution.   For the solution it is beneficial to review eqs. (6.19)–(6.25) of Section 6.2.

A rotation of the reference configuration by tensor $\mathbf{Q}$ transforms a typical point $\mathbf{X}$ into position $\mathbf{X}^\star = \mathbf{QX}$. Consequently, fiber direction $\mathbf{a}_0$ transforms into the new fiber direction $\mathbf{a}_0^\star = \mathbf{Qa}_0$ so that $\mathbf{a}_0^\star \otimes \mathbf{a}_0^\star$ becomes $\mathbf{Qa}_0 \otimes \mathbf{a}_0\mathbf{Q}^\mathrm{T}$. Now, after a subsequent motion of the rotated reference configuration, $\mathbf{X}^\star$ maps into position $\mathbf{x}$. Thus, the deformation gradient $\mathbf{F}^\star$ and the strain measure $\mathbf{C}^\star = \mathbf{F}^{\star\mathrm{T}}\mathbf{F}^\star$ relative to the rotated reference configuration are $\mathbf{F}^\star = \mathbf{FQ}^\mathrm{T}$ (compare with (6.22)) and $\mathbf{C}^\star = \mathbf{F}^{\star\mathrm{T}}\mathbf{F}^\star = \mathbf{QF}^\mathrm{T}\mathbf{FQ}^\mathrm{T} = \mathbf{QCQ}^\mathrm{T}$, respectively.

We say that a hyperelastic material is transversely isotropic relative to a reference configuration if the identity $\Psi(\mathbf{C}, \mathbf{a}_0 \otimes \mathbf{a}_0) = \Psi(\mathbf{C}^\star, \mathbf{a}_0^\star \otimes \mathbf{a}_0^\star)$ is satisfied for all proper orthogonal tensors $\mathbf{Q}$. Hence, restriction (6.202) follows directly. Note that in view of (6.202), Ψ may be seen as a scalar-valued *isotropic* tensor function of the two tensor variables $\mathbf{C}$ and $\mathbf{a}_0 \otimes \mathbf{a}_0$.   ∎

According to (6.27), an isotropic hyperelastic material may be represented by the first three invariants I_1, I_2, I_3 of either $\mathbf{C}$ or $\mathbf{b}$, characterized in (5.89)–(5.91). These invariants can be used to fulfil requirement (6.25), i.e. $\Psi(\mathbf{C}) = \Psi(\mathbf{QCQ}^\mathrm{T})$ for all $(\mathbf{Q}, \mathbf{C})$. Following SPENCER [1971, 1984], two additional (new) scalars, I_4 and I_5, are necessary to form the *integrity bases* of the tensors $\mathbf{C}$ and $\mathbf{a}_0 \otimes \mathbf{a}_0$ and to satisfy relation (6.202). They are the so-called **pseudo-invariants** of $\mathbf{C}$ *and* $\mathbf{a}_0 \otimes \mathbf{a}_0$, which are given by

$$I_4(\mathbf{C}, \mathbf{a}_0) = \mathbf{a}_0 \cdot \mathbf{Ca}_0 = \lambda^2 \quad , \qquad I_5(\mathbf{C}, \mathbf{a}_0) = \mathbf{a}_0 \cdot \mathbf{C}^2\mathbf{a}_0 \quad . \tag{6.203}$$

The two pseudo-invariants I_4, I_5 arise directly from the anisotropy and contribute to the free energy. They describe the properties of the fiber family and its interaction with the other material constituents. Note that invariant I_4 is equal to the square of the stretch λ in the fiber direction $\mathbf{a}_0$ (compare with eq. (6.200)).

For the definition of the *integrity bases* and the related *theory of invariants* see the lecture notes by SCHUR [1968], the articles by SPENCER [1971] and ZHENG [1994, and references therein]. For applications in continuum mechanics the reader is referred to the works by RIVLIN [1970], BETTEN [1987a, Chapter D and 1987b], TRUESDELL and NOLL [2004] among others. A brief review of the theory of invariants may also be found in SCHRÖDER [1996].

For a transversely isotropic material, the free energy can finally be written in terms of the five independent scalar invariants, and eq. (6.27)$_1$, valid for isotropic material response, and may consequently be expanded according to

$$\Psi = \Psi \left[I_1(\mathbf{C}), I_2(\mathbf{C}), I_3(\mathbf{C}), I_4(\mathbf{C}, \mathbf{a}_0), I_5(\mathbf{C}, \mathbf{a}_0) \right] \quad . \tag{6.204}$$

The free energy (6.204) provides a fundamental basis for deriving the associated constitutive equations.

Constitutive equations in terms of invariants. In order to derive the constitutive equations we apply (6.13)$_2$. Then, by use of the chain rule, the second Piola-Kirchhoff stress tensor $\mathbf{S}$ is given as a function of the five scalar invariants, i.e.

$$\mathbf{S} = 2 \frac{\partial \Psi (\mathbf{C}, \mathbf{a}_0 \otimes \mathbf{a}_0)}{\partial \mathbf{C}} = 2 \sum_{a=1}^{5} \frac{\partial \Psi (\mathbf{C}, \mathbf{a}_0 \otimes \mathbf{a}_0)}{\partial I_a} \frac{\partial I_a}{\partial \mathbf{C}} \quad , \tag{6.205}$$

in which $\partial I_1 / \partial \mathbf{C}$ and $\partial I_2 / \partial \mathbf{C}$, $\partial I_3 / \partial \mathbf{C}$ are given by (6.30) and (6.31), respectively.

The remaining derivatives follow from (6.203) and have the forms

$$\frac{\partial I_4}{\partial \mathbf{C}} = \mathbf{a}_0 \otimes \mathbf{a}_0 \qquad \text{or} \qquad \frac{\partial I_4}{\partial C_{AB}} = a_{0\,A} a_{0\,B} \quad , \tag{6.206}$$

$$\frac{\partial I_5}{\partial \mathbf{C}} = \mathbf{a}_0 \otimes \mathbf{C} \mathbf{a}_0 + \mathbf{a}_0 \mathbf{C} \otimes \mathbf{a}_0$$

$$\text{or} \qquad \frac{\partial I_5}{\partial C_{AB}} = a_{0\,A} C_{BC} a_{0\,C} + a_{0\,B} C_{AC} a_{0\,C} \quad . \tag{6.207}$$

Finally, (6.205) reads, with eqs. (6.30), (6.31), (6.206) and (6.207),

$$\mathbf{S} = 2 \left[\left(\frac{\partial \Psi}{\partial I_1} + I_1 \frac{\partial \Psi}{\partial I_2} \right) \mathbf{I} - \frac{\partial \Psi}{\partial I_2} \mathbf{C} + I_3 \frac{\partial \Psi}{\partial I_3} \mathbf{C}^{-1} \right.$$

$$\left. + \frac{\partial \Psi}{\partial I_4} \mathbf{a}_0 \otimes \mathbf{a}_0 + \frac{\partial \Psi}{\partial I_5} (\mathbf{a}_0 \otimes \mathbf{C} \mathbf{a}_0 + \mathbf{a}_0 \mathbf{C} \otimes \mathbf{a}_0) \right] \quad , \tag{6.208}$$

which extends the constitutive equation (6.32) by the addition of the last two terms.

Using arguments similar to those used for the derivation of the spatial version of the stress relation (6.34), namely a push-forward operation on the material stress tensor $\mathbf{S}$ by the motion χ, we arrive, using (6.199) and (6.203)$_1$, at

$$\sigma = 2J^{-1} \left[I_3 \frac{\partial \Psi}{\partial I_3} \mathbf{I} + \left(\frac{\partial \Psi}{\partial I_1} + I_1 \frac{\partial \Psi}{\partial I_2} \right) \mathbf{b} - \frac{\partial \Psi}{\partial I_2} \mathbf{b}^2 \right.$$

$$\left. + I_4 \frac{\partial \Psi}{\partial I_4} \mathbf{a} \otimes \mathbf{a} + I_4 \frac{\partial \Psi}{\partial I_5} (\mathbf{a} \otimes \mathbf{b} \mathbf{a} + \mathbf{a} \mathbf{b} \otimes \mathbf{a}) \right] \quad . \tag{6.209}$$

Recall that the unit vector $\mathbf{a}(\mathbf{x}, t)$ denotes the fiber direction in the deformed configuration while $\mathbf{b} = \mathbf{F} \mathbf{F}^{\mathrm{T}}$ is the (second-order) left Cauchy-Green tensor. Observe the similar structure of the last two terms in the stress relation (6.209) to that presented in (6.208).

The associated elasticity tensors in the material and spatial descriptions follow by means of expressions (6.154) and (6.159), respectively (for an explicit derivation compare with the work of WEISS et al. [1996]). For implementations of large strain transversely isotropic models in a finite element program see WEISS et al. [1996], SCHRÖDER [1996], BONET and BURTON [1998] among others.

Incompressible transversely isotropic materials.   We now consider transversely isotropic materials with an *incompressible isotropic* matrix material.

Firstly, we study the case in which the embedded fibers are *extensible*. Since we assume incompressibility of the isotropic matrix material, i.e. $I_3 = 1$, we are able to postulate a free energy in terms of the remaining four independent invariants. Because of the incompressibility constraint $I_3 = 1$ the free energy Ψ is enhanced by an indeterminate *Lagrange multiplier* $p/2$ which is identified as a reaction pressure. In view of (6.62) we have the assumption

$$\Psi = \Psi[I_1(\mathbf{C}), I_2(\mathbf{C}), I_4(\mathbf{C}, \mathbf{a}_0), I_5(\mathbf{C}, \mathbf{a}_0)] - \frac{1}{2}p(I_3 - 1) \ . \tag{6.210}$$

The associated stress relations given in the reference and current configurations are basically those presented by (6.63) and (6.64) supplemented by the fourth and fifth term in eqs. (6.208) and (6.209), respectively.

Secondly, we study an incompressible isotropic matrix material which is continuously reinforced throughout by *inextensible* fibers. This means that $\lambda = 1$ and, in view of (6.203)$_1$, the fourth invariant is equal to one. With this additional internal constraint, the free energy Ψ is a function only of I_1, I_2, which are responsible for the hyperelastic isotropic matrix material, and I_5, which is responsible for the fibers. By adding the term $q(I_4 - 1)/2$ to the free energy Ψ we obtain the function

$$\Psi = \Psi[I_1(\mathbf{C}), I_2(\mathbf{C}), I_5(\mathbf{C}, \mathbf{a}_0)] - \frac{1}{2}p(I_3 - 1) - \frac{1}{2}q(I_4 - 1) \ , \tag{6.211}$$

where $q/2$ is an additional indeterminate *Lagrange multiplier*.

The associated stress relations in the Lagrangian and Eulerian descriptions for the transversely isotropic materials with an *incompressible isotropic* matrix material and *inextensible* fibers (with direction $\mathbf{a}_0$) are the extended constitutive equations (6.63) and (6.65), i.e.

$$\mathbf{S} = -p\mathbf{C}^{-1} - q\mathbf{a}_0 \otimes \mathbf{a}_0 + 2\left(\frac{\partial\Psi}{\partial I_1} + I_1\frac{\partial\Psi}{\partial I_2}\right)\mathbf{I} - 2\frac{\partial\Psi}{\partial I_2}\mathbf{C}$$

$$+2\frac{\partial\Psi}{\partial I_5}\left(\mathbf{a}_0 \otimes \mathbf{C}\mathbf{a}_0 + \mathbf{a}_0\mathbf{C} \otimes \mathbf{a}_0\right) \ , \tag{6.212}$$

$$\sigma = -p\mathbf{I} - q\mathbf{a} \otimes \mathbf{a} + 2\frac{\partial \Psi}{\partial I_1}\mathbf{b} - 2\frac{\partial \Psi}{\partial I_2}\mathbf{b}^{-1} + 2\frac{\partial \Psi}{\partial I_5}(\mathbf{a} \otimes \mathbf{ba} + \mathbf{ab} \otimes \mathbf{a}) \quad . \quad (6.213)$$

Note that the indeterminate terms $-q\mathbf{a}_0 \otimes \mathbf{a}_0$ and $-q\mathbf{a} \otimes \mathbf{a}$ are identified as fiber reaction stresses which respond to the inextensibility constraint $I_4 = 1$.

EXERCISES

1. Starting from the pseudo-invariants I_4 and I_5, i.e. eqs. (6.203), show their derivative with respect to $\mathbf{C}$, eqs. (6.206) and (6.207).

2. We characterize a transversely isotropic material by the *decoupled* free energy in the form

$$\Psi = \Psi(\mathbf{C}, \mathbf{a}_0 \otimes \mathbf{a}_0) = \Psi_{\mathrm{vol}}(J) + \Psi_{\mathrm{iso}}(\bar{I}_1, \bar{I}_2, \bar{I}_4, \bar{I}_5) \quad , \quad (6.214)$$

where Ψ_{vol} and Ψ_{iso} are the volumetric and isochoric contributions to the hyperelastic response (recall Section 6.4). The modified invariants $\bar{I}_1, \bar{I}_2$ are given according to eqs. (6.109) and (6.110), while $\bar{I}_3 = \det \bar{\mathbf{C}} = 1$ (note that $\bar{I}_1, \bar{I}_2, \bar{I}_3$ are the modified principal invariants of the modified tensor $\bar{\mathbf{C}} = J^{-2/3}\mathbf{C}$). The remaining modified pseudo-invariants are expressed by $\bar{I}_4 = J^{-2/3}I_4$ and $\bar{I}_5 = J^{-4/3}I_5$.

(a) Having in mind the free energy (6.214) and the derivatives (6.206) and (6.207), show that the constitutive equation $\mathbf{S} = Jp\mathbf{C}^{-1} + J^{-2/3}\mathbb{P} : \bar{\mathbf{S}}$ specializes to

$$\bar{\mathbf{S}} = 2\frac{\partial \Psi_{\mathrm{iso}}(\bar{I}_1, \bar{I}_2, \bar{I}_4, \bar{I}_5)}{\partial \bar{\mathbf{C}}} = 2\sum_{\substack{a=1 \\ a\neq 3}}^{5} \frac{\partial \Psi_{\mathrm{iso}}}{\partial \bar{I}_a} \frac{\partial \bar{I}_a}{\partial \bar{\mathbf{C}}}$$

$$= \bar{\gamma}_1\mathbf{I} + \bar{\gamma}_2\bar{\mathbf{C}} + \bar{\gamma}_4\mathbf{a}_0 \otimes \mathbf{a}_0 + \bar{\gamma}_5(\mathbf{a}_0 \otimes \bar{\mathbf{C}}\mathbf{a}_0 + \mathbf{a}_0\bar{\mathbf{C}} \otimes \mathbf{a}_0) \quad , \quad (6.215)$$

with the response coefficients

$$\bar{\gamma}_4 = 2\frac{\partial \Psi_{\mathrm{iso}}}{\partial \bar{I}_4} \quad , \qquad \bar{\gamma}_5 = 2\frac{\partial \Psi_{\mathrm{iso}}}{\partial \bar{I}_5}$$

for the fictitious second Piola-Kirchhoff stress $\bar{\mathbf{S}}$. Note that the coefficients $\bar{\gamma}_1$ and $\bar{\gamma}_2$ reflect the isotropic stress response, as given in eqs. (6.116).

(b) By recalling Section 6.6, a closed form expression for the elasticity tensor $\mathbb{C}$ in the material description is given by relation (6.162), with contributions $\mathbb{C}_{\mathrm{vol}}$, i.e. (6.166)$_4$, and $\mathbb{C}_{\mathrm{iso}}$, i.e. (6.168).

By use of the important property (1.256) and the constitutive equation for the fictitious second Piola-Kirchhoff stress (6.215)$_3$, show that the fictitious elasticity tensor $\overline{\mathbb{C}}$ in the *material description* takes on the form

$$J^{4/3}\overline{\mathbb{C}} = 2\frac{\partial\overline{\mathbf{S}}}{\partial\overline{\mathbf{C}}} = 4\frac{\partial^2\Psi_{\mathrm{iso}}(\overline{I}_1,\overline{I}_2,\overline{I}_4,\overline{I}_5)}{\partial\overline{\mathbf{C}}\partial\overline{\mathbf{C}}}$$

$$= \overline{\delta}_1\mathbf{I}\otimes\mathbf{I} + \overline{\delta}_2(\mathbf{I}\otimes\overline{\mathbf{C}} + \overline{\mathbf{C}}\otimes\mathbf{I}) + \overline{\delta}_3\overline{\mathbf{C}}\otimes\overline{\mathbf{C}} + \overline{\delta}_4\mathbb{I}$$

$$+\overline{\delta}_5(\mathbf{I}\otimes\mathbf{a}_0\otimes\mathbf{a}_0 + \mathbf{a}_0\otimes\mathbf{a}_0\otimes\mathbf{I})$$

$$+\overline{\delta}_6\left(\overline{\mathbf{C}}\otimes\mathbf{a}_0\otimes\mathbf{a}_0 + \mathbf{a}_0\otimes\mathbf{a}_0\otimes\overline{\mathbf{C}}\right)$$

$$+\overline{\delta}_7\mathbf{a}_0\otimes\mathbf{a}_0\otimes\mathbf{a}_0\otimes\mathbf{a}_0 + \overline{\delta}_8\left(\mathbf{I}\otimes\frac{\partial\overline{I}_5}{\partial\overline{\mathbf{C}}} + \frac{\partial\overline{I}_5}{\partial\overline{\mathbf{C}}}\otimes\mathbf{I}\right)$$

$$+\overline{\delta}_9\left(\overline{\mathbf{C}}\otimes\frac{\partial\overline{I}_5}{\partial\overline{\mathbf{C}}} + \frac{\partial\overline{I}_5}{\partial\overline{\mathbf{C}}}\otimes\overline{\mathbf{C}}\right) + \overline{\delta}_{10}\left(\frac{\partial\overline{I}_5}{\partial\overline{\mathbf{C}}}\otimes\frac{\partial\overline{I}_5}{\partial\overline{\mathbf{C}}}\right)$$

$$+\overline{\delta}_{11}\left(\mathbf{a}_0\otimes\mathbf{a}_0\otimes\frac{\partial\overline{I}_5}{\partial\overline{\mathbf{C}}} + \frac{\partial\overline{I}_5}{\partial\overline{\mathbf{C}}}\otimes\mathbf{a}_0\otimes\mathbf{a}_0\right) + \overline{\delta}_{12}\frac{\partial^2\overline{I}_5}{\partial\overline{\mathbf{C}}\partial\overline{\mathbf{C}}} \quad,$$

with the fourth-order unit tensor $\mathbb{I}$ defined by (1.160) and the coefficients $\overline{\delta}_a$, $a = 5, \ldots, 12$, by

$$\overline{\delta}_5 = 4\left(\frac{\partial^2\Psi_{\mathrm{iso}}}{\partial\overline{I}_1\partial\overline{I}_4} + \overline{I}_1\frac{\partial^2\Psi_{\mathrm{iso}}}{\partial\overline{I}_2\partial\overline{I}_4}\right) \quad , \qquad \overline{\delta}_6 = -4\frac{\partial^2\Psi_{\mathrm{iso}}}{\partial\overline{I}_2\partial\overline{I}_4} \quad ,$$

$$\overline{\delta}_7 = 4\frac{\partial^2\Psi_{\mathrm{iso}}}{\partial\overline{I}_4\partial\overline{I}_4} \quad , \qquad \overline{\delta}_8 = 4\left(\frac{\partial^2\Psi_{\mathrm{iso}}}{\partial\overline{I}_1\partial\overline{I}_5} + \overline{I}_1\frac{\partial^2\Psi_{\mathrm{iso}}}{\partial\overline{I}_2\partial\overline{I}_5}\right) \quad ,$$

$$\overline{\delta}_9 = -4\frac{\partial^2\Psi_{\mathrm{iso}}}{\partial\overline{I}_2\partial\overline{I}_5} \quad , \qquad \overline{\delta}_{10} = 4\frac{\partial^2\Psi_{\mathrm{iso}}}{\partial\overline{I}_5\partial\overline{I}_5} \quad ,$$

$$\overline{\delta}_{11} = 4\frac{\partial^2\Psi_{\mathrm{iso}}}{\partial\overline{I}_4\partial\overline{I}_5} \quad , \qquad \overline{\delta}_{12} = 4\frac{\partial\Psi_{\mathrm{iso}}}{\partial\overline{I}_5} \quad .$$

Note that the coefficients $\overline{\delta}_a$, $a = 1, \ldots, 4$, were given previously in relations (6.195) and reflect the isotropic contributions.

6.8 Composite Materials with Two Families of Fibers

In the following we discuss appropriate constitutive equations for the *finite elastic* response of fiber-reinforced composites in which the matrix material is reinforced by *two* families of fibers. We assume that the fibers are continuously distributed throughout the material so that the continuum theory of fiber-reinforced composites is the constitutive theory of choice.

There are many different fibers and matrix materials now in use for composite materials. Examples of specific fibers for structural applications are *boron* and *glass*. The latter is an important engineering fiber with high strength and low cost. Further examples are *carbon* and *graphite* (the difference is in the carbon content), the organic fiber *aramid* and the ceramic fibers *silicon carbide* and *alumina* among others. Many specific matrix materials are available for the use in composites; for example, *thermoplastic polymers*, *thermoset polymers*, *metals* (such as aluminum, titanium and copper) and *ceramics*.

Vast numbers of applications in industry are concerned with composite materials, such as the finite elastic response of belts and high pressure tubes, steel reinforced rubber used in tyres, and integrated circuits used in electronic computing devices. Typical medical applications are lightweight wheelchairs and implant devices such as hip joints (see also the textbook by HERAKOVICH [1998, Chapter 1]). The five-volume encyclopedia of composites edited by LEE [1990, 1991] includes a detailed account of special types of fiber, matrix materials and composites as engineering materials. Typical engineering properties, manufacturing and fabrication processes and details on how to use composite materials for different applications are also provided.

However, it is important to note that numerous organisms such as the human body, animals and plants are heterogeneous systems of various composite biomaterials. The textbooks by FUNG [1990, 1993, 1997] are concerned with the biomechanics of various biomaterials, soft tissues and organs of the human body. One important example of a fibre-reinforced biomaterial is the artery. The layers of the arterial wall are composed mainly of an isotropic matrix material (associated with the *elastin*) and two families of fibers (associated with the *collagen*) which are arranged in symmetrical spirals (for arterial histology see RHODIN [1980]). For mechanical properties and constitutive equations of arterial walls, see the reviews by, for example, HAYASHI [1993], HUMPHREY [1995] and the data book edited by ABÉ et al. [1996, Chapter 2]. A simple finite element simulation of the orthotropic biomechanical behavior of the arterial wall is provided by HOLZAPFEL et al. [1996d, 1996e] and HOLZAPFEL and WEIZSÄCKER [1998]. For a review of finite element models for arterial wall mechanics, see the article by SIMON et al. [1993].

Free energy and constitutive equations. We may now consider a body built up of a matrix material with two families of fibers each of which is unidirectional with preferred direction. The matrix material is assumed to be hyperelastic. The preferential fiber directions in the reference and the current configuration are denoted by the unit vector fields $\mathbf{a}_0, \mathbf{g}_0$ and $\mathbf{a}, \mathbf{g}$, respectively. By analogy with relation (6.201) we may postulate the free energy

$$\Psi = \Psi(\mathbf{C}, \mathbf{A}_0, \mathbf{G}_0) \tag{6.216}$$

per unit reference volume. For notational simplicity we have introduced the abbreviations $\mathbf{A}_0 = \mathbf{a}_0 \otimes \mathbf{a}_0$ and $\mathbf{G}_0 = \mathbf{g}_0 \otimes \mathbf{g}_0$, frequently referred to as **structural tensors**.

The free energy must be unchanged if the fiber-reinforced composite (i.e. a hyperelastic (matrix) material with two families of fibers) in the reference configuration undergoes a rotation described by the proper orthogonal tensor $\mathbf{Q}$. Using arguments similar to those used for a single fiber family (see the previous Section 6.7), the requirement for this type of composite is, in view of (6.202), given by

$$\Psi(\mathbf{C}, \mathbf{A}_0, \mathbf{G}_0) = \Psi(\mathbf{QCQ}^\mathrm{T}, \mathbf{QA}_0\mathbf{Q}^\mathrm{T}, \mathbf{QG}_0\mathbf{Q}^\mathrm{T}) \quad , \tag{6.217}$$

which holds for all tensors $\mathbf{Q}$ (recall Example 6.9). Here, Ψ is a scalar-valued *isotropic* tensor function of the three tensor variables $\mathbf{C}$, $\mathbf{A}_0$ and $\mathbf{G}_0$.

According to SPENCER [1971, 1984], requirement (6.217) is satisfied if Ψ is a function of the set of invariants

$$\left.\begin{array}{ccc} I_1(\mathbf{C}) \quad , & I_2(\mathbf{C}) \quad , & I_3(\mathbf{C}) \quad , \\[2mm] I_4(\mathbf{C}, \mathbf{a}_0) \quad , & I_5(\mathbf{C}, \mathbf{a}_0) \quad , & \\[2mm] I_6(\mathbf{C}, \mathbf{g}_0) = \mathbf{g}_0 \cdot \mathbf{C}\mathbf{g}_0 \quad , & I_7(\mathbf{C}, \mathbf{g}_0) = \mathbf{g}_0 \cdot \mathbf{C}^2\mathbf{g}_0 \quad , & \\[2mm] I_8(\mathbf{C}, \mathbf{a}_0, \mathbf{g}_0) = \mathbf{a}_0 \cdot \mathbf{C}\mathbf{g}_0 \quad , & I_9(\mathbf{a}_0, \mathbf{g}_0) = (\mathbf{a}_0 \cdot \mathbf{g}_0)^2 \quad . \end{array}\right\} \tag{6.218}$$

The three invariants I_1, I_2, I_3 are identical to those from the isotropic theory presented in eqs. (5.89)–(5.91). The pseudo-invariants I_4, I_5 are given by eq. (6.203) and characterize one family of fibers with direction $\mathbf{a}_0$. The pseudo-invariants $I_4, \ldots, I_9$ are associated with the anisotropy generated by the two families of fibers. The dot product $\mathbf{a}_0 \cdot \mathbf{g}_0$ is a geometrical constant determining the cosine of the angle between the two fiber directions in the reference configuration. Therefore, the invariant I_9 does not depend on the deformation and is subsequently no longer considered. Note that I_4 and I_6 are equal to the squares of the stretch in the fiber directions $\mathbf{a}_0$ and $\mathbf{g}_0$, respectively.

The constitutive equation for the second Piola-Kirchhoff stress $\mathbf{S}$ follows from the postulated free energy (6.216) by differentiation with respect to $\mathbf{C}$. By means of the chain rule, $\mathbf{S}$ is given as a function of the remaining eight scalar invariants in the form

$$\mathbf{S} = 2\frac{\partial\Psi(\mathbf{C}, \mathbf{A}_0, \mathbf{G}_0)}{\partial\mathbf{C}} = 2\sum_{a=1}^{8}\frac{\partial\Psi(I_1, \ldots, I_8)}{\partial I_a}\frac{\partial I_a}{\partial\mathbf{C}} \quad , \tag{6.219}$$

in which $\partial I_1/\partial\mathbf{C}, \ldots, \partial I_3/\partial\mathbf{C}$ and $\partial I_4/\partial\mathbf{C}, \partial I_5/\partial\mathbf{C}$ are given by eqs. (6.30), (6.31) and (6.206), (6.207), respectively.

The remaining derivatives of the invariants follow from (6.218)$_6$–(6.218)$_8$ and have the forms

$$\frac{\partial I_6}{\partial\mathbf{C}} = \mathbf{G}_0 \qquad \text{or} \qquad \frac{\partial I_6}{\partial C_{AB}} = g_{0\,A}g_{0\,B} = G_{AB} \quad , \tag{6.220}$$

$$\frac{\partial I_7}{\partial \mathbf{C}} = \mathbf{g}_0 \otimes \mathbf{Cg}_0 + \mathbf{g}_0 \mathbf{C} \otimes \mathbf{g}_0$$

$$\text{or} \quad \frac{\partial I_7}{\partial C_{AB}} = g_{0\,A} C_{BC} g_{0\,C} + g_{0\,B} C_{AC} g_{0\,C} \quad , \tag{6.221}$$

$$\frac{\partial I_8}{\partial \mathbf{C}} = \frac{1}{2}(\mathbf{a}_0 \otimes \mathbf{g}_0 + \mathbf{g}_0 \otimes \mathbf{a}_0)$$

$$\text{or} \quad \frac{\partial I_8}{\partial C_{AB}} = \frac{1}{2}(a_{0\,A} g_{0\,B} + g_{0\,A} a_{0\,B}) \quad . \tag{6.222}$$

Using arguments similar to those used for deriving the stress relation (6.208), we obtain the explicit expressions

$$\begin{aligned}
\mathbf{S} = 2 \Bigg[& \left(\frac{\partial \Psi}{\partial I_1} + I_1 \frac{\partial \Psi}{\partial I_2} \right) \mathbf{I} - \frac{\partial \Psi}{\partial I_2} \mathbf{C} + I_3 \frac{\partial \Psi}{\partial I_3} \mathbf{C}^{-1} \\
& + \frac{\partial \Psi}{\partial I_4} \mathbf{A}_0 + \frac{\partial \Psi}{\partial I_5} (\mathbf{a}_0 \otimes \mathbf{Ca}_0 + \mathbf{a}_0 \mathbf{C} \otimes \mathbf{a}_0) + \frac{\partial \Psi}{\partial I_6} \mathbf{G}_0 \\
& + \frac{\partial \Psi}{\partial I_7} (\mathbf{g}_0 \otimes \mathbf{Cg}_0 + \mathbf{g}_0 \mathbf{C} \otimes \mathbf{g}_0) + \frac{1}{2} \frac{\partial \Psi}{\partial I_8} (\mathbf{a}_0 \otimes \mathbf{g}_0 + \mathbf{g}_0 \otimes \mathbf{a}_0) \Bigg] \quad . \tag{6.223}
\end{aligned}$$

The anisotropic stress response (6.223) extends relation (6.208) by the addition of the last three terms. Note that $(\partial \Psi / \partial I_4) \mathbf{A}_0$ and $(\partial \Psi / \partial I_6) \mathbf{G}_0$ characterize the (decoupled) stress contributions arising only from the fibers.

Orthotropic hyperelastic materials. If $\mathbf{a}_0 \cdot \mathbf{g}_0 = 0$, the two families of fibers have orthogonal directions. Then, the material is said to be **orthotropic** in the reference configuration with respect to the planes normal to the fibers and the surface in which the fibers lie. Since now two directions in space $(\mathbf{a}_0, \mathbf{g}_0)$ are preferred with respect to the mechanical response of the composite, the remaining third direction orthogonal to the fiber plane is also a preferred direction. The mechanical response in the third direction is governed by the matrix material. The list of invariants (6.218) reduces to the first seven and the free energy has the form $\Psi = \Psi(I_1, \dots, I_7)$.

A further special case may be found under the assumption that the isotropic matrix material is *incompressible*, i.e. $I_3 = 1$. Additionally, the families of fibers may also be *inextensible* in the two fiber directions $\mathbf{a}_0$ and $\mathbf{g}_0$, consequently $I_4 = 1$ and $I_6 = 1$. For this case a suitable Helmholtz free-energy function is given by

$$\begin{aligned}
\Psi = \Psi[&I_1(\mathbf{C}), I_2(\mathbf{C}), I_5(\mathbf{C}, \mathbf{a}_0), I_7(\mathbf{C}, \mathbf{g}_0)] \\
&- \frac{1}{2}p(I_3 - 1) - \frac{1}{2}q(I_4 - 1) - \frac{1}{2}r(I_6 - 1) \quad , \tag{6.224}
\end{aligned}$$

with the indeterminate *Lagrange multipliers* $p/2, q/2, r/2$.

The constitutive equations for an orthotropic material composed of an *incompressible isotropic* matrix material and *inextensible* fibers (with directions $\mathbf{a}_0$ and $\mathbf{g}_0$) depend only on the invariants I_1, I_2 and I_5, I_7. The constitutive equations in the Lagrangian and Eulerian descriptions, extending eqs. (6.212) and (6.213), follow from (6.224) and are

$$
\mathbf{S} = 2 \sum_{\substack{a=1 \\ a \neq 3,4,6}}^{7} \frac{\partial \Psi(I_1, I_2, I_5, I_7)}{\partial I_a} \frac{\partial I_a}{\partial \mathbf{C}} - \frac{\partial}{\partial \mathbf{C}} \left[p(I_3 - 1) + q(I_4 - 1) + r(I_6 - 1) \right]
$$

$$
= -p\mathbf{C}^{-1} - q\mathbf{A}_0 - r\mathbf{G}_0 + 2 \left(\frac{\partial \Psi}{\partial I_1} + I_1 \frac{\partial \Psi}{\partial I_2} \right) \mathbf{I} - 2 \frac{\partial \Psi}{\partial I_2} \mathbf{C}
$$

$$
+ 2 \frac{\partial \Psi}{\partial I_5} (\mathbf{a}_0 \otimes \mathbf{C}\mathbf{a}_0 + \mathbf{a}_0\mathbf{C} \otimes \mathbf{a}_0) + 2 \frac{\partial \Psi}{\partial I_7} (\mathbf{g}_0 \otimes \mathbf{C}\mathbf{g}_0 + \mathbf{g}_0\mathbf{C} \otimes \mathbf{g}_0) \quad , \tag{6.225}
$$

$$
\boldsymbol{\sigma} = 2\mathbf{F} \frac{\partial \Psi}{\partial \mathbf{C}} \mathbf{F}^{\mathrm{T}} = -p\mathbf{I} - q\mathbf{A} - r\mathbf{G} + 2 \frac{\partial \Psi}{\partial I_1} \mathbf{b} - 2 \frac{\partial \Psi}{\partial I_2} \mathbf{b}^{-1}
$$

$$
+ 2 \frac{\partial \Psi}{\partial I_5} (\mathbf{a} \otimes \mathbf{b}\mathbf{a} + \mathbf{a}\mathbf{b} \otimes \mathbf{a}) + 2 \frac{\partial \Psi}{\partial I_7} (\mathbf{g} \otimes \mathbf{b}\mathbf{g} + \mathbf{g}\mathbf{b} \otimes \mathbf{g}) \quad , \tag{6.226}
$$

respectively. Herein, the indeterminate terms $-p\mathbf{C}^{-1}$, $-q\mathbf{A}_0$, $-r\mathbf{G}_0$ (and $-p\mathbf{I}$, $-q\mathbf{A}$, $-r\mathbf{G}$) are identified as reaction stresses associated with the constraints $I_3 = 1$, $I_4 = 1$, $I_6 = 1$, with the pressure-like quantity p and the fiber tensions q, r, respectively. In eq. $(6.226)_2$ we have introduced the definitions $\mathbf{A} = \mathbf{a} \otimes \mathbf{a}$ and $\mathbf{G} = \mathbf{g} \otimes \mathbf{g}$ of the structural tensors $\mathbf{A}$ and $\mathbf{G}$. Recall that $\mathbf{a}(\mathbf{x}, t)$, $\mathbf{g}(\mathbf{x}, t)$ denote the fiber directions in the deformed configuration while $\mathbf{b} = \mathbf{F}\mathbf{F}^{\mathrm{T}}$ is the spatial strain tensor of second-order.

However, if the two families of fibers are mechanically equivalent – and not necessarily orthogonal – then the material is said to be **locally orthotropic** in the reference configuration with respect to the mutually orthogonal planes which bisect the two fiber families (with directions $\mathbf{a}_0$ and $\mathbf{g}_0$) and the surface in which the fibers lie. Then, Ψ is a function of the first eight invariants listed in (6.218) and is symmetric with respect to interchanges of $\mathbf{a}_0$ and $\mathbf{g}_0$.

It can finally be shown that for a *locally orthotropic* material Ψ can be expressed as a function of the seven invariants

$$
\left.
\begin{aligned}
&I_1(\mathbf{C}) \quad , \qquad I_2(\mathbf{C}) \quad , \qquad I_3(\mathbf{C}) \quad , \qquad I_8(\mathbf{C}, \mathbf{a}_0, \mathbf{g}_0) \quad , \\
&I_9(\mathbf{C}, \mathbf{a}_0, \mathbf{g}_0) \quad , \qquad I_{10}(\mathbf{C}, \mathbf{a}_0, \mathbf{g}_0) \quad , \qquad I_{11}(\mathbf{C}, \mathbf{a}_0, \mathbf{g}_0)
\end{aligned}
\right\} \tag{6.227}
$$

(see SPENCER [1984]), with I_1, I_2, I_3 and I_8 given by eqs. (5.89)–(5.91) and $(6.218)_8$, and the definitions

$$
I_9 = I_4 + I_6 \quad , \qquad I_{10} = I_4 I_6 \quad , \qquad I_{11} = I_5 + I_7 \tag{6.228}
$$

for the remaining three pseudo-invariants.

EXERCISES

1. Consider a *locally orthotropic* material with the free energy Ψ expressed as a function of the invariants presented by (6.227) using (6.228). Assume an *incompressible isotropic* matrix material and two families of *inextensible* fibers.

   Show that the constitutive equation for the Cauchy stress tensor $\boldsymbol{\sigma}$ is given by

   $$\boldsymbol{\sigma} = -p\mathbf{I} - q\mathbf{A} - r\mathbf{G} + 2\frac{\partial\Psi}{\partial I_1}\mathbf{b} - 2\frac{\partial\Psi}{\partial I_2}\mathbf{b}^{-1} + \frac{\partial\Psi}{\partial I_8}(\mathbf{a}\otimes\mathbf{g} + \mathbf{g}\otimes\mathbf{a})$$

   $$+ 2\frac{\partial\Psi}{\partial I_{11}}(\mathbf{a}\otimes\mathbf{ba} + \mathbf{ab}\otimes\mathbf{a} + \mathbf{g}\otimes\mathbf{bg} + \mathbf{gb}\otimes\mathbf{g}) \quad,$$

   where the first three terms characterize reaction stresses.

2. We characterize a compressible composite with two families of fibers by the *decoupled* representation of the free energy

   $$\Psi = \Psi(\mathbf{C}, \mathbf{A}_0, \mathbf{G}_0) = \Psi_{\mathrm{vol}}(J) + \Psi_{\mathrm{iso}}(\bar{I}_1, \bar{I}_2, \bar{I}_4, \ldots, \bar{I}_8) \qquad (6.229)$$

   (compare also with Exercise 2 on p. 271), with the volumetric and isochoric parts Ψ_{vol} and Ψ_{iso}, and the modified invariants $\bar{I}_1, \bar{I}_2$ given by eqs. (6.109) and (6.110) ($\bar{I}_3 = \det\overline{\mathbf{C}} = 1$). The remaining modified pseudo-invariants are $\bar{I}_a = J^{-2/3}I_a$, $a = 4, 6, 8$, and $\bar{I}_a = J^{-4/3}I_a$, $a = 5, 7$.

   Use the free energy (6.229) to particularize the fictitious second Piola-Kirchhoff stress $\overline{\mathbf{S}}$ which appears in the constitutive equation for $\mathbf{S} = Jp\mathbf{C}^{-1} + J^{-2/3}\mathbb{P}:\overline{\mathbf{S}}$. Show that

   $$\overline{\mathbf{S}} = 2\frac{\partial\Psi_{\mathrm{iso}}(\bar{I}_1, \bar{I}_2, \bar{I}_4, \ldots, \bar{I}_8)}{\partial\overline{\mathbf{C}}} = 2\sum_{\substack{a=1 \\ a\neq 3}}^{8}\frac{\partial\Psi_{\mathrm{iso}}}{\partial\bar{I}_a}\frac{\partial\bar{I}_a}{\partial\overline{\mathbf{C}}} \quad,$$

   with the explicit expressions

   $$\frac{\partial\bar{I}_1}{\partial\overline{\mathbf{C}}} = \mathbf{I} \quad, \qquad \frac{\partial\bar{I}_2}{\partial\overline{\mathbf{C}}} = \bar{I}_1\mathbf{I} - \overline{\mathbf{C}} \quad, \qquad \frac{\partial\bar{I}_4}{\partial\overline{\mathbf{C}}} = \mathbf{A}_0 \quad,$$

   $$\frac{\partial\bar{I}_5}{\partial\overline{\mathbf{C}}} = \mathbf{a}_0\otimes\overline{\mathbf{C}}\mathbf{a}_0 + \mathbf{a}_0\overline{\mathbf{C}}\otimes\mathbf{a}_0 \quad, \qquad \frac{\partial\bar{I}_6}{\partial\overline{\mathbf{C}}} = \mathbf{G}_0 \quad,$$

   $$\frac{\partial\bar{I}_7}{\partial\overline{\mathbf{C}}} = \mathbf{g}_0\otimes\overline{\mathbf{C}}\mathbf{g}_0 + \mathbf{g}_0\overline{\mathbf{C}}\otimes\mathbf{g}_0 \quad, \qquad \frac{\partial\bar{I}_8}{\partial\overline{\mathbf{C}}} = \frac{1}{2}(\mathbf{a}_0\otimes\mathbf{g}_0 + \mathbf{g}_0\otimes\mathbf{a}_0) \quad.$$

6.9 Constitutive Models with Internal Variables

Many materials used in the fields of engineering and physics are inelastic. It turns out that the constitutive models introduced hitherto are not adequate to describe this class of materials, for which every admissible process is dissipative. Within the remaining sections of this chapter we study inelastic materials and, based on the concept of internal variables, we derive constitutive models for viscoelastic materials and hyperelastic materials with isotropic damage.

Concept of internal variables.   The current thermodynamic state of thermoelastic materials can be determined solely by the current values of the deformation gradient $\mathbf{F}$ and the temperature Θ. Variables such as $\mathbf{F}$ or Θ are *measurable* and *controllable* quantities and are *accessible* to direct observation. In practice these type of variables are usually called **external variables**.

The current thermodynamic state of materials that involve dissipation can be determined by a finite number of so-called **internal variables**, or in the literature sometimes called **hidden variables** (hidden to the eyes of external observers). These additional thermodynamic state variables, which we denote collectively by $\boldsymbol{\xi}$, are supposed to describe aspects of the internal structure of materials associated with irreversible (dissipative) effects. Note that strain (stress) and temperature (entropy) depend on these internal variables. The evolution of internal variables replicates indirectly the history of the deformation, and hence they are often also termed **history variables**. Materials that involve dissipative effects we refer to as **dissipative materials** or **materials with dissipation**.

Hence, the **concept of internal variables** postulates that the current thermodynamic state at a point of a dissipative material is specified by the triple $(\mathbf{F}, \Theta, \boldsymbol{\xi})$ (the current thermodynamic state may be imagined as a *fictitious* state of thermodynamic equilibrium). Then, the current thermodynamic state is represented in a finite-dimensional state space and described by the current values (and *not* by their past history) of the deformation gradient, the temperature and the finite number of internal variables.

The nature of internal variables may be physical, describing the physical structure of materials. In the course of phenomenological experiments one may be able to *identify* internal variables; however, they are certainly *not controllable* or *observable*.

We use the internal variables as phenomenological variables which are constructed mathematically. They are *mechanical* (or thermal, or even chemical or electrical . . .) state variables describing structural properties within a macroscopic framework, such as the 'dashpot displacements' in viscoelastic models, damage, inelastic strains, dislocation densities, point-defects and so on. Hence, here we introduce both external and internal variables as macroscopic quantities without referring to the internal microstructure of the material.

The concept of internal variables serves as a profound basis for the development of constitutive equations for dissipative materials studied in the following section.

Constitutive equations and internal dissipation.   The existence of non-equilibrium states that do evolve with time is an essential feature of inelastic materials.

Two typical examples of irreversible processes known from classical mechanics which govern non-equilibrium states are **relaxation** and **creep**. Relaxation (and creep) is the time-dependent return to the (new) equilibrium state after a disturbance. In general, stress will decrease with time at a *fixed (constant) strain*, which is referred to as *relaxation*, while during a *creeping* process strain will increase with time at a *fixed (constant) stress*. For an illustration of these two simple processes, see Figure 6.5. The (strain or stress) response of removing a strain or stress is called **recovery**. A viscoelastic behavior of a material is characterized by **hysteresis**. The term 'hysteresis' means that the loading and unloading curves do not coincide. It represents the non-recoverable energy when a material is loaded to a point and then unloaded.

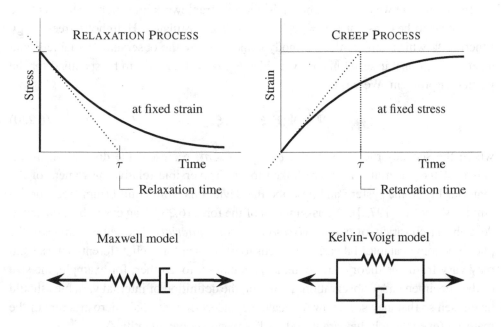

Figure 6.5 Maxwell and Kelvin-Voigt models associated with relaxation and creep behavior.

The **Maxwell model** (a dashpot is arranged in series with a spring) and the **Kelvin-Voigt model** (a dashpot is arranged in parallel with a spring), two mechanical models known from linear viscoelasticity, are frequently used to discuss relaxation and creep behavior. These models combine 'viscous' (or fluid-like) with 'elastic' (or solid-like) behavior. Under the action of a constant deformation (strain), the Maxwell model is supposed to produce instantaneously a stress response by the **spring** which is followed by an exponential stress relaxation due to the **dashpot**. On the other hand, the Kelvin-Voigt model is supposed to produce no immediate deformation for a constant load (stress). However, in a Kelvin-Voigt model a deformation (strain) will be created with time according to an exponential function. Within the realm of non-equilibrium thermodynamics the viscoelastic deformation mechanisms of these material models are not reversible.

The rate of decay of the stress and strain in a viscoelastic process is characterized by the so-called **relaxation time** $\tau \in (0, \infty)$, with dimension of time, known from linear viscoelasticity. The parameter τ associated with a creeping process is often referred to as the **retardation time**.

The constitutive equations introduced hitherto are no longer sufficient to describe dissipative materials. The vast majority of constitutive models that are used to approximate the physical behavior of real nonlinear inelastic materials are developed on the basis of internal variables.

In this chapter we remain within an isothermal framework, in which the temperature is assumed to be constant ($\Theta = \Theta_0$). Hence, we postulate a Helmholtz free-energy function Ψ which defines the thermodynamic state by the observable variable $\mathbf{F}$ and a set of additional internal history variables $\boldsymbol{\xi}_\alpha$, $\alpha = 1, \ldots, m$, to be specified for the particular problem. We write

$$\Psi = \Psi(\mathbf{F}, \boldsymbol{\xi}_1, \ldots, \boldsymbol{\xi}_m) \ , \tag{6.230}$$

where the second-order tensors $\boldsymbol{\xi}_\alpha$, $\alpha = 1, \ldots, m$, represent the dissipation mechanism of the material. They are linked to the irreversible relative movement of the material inside the system and describe the deviations from equilibrium (see, for example, VALANIS [1972]). An assumption of the form (6.230) can easily be adjusted to describe a rich variety of *porous, viscous* or *plastic materials*. The actual number of the phenomenological internal variables needs to be chosen for each different material and may vary from one theory and (boundary) condition to another; for example, the size of the specimen under observation. However, the definition of internal variables should be chosen so that they somehow replicate the underlying internal microstructure of the material (even though they are introduced as macroscopic quantities).

In general, the internal variables may take on *scalar, vector* or *tensor* values. Here the internal variables are all denoted by second-order tensors.

In order to particularize the Clausius-Planck inequality of the form (4.154) to the free energy Ψ at hand, we must differentiate (6.230) with respect to time. By means of the chain rule we obtain $\dot{\Psi}(\mathbf{F}, \boldsymbol{\xi}_1, \ldots, \boldsymbol{\xi}_m) = \partial\Psi/\partial\mathbf{F} : \dot{\mathbf{F}} + \sum_{\alpha=1}^{m} \partial\Psi/\partial\boldsymbol{\xi}_\alpha : \dot{\boldsymbol{\xi}}_\alpha$, and finally, with the expression for the stress power $w_{\text{int}} = \mathbf{P} : \dot{\mathbf{F}}$ per unit reference volume, we find from (4.154) that

$$\mathcal{D}_{\text{int}} = \left(\mathbf{P} - \frac{\partial \Psi(\mathbf{F}, \boldsymbol{\xi}_1, \ldots, \boldsymbol{\xi}_m)}{\partial \mathbf{F}} \right) : \dot{\mathbf{F}} - \sum_{\alpha=1}^{m} \frac{\partial \Psi(\mathbf{F}, \boldsymbol{\xi}_1, \ldots, \boldsymbol{\xi}_m)}{\partial \boldsymbol{\xi}_\alpha} : \dot{\boldsymbol{\xi}}_\alpha \geq 0 \ . \quad (6.231)$$

In order to satisfy $\mathcal{D}_{\text{int}} \geq 0$ we apply the Coleman-Noll procedure. For arbitrary choices of the tensor variable $\dot{\mathbf{F}}$, we deduce a physical expression for the first Piola-Kirchhoff stress $\mathbf{P}$ and a remainder inequality governing the non-negativeness of the internal dissipation $\mathcal{D}_{\text{int}}$ (required by the second law of thermodynamics). We have

$$\mathbf{P} = \frac{\partial \Psi(\mathbf{F}, \boldsymbol{\xi}_1, \ldots, \boldsymbol{\xi}_m)}{\partial \mathbf{F}} \quad , \qquad \mathcal{D}_{\text{int}} = \sum_{\alpha=1}^{m} \Xi_\alpha : \dot{\boldsymbol{\xi}}_\alpha \geq 0 \ , \qquad (6.232)$$

which must hold at every point of the continuum body and for all times during a thermodynamic process. In $(6.232)_2$ we have defined the internal (second-order) tensor variables $\Xi_\alpha, \alpha = 1, \ldots, m$, which are related (conjugate) to $\boldsymbol{\xi}_\alpha$ through the **internal constitutive equations**

$$\Xi_\alpha = - \frac{\partial \Psi(\mathbf{F}, \boldsymbol{\xi}_1, \ldots, \boldsymbol{\xi}_m)}{\partial \boldsymbol{\xi}_\alpha} \quad , \qquad \alpha = 1, \ldots, m \ . \qquad (6.233)$$

The additional constitutive equations (6.233) restrict the free energy Ψ and relate the gradient of the free energy Ψ with respect to the internal variables $\boldsymbol{\xi}_\alpha$ to the associated internal variables $\Xi_\alpha, \alpha = 1, \ldots, m$. Note that the presence of additional variables in the free energy (6.230) justifies additional constitutive equations. A physical motivation of restriction (6.233) may be given by several examples, one of which, stemming from linear viscoelasticity, is presented on p. 286; in particular, see eq. (6.251).

In constitutive equations $(6.232)_1$ and (6.233) the tensor variables $\mathbf{F}$ and $\boldsymbol{\xi}_\alpha$ are associated with the thermodynamic forces $\mathbf{P}$ and Ξ_α, respectively. A constitutive model which is characterized by the set of equations (6.231)–(6.233) is called an **internal variable model**.

For the case in which the internal variables $\boldsymbol{\xi}_\alpha$ are not needed to characterize the thermodynamic state of a system, then, the internal dissipation $\mathcal{D}_{\text{int}}$ in $(6.232)_2$ is zero (the material is considered to be perfectly elastic) and all relations from previous sections of this chapter may be applied. In order to describe materials without dissipative character, the set of equations (6.231)–(6.233) simply reduces to (6.3) and $(6.1)_1$.

Evolution equations and thermodynamic equilibrium.   The derived set of equations (6.232) and (6.233) must be complemented by a kinetic relation, which describes the evolution of the involved internal variable $\boldsymbol{\xi}_\alpha$ and the associated dissipation mechanism. Consequently, suitable **equations of evolution** (rate equations) are required in order to describe the way an irreversible process evolves.

The only restriction on these equations is thermodynamic admissibility, i.e. the satisfaction of the fundamental inequality $(6.232)_2$ characterizing local entropy production. The missing equations for the evolution of the internal variables $\boldsymbol{\xi}_\alpha$ may be written, for example, as

$$\dot{\boldsymbol{\xi}}_\alpha(t) = \boldsymbol{\mathcal{A}}_\alpha(\mathbf{F}, \boldsymbol{\xi}_1, \dots, \boldsymbol{\xi}_m) \ , \qquad \alpha = 1, \dots, m \ . \tag{6.234}$$

The evolution of the system is described by $\boldsymbol{\mathcal{A}}_\alpha$, $\alpha = 1, \dots, m$, which are tensor-valued functions of $1 + m$ tensor variables.

Every system will tend towards a state of **thermodynamic equilibrium**, which implies that the observable and internal variables reach equilibrium under a prescribed stress or strain; they remain constant at any particle of the system with time. Hence, the behavior at the equilibrium state may be considered as a limiting case and does not depend upon time.

In view of eq. (6.234), the definition of an equilibrium state now requires the additional conditions

$$\boldsymbol{\mathcal{A}}_\alpha(\mathbf{F}, \boldsymbol{\xi}_1, \dots, \boldsymbol{\xi}_m) = \mathbf{0} \ , \qquad \alpha = 1, \dots, m \ . \tag{6.235}$$

Hence, $\dot{\boldsymbol{\xi}}_\alpha$ may be seen as the rate of change with which $\boldsymbol{\xi}_\alpha(t)$ tends toward its equilibrium.

In an elastic continuum, every state is an equilibrium state. The internal dissipation $\mathcal{D}_{\text{int}}$ at equilibrium is zero, which characterizes, for instance, a perfectly elastic material, as pointed out in Section 6.1.

6.10 Viscoelastic Materials at Large Strains

Many materials of practical interest appear to behave in a markedly viscoelastic manner over a certain range of stresses and times. The mechanical behavior of, for example, *thermoplastic elastomers* (actually rubber-like materials) or some other types of *natural* and *synthetic polymers* are associated with relaxation and/or creep phenomena, which are important design factors (see, for example, McCRUM et al. [1997], SPERLING [1992] and WARD and HADLEY [1993]). Problems that involve relaxation and/or creep effects determine irreversible processes and belong to the realm of *non-equilibrium thermodynamics*. For a detailed introduction of the linear and nonlinear theory of viscoelasticity the reader is referred to the book by CHRISTENSEN [1982]. Experimental investigations are documented by, for example, SULLIVAN [1986], LION [1996] and MIEHE and KECK [2000].

In the following we characterize the thermodynamic state of such problems explicitly by means of an internal variable model as introduced in the previous section. A description solely via external variables is also possible, but it emerges that such types of formulation are not preferred for numerical realizations using the finite element method.

Numerous viscoelastic materials can often not be modeled adequately within limits by means of a linear theory. Here we postulate a three-dimensional **viscoelastic model** suitable for finite strains and small perturbations away from the equilibrium state. In contrast to several theories of viscoelasticity (see, for example, the pioneering paper by GREEN and TOBOLSKY [1946]) the present phenomenological approach is *not* restricted to isotropy. For theories that account for finite perturbations away from the equilibrium state, the reader is referred to, for example, KOH and ERINGEN [1963], HAUPT [1993a, b] and REESE and GOVINDJEE [1998a].

Additionally we follow a phenomenological approach that does not consider the underlying molecular structure of the physical object.

Structure of the free energy with internal variables. In particular, we choose an approach which applies the concept of internal variables motivated by SIMO [1987] and followed by, for example, GOVINDJEE and SIMO [1992b, 1993], HOLZAPFEL [1996a], KALISKE and ROTHERT [1997], SIMO and HUGHES [1998, Chapter 10] and HOLZAPFEL and GASSER [2001].

Our study is based on the theory of compressible hyperelasticity within the isothermal regime, as discussed in Section 6.4. We postulate a *decoupled* representation of the Helmholtz free-energy function Ψ. The free energy uses the multiplicative decomposition of the deformation gradient into dilational and volume-preserving parts. Our present approach is in contrast to that which uses the multiplicative decomposition of the deformation gradient into elastic (rate-independent) and permanent (viscous) parts (see SIDOROFF [1974] and LUBLINER [1985] among others).

The change of Ψ within an isothermal viscoelastic process from the reference to the current configuration is given as

$$\Psi(\mathbf{C}, \boldsymbol{\Gamma}_1, \ldots, \boldsymbol{\Gamma}_m) = \Psi_{\text{vol}}^{\infty}(J) + \Psi_{\text{iso}}^{\infty}(\overline{\mathbf{C}}) + \sum_{\alpha=1}^{m} \Upsilon_\alpha(\overline{\mathbf{C}}, \boldsymbol{\Gamma}_\alpha) \ , \tag{6.236}$$

valid for some closed time interval $t \in [0, T]$ of interest. We assume that each contribution to the free energy Ψ must satisfy the normalization condition (6.4), i.e.

$$\Psi_{\text{vol}}^{\infty}(1) = 0 \ , \qquad \Psi_{\text{iso}}^{\infty}(\mathbf{I}) = 0 \ , \qquad \sum_{\alpha=1}^{m} \Upsilon_\alpha(\mathbf{I}, \mathbf{I}) = 0 \ . \tag{6.237}$$

A material which is characterized by the free energy (6.236) for any point and time we call a **viscoelastic material**.

The first two terms in (6.236), i.e. $\Psi_{\text{vol}}^{\infty}(J)$ and $\Psi_{\text{iso}}^{\infty}(\overline{\mathbf{C}})$, are strain-energy functions per unit reference volume and characterize the *equilibrium state* of the solid. They can be identified as the terms presented by eq. (6.85) describing the *volumetric elastic response* and the *isochoric elastic response* as $t \to \infty$, respectively. In fact, the superscript $(\bullet)^{\infty}$ characterizes functions which represent the hyperelastic behavior of sufficiently slow processes.

The additional third term in (6.236), i.e. the 'dissipative' potential $\sum_{\alpha=1}^{m} \Upsilon_\alpha$, is responsible for the viscoelastic contribution and extends the decoupled strain-energy function (6.85) to the viscoelastic regime. The scalar-valued functions $\Upsilon_\alpha, \alpha = 1, \ldots, m$, represent the so-called **configurational free energy** of the viscoelastic solid and characterize the *non-equilibrium state*, i.e. the behavior of relaxation and creep.

Motivated by experimental data we assume a time-dependent change of the system caused purely by *isochoric* deformations. Hence, the volumetric response remains *fully elastic* and the configurational free energy is a function of the modified right Cauchy-Green tensor $\overline{\mathbf{C}}$ and a set of strain-like internal variables (history variables) not accessible to direct observation, here denoted by $\mathbf{\Gamma}_\alpha, \alpha = 1, \ldots, m$. Each hidden tensor variable $\mathbf{\Gamma}_\alpha$ characterizes the relaxation and/or creep behavior of the material. They are considered to be (inelastic) strains *akin* to the strain measure $\overline{\mathbf{C}}$, with $\mathbf{\Gamma}_\alpha = \mathbf{I}$, $\alpha = 1, \ldots, m$, at the (stress-free) reference configuration. The viscoelastic behavior is, in particular, modeled by $\alpha = 1, \ldots, m$ viscoelastic processes with corresponding *relaxation times* (or *retardation times*) $\tau_\alpha \in (0, \infty), \alpha = 1, \ldots, m$.

Note that the set of $1+m$ tensor variables $(\mathbf{C}, \mathbf{\Gamma}_1, \ldots, \mathbf{\Gamma}_m)$ completely characterizes the isothermal viscoelastic state.

Decoupled volumetric-isochoric stress response.   In order to obtain the associated constitutive equations describing viscoelastic behavior at finite strains we specify postulate (6.236).

Following arguments analogous to those which led from (6.230) to eqs. (6.232) and (6.233), we obtain physical expressions for the (symmetric) second Piola-Kirchhoff stress $\mathbf{S}$ and the non-negative internal dissipation (local entropy production) $\mathcal{D}_{\text{int}}$ in the forms

$$\mathbf{S} = 2\frac{\partial \Psi(\mathbf{C}, \mathbf{\Gamma}_1, \ldots, \mathbf{\Gamma}_m)}{\partial \mathbf{C}} \quad , \qquad \mathcal{D}_{\text{int}} = -\sum_{\alpha=1}^{m} 2\frac{\partial \Upsilon_\alpha(\overline{\mathbf{C}}, \mathbf{\Gamma}_\alpha)}{\partial \mathbf{\Gamma}_\alpha} : \frac{1}{2}\dot{\mathbf{\Gamma}}_\alpha \geq 0 \quad . \quad (6.238)$$

Starting from the decoupled free energy (6.236), a straightforward computation leads to an additive split of $\mathbf{S}$, as already derived for purely elastic compressible hyperelastic materials (see Section 6.4). We have

$$\mathbf{S} = 2\frac{\partial \Psi(\mathbf{C}, \mathbf{\Gamma}_1, \ldots, \mathbf{\Gamma}_m)}{\partial \mathbf{C}} = \mathbf{S}_{\text{vol}}^\infty + \mathbf{S}_{\text{iso}} \quad , \tag{6.239}$$

with the definition

$$\mathbf{S}_{\text{iso}} = \mathbf{S}_{\text{iso}}^\infty + \sum_{\alpha=1}^{m} \mathbf{Q}_\alpha \tag{6.240}$$

of the isochoric contributions. In eqs. $(6.239)_2$ and (6.240) the quantities

$$\mathbf{S}_{\text{vol}}^\infty = J\frac{\mathrm{d}\Psi_{\text{vol}}^\infty(J)}{\mathrm{d}J}\mathbf{C}^{-1} \quad , \qquad \mathbf{S}_{\text{iso}}^\infty = J^{-2/3}\mathbb{P} : 2\frac{\partial \Psi_{\text{iso}}^\infty(\overline{\mathbf{C}})}{\partial \overline{\mathbf{C}}} \tag{6.241}$$

determine volumetric and isochoric contributions, which we take to be fully elastic. In relation $(6.241)_2$ the (fourth-order) projection tensor $\mathbb{P} = \mathbb{I} - \frac{1}{3}\mathbf{C}^{-1} \otimes \mathbf{C}$ furnishes the deviatoric operator in the *Lagrangian description*. Note that for these elastic contributions we may apply the framework of compressible hyperelasticity and adopt relations (6.88)–(6.91) by using $\Psi_{\mathrm{vol}}^{\infty}$ and $\Psi_{\mathrm{iso}}^{\infty}$ instead of Ψ_{vol} and Ψ_{iso}.

In (6.240) we have introduced additional internal tensor variables $\mathbf{Q}_\alpha$, $\alpha = 1, \ldots, m$, which may be interpreted as **non-equilibrium stresses** in the sense of non-equilibrium thermodynamics. Note that the symbol $\mathbf{Q}$ has already been used and must not be confused with the orthogonal tensor. As can be seen from (6.240) the isochoric second Piola-Kirchhoff stress is decomposed into an **equilibrium part** and a **non-equilibrium part** characterized by the elastic response of the system $\mathbf{S}_{\mathrm{iso}}^{\infty}$ and the viscoelastic response $\sum_{\alpha=1}^{m} \mathbf{Q}_\alpha$, respectively.

By analogy with (6.90) we have defined the relationship

$$\mathbf{Q}_\alpha = 2\frac{\partial \Upsilon_\alpha(\overline{\mathbf{C}}, \Gamma_\alpha)}{\partial \mathbf{C}} = J^{-2/3}(\mathbb{I} - \frac{1}{3}\mathbf{C}^{-1} \otimes \mathbf{C}) : \overline{\mathbf{Q}}_\alpha$$

$$= J^{-2/3}\mathrm{Dev}\overline{\mathbf{Q}}_\alpha = J^{-2/3}\mathbb{P} : \overline{\mathbf{Q}}_\alpha \ , \qquad \alpha = 1, \ldots, m \qquad (6.242)$$

for the second-order tensors $\mathbf{Q}_\alpha$, with the definition

$$\overline{\mathbf{Q}}_\alpha = 2\frac{\partial \Upsilon_\alpha(\overline{\mathbf{C}}, \Gamma_\alpha)}{\partial \overline{\mathbf{C}}} \ , \qquad \alpha = 1, \ldots, m \qquad (6.243)$$

of the so-called **fictitious non-equilibrium stresses** $\overline{\mathbf{Q}}_\alpha$. As can be seen from (6.242), $\mathbf{Q}_\alpha$ is the deviatoric projection of $\overline{\mathbf{Q}}_\alpha$ times $J^{-2/3}$, with projection tensor $\mathbb{P}$.

Motivated by the (mechanical) equilibrium equations for the linear viscoelastic solid (see the following Example 6.10, in particular, eq. (6.251)), we conclude further that $\mathbf{Q}_\alpha$ are variables related (conjugate) to Γ_α, $\alpha = 1, \ldots, m$, with the internal constitutive equations

$$\mathbf{Q}_\alpha = -2\frac{\partial \Upsilon_\alpha(\overline{\mathbf{C}}, \Gamma_\alpha)}{\partial \Gamma_\alpha} \ , \qquad \alpha = 1, \ldots, m \ . \qquad (6.244)$$

These conditions restrict the configurational free energy $\sum_{\alpha=1}^{m} \Upsilon_\alpha$ in view of $(6.242)_1$. Hence, the internal dissipation $\mathcal{D}_{\mathrm{int}}$ in eq. $(6.238)_2$ equivalently reads $\mathcal{D}_{\mathrm{int}} = \sum_{\alpha=1}^{m} \mathbf{Q}_\alpha : \dot{\Gamma}_\alpha/2 \geq 0$.

The condition for *thermodynamic equilibrium* (compare with eq. (6.235)) implies that for $t \to \infty$ the stresses in eq. (6.240) reach equilibrium, which means that $\mathbf{Q}_\alpha = -2\partial \Upsilon_\alpha/\partial \Gamma_\alpha|_{t\to\infty} \equiv \mathbf{O}$, $\alpha = 1, \ldots, m$, and hence, $\mathbf{Q}_\alpha$ characterize the current '*distance from equilibrium*'. Consequently, the dissipation at equilibrium is zero as seen from $(6.238)_2$ and (6.244). In other words, at thermodynamic equilibrium the material responds as perfectly elastic; general finite elasticity is recovered.

EXAMPLE 6.10   By using a simple *spring-and-dashpot model* find a meaningful rheological interpretation for the phenomenological viscoelastic constitutive model presented. Start with a one-dimensional and linear approach. Derive physically motivated evolution equations for the internal variables.

Solution.   To begin with linear geometry consider a *rheological model*, as illustrated in Figure 6.6. It is a one-dimensional **generalized Maxwell model** with a free **spring** on one end and an arbitrary number m of **Maxwell elements** arranged in parallel.

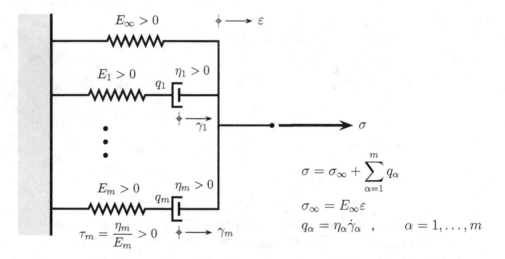

$$\sigma = \sigma_\infty + \sum_{\alpha=1}^{m} q_\alpha$$

$$\sigma_\infty = E_\infty \varepsilon$$

$$q_\alpha = \eta_\alpha \dot{\gamma}_\alpha , \qquad \alpha = 1, \ldots, m$$

Figure 6.6 Rheological model.

The viscoelastic model in Figure 6.6, which we call temporarily a **mechanical device**, displays both *relaxation* and *creep behavior*. It is a suitable simple model to represent quantitatively the mechanical behavior of real viscoelastic materials. The mechanical device is assumed to have unit area and unit length so that stresses and strains are to be interpreted as forces and extensions (contractions), respectively.

We assume that the solid behavior is modeled by a set of springs responding linear elastically according to **Hooke's law**. The stiffnesses of the free spring on one end and the spring for the so-called α-Maxwell element are determined by **Young's moduli** $E_\infty > 0$ and $E_\alpha > 0$, $\alpha = 1, \ldots, m$, respectively. The flow behavior is modeled by a **Newtonian viscous fluid** responding like a **dashpot**. The viscous fluid of the α-Maxwell element is specified adequately by the material constant $\eta_\alpha > 0$, called the **viscosity**. Based on physics all these parameters are positive.

Let σ be the total stress applied to the *generalized Maxwell model* and ε be an *external variable* which measures the total linear strain due to the stress. By equilibrium, the total stress applied to the device is found to be

$$\sigma = \sigma_\infty + \sum_{\alpha=1}^{m} q_\alpha \qquad (6.245)$$

(see Figure 6.6), where the definition of the stress at equilibrium, i.e. $\sigma_\infty = E_\infty \varepsilon$, is to be used. The *internal variables* q_α, $\alpha = 1, \ldots, m$, are the non-equilibrium stresses in the dashpot of the α-Maxwell element characterizing the dissipation mechanism of the viscoelastic model.

The stresses q_α, $\alpha = 1, \ldots, m$, acting on each dashpot are related to the associated internal variables γ_α, which we interpret as (inelastic) strains on each dashpot. In particular, for a Newtonian viscous fluid, q_α are set to be proportional to the current '*distance from equilibrium*', i.e. the *strain rates* $\dot\gamma_\alpha$. We adopt the linear constitutive equation by *Newton*, i.e. $q_\alpha = \eta_\alpha \dot\gamma_\alpha$, $\alpha = 1, \ldots, m$. On the other hand, the stress in the spring of the α-Maxwell element is determined by $q_\alpha = E_\alpha(\varepsilon - \gamma_\alpha)$ (see Figure 6.6). Consequently, the stresses (not necessarily at equilibrium) acting on each dashpot is

$$q_\alpha = \eta_\alpha \dot\gamma_\alpha = E_\alpha(\varepsilon - \gamma_\alpha) \ , \qquad \alpha = 1, \ldots, m \ . \qquad (6.246)$$

Hence, time differentiation of $(6.246)_2$, i.e. $\dot q_\alpha = E_\alpha(\dot\varepsilon - \dot\gamma_\alpha)$, implies by means of $(6.246)_1$ the important evolution equations

$$\dot q_\alpha + \frac{q_\alpha}{\tau_\alpha} = \overline{E_\alpha \varepsilon} \ , \qquad \alpha = 1, \ldots, m \qquad (6.247)$$

for the internal variables within the one-dimensional and linear regime, where the definition of the *relaxation time* (or *retardation time*) $\tau_\alpha = \eta_\alpha / E_\alpha > 0$, $\alpha = 1, \ldots, m$, is to be used.

Since q_α and $\dot\gamma_\alpha$ are the stresses and the strain rates acting on each dashpot, we are in the position to define the *rate of work dissipated* within the considered device. By means of $(6.246)_1$, the internal dissipation takes on the form

$$\mathcal{D}_{\text{int}} = \sum_{\alpha=1}^{m} q_\alpha \dot\gamma_\alpha = \sum_{\alpha=1}^{m} \eta_\alpha(\dot\gamma_\alpha)^2 \geq 0 \ , \qquad (6.248)$$

which is always non-negative, since $\eta_\alpha > 0$. It disappears at equilibrium.

We now define the strain energy $\psi(\varepsilon, \gamma_1, \ldots, \gamma_m) = \psi_\infty(\varepsilon) + \sum_{\alpha=1}^{m} \upsilon_\alpha(\varepsilon, \gamma_\alpha)$, with the quadratic forms $\psi_\infty(\varepsilon) = \frac{1}{2} E_\infty \varepsilon^2$ and $\upsilon_\alpha(\varepsilon, \gamma_\alpha) = \frac{1}{2} E_\alpha(\varepsilon - \gamma_\alpha)^2$, and the normalization conditions $\psi_\infty(0) = 0$ and $\upsilon_\alpha(0, 0) = 0$, $\alpha = 1, \ldots, m$. The physically motivated strain energy ψ determines the energy *stored* elastically in the springs of the device, as illustrated in Figure 6.6. The strain energy $\upsilon_\alpha = \upsilon_\alpha(\varepsilon, \gamma_\alpha)$ is responsible for the viscoelastic contribution and is related to the α-*relaxation (retardation) process* with relaxation (retardation) time $\tau_\alpha \in (0, \infty)$.

Differentiation of ψ with respect to the total strain ε gives the total stress σ applied to the device. On comparison with (6.245) we conclude that

$$\frac{\partial \psi(\varepsilon, \gamma_1, \ldots, \gamma_m)}{\partial \varepsilon} = \sigma_\infty(\varepsilon) + \sum_{\alpha=1}^{m} q_\alpha(\varepsilon, \gamma_\alpha) = \sigma \quad , \tag{6.249}$$

where the physical expressions

$$\sigma_\infty = \frac{d\psi_\infty(\varepsilon)}{d\varepsilon} = E_\infty \varepsilon \quad , \qquad q_\alpha = \frac{\partial v_\alpha(\varepsilon, \gamma_\alpha)}{\partial \varepsilon} = E_\alpha(\varepsilon - \gamma_\alpha) \tag{6.250}$$

for the stress at equilibrium $\sigma_\infty(\varepsilon)$ and the non-equilibrium stresses $q_\alpha(\varepsilon, \gamma_\alpha)$, $\alpha = 1, \ldots, m$, are to be used.

Finally, the derivative of ψ with respect to the internal variables γ_α gives with (6.250)$_2$ (or (6.246)) the associated non-equilibrium stresses q_α in the dashpots. The resulting internal constitutive equations read

$$-\frac{\partial v_\alpha(\varepsilon, \gamma_\alpha)}{\partial \gamma_\alpha} = E_\alpha(\varepsilon - \gamma_\alpha) = q_\alpha \quad , \qquad \alpha = 1, \ldots, m \quad , \tag{6.251}$$

which, when substituted into (6.248)$_1$, gives the internal dissipation $\mathcal{D}_{\text{int}}$ expressed through the strain energy, i.e. $\mathcal{D}_{\text{int}} = -\sum_{\alpha=1}^{m} (\partial v_\alpha / \partial \gamma_\alpha) \dot{\gamma}_\alpha$.

Note that the general stress relation (6.239) with definition (6.240) may be identified as the three-dimensional and nonlinear version of the linear rheological model (6.249), which, in view of Figure 6.6, decomposes the stresses in *equilibrium* and *non-equilibrium* parts. In addition, the internal constitutive equations (6.244) and definition (6.242)$_1$ may be considered as the three-dimensional generalization of (6.251) and (6.250)$_2$ and also its extension to the finite strain regime. ∎

Evolution equations and their solutions.   In order to describe the way a viscoelastic process evolves it is necessary to specify complementary *equations of evolution* so that the local entropy production, i.e. the inequality (6.238)$_2$, is satisfied. In particular, we look for a law which governs the internal variables $\mathbf{Q}_\alpha$, $\alpha = 1, \ldots, m$, introduced as *isochoric* non-equilibrium stresses. We require that the evolution equations have a physical basis and provide a good approximation to the observed physical behavior of real materials in the large strain regime. In addition, we require that they are suitable for efficient time integration algorithms that are accessible for use within a finite element procedure.

We motivate the evolution equations for the three-dimensional and nonlinear deformation regime by reference to the relationship (6.247). Having this in mind, an obvious choice of appropriate (linear) evolution equations for each of the internal variables has the form

$$\dot{\mathbf{Q}}_\alpha + \frac{\mathbf{Q}_\alpha}{\tau_\alpha} = \dot{\mathbf{S}}_{\text{iso}\,\alpha} \ , \qquad \alpha = 1, \ldots, m \ , \tag{6.252}$$

where (6.252) is valid for some semi-open time interval $t \in (0, T]$, in which the value 0 is not included in the interval. Here we employ a superposed dot to designate the material time derivative as usual. The values $\mathbf{Q}_\alpha|_{t=0} = \mathbf{0}$, $\alpha = 1, \ldots, m$, for the internal variables at initial time $t = 0$ are assumed to be zero, since we agreed to start from a stress-free reference configuration.

In the linear differential equations (6.252), the tensors $\mathbf{S}_{\text{iso}\,\alpha}$ characterize isochoric second Piola-Kirchhoff stresses corresponding to the strain energies $\Psi_{\text{iso}\,\alpha}(\overline{\mathbf{C}})$ of the system (with $\Psi_{\text{iso}\,\alpha}(\mathbf{I}) = 0$ in the reference configuration) responsible for the α-relaxation (retardation) process with relaxation (retardation) time $\tau_\alpha \in (0, \infty)$, $\alpha = 1, \ldots, m$. The definition of the material variables $\mathbf{S}_{\text{iso}\,\alpha}$ is based on structure (6.90) and has the form

$$\mathbf{S}_{\text{iso}\,\alpha} = J^{-2/3}\mathbb{P} : \overline{\mathbf{S}}_\alpha \ , \qquad \overline{\mathbf{S}}_\alpha = 2\frac{\partial \Psi_{\text{iso}\,\alpha}(\overline{\mathbf{C}})}{\partial \overline{\mathbf{C}}} \ , \qquad \alpha = 1, \ldots, m \ , \tag{6.253}$$

where (6.253)$_2$ define the constitutive equations for the fictitious second Piola-Kirchhoff stresses $\overline{\mathbf{S}}_\alpha$. A particular stress $\mathbf{S}_{\text{iso}\,\alpha}$ depends only on the *external variable* $\overline{\mathbf{C}}$, i.e. the modified right Cauchy-Green tensor introduced in (6.79)$_2$.

The linear equations (6.252) are straightforward generalizations of eqs. (6.247), which are physically based. Both tensor quantities $\mathbf{Q}_\alpha$ and $\mathbf{S}_{\text{iso}\,\alpha}$ contribute to the isochoric response of the system. It emerges that the structure (6.252) introduced here is suitable for efficient time integration algorithms as we discuss later in this section.

For the case of non-constant relaxation times τ_α, the convenient concept of **'modified' time** is used in order to obtain linear evolution equations of the type (6.252). Within this concept, τ_α is kept fixed during the process; for more details see, for example, KNAUSS and EMRI [1981] and GOVINDJEE and SIMO [1993].

Fairly simple closed form solutions of the linear evolution equations (6.252), which are valid for some semi-open time interval $t \in (0, T]$, are given by the **convolution integrals**

$$\mathbf{Q}_\alpha = \exp(-T/\tau_\alpha)\mathbf{Q}_{\alpha\,0+}$$
$$+ \int_{t=0+}^{t=T} \exp[-(T-t)/\tau_\alpha]\dot{\mathbf{S}}_{\text{iso}\,\alpha}(t)\mathrm{d}t \ , \qquad \alpha = 1, \ldots, m \tag{6.254}$$

(the proof is omitted). The instantaneous response $\mathbf{Q}_{\alpha\,0+}$ (set of initial conditions) is given by

$$\mathbf{Q}_{\alpha\,0+} = J_{0+}^{-2/3}\mathbb{P}_{0+} : 2\frac{\partial \Upsilon_\alpha(\overline{\mathbf{C}}_{0+}, \mathbf{\Gamma}_{\alpha\,0+})}{\partial \overline{\mathbf{C}}_{0+}} \ , \qquad \alpha = 1, \ldots, m \tag{6.255}$$

(compare with eqs. (6.242)$_4$, (6.243)), with J_{0^+}, $\overline{\mathbf{C}}_{0^+}$, $\boldsymbol{\Gamma}_{\alpha 0^+}$, $\mathbb{P}_{0^+} = \mathbb{I} - \frac{1}{3}\mathbf{C}_{0^+}^{-1} \otimes \mathbf{C}_{0^+}$ defining, respectively, the volume ratio, the modified right Cauchy-Green tensor, the internal variables and the projection tensor at time $t = 0^+$.

The evolution of the internal variables is governed mainly by the strain-energy functions $\Psi_{\mathrm{iso}\,\alpha}$ via relation (6.253). However, if a viscoelastic medium such as a thermoplastic elastomer is composed of identical polymer chains we can motivate the assumption that $\Psi_{\mathrm{iso}\,\alpha}$ is replaceable by the strain-energy function $\Psi_{\mathrm{iso}}^\infty$, which is responsible for the isochoric elastic response as $t \to \infty$. We adopt the expression

$$\Psi_{\mathrm{iso}\,\alpha}(\overline{\mathbf{C}}) = \beta_\alpha^\infty \Psi_{\mathrm{iso}}^\infty(\overline{\mathbf{C}}) \ , \qquad \alpha = 1, \ldots, m \tag{6.256}$$

(which is due to GOVINDJEE and SIMO [1992b]), where $\beta_\alpha^\infty \in [0, \infty)$ are given non-dimensional **strain-energy factors** associated with $\tau_\alpha \in (0, \infty)$, $\alpha = 1, \ldots, m$. Consequently, the stresses $\mathbf{S}_{\mathrm{iso}\,\alpha}$, $\alpha = 1, \ldots, m$, as introduced in (6.253), may be replaced by means of (6.256) and (6.241)$_2$. We write

$$\mathbf{S}_{\mathrm{iso}\,\alpha} = J^{-2/3}\mathbb{P} : 2\beta_\alpha^\infty \frac{\partial \Psi_{\mathrm{iso}}^\infty(\overline{\mathbf{C}})}{\partial \overline{\mathbf{C}}} = \beta_\alpha^\infty \mathbf{S}_{\mathrm{iso}}^\infty(\overline{\mathbf{C}}) \ , \qquad \alpha = 1, \ldots, m \ . \tag{6.257}$$

Hence, the material time derivative of the stress tensors, $\dot{\mathbf{S}}_{\mathrm{iso}\,\alpha}$, which govern evolution equations (6.252), are replaced by $\beta_\alpha^\infty \dot{\mathbf{S}}_{\mathrm{iso}}^\infty$.

In summary: the phenomenological viscoelastic model valid over any range of strains is described by constitutive equations (6.239)–(6.243), evolution equations (6.252) with solutions (6.254) and replacements (6.257). Note that with reference to assumption (6.256) the model problem is completely determined by the specification of only two scalar-valued functions, namely $\Psi_{\mathrm{vol}}^\infty(J)$ and $\Psi_{\mathrm{iso}}^\infty(\overline{\mathbf{C}})$, a crucial advantage of the introduced finite strain viscoelastic model.

It is important to emphasize that the described constitutive model fits within the framework of so-called **simple materials** with memory, which are expressed in the general form by $\mathbf{S}(t) = \mathfrak{S}_{s=0^+}^\infty [\mathbf{C}(t - s), \mathbf{C}(t)]$ (see, for example, MALVERN [1969, p. 400, eq. (6.7.62)]), with $T = t - s$, where $\mathfrak{S}$ is a (response) functional depending on the history of $\mathbf{C}$ from $T = -\infty$ to $T = t$. For further discussion of this issue the interested reader is referred to MALVERN [1969, Section 6.7, and references therein].

Time integration algorithm.   The total second Piola-Kirchhoff stress tensor $\mathbf{S}$ is computed according to relations (6.239)–(6.241) with the *volumetric* and *isochoric* (elastic) response $\mathbf{S}_{\mathrm{vol}}^\infty$ and $\mathbf{S}_{\mathrm{iso}}^\infty$ and the contribution due to the *non-equilibrium stresses* $\sum_{\alpha=1}^m \mathbf{Q}_\alpha$ as given by the convolution integrals in the form of (6.254).

For the solutions of the crucial *Cauchy's equations of motion* (see the local forms (4.53) and (4.68)) the stress tensor is required. The main goal of the following is to outline an appropriate **update algorithm** for the total stresses suitable for implementation

in a finite element program. The update procedure is realized in the reference configuration, and hence the objectivity requirement based on a Euclidean transformation is
trivially satisfied (see Chapter 5). The key of the update algorithm is the *numerical
integration* of the convolution integrals (6.254).

In order to obtain the *algorithmic update* of the second Piola-Kirchhoff stress **S**
we consider a partition (time discretization) $\bigcup_{n=0}^{M}[t_n, t_{n+1}]$ of the closed time interval
$t \in [0^+, T]$ of interest, where $0^+ = t_0 < \ldots < t_{M+1} = T$. We now concentrate
attention on a typical closed **time sub-interval** $[t_n, t_{n+1}]$, with

$$\Delta t = t_{n+1} - t_n \tag{6.258}$$

characterizing the associated **time increment**.

Assume now that up to a certain time t_n the stress $\mathbf{S}_n$ satisfies the equilibrium
equation and that the displacement field $\mathbf{u}_n$, the tensor variables

$$\left. \begin{array}{ll} \mathbf{F}_n = \mathbf{I} + \mathrm{Grad}\mathbf{u}_n \quad, & J_n = \det\mathbf{F}_n \quad, \\[2mm] \mathbf{C}_n = \mathbf{F}_n^{\mathrm{T}}\mathbf{F}_n \quad, & \overline{\mathbf{C}}_n = J_n^{-2/3}\mathbf{C}_n \end{array} \right\} \tag{6.259}$$

(see $(2.45)_2$, (2.52), $(2.66)_1$, $(6.79)_2$), and the stress $\mathbf{S}_n$ (determined via the associated
constitutive equation) are specified uniquely by the given motion $\boldsymbol{\chi}_n$ at time t_n.

Within a *strain-driven* type of numerical procedure, the aim is now to advance the
solution to time $t_{n+1} = \Delta t + t_n$ and *update* all relevant quantities. At first we make
an *initial guess* for $\boldsymbol{\chi}_{n+1}$, known as a **trial solution**, and update the prescribed loads.
Within a classical solution technique, such as *Newton's method*, the new motion $\boldsymbol{\chi}_{n+1}$
at time t_{n+1} is corrected iteratively until the balance principles are satisfied within a
given tolerance of accuracy. To check equilibrium at time t_{n+1} the tensor variables

$$\left. \begin{array}{ll} \mathbf{F}_{n+1} = \mathbf{I} + \mathrm{Grad}\mathbf{u}_{n+1} \quad, & J_{n+1} = \det\mathbf{F}_{n+1} \quad, \\[2mm] \mathbf{C}_{n+1} = \mathbf{F}_{n+1}^{\mathrm{T}}\mathbf{F}_{n+1} \quad, & \overline{\mathbf{C}}_{n+1} = J_{n+1}^{-2/3}\mathbf{C}_{n+1} \end{array} \right\} \tag{6.260}$$

have to be computed. This process is straightforward since the new motion $\boldsymbol{\chi}_{n+1}$ with
the updated displacement field $\mathbf{u}_{n+1}$ is considered to be given. The remaining second
Piola-Kirchhoff stress at time t_{n+1} is again determined uniquely via the associated
constitutive equation. In particular, the so-called **algorithmic stress** at time t_{n+1} reads
as

$$\mathbf{S}_{n+1} = (\mathbf{S}_{\mathrm{vol}}^{\infty} + \mathbf{S}_{\mathrm{iso}}^{\infty} + \sum_{\alpha=1}^{m}\mathbf{Q}_{\alpha})|_{n+1} \quad . \tag{6.261}$$

Since all required strain measures at t_{n+1} are known, the first two stress contributions,
i.e. $\mathbf{S}_{\mathrm{vol}\,n+1}^{\infty}$ and $\mathbf{S}_{\mathrm{iso}\,n+1}^{\infty}$, are determined via (6.241), which, in the present notation,
reads as

$$\mathbf{S}_{\mathrm{vol}\,n+1}^{\infty} = 2\frac{\partial\Psi_{\mathrm{vol}}^{\infty}(J_{n+1})}{\partial\mathbf{C}_{n+1}} \quad, \qquad \mathbf{S}_{\mathrm{iso}\,n+1}^{\infty} = 2\frac{\partial\Psi_{\mathrm{iso}}^{\infty}(\overline{\mathbf{C}}_{n+1})}{\partial\mathbf{C}_{n+1}} \quad . \tag{6.262}$$

The third term in (6.261), which is the viscoelastic stress contribution $\sum_{\alpha=1}^{m} \mathbf{Q}_{\alpha n+1}$ based on (6.254), remains to be evaluated.

The following derivation is related to the approach by SIMO [1987], which bypasses the need for **incremental objectivity** as proposed by HUGHES and WINGET [1980]. Incremental objectivity requires that the algorithmic constitutive equations must be objective (frame-indifferent) during a superimposed (time-dependent) rigid-body motion. Incremental objectivity represents the numerical version of the principle of material frame-indifference, as introduced in Section 5.4.

We now split the convolution integral (6.254) into the form of

$$\int_{0+}^{t_{n+1}} (\bullet) \mathrm{d}t = \int_{0+}^{t_n} (\bullet) \mathrm{d}t + \int_{t_n}^{t_{n+1}} (\bullet) \mathrm{d}t \quad . \tag{6.263}$$

Hence, the internal variables $\mathbf{Q}_\alpha$, $\alpha = 1, \ldots, m$, at t_{n+1} are given by

$$\mathbf{Q}_{\alpha n+1} = \exp(-t_{n+1}/\tau_\alpha)\mathbf{Q}_{\alpha 0+} + \int_{t=0+}^{t=t_n} \exp[-(t_{n+1} - t)/\tau_\alpha]\dot{\mathbf{S}}_{\mathrm{iso}\,\alpha}(t)\mathrm{d}t$$

$$+ \int_{t=t_n}^{t=t_{n+1}} \exp[-(t_{n+1} - t)/\tau_\alpha]\dot{\mathbf{S}}_{\mathrm{iso}\,\alpha}(t)\mathrm{d}t \quad , \qquad \alpha = 1, \ldots, m \quad . \tag{6.264}$$

In order to simplify (6.264) we apply relation (6.258) to all three terms. For the first two terms we use the standard property

$$\exp[-(\Delta t + \beta)/\tau_\alpha] = \exp(-\Delta t/\tau_\alpha)\exp(-\beta/\tau_\alpha) \tag{6.265}$$

for the exponential function, for any constants Δt, τ_α and parameter β which takes on values t_n and $t_n - t$. In addition to (6.265) we use the second-order accurate **mid-point rule** on the third term $\int_{t_n}^{t_{n+1}} (\bullet) \mathrm{d}t$ of eq. (6.264), which means that the time variable t is approximated by $(t_{n+1} + t_n)/2$. We deduce from (6.264) that

$$\mathbf{Q}_{\alpha n+1} = \exp(-\Delta t/\tau_\alpha)[\exp(-t_n/\tau_\alpha)\mathbf{Q}_{\alpha 0+} + \int_{t=0+}^{t=t_n} \exp[-(t_n - t)/\tau_\alpha]\dot{\mathbf{S}}_{\mathrm{iso}\,\alpha}(t)\mathrm{d}t]$$

$$+\exp(-\Delta t/2\tau_\alpha) \int_{t=t_n}^{t=t_{n+1}} \dot{\mathbf{S}}_{\mathrm{iso}\,\alpha}(t)\mathrm{d}t \quad , \qquad \alpha = 1, \ldots, m \quad . \tag{6.266}$$

Note that the terms within the brackets are $\mathbf{Q}_\alpha$ at time t_n (compare with eq. (6.254)). By solving the last term in (6.266) and by means of assumption (6.257), we may write

$$\mathbf{Q}_{\alpha n+1} = \exp(2\xi_\alpha)\mathbf{Q}_{\alpha n}$$

$$+\exp(\xi_\alpha)\beta_\alpha^\infty (\mathbf{S}_{\mathrm{iso}\,n+1}^\infty - \mathbf{S}_{\mathrm{iso}\,n}^\infty) \quad , \qquad \xi_\alpha = -\frac{\Delta t}{2\tau_\alpha} \quad , \tag{6.267}$$

for $\alpha = 1, \ldots, m$.

After rearranging eqs. $(6.267)_1$ we arrive finally at a second-order accurate **recurrence update formula** for the internal stresses in a simple format, namely

$$\mathbf{Q}_{\alpha\, n+1} = \beta_\alpha^\infty \exp(\xi_\alpha) \mathbf{S}_{\mathrm{iso}\, n+1}^\infty + \mathcal{H}_{\alpha\, n} \,, \qquad \alpha = 1, \ldots, m \,, \tag{6.268}$$

$$\mathcal{H}_{\alpha\, n} = \exp(\xi_\alpha)[\exp(\xi_\alpha)\mathbf{Q}_{\alpha\, n} - \beta_\alpha^\infty \mathbf{S}_{\mathrm{iso}\, n}^\infty] \,, \tag{6.269}$$

and with definition $(6.267)_2$ of the dimensionless parameters ξ_α. In recurrence relation (6.268) we have introduced the *(algorithmic) history term* $\mathcal{H}_{\alpha\, n}$, $\alpha = 1, \ldots, m$. This term is determined by the internal history variables $\mathbf{Q}_{\alpha\, n}$ and $\mathbf{S}_{\mathrm{iso}\, n}^\infty$, which are known from the previous step serving as an 'initial' data base.

The recurrence update formula of the type (6.268) was proposed by TAYLOR et al. [1970, and references therein]. Instead of the time integration algorithm outlined above we may use other algorithmic updates for the total stresses. For a slightly different structure see, for example, SIMO [1987] and co-workers, and HOLZAPFEL [1996a]. For an application of the described viscoelastic model to fiber-reinforced composites at finite strains see the recent paper by HOLZAPFEL and GASSER [2001].

Elasticity tensor in material description. The importance of *consistent linearized tangent moduli* in the solution of nonlinear problems by incremental/iterative techniques of Newton's type was emphasized in Section 6.6.

We now determine the consistent linearization of the constitutive model presented above, noting the algorithmic stress (6.261) with relations (6.262), (6.268), (6.269). In view of definition (6.154) and the decoupled stress relation (6.261), the associated *algorithmic elasticity tensor* in the material description at time t_{n+1} may be written in the form

$$\mathbb{C}_{n+1} = (\mathbb{C}_{\mathrm{vol}}^\infty + \mathbb{C}_{\mathrm{iso}}^\infty + \sum_{\alpha=1}^m \mathbb{C}_{\mathrm{vis}}^\alpha)|_{n+1} \,, \tag{6.270}$$

where the first two contributions to $\mathbb{C}_{n+1}$ are given by (6.163), which, in the present notation, reads as

$$\mathbb{C}_{\mathrm{vol}\, n+1}^\infty = 2 \frac{\partial \mathbf{S}_{\mathrm{vol}\, n+1}^\infty}{\partial \mathbf{C}_{n+1}} \,, \qquad \mathbb{C}_{\mathrm{iso}\, n+1}^\infty = 2 \frac{\partial \mathbf{S}_{\mathrm{iso}\, n+1}^\infty}{\partial \mathbf{C}_{n+1}} \,. \tag{6.271}$$

Explicit expressions for (6.271) are found in (6.166) and (6.168).

The third (viscoelastic) contribution to $\mathbb{C}_{n+1}$, i.e. $\sum_{\alpha=1}^m \mathbb{C}_{\mathrm{vis}}^\alpha$ at t_{n+1}, is derived using the expression for the internal stresses (6.268). Note that the derivative of the (algorithmic) history term $\mathcal{H}_{\alpha\, n}$, $\alpha = 1, \ldots, m$, (which defines quantities at t_n) with respect to $\mathbf{C}_{n+1}$ is zero. Hence, the third contribution to the elasticity tensor is

$$\begin{aligned} \mathbb{C}_{\mathrm{vis}\, n+1}^\alpha &= 2 \frac{\partial \mathbf{Q}_{\alpha\, n+1}}{\partial \mathbf{C}_{n+1}} \\ &= \delta_\alpha \mathbb{C}_{\mathrm{iso}\, n+1}^\infty \,, \qquad \delta_\alpha = \beta_\alpha^\infty \exp(\xi_\alpha) \,, \qquad \alpha = 1, \ldots, m \,. \end{aligned} \tag{6.272}$$

Remarkably, the viscoelastic contribution to the algorithmic elasticity tensor may be expressed as the **viscoelastic factors** δ_α, $\alpha = 1, \ldots, m$, governing the time-dependent part, and by $\mathbb{C}^\infty_{\text{iso}\,n+1}$, which is associated with the isochoric elastic response as $t \to \infty$.

Hence, relation (6.270) may be rewritten by means of (6.272) as

$$\mathbb{C}_{n+1} = \mathbb{C}^\infty_{\text{vol}\,n+1} + (1+\delta)\mathbb{C}^\infty_{\text{iso}\,n+1} \quad , \qquad \delta = \sum_{\alpha=1}^{m} \delta_\alpha \quad . \tag{6.273}$$

If $\delta = \sum_{\alpha=1}^{m} \beta^\infty_\alpha \exp(\xi_\alpha)$ tends to zero, the algorithmic elasticity tensor reduces to (6.271) and compressible finite hyperelasticity is recovered.

The implementation of the viscoelastic model described above into a finite element program is based on the derived algorithmic stress (6.261) and the associated algorithmic elasticity tensor (6.273). The algorithmic update is carried out at each Gauss point of a finite element. The presented formulation only needs a particularization of the two strain-energy functions $\Psi^\infty_{\text{vol}}(J)$ and $\Psi^\infty_{\text{iso}}(\overline{\mathbf{C}})$. If we use a strain energy which is expressed in principal stretches, then the algorithmic update is actually performed on the principal values.

For an efficient computational application of the iterative process, see the works by SIMO [1987], GOVINDJEE and SIMO [1992b], HOLZAPFEL [1996a] and HOLZAPFEL and GASSER [2001].

EXERCISES

1. Using the relations

$$\dot{\mathbf{Q}}_\alpha + \frac{\mathbf{Q}_\alpha}{\tau_\alpha} = \exp(-t/\tau_\alpha)\frac{\mathrm{D}}{\mathrm{D}t}[\exp(t/\tau_\alpha)\mathbf{Q}_\alpha] \quad , \qquad \alpha = 1, \ldots, m$$

and integration over time interval $t \in (0, T]$ obtain the convolution integrals (6.254), which are the closed form solutions of the linear differential equations (6.252).

2. *Relaxation test.* Assume a thin sheet of incompressible material in the undeformed configuration which may undergo viscoelastic deformations in the large strain domain. The sheet is stretched to $\lambda_1 = \lambda$ ($\lambda_2 = \lambda_3 = \lambda^{-1/2}$) in one direction (simple tension) and is *fixed* subsequently at this elongation. The deformation is assumed to be homogeneous.

   The viscoelastic behavior of the material is based on a phenomenological Maxwell-type model with a free spring at one end and one Maxwell element arranged in parallel ($m = 1$). The underlying strain energy is due to Ogden, as introduced in Section 6.5. We assume the six-parameter model and use the typical values according to (6.121).

(a) For a certain closed time interval $t \in [0, T]$ compute an explicit expression for the evolution of the remaining non-vanishing internal stress $Q = Q(\lambda, t)$ and the Cauchy stress $\sigma = \sigma(\lambda, t)$ (the stresses in the transverse directions are zero). Discuss the internal dissipation $\mathcal{D}_{\text{int}}$ along the time interval and determine thermodynamic equilibrium.

(b) For $t \in [0, T]$ plot the stress decay (Q, σ) at a stretch ratio $\lambda = 5$, with the strain-energy factor $\beta_1^\infty = 1$ and the relaxation time $\tau = 10s$. Give a physical interpretation of the relaxation time τ.

3. *Creep test.*   Consider the specimen and the viscoelastic constitutive model as described in the previous exercise, but now let a Cauchy stress σ be applied in one direction up to a certain value σ_0 and then be held constant. The stress causes a homogeneous deformation characterized by the stretch ratios $\lambda_1 = \lambda$, $\lambda_2 = \lambda_3 = \lambda^{-1/2}$ (simple tension). The underlying strain energy is of neo-Hookean type.

(a) Derive the stress relation in the form of a nonlinear differential equation of first-order.

(b) Solve the differential equation by means of Newton's method and use the derived recurrence update formula, or alternatively solve it with the *Runge-Kutta* method. For this purpose write a computer program or simply use some commercially available mathematical software-package. For a certain time domain $t \in [0, T]$ plot the stress evolution (Q, σ), and the stretch evolution λ (for a fixed σ_0), $\beta_1^\infty = 1$ and $\tau = 10s$.

6.11   Hyperelastic Materials with Isotropic Damage

Continuum damage theories are either **micromechanical** or **phenomenological** in nature. Microscopic approaches are certainly the best, but necessitate a strong physical background, and models are mathematically complex and often difficult to identify.

The aim of this section is to formulate a three-dimensional and rate-independent isotropic damage model for the large strain domain describing the Mullins effect. We employ **continuum damage mechanics** (often abbreviated as CDM) and follow a purely phenomenological approach, which leads to damage models describing the *macroscopic* constitutive behavior of materials containing distributed microcracks. The material model introduced is suitable for numerical procedures.

For additional information on the subject of constitutive models within the context of continuum damage mechanics, see the monographs by, for example, KACHANOV

[1986], LEMAITRE [1996] and KRAJCINOVIC [1996, Chapter 4] and the review article by DE SOUZA NETO et al. [1998], which also describes techniques for the numerical simulation of (isotropic) internal damage in finitely deformed solids.

Mullins effect.   Many rubber-like materials consist of a cross-linked elastomer substance with a distribution of small carbon particles as fillers (see the account of filled elastomers in MARK and ERMAN [1988, Chapter 20]). A piece of filler-loaded rubber subjected to a series of loadings typically displays pronounced (strain-induced) **stress softening** associated with damage, known as the **Mullins effect** (this effect was pointed out in the early pioneering work of MULLINS [1947]; for a detailed description see, for example, JOHNSON and BEATTY [1993a, b]). Nearly all practical engineering rubbers contain carbon particles as fillers and exhibit a certain degree of Mullins effect, which is regarded as essentially being caused by the fillers.

In order to explain the main features of this stress softening phenomenon we consider a strain-controlled cyclic tension test of a piece of filled rubber with two different strain amplitudes and neglect viscoelastic effects (slow strain rates). The cyclic loading and unloading process starts from its unstressed (initial) virgin state 0 and follows a path A, which we call the **primary loading path** (see Figure 6.7). After subsequent unloading initiated from any point 1 on the primary loading path the piece of rubber follows path B and completely returns to the unstressed state 0 (for real rubber this will hardly ever occur). Note that after the test piece has been subjected to a load up to the point 1 the initial properties of the virgin material containing fillers are changed permanently (see MULLINS [1947]). The first loading and unloading cycle involves dissipation which is represented by the area between the curves A and B (hysteresis behavior). The area is a measure for non-recoverable energy.

When the material is re-loaded the stress-strain behavior follows the path B again and if a strain beyond the point 1 (at which unloading began) is applied, the path D is activated. It is a continuation of the primary loading path A. For additional unloading which begins at any point 2 of the primary loading path the rubber is retraced back to the unstressed state 0 along the path C. Note that the shape of the second stress-strain cycle differs significantly from the first one. Path C retraces the piece of rubber back to 2 on re-loading and the primary loading path is activated again.

In summary: the stiffness of rubbers containing reinforcing fillers such as carbon black decreases as a result of extensional loading and unloading. The material properties associated with the initial extension of rubber compounds may differ significantly from those associated with successive deformation. With reference to Figure 6.7 we recognize that for a given strain level the stress required on the first unloading and re-loading path B (and also on the second unloading and re-loading path C) is *less* than that on the primary loading path A, and is the essential feature of the Mullins effect.

There are a few theories in the literature aimed at explaining the microscopic dam-

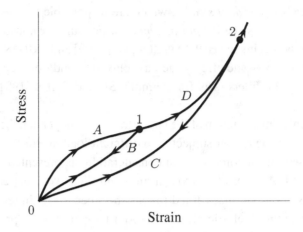

Figure 6.7 Cyclic tension test displaying Mullins effect.

age mechanism. One class of theory is based on the idea that the internal damage is caused by debonding of rubber molecular chains attached between the filler particles (see BUECHE [1960, 1961]). The higher the (macroscopic) deformation of the rubber the higher the strain-induced damage. Based on statistical arguments the stress softening effect may be predicted for a fixed level of strain due to previous higher strains in the material.

The other class of theory for explaining the microscopic damage mechanism goes back to MULLINS and TOBIN [1957] and MULLINS [1969]. They proposed that initially the filler-loaded rubber exists in a so-called *hard phase* which degrades into a *soft phase* with increasing strain. The transition between the two phases is characterized by a damage parameter which is associated with a strain-amplification function. JOHNSON and BEATTY [1993a, b] have adopted the two-phase approach, which shows good agreement with experimental data in simple tension.

The Mullins effect was observed experimentally in uniaxial cyclic extension tests performed by, for example, MULLINS [1947], MULLINS and TOBIN [1957] (who provided experimental data for loading), BUECHE [1961], MULLINS and TOBIN [1965] (experimental data obtained from the elastic and swelling behavior of filler-loaded rubbers), HARWOOD et al. [1965], HARWOOD and PAYNE [1966a, b], STERN [1967] and MULLINS [1969].

It is important to emphasize that in practice several other (inelastic) effects arise under extensional loading and unloading and we address these in brief. From the experimental observation it is known that, besides the Mullins effect (which is an idealized phenomenon), the shape of the stress-strain curves is essentially rate and temperature dependent. In addition, the shape of a piece of carbon-black filled rubber after unloading differs significantly from its virgin shape. This interesting effect caused by reinforcement leads to *residual strains*, which are responsible for the change of shape (in the rubber industry also called *permanent set*). For suitable constitutive models incorporating residual strains see HOLZAPFEL et al. [1999] and OGDEN and ROXBURGH [1999b]. Additional inelastic effects such as relaxation and/or creep depend strongly on the content of solid fillers (see, for example, SO and CHEN [1991, and references therein]).

An experimental investigation of the large strain time-dependent behavior of carbon-black filled Chloroprene rubber subjected to different loading conditions and a constitutive model based on micromechanical considerations is presented by BERGSTRÖM and BOYCE [1998]. A series of uniaxial strain-controlled cyclic experiments on cylindrical specimens of carbon-black filled rubbers between 100% in tension and 30% in compression, including inelastic effects at room temperature, is presented by LION [1996] and MIEHE and KECK [2000]. Phenomenological material models for filled rubbery polymers are also proposed therein. A cyclic tension process with three different strain amplitudes and 12 loading cycles were performed on a virgin specimen. The process shows clearly that the magnitude of the resulting stress softening depends on the number of loading cycles and the strain amplitude. The experimental studies also consider relaxation periods in tension and compression at constant strain. The work of MIEHE and KECK [2000] also provides experimental stress-deformation curves of pre-damaged specimens under monotonic tension and compression and cyclic tension/compression. For an experimental investigation of a filler-loaded tread compound at different temperature levels see LION [1997a].

Damage model in coupled material description.   Now we are concerned with the continuum formulation of the (ideal) Mullins effect and neglect rate and temperature dependency as well as residual strains. Additionally, in the phenomenological model we do not consider the presence of carbon-black fillers. We choose a *Lagrangian* description and express the relevant equations in terms of the right Cauchy-Green tensor.

Consider an isothermal elastic process, and postulate a Helmholtz free-energy function Ψ in the *coupled* form

$$\Psi = \Psi(\mathbf{C}, \zeta) = (1 - \zeta)\Psi_0(\mathbf{C}) \ , \tag{6.274}$$

where Ψ_0 is the **effective strain-energy function** of the hypothetical *undamaged* material, with the normalization condition $\Psi_0(\mathbf{I}) = 0$ and the restriction $\Psi_0(\mathbf{C}) \geq 0$. The factor $(1 - \zeta)$ is known as the **reduction factor**, first proposed by KACHANOV [1958] who modeled the creep rupture of metals as a uniaxial problem (*Kachanov* actually introduced $(1 - \zeta)$ as the 'integrity' parameter). Here the *internal variable* $\zeta \in [0, 1]$ is a *scalar*, referred to as the **damage variable**. The damage variable describes an (ideally) isotropic damage process and is related to the ultimate failure of the material. Note that the strain energy Ψ_0 is assumed to be objective as usual.

Of course this simple type of damage model has limited use in practice, but describes both the *dissipation mechanisms* and the *irreversible rearrangements of the structure*. To refine the model more general (tensorial) forms are needed, especially to describe anisotropic damage.

In order to obtain the stress relation we differentiate first (6.274) with respect to time. Using the chain rule we find that

$$\dot{\Psi} = (1 - \zeta)\frac{\partial \Psi_0(\mathbf{C})}{\partial \mathbf{C}} : \dot{\mathbf{C}} - \Psi_0(\mathbf{C})\dot{\zeta} \quad , \tag{6.275}$$

with the rate of change of the right Cauchy-Green tensor and the damage variable, i.e. $\dot{\mathbf{C}}$, (2.168), and $\dot{\zeta}$, respectively.

As a particularization of the Clausius-Planck inequality, as given in (4.154), we obtain, by means of (6.275),

$$\mathcal{D}_{\text{int}} = \left(\mathbf{S} - (1 - \zeta)2\frac{\partial \Psi_0(\mathbf{C})}{\partial \mathbf{C}}\right) : \frac{\dot{\mathbf{C}}}{2} + \Psi_0(\mathbf{C})\dot{\zeta} \geq 0 \quad , \tag{6.276}$$

and therefore the second Piola-Kirchoff stress tensor $\mathbf{S}$ and the *non-negative* internal dissipation $\mathcal{D}_{\text{int}}$ are

$$\mathbf{S} = (1 - \zeta)\mathbf{S}_0 \qquad \text{with} \qquad \mathbf{S}_0 = 2\frac{\partial \Psi_0(\mathbf{C})}{\partial \mathbf{C}} \quad , \tag{6.277}$$

$$\mathcal{D}_{\text{int}} = f\dot{\zeta} \geq 0 \qquad \text{with} \qquad f = \Psi_0(\mathbf{C}) \geq 0 \quad . \tag{6.278}$$

In constitutive equation (6.277) the quantity $\mathbf{S}_0$ denotes the **effective second Piola-Kirchhoff stress tensor**. The dissipation inequality (6.278) clearly shows that damage is a dissipative process, the quantity f therein denotes the **thermodynamic force** which governs the damage evolution. In continuum damage mechanics the thermodynamic quantity f has the meaning of the effective strain energy Ψ_0 released per unit reference volume.

The thermodynamic force f is related (conjugate) to the internal variable ζ according to

$$f = \Psi_0(\mathbf{C}) = -\frac{\partial \Psi}{\partial \zeta} \quad , \tag{6.279}$$

see relation (6.274). Therefore, instead of controlling the damage process by the internal variable ζ we can equivalently use its conjugate quantity, i.e. the effective strain energy Ψ_0.

The evolution of f is given by

$$\dot{f} = 2\frac{\partial \Psi_0(\mathbf{C})}{\partial \mathbf{C}} : \frac{\dot{\mathbf{C}}}{2} = \mathbf{S}_0 : \frac{\dot{\mathbf{C}}}{2} \quad , \tag{6.280}$$

recognizing that $\dot{f}$ characterizes the **effective stress power** per unit reference volume according to (4.113).

As noted above we assume that the total damage accumulation is based on *Mullins* type (stress) softening of the material. For a strain-controlled cyclic loading process with a fixed strain amplitude this type of phenomenological damage occurs only within the first cycle. We adopt the smooth function $\zeta = \zeta(\alpha)$ as the damage variable, with conditions $\zeta(0) = 0$ and $\zeta(\infty) \in [0,1]$. The phenomenological variable α describes the **discontinuous damage**. A constitutive particularization of the damage variable ζ may, according to MIEHE [1995a], be given by

$$\zeta = \zeta(\alpha) = \zeta_\infty \left[1 - \exp(-\alpha/\iota) \right] \quad , \tag{6.281}$$

where ζ_∞ describes the dimensionless **maximum** (possible) **damage** and ι is referred to as the **damage saturation parameter**.

We now aim to determine the *discontinuous damage variable* α over the past history up to the current time, i.e. the history time interval $[0, t]$. To control a discontinuous damage process we use the evolution of the effective strain-energy function Ψ_0. Within the closed time interval $[0, t]$ we take the phenomenological variable α to be related to the *maximum value* of Ψ_0 and write

$$\alpha(t) = \max_{s \in [0,t]} \Psi_0(s) \quad , \tag{6.282}$$

where $s \in [0, t]$ denotes the history variable. Thus, it emerges that α is the **maximum thermodynamic force** with the same dimension as the effective strain energy per unit reference volume. Definition (6.282) was employed by, for example, DE SOUZA NETO et al. [1994] and by MIEHE [1995a], while the rate-independent damage model of SIMO [1987] invoked the *principle of strain-equivalence* (see the chapter on damage mechanics in the book by LEMAITRE and CHABOCHE [1990, Chapter 7]). Other strain-based models for the finite strain domain are due to GOVINDJEE and SIMO [1991, 1992a, b], JOHNSON and BEATTY [1993a, b] among others.

Relation (6.282) generalizes the one-dimensional damage model (for small strains) proposed by GURTIN and FRANCIS [1981b]. It uses the hypothesis that the current state of damage is characterized by the maximum axial strain attained in the history of deformation. Here we do not recall the many important works in the area of small strain damage mechanics.

A straightforward refinement of the isotropic damage model may be obtained by including damage effects governed by *continuous damage accumulation*. In particular, the work of MIEHE [1995a], which is exclusively presented in an *Eulerian* setting, describes continuous damage accumulation. It is based on the arc-length of the effective strain-energy function.

A fairly simple and efficient energy-based damage model to describe the main features of the Mullins effect in filled rubber was proposed by OGDEN and ROXBURGH

[1999a]. The material model is composed of a (classical) strain-energy function describing the primary loading path from the unstressed virgin state and an additive damage function responsible for the unloading path which is initiated from any point on the primary loading path. The formulation is based on the **concept of pseudo-elasticity** in which the material is treated as one elastic material in loading and another elastic material in unloading. This idea was used by, for example, FUNG et al. [1979] within the context of modelling arterial walls. It has the significant advantage of convenient and simple description of the stress-strain relationships in cyclic loading and their numerical (finite element) realization.

Damage criterion and damage evolution.   We now define a **damage criterion** in the *strain space* at any time of the loading process in the form

$$\phi(\mathbf{C}, \alpha) = f(\mathbf{C}) - \alpha \leq 0 \ , \tag{6.283}$$

with the **damage function** ϕ. For $\phi < 0$ no evolution of damage occurs. The second possible situation is $\phi = 0$ which characterizes the so-called **damage surface** with normal

$$\mathbf{N} = \frac{\partial \phi(\mathbf{C}, \alpha)}{\partial \mathbf{C}} = \frac{\partial f(\mathbf{C})}{\partial \mathbf{C}} \ . \tag{6.284}$$

For a fixed α, the damage surface delimits the strain space in which the behavior of the material is considered to be fully elastic (no damage accumulation occurs). A representation of the damage surface with the associated normal in the principal strain space is shown in Figure 6.8.

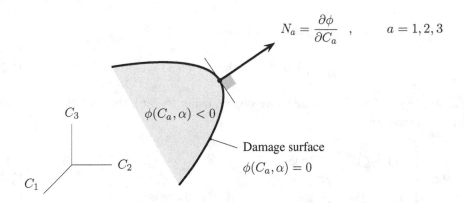

Figure 6.8 inset:
$$N_a = \frac{\partial \phi}{\partial C_a} \ , \qquad a = 1, 2, 3$$

C_3

$\phi(C_a, \alpha) < 0$

C_2

C_1

Damage surface

$\phi(C_a, \alpha) = 0$

Figure 6.8 Illustration of the damage criterion in the principal strain space.

Hence, a double contraction of the two tensors $\mathbf{N}$ and $\dot{\mathbf{C}}$ gives, using $(6.284)_2$, the scalar $\mathbf{N} : \dot{\mathbf{C}} = \partial f(\mathbf{C})/\partial \mathbf{C} : \dot{\mathbf{C}} = \dot{f}(\mathbf{C})$. Borrowing the terminology from *strain space plasticity* (see NAGHDI and TRAPP [1975] and SIMO and HUGHES [1998, p. 84]), at $\phi = 0$ we must distinguish between

$$\phi = 0 \qquad \text{and} \qquad \begin{cases} \dot{f} < 0 \ , \\ \dot{f} = 0 \ , \\ \dot{f} > 0 \ , \end{cases} \qquad (6.285)$$

describing *unloading*, *neutral loading* and *loading*, respectively.

Finally, the evolution of the maximum thermodynamic force α, i.e. (6.282), is given, with $(6.280)_2$, by

$$\dot{\alpha} = \begin{cases} \dot{f} = \mathbf{S}_0 : \dfrac{\dot{\mathbf{C}}}{2} & \text{if} \quad \phi = 0 \quad \text{and} \quad \dot{f} > 0 \ , \\ 0 & \text{otherwise} \ . \end{cases} \qquad (6.286)$$

Advancement of damage only occurs for the case of loading, and the initial condition for α is zero. The evolution equation (6.286), which clearly shows the discontinuous property of this damage model, corresponds to those given in GURTIN and FRANCIS [1981b], SIMO [1987] and MIEHE [1995a].

Coupled representation of the elasticity tensor.   According to relations (6.153) and (6.154) we derive the symmetric fourth-order elasticity tensor $\mathbb{C}$ in the *material description*. Starting with the constitutive equation (6.277), we have

$$\dot{\mathbf{S}} = (1 - \zeta)2\frac{\partial \mathbf{S}_0}{\partial \mathbf{C}} : \frac{1}{2}\dot{\mathbf{C}} - \mathbf{S}_0\dot{\zeta} \ , \qquad (6.287)$$

where the rate of damage $\dot{\zeta}$ takes on the form

$$\dot{\zeta} = \zeta'(\alpha)\dot{\alpha} \qquad \text{with} \qquad \zeta'(\alpha) = \frac{\partial \zeta}{\partial \alpha} \ . \qquad (6.288)$$

Substituting (6.288) into (6.287), we obtain, using (6.286), the evolution of the stress tensor in the form

$$\dot{\mathbf{S}} = \begin{cases} [(1 - \zeta)\mathbb{C}_0 - \zeta'(\alpha)\mathbf{S}_0 \otimes \mathbf{S}_0] : \dfrac{\dot{\mathbf{C}}}{2} & \text{if} \quad \phi = 0 \quad \text{and} \quad \dot{f} > 0 \ , \\ (1 - \zeta)\mathbb{C}_0 : \dfrac{\dot{\mathbf{C}}}{2} & \text{otherwise} \ , \end{cases} \qquad (6.289)$$

with the **effective elasticity tensor** $\mathbb{C}_0$ in the *material description*. For an *undamaged material*, $\mathbb{C}_0$ is defined to be

$$\mathbb{C}_0 = 2\frac{\partial \mathbf{S}_0(\mathbf{C})}{\partial \mathbf{C}} \ . \qquad (6.290)$$

By comparing $(6.289)_1$ with (6.153) we find that the terms within the bracket characterize the elasticity tensor $\mathbb{C}$ in the material description. The second term in the rate equation $(6.289)_1$ governs the damage that causes stress softening of the material.

Damage model in decoupled material description.   To complete our considerations of finite strain elasticity with isotropic damage in the sense of decoupled volumetric-isochoric response (introduced due to a multiplicative split of the deformation gradient (6.79)), we postulate finally a *decoupled* representation of $\Psi = \Psi(\mathbf{C}, \zeta)$ in accordance with (6.85) in the form

$$\Psi(\mathbf{C}, \zeta) = \Psi_{\mathrm{vol}}(J) + (1 - \zeta)\Psi_{0\,\mathrm{iso}}(\overline{\mathbf{C}}) \ . \tag{6.291}$$

Here, Ψ_{vol} is a strictly convex function (with the minimum at $J = 1$) which describes the *volumetric* elastic response. The second function $\Psi_{0\,\mathrm{iso}}$ denotes the **isochoric effective strain energy** of the *undamaged* material, which describes the *isochoric* elastic response. Hence, the damage phenomenon is assumed to affect only the isochoric part of the deformation, as proposed by, for example, SIMO [1987]. We require that $\Psi_{\mathrm{vol}}(J) = 0$ and $\Psi_{0\,\mathrm{iso}}(\overline{\mathbf{C}}) = 0$ hold *if and only if* $J = 1$ and $\overline{\mathbf{C}} = \mathbf{I}$, respectively.

Consider the structure (6.291), the purely volumetric contribution to the stress and the elasticity tensor are presented by eqs. (6.89) and (6.166), respectively. The *isochoric* contribution to the stress is, by analogy with (6.277), given by

$$\mathbf{S}_{\mathrm{iso}} = (1 - \zeta)\mathbf{S}_{0\,\mathrm{iso}} \qquad \text{with} \qquad \mathbf{S}_{0\,\mathrm{iso}} = 2\frac{\partial \Psi_{0\,\mathrm{iso}}(\overline{\mathbf{C}})}{\partial \mathbf{C}} \ . \tag{6.292}$$

The isochoric contribution to the elasticity tensor includes damage and is, by analogy with (6.289), given by

$$\dot{\mathbf{S}}_{\mathrm{iso}} = \begin{cases} [(1 - \zeta)\mathbb{C}_{0\,\mathrm{iso}} - \zeta'(\alpha)\mathbf{S}_{0\,\mathrm{iso}} \otimes \mathbf{S}_{0\,\mathrm{iso}}] : \dfrac{\dot{\mathbf{C}}}{2} & \text{if} \quad \phi = 0 \quad \text{and} \quad \dot{f} > 0 \ , \\[2mm] (1 - \zeta)\mathbb{C}_{0\,\mathrm{iso}} : \dfrac{\dot{\mathbf{C}}}{2} & \text{otherwise} \ , \end{cases} \tag{6.293}$$

with the isochoric part $\mathbb{C}_{0\,\mathrm{iso}}$ of the *effective elasticity tensor* in the material description. For an *undamaged material*, $\mathbb{C}_{0\,\mathrm{iso}}$ is defined to be

$$\mathbb{C}_{0\,\mathrm{iso}} = 2\frac{\partial \mathbf{S}_{0\,\mathrm{iso}}}{\partial \mathbf{C}} \ . \tag{6.294}$$

Explicit forms of $(6.292)_2$ and (6.294) are given by (6.90) and (6.168), respectively.

By analogy with eq. $(6.278)_2$ the thermodynamic force f has here the meaning of the isochoric effective strain energy $\Psi_{0\,\mathrm{iso}}(\overline{\mathbf{C}})$ of the undamaged material. Within the decoupled framework of volumetric-isochoric response, eqs. $(6.278)_2$ and $(6.280)_2$ take on the forms

$$f = \Psi_{0\,\mathrm{iso}}(\overline{\mathbf{C}}) \geq 0 \qquad \text{and} \qquad \dot{f} = \mathbf{S}_{0\,\mathrm{iso}} : \frac{\dot{\mathbf{C}}}{2} \ . \tag{6.295}$$

The applicability of the constitutive damage model thus described is limited to sufficiently slow processes (viscous effects are not considered). However, the damage model may easily be combined with the viscoelastic model as proposed in the last section. A suitable *decoupled* free energy for characterizing finite-strain viscoelastic damage mechanisms might be given, with reference to (6.236) and (6.291), as

$$\Psi(\mathbf{C}, \zeta, \boldsymbol{\Gamma}_1, \ldots, \boldsymbol{\Gamma}_m) = \Psi_{\mathrm{vol}}^{\infty}(J) + (1-\zeta)[\Psi_{0\,\mathrm{iso}}^{\infty}(\overline{\mathbf{C}}) + \sum_{\alpha=1}^{m} \Upsilon_{\alpha 0}(\overline{\mathbf{C}}, \boldsymbol{\Gamma}_\alpha)].$$

Here, $\Psi_{0\,\mathrm{iso}}^{\infty}(\overline{\mathbf{C}})$ and $\sum_{\alpha=1}^{m} \Upsilon_{\alpha 0}(\overline{\mathbf{C}}, \boldsymbol{\Gamma}_\alpha)$ denote, respectively, the strain energy and the configurational free energy (per unit reference volume) for the hyperelastic *undamaged* material. Both functions are associated with the isochoric response.

EXERCISES

1. *Pure shear with isotropic damage.* Consider a thin sheet of (incompressible) hyperelastic material which is subjected to a homogeneous pure shear deformation with the kinematic relation $\lambda_1 = \lambda, \lambda_2 = 1, \lambda_3 = 1/\lambda$ (compare with Exercise 1(b) on p. 226). The stress state of this mode of deformation is characterized by σ_1, σ_2 and $\sigma_3 = 0$ (recall eqs. (6.76), (6.77)). The material is supposed to undergo stress softening of Mullins type.

   (a) Based on the isotropic damage model introduced, compute the loading path up to $\lambda = 2$ followed by the unloading path back to $\lambda = 1$. Apply the constitutive particularization of the phenomenological damage variable $\zeta = \zeta(\alpha)$, as given in eq. (6.281), and use the strain-energy function of the Mooney-Rivlin form, i.e. $\Psi = c_1(I_1 - 3) + c_2(I_2 - 3)$, with the ratio of the material constants $c_1/c_2 = 7$.

   (b) Plot the loading and unloading path in the form of the two functions $\sigma_1 = \sigma_1(\lambda)$ and $\sigma_2 = \sigma_2(\lambda)$ with the Mooney-Rivlin parameters $c_1 = 0.4375\mu$, $c_2 = 0.0625\mu$ and the shear modulus in the reference configuartion, i.e. $\mu = 4.225 \cdot 10^5 \mathrm{N/m}^2$. In addition, take $\zeta_\infty = 0.8$ for the maximum damage and $\iota = 0.3 \cdot 10^6 \mathrm{N/m}^2$ for the damage saturation parameter.

2. *Equibiaxial deformation with isotropic damage.* Consider an equibiaxial deformation of an incompressible material, which may be modeled by the strain energy due to Ogden. Recall the stress strain relation derived in Example 6.6 on p. 239, i.e. $\sigma = \sum_{p=1}^{N} \mu_p(\lambda^{\alpha_p} - \lambda^{-2\alpha_p})$. During a load cycle $\lambda = 1 \to 3 \to 1 \to 5 \to 1$ the material accumulates damage of Mullins type.

   Plot the load cycle $\sigma = \sigma(\lambda)$ with the typical values of the constants $\alpha_p, \mu_p, p = 1, \ldots, 3$, for Ogden's model given by (6.121). Assume the maximum damage $\zeta_\infty = 0.8$ and the damage saturation parameter $\iota = 1.0 \cdot 10^6 \mathrm{N/m}^2$.

7 Thermodynamics of Materials

Thermodynamics is the science of energy, which studies processes in systems outside the thermodynamic equilibrium state. The term 'thermodynamics' comes from the Greek words $\theta\acute{\epsilon}\rho\mu\eta$ and $\delta\acute{\upsilon}\nu\alpha\mu\iota\varsigma$ meaning 'heat' and 'force' (or 'power'), respectively. Thermodynamics has long been a fundamental part of engineering. It constitutes a concept of great generality which is based on a few, simple hypotheses. Today the name 'thermodynamics' is interpreted as including all aspects of *energy*.

There are two principle ways of dealing with thermodynamics: long ago it was recognized that in the real world physical objects are compositions of molecules which are formed by atoms and even smaller subatomic particles. The special field in which the laws of classical mechanics (or quantum mechanics) are applied to large groups of individual particles (molecules, atoms) is called **statistical thermodynamics**. This approach investigates the correlation between the average behavior of particles and the macroscopic properties of a system on a statistical basis. The traditional classical, or phenomenological, approach to the study of thermodynamics, in which the molecular structure of a physical object is disregarded (the object is considered as continuous matter with no microscopic holes) is called **classical thermodynamics** or **(phenomenological) continuum thermodynamics**. It may be viewed as a unified field theory of *mechanics* and *thermodynamics* in which all thermodynamic state variables depend on position and time.

The essential feature of continuum thermodynamics is the derivation of *constitutive equations* (for the *stress tensor*, the *entropy*, the *heat flux vector*) from the basic physical principles of thermodynamics representing the individual (mechanical and thermodynamic) material properties of matter. However, for the formulation of constitutive equations we have several possible choices of independent and dependent variables. In this chapter the list of independent variables is supplemented by non-mechanical variables such as *temperature*, *entropy* and their gradients. We shall combine the (isothermal) constitutive theory of *finite (visco)elasticity*, as introduced in Chapter 6, with the theory of *heat conduction* under *transient* conditions. Solutions of the resulting **coupled thermomechanical problem** are able to describe the interaction between the *mechanical field* and the *thermal field*. In this chapter we study the thermodynamics of continuous media and, in particular, two different classes of constitutive models within the nonlinear constitutive theory of **finite thermoelasticity** and **finite thermoviscoelasticity**.

The only materials undergoing finite strains and temperature changes relative to an equilibrium state are biological soft tissues and rubber-like materials. As known from statistical thermodynamics of rubber elasticity, extended rubber chains tend to return to a less-ordered curled up-state which is characterized by a higher conformation entropy. The thermomechanical behavior of solid polymers is almost entirely based on an entropy concept. Therefore, in this chapter we start out with the aim of reviewing the crucial difference between rubber and metal within a thermodynamic context. In order to describe the three-dimensional network of rubber by means of the Helmholtz free-energy function, some insights in the statistical thermodynamics of rubber and the (molecular) network theory are presented briefly. We restrict attention to the *Gaussian* statistical theory, which enables us to characterize the thermoelastic behavior of a (molecular) network within small strains.

Within this statistical context, the neo-Hookean model, as derived in Section 6.5, is motivated. In the subsequent sections, quite independently of the network theory, we follow an approach to a macroscopic continuum formulation of *thermoelastic* and *thermoviscoelastic materials* by making use of continuum particles. We introduce a constitutive model for the thermoelastic behavior of materials and present a thermodynamic extension of the classical strain-energy function originally proposed by *Ogden*. The material model is set up in order to reproduce the realistic physical stress-strain-temperature response of rubber-like materials.

Moreover, a study of one-dimensional problems of finite thermoelasticity is presented. Distinctive attention is paid to the so-called *thermoelastic inversion phenomena*, a remarkable property of rubber-like materials.

The last section in this chapter is concerned with the study of thermodynamics in terms of internal variables. A constitutive model for highly deformable media that accounts for several thermomechanical coupling effects is examined. The proposed phenomenological model is capable of describing relaxation and/or creep phenomena within the thermomechanical regime.

7.1   Physical Preliminaries

As a basis for our next studies we present an introductory review of some of the interesting physical aspects of the thermoelastic behavior of amorphous solid polymers, that are chemically cross-linked, for example, by sulphur bridges.

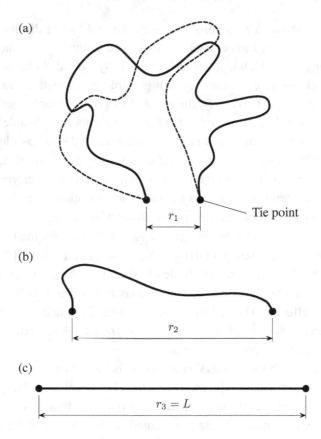

(a)

r_1

Tie point

(b)

r_2

(c)

$r_3 = L$

Figure 7.1 Single polymer chain lying between two tie points with various distances r. Two possible conformations are shown in (a). The number of possible conformations decreases with increasing end-to-end distance (b). Fully extended chain showing the limiting case with only one possible conformation (c).

Statistical concept.   Based on several physical techniques we know that amorphous polymers are composed of bundles of long-chain molecules (which may be imagined as strings) having a high degree of flexibility. Figure 7.1 shows a model for a single polymer chain lying between successive points of cross-linkage, which we call **tie points**. The single polymer chain forms a typical segment in the coherent and three-dimensional network of rubber. This model is due to GUTH and MARK [1935] and KUHN [1938, 1946]. The distance between the tie points of the chain molecule, denoted by r, we call subsequently **end-to-end distance** (or **separation**). The distance r is a parameter that characterizes a molecular **conformation**. The name conformation comes from chemistry and refers to different shapes (arrangements) of a chain molecule. The most powerful physical technique now available for determining conformations of chain molecules is *small-angle neutron scattering* (for more details see, for example, SPERLING [1992, Section 5.2]).

Figure 7.1(a) shows a polymer chain with the end-to-end distance r_1 of the tie points, which are assumed to be fixed in space. The distance r_1 is much smaller than the **contour length** L which is the length of the fully extended chain. Consequently, the chain may take on an enormous range of possible conformations, two of which are shown in Figure 7.1(a). Obviously, the number of possible conformations decreases for a larger end-to-end distance, and in the limit the number of possible conformations diminishes to only one if the chain is in its most extended state, as illustrated in Figure 7.1(c). Then the value of the end-to-end distance reaches its maximum, i.e. r_3, and equals the contour length L, and the chain is straight. The most crumpled conformation occurs when r tends to zero, and the tie points coincide. Clearly the end-to-end distance r of a chain characterizes the molecular conformation.

The *statistical theory* of rubber elasticity (see, for example, the notable works by TRELOAR [1943a, b], JAMES and GUTH [1943, 1949] and FLORY [1953, Section XI-3, and references therein]), which is basically set up on these concepts makes use of the idea to express the number of conformations that a chain molecule can assume as an '**entropic effect**'. The chains occur in randomly coiled conformations in the unstretched state, as seen in Figure 7.1(a), and as the chains are extended the number of conformations and the entropy decrease.

We proceed now to model this characteristic behavior and to analyze the conformations of a chain molecule. However, just for clarity, consider first the problem in *one dimension* and project the conformation on one coordinate axis, say the x_1-axis (the chain may be imagined as being constrained artificially so that the tie points lie on the x_1-axis). The conformations of the individual chains are distributed in a random manner. The **probability** $p(x_1)dx_1$ that the end-to-end distance of a chain lies in the interval between x_1 and $x_1 + dx_1$ is expressed by the **Gaussian distribution function**

$$p(x_1)dx_1 = \frac{b}{\pi^{1/2}}\exp(-b^2x_1^2)dx_1 \quad , \tag{7.1}$$

where $p(x_1)$ is the **probability density** (per unit length) and b is a parameter of the model. This parameter is a measure of a representative length, as we see later in Example 7.1, p. 313. For an explicit derivation of the Gaussian distribution function (7.1) for a chain in one dimension the interested reader is referred to the classical book by FLORY [1953, Appendix A of Chapter X].

The Gaussian function is a bell-shaped curve, which provides numerous applications in engineering practice and statistics. The most probable value of x_1, i.e. the maximum of the Gaussian function, may be found by differentiating eq. (7.1) with respect to x_1 and occurs at $x_1 = 0$. The probability decreases monotonically as x_1^2 increases. Note that the entropy may be interpreted as a quantitative measure of probability (microscopic randomness and disorder) by using a fundamental finding which is due to *Boltzmann* and *Planck*. This will be made clear in Section 7.2.

For a more comprehensive survey of the general concepts of statistical mechanics, a terminology which was introduced by *Gibbs*, the reader may be referred to the books by FLORY [1969], WEINER [1983], CALLEN [1985, Part II], MARK and ERMAN [1988] and TRELOAR [2005]; see also the review paper by GUTH [1966].

Rubber versus 'hard' solids.   One of the remarkable differences between *rubber* and *'hard' solids*, such as *metals, glasses, ceramics, crystals*, etc. lies in the effect of temperature. The following crucial physical properties, explored and quantified in a set of experiments by *Joule*, exhibit the distinctive behavior of rubber (see JOULE [1859, p. 105]):

(i) a piece of vulcanized rubber subjected to a weight produces a slight *cooling effect* in the very low strain range and changes to a *heating effect* by increasing the weight, and

(ii) rubber will *contract* its length under tension when its temperature is raised (it is not very known that healthy human and animal arteries also *shrink* upon heating, a phenomenon that was pointed out in the early work of ROY [1880–1882] for the first time).

These results are based on a (previous) simple qualitative observation by GOUGH [1805] that a rapidly stretched *rubber band* (adiabatic straining) brought into slight contact with the lips as a sensitive detector feels warm. On the other hand a stretched rubber band in thermodynamic equilibrium feels cold after releasing the tension. This (thermoelastic) coupling phenomenon entered the literature as the so-called **Gough-Joule effect**. Note that the behavior of a metallic spring is in striking contrast to a rubber band. A metallic spring cools continuously on elastic stretching. This is the opposite behavior of a rubber band which warms on stretching, a remarkable experimental observation (see Box 1).

The properties of rubber are well-known above the **glass transition temperature** (see, for example, CYR [1988], WARD and HADLEY [1993]), and are characterized by

(i) extremely long-range extensibility, typically 300–500% extension for vulcanized natural rubber (i.e., without carbon black or other reinforcing fillers) and even more for synthetic rubber (generating low mechanical stresses), accompanied by

(ii) full recovery to the initial dimensions without mechanical and thermal hysteresis within the lower temperature domain of the 'rubbery' region.

Below the glass transition temperature, flexibility and mobility of the chains are so reduced that rubber behaves like a brittle glass (the material consists of rigid crystals). The glass transition temperature, for example, for *natural rubber* and *butyl rubber* is $-73°C$ (see MARK and ERMAN [1988, Chapter 2]).

Based on experimental observations by ANTHONY et al. [1942] (see also TRELOAR [2005, Chapters 2, 13]) the retractive force in a *real rubber* is approximately 90% based on entropy (for additional information the reader is referred to the papers by SHEN and CROUCHER [1975] and CHADWICK and CREASY [1984]).

Polymers	Metals
Entropic elasticity	Energetic elasticity
⇒ Total stress is entirely caused by a change in entropy with deformation.	⇒ Entropy does not change with deformation at all.
⇒ Internal energy does not change with deformation at all.	⇒ Total stress is internal energy driven, which changes rapidly with deformation.
⇒ Elasticity arises through entropic straightening of a polymer chain, followed by recoiling into a conformation of maximum entropy, see Figure 7.1 (iso-volumetric phenomena).	⇒ Elasticity arises through energetic increases due to distance changes between atoms against atomic attractive forces, followed by removing the interatomic forces back to its initial dimensions (in general, substantial volume changes accompany deformation).
'Polymers are entropic'	'Metals are energetic'
⇒ A piece of rubber warms on stretching.	⇒ An elastic metallic spring cools on stretching.
⇒ A rubber band under constant tensile force substantially will shrink upon heating and expand upon cooling.	⇒ An elastic metallic spring under constant tensile force will expand upon heating and shrink upon cooling.
From a thermodynamic point of view, the work done in elongating a rubber band is unlike the work produced by stretching a coiled elastic metallic spring.	

Box 1   Composition of the concepts in polymers (ideal rubbers) and 'hard' solids (metals, glasses, ceramics, crystals, etc.) in the elastic range.

For an *ideal rubber*, for which by definition one property is incompressibility (the volume remains constant (locally and globally) during a mechanical process), the retractive force is, however, purely determined by changes in entropy and the internal energy does not change with deformation at all, i.e. a significant characteristic of rubber elasticity. We term this type of rubber-like material **'entropic elastic'**.

However, elasticity of metals, glasses, ceramics or crystals arises basically through removing atoms from their equilibrium positions, accompanied by rapid internal energy changes, while the entropy does not change at all (see, for example, HILL [1975] and ERICKSEN [1977]). These materials with a regular atomic structure and usually with high strength are typically called **'energetic elastic'**. They exhibit, in general, substantial volume changes on deformations and stand in sharp contrast to rubber-like solids. For a general overview, the corresponding concepts in *polymers* (ideal rubber) and *'hard' solids* (metals) are summarized in Box 1.

Natural rubber (cis-polyisoprene) does not recover completely. In order to achieve dimensional stability and deformations which are completely reversible in the 'rubbery' state, vulcanization of the rubber (typically done for commercial products) is required. Within a vulcanization process polymer chains are chemically connected to other chains at different locations to produce a cross-linked monolithic three-dimensional network (see SPERLING [1992]). On the other hand crystallization occurring in the highly stretched rubber is influenced by vulcanization. Crystallization is lower for a higher concentration of sulfur used in the vulcanization process (see TRELOAR [2005, pp. 16-23]).

It is important to note that the material properties of highly stretched crystalline rubber become anisotropic and that the heat of crystallization is much larger than produced by the mentioned thermoelastic *Gough-Joule effect*. Moreover, due to frictional losses during the deformation process an additional heat is generated. In addition, it is mentioned that real networks contain defects (see, for example, MULLINS and THOMAS [1960] and SCANLAN [1960]). Crystallinity effects and network imperfections lie outside the scope of this text.

7.2   Thermoelasticity of Macroscopic Networks

A typical vulcanized rubber may be considered as the assembly of long-chain molecules. Each chain is attached at both ends and thus produces one giant molecule, which we call the **(molecular) network**. From the irregular three-dimensional network we draw conclusions regarding its material properties. In the following, motivated by statistical thermodynamics, we describe the material properties of rubber through the Helmholtz free-energy function Ψ.

The freely jointed chain in three dimensions.   We assume that the molecular network contains N chains per unit volume which is often referred to as the **network density**.

We consider first a representative single polymer chain in space **detached** from the network, which means that the chain is taken *out* of the network. Our aim is to compute the entropy of this chain and to study its thermodynamic behavior. The chain with contour length L is cross-linked at the tie points O and P of the network (see Figure 7.2). One end of the chain is attached to the fixed origin O of the x_1, x_2, x_3-coordinate system. The other end is given by the **end-to-end vector** $\mathbf{r} = x_1 \mathbf{e}_1 + x_2 \mathbf{e}_2 + x_3 \mathbf{e}_3$, pointing to P and characterizing a certain number of different shapes.

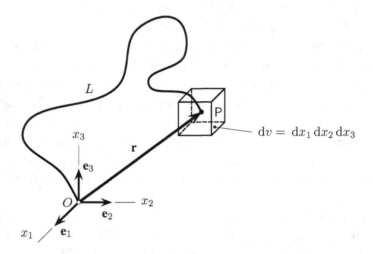

Figure 7.2 A representative single polymer chain OP detached from the network.

Further, we assume a so-called **Gaussian chain** which is defined so that the distance between the tie points (chain ends) O and P, i.e. $r = |\mathbf{r}|$, is considerably less than the contour length L, i.e.

$$r \ll L \ . \tag{7.2}$$

Hence, we follow the context of the **Gaussian statistical theory** of elasticity which is valid for problems where only small strains are involved.

The contour length L of a single chain is commonly considered to be an assembly of n (statistical) segments joined together, each of length l so that $L = nl$. We suppose that there is no correlation between the directions of the successive segments. Based on this simple mechanical model we may determine the so-called **mean value** $\bar{r}$ of the end-to-end distance r for this **freely jointed chain** observed at one instant of time. This is given by

$$\overline{r^2} = nl^2 \quad , \tag{7.3}$$

where the mean-square value $\overline{r^2}$ denotes the average over r^2. For a more detailed explanation the interested reader is referred to the textbooks by, for example, MCCRUM et al. [1997, Section 2.8] or WARD and HADLEY [1993, Section 3.3] and left to study Exercise 1 on p. 320.

We wish to calculate the probability that the tie point P lies within the infinitesimal volume element of size $dv = dx_1 dx_2 dx_3$ at P (see Figure 7.2). By analogy with eq. (7.1) we may introduce the probability densities $p(x_2)$ and $p(x_3)$ which are associated with the x_2-axis and the x_3-axis, respectively. In addition, it is possible to show that $p(x_1)$ depends only on x_1 ($p(x_2)$ only on x_2 and $p(x_3)$ only on x_3) provided that n is large and x_1, x_2, x_3 are much smaller than the contour length L of the chain.

Generalizing the relation (7.1) to *three dimensions* we may find the probability density $p(x_1, x_2, x_3)$, now per unit volume, that the tie point P of the freely jointed chain occurs in the infinitesimal volume element dv (see Figure 7.2). Under the restrictions considered it is the product of the independent probability densities according to

$$p(x_1, x_2, x_3)dx_1 dx_2 dx_3 = p(x_1)p(x_2)p(x_3)dx_1 dx_2 dx_3$$

$$= \frac{b^3}{\pi^{3/2}}\exp(-b^2 r^2)dx_1 dx_2 dx_3 \quad , \tag{7.4}$$

where $r^2 = x_1^2 + x_2^2 + x_3^2$ is the square of the distance between the tie points O and P for this detached Gaussian chain and parameter b denotes a measure of a representative length. As for the one-dimensional case the maximum of the Gaussian function (7.4) occurs at $r = 0$. The Gaussian distribution function, as given in (7.4), represents a sufficiently accurate solution to the stochastic problem in question.

EXAMPLE 7.1   In this example consider a freely jointed Gaussian chain with tie points O and P and calculate all possible conformations of the chain at any given value of the end-to-end distance $r = |\mathbf{r}|$, *irrespective of direction*. Furthermore, show that the measure of the representative length b controls the mean-square value $\overline{r^2}$ according to

$$b^2 = \frac{3}{2\overline{r^2}} \tag{7.5}$$

(compare also with FLORY [1956, Section X-1b]). The mean-square value $\overline{r^2}$ is defined to be $\int_0^\infty r^2 p(r)dv / \int_0^\infty p(r)dv$, with the probability density p given by eq. (7.4)$_2$.

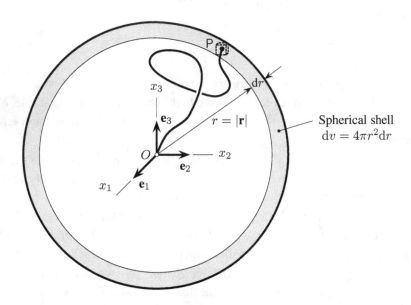

Figure 7.3 Spherical shell which defines all possible conformations of a representative Gaussian chain OP irrespective of direction.

Solution.   The restriction to a particular direction in space as considered in the previous analysis is not appropriate anymore, so we take into account all directions of the vector $\mathbf{r}$ equally. Doing so, the tie point P does not move within the infinitesimal rectangular block, rather within an infinitesimal volume dv of a spherical shell which is

$$dv = 4\pi r^2 dr \quad . \tag{7.6}$$

The infinitesimal volume is defined between the inner radius r and the outer radius $r + dr$ from the other tie point, which is fixed at the origin O of the coordinate system (see Figure 7.3). The required probability $p(r)dr$ that the chain length lies in the interval between r and $r + dr$, is, by means of $(7.4)_2$,

$$p(r)dv = p(r)4\pi r^2 dr = \frac{b^3}{\pi^{3/2}}4\pi r^2 \exp(-b^2 r^2)dr \quad . \tag{7.7}$$

The function (7.7) represents the Gaussian distribution of the distance r for a set of free chains. In the r-distribution function, no restriction on the direction of the vector $\mathbf{r}$ is involved. The maximum of function (7.7), i.e. the most probable value of r, is obtained by differentiation of $p(r)$ with respect to r. This maximum occurs at $r = 1/b$.

In order to compute the important mean-square value of r, i.e. $\overline{r^2}$, we find after some manipulations the analytical solution

$$\overline{r^2} = \frac{\int\limits_0^\infty r^2 p(r) 4\pi r^2 dr}{\int\limits_0^\infty p(r) 4\pi r^2 dr} = \frac{\int\limits_0^\infty r^4 \exp(-b^2 r^2) dr}{\int\limits_0^\infty r^2 \exp(-b^2 r^2) dr}$$

$$= \frac{3\sqrt{\pi}/8b^5}{\sqrt{\pi}/4b^3} = \frac{3}{2b^2} \quad, \tag{7.8}$$

which proves (7.5). We know from eq. (7.3) that the mean-square value $\overline{r^2}$ depends on the (statistical) segments n and their lengths l (recall that nl is the contour length L of the chains). Hence, we conclude that b is a measure of a representative length.   ∎

The entropy of a single chain.   On the basis of the statistical concept of thermo-dynamics we now determine the entropy η_i of a representative single Gaussian chain i whose ends are located at specified points in space. The chain is assumed to be taken out (detached) from the network. We apply **Boltzmann's equation** (or what *Einstein* called the **Boltzmann principle**) relating thermodynamic entropy and the probability of a thermodynamic state (molecular conformations). Hence, from the statistical point of view, the entropy η_i of a single chain is defined to be proportional to the logarithm of the probability density $p(r)$ and varies with the end-to-end distance r according to Boltzmann's equation, which is given in the form

$$\eta_i = a + k \ln p(r) \quad, \tag{7.9}$$

where a denotes a constant entropy with respect to a reference level, which need not be specified here in more detail. The universal constant of proportionality $k = 1.38 \cdot 10^{-23} \text{Nm/K}$ denotes **Boltzmann's constant**.

The famous relation between entropy and probability was published by the Aus-trian physicist *Boltzmann* in 1877 (at this time he worked as a professor for physics in Graz). The term 'Boltzmann's constant' for k and the mathematical formulation of the principle in the form of $S = k \log W$ due to to *Planck*. This form was carved on *Boltzmann's* gravestone at the 'Zentralfriedhof' in Vienna in 1933.

Substituting $(7.7)_2$ into (7.9) we obtain finally

$$\eta_i = c - kb^2 r^2 \quad, \tag{7.10}$$

where the constant $c = a + k \ln(b^3/\pi^{3/2})$ incorporates the constant entropy a. The mea-sure of the representative length b for this Gaussian chain detached from the network is given by

$$b^2 = \frac{3}{2\overline{r^2}_{\text{out}}} \tag{7.11}$$

(see eq. (7.5)), where $\overline{r^2}_{\text{out}}$ denotes the mean-square value of the end-to-end distance of this un-cross-linked free chain *out* of the network. It is an *intrinsic* property of the chain molecule and is independent of volume changes. As can be seen from expression (7.10) the entropy tends to its largest value for $r \to 0$, as expected (see the considerations in the previous section).

The elasticity of a molecular network.   In order to determine the elasticity of a molecular network, we choose, without loss of generality, the case of a homogeneous deformation state of a rubber block given by the principal stretches λ_a, $a = 1, 2, 3$. Further, we introduce two crucial assumptions:

  (i) there is no change in volume on deformation, the material is idealized as totally incompressible *(incompressibility assumption)*, i.e. $\lambda_1 \lambda_2 \lambda_3 = 1$,

 (ii) changes in the length and orientation of lines marked on chains in a network are identical to changes in lines marked on the corresponding dimensions of the macroscopic rubber sample *(affine motion assumption)*.

   Thus, we refer to Figure 7.4, which shows the affine motion of a representative Gaussian chain with one end at the origin.

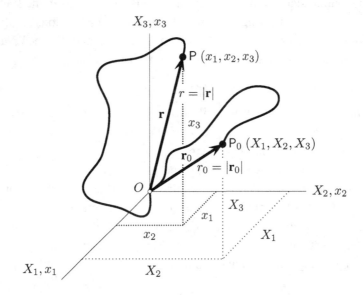

Figure 7.4 Undeformed and deformed configurations of a representative Gaussian chain.

   In the undeformed configuration of a network the end-to-end distance of the chain OP_0 is characterized by the vector $\mathbf{r}_0$, with material coordinates X_A, $A = 1, 2, 3$, and length $r_0 = |\mathbf{r}_0|$. Since we consider the affine motion assumption all N chains deform like the representative chain shown in Figure 7.4 and vector $\mathbf{r}_0$ becomes $\mathbf{r}$ after deformation, with spatial coordinates x_a, $a = 1, 2, 3$, and length $r = |\mathbf{r}|$. In the course of the motion the tie point P_0 is displaced to P. Because of the affine motion assumption we may write

$$x_1 = \lambda_1 X_1 \ , \qquad x_2 = \lambda_2 X_2 \ , \qquad x_3 = \lambda_3 X_3 \ . \qquad (7.12)$$

The change in the end-to-end distance of the chain due to the deformation produces a difference of entropy between the state before deformation (with $\lambda_a = 1$, $a = 1, 2, 3$), i.e. $c - kb^2(X_1^2 + X_2^2 + X_3^2)$ (see eq. (7.10)), and the state after deformation, i.e. $c - kb^2(\lambda_1^2 X_1^2 + \lambda_2^2 X_2^2 + \lambda_3^2 X_3^2)$. Hence, the entropy change in the chain caused by the deformation of that individual chain i, denoted by $\Delta\eta_i$, is therefore

$$\Delta\eta_i = [c - kb^2(\lambda_1^2 X_1^2 + \lambda_2^2 X_2^2 + \lambda_3^2 X_3^2)] - [c - kb^2(X_1^2 + X_2^2 + X_3^2)]$$

$$= -kb^2[(\lambda_1^2 - 1)X_1^2 + (\lambda_2^2 - 1)X_2^2 + (\lambda_3^2 - 1)X_3^2] \ . \qquad (7.13)$$

The constant c has no physical relevance since we are only concerned with the change of entropy.

In order to reform a network all the detached chains are transferred back into the rubber specimen and cross-linked. Of course, the end-to-end distances of (detached) chains *out* of the network are not the same as the end-to-end distances of (cross-linked) chains *in* the network. Therefore, we introduce the mean-square value of the end-to-end distance r for the whole assembly of chains *in* the specimen, denoted by $\overline{r^2}_{\mathrm{in}}$. In contrast to the mean-square value $\overline{r^2}_{\mathrm{out}}$, $\overline{r^2}_{\mathrm{in}}$ is not an intrinsic property of the chain molecule. Since some constraints must be applied to the detached chains in order to reform a network, the mean-square value $\overline{r^2}_{\mathrm{in}}$ differs from $\overline{r^2}_{\mathrm{out}}$. In particular, the value $\overline{r^2}_{\mathrm{in}}$ depends on the volume of the rubber and changes by heating (or cooling) (compare also with MCCRUM et al. [1997, Chapter 3]).

Our next aim is to compute the entropy change of a network of such chains generated by the macroscopic deformation state. It is the sum of the entropy changes of all N chains in a unit volume in the network, which we denote by $\Delta\eta$. Since we have assumed affine motion, all the chains have the same given intrinsic property b and imposed λ_a, $a = 1, 2, 3$. With eq. (7.13) we may write

$$\Delta\eta = \sum_{i=1}^{N} \Delta\eta_i$$

$$= -kb^2[(\lambda_1^2 - 1)\sum_{1}^{N} X_1^2 + (\lambda_2^2 - 1)\sum_{1}^{N} X_2^2 + (\lambda_3^2 - 1)\sum_{1}^{N} X_3^2] \ . \qquad (7.14)$$

We claim that

$$\sum_{1}^{N} r^2_{0\mathrm{in}} = N\overline{r^2}_{0\mathrm{in}} \qquad (7.15)$$

holds, where $\overline{r_{0\text{in}}^2}$ is the mean-square value of the end-to-end distance $r_{0\text{in}}$ of the assembled chains *in* the specimen in the undeformed state.

Since the vector $\mathbf{r}_0$ has no preferred direction in the undeformed state (which is isotropic), we may write $\sum_1^N X_1^2 = \sum_1^N X_2^2 = \sum_1^N X_3^2$. But since $\sum_1^N X_1^2 + \sum_1^N X_2^2 + \sum_1^N X_3^2 = \sum_1^N r_{0\text{in}}^2$ we deduce that $\sum_1^N X_1^2 = \sum_1^N X_2^2 = \sum_1^N X_3^2 = 1/3 \sum_1^N r_{0\text{in}}^2$, and finally, from (7.15)

$$\sum_1^N X_1^2 = \sum_1^N X_2^2 = \sum_1^N X_3^2 = \frac{1}{3} N \overline{r_{0\text{in}}^2} \quad . \tag{7.16}$$

Combining (7.16) and (7.11) with (7.14) we find the entropy change of the network independent of the parameter b, i.e.

$$\Delta\eta = -\frac{1}{2} N k \frac{\overline{r_{0\text{in}}^2}}{\overline{r^2}_{\text{out}}} (\lambda_1^2 + \lambda_2^2 + \lambda_3^2 - 3) \quad . \tag{7.17}$$

The term $\overline{r_{0\text{in}}^2}/\overline{r^2}_{\text{out}}$ accounts for the different end-to-end distances of chains in the network and detached from the network.

We have learnt in the previous section that for an ideal rubber the internal energy e does not change with deformation at all. Hence, from the Legendre transformation (4.152) it follows that for an isothermal process the change in the Helmholtz free-energy function Ψ is $\Delta\Psi = -\Theta\Delta\eta$. As a consequence of the Gaussian statistical theory of a molecular network, using the fundamental expression eq. (7.17), we find finally that

$$\Psi = \frac{1}{2} N k \Theta \frac{\overline{r_{0\text{in}}^2}}{\overline{r^2}_{\text{out}}} (\lambda_1^2 + \lambda_2^2 + \lambda_3^2 - 3) \quad . \tag{7.18}$$

According to (6.4) we have assumed that the free energy is zero in the undeformed configuration. This important result shows that, within the scope of the Gaussian statistical theory, the only quantities pertaining to the molecular network are the total number of chains N contained in the network (per unit volume) and $\overline{r_{0\text{in}}^2}/\overline{r^2}_{\text{out}}$.

On comparison with relation (6.128) it emerges that (7.18) represents the simple *neo-Hookean model* with the physical parameter μ, known as the shear modulus, which is proportional to the concentration of network chains N, given as

$$\mu = N k \Theta \frac{\overline{r_{0\text{in}}^2}}{\overline{r^2}_{\text{out}}} \tag{7.19}$$

(see, in addition, FLORY [1956] or TRELOAR [2005, p. 114]). A simple method for determining the number of chains N per unit volume is to measure the shear modulus μ for rubber.

For high chain extensions the end-to-end distance r is close to or equal to the contour length L. Therefore, condition (7.2) cannot be satisfied anymore and the Gaussian statistical theory becomes increasingly inadequate for the finite strain domain. In order to account for the finite extensibility of chains some significant refinements must be taken into account (SPERLING [1992, Section 9.10]). The more accurate **non-Gaussian statistical theory** is required (see, for example, MARK and ERMAN [1988, Chapter 13] and TRELOAR [2005, Chapter 6]). Within the non-Gaussian statistical theory the finite extensibility of chains is considered in the form of correction terms leading to a more realistic form of the distribution function which is valid over the whole range of r-values up to the maximum or fully extended length.

One example of such a refined theory is based on the **Langevin distribution function**. The exact treatment of the freely jointed chain is considered by KUHN and GRÜN [1942], JAMES and GUTH [1943], and summarized by FLORY [1953, Appendix B of Chapter X]. In this type of refined theory the Gaussian distribution function is included as a special case.

Other examples are phenomenologically motivated and based on mathematical arguments (see the material models introduced in Section 6.5).

EXAMPLE 7.2   At a given temperature Θ we consider a thermodynamic process in a closed system within some closed time interval $t \in [0, T]$, in which the values 0 and T denote the initial (reference) and the final time, respectively. Assume that the closed system is *thermally isolated* and *conservative*. During the thermodynamic process a unit cube of ideal (incompressible) rubber deforms homogeneously to a parallelepiped with sides of length λ_1, λ_2 and $\lambda_3 = (\lambda_1\lambda_2)^{-1}$, i.e. the principal stretches. The cross-linked network of the rubber cube consists of N Gaussian chains per unit volume.

Based on the Gaussian statistical theory find the heat generated (or destroyed) and the induced total entropy change of the rubber block due to the homogeneous deformation. Specify the problem with the values $\lambda_1 = 2$, $\lambda_2 = 3$, $\lambda_3 = 1/6$ and $N = 3.0 \cdot 10^{21} \mathrm{m}^{-3}$. The temperature is assumed to be $\Theta = 293.15\mathrm{K}\ (= 20°\mathrm{C})$, where K denotes the '*Kelvin temperature*' and °C the '*Celsius temperature*'. The term $\overline{r_{0\mathrm{in}}^2}/\overline{r^2}_{\mathrm{out}}$ is equal to 1.

Solution.   Since rubber is incompressible, the change in internal energy, i.e. $\dot{\mathcal{E}}$, during deformation is zero. Hence, the first law of thermodynamics (4.122) reads

$$\dot{\mathcal{E}}(t) = \mathcal{P}_{\mathrm{int}}(t) + \mathcal{Q}(t) = 0 \ . \tag{7.20}$$

The expressions for the *thermal power* $\mathcal{Q}(t) = \int_{\Omega_0} R dV$ (the system is thermally isolated, i.e. thermal energy can not enter or leave the boundary ($Q_\mathrm{N} = 0$), no heat transfer) and the *stress power* $\mathcal{P}_{\mathrm{int}} = \mathrm{D}/\mathrm{D}t \int_{\Omega_0} \Psi dV$ (the system is conservative) are adopted from (4.118) and (4.116), respectively. After integration over time interval $t \in [0, T]$ we find from the last equation (7.20), by means of (7.18), the particularized first law of thermodynamics, i.e.

$$\int_{t=0}^{t=T} R \mathrm{d}t = -\Psi = -\frac{1}{2}Nk\Theta\frac{\overline{r^2}_{0\text{in}}}{r^2_{\text{out}}}(\lambda_1^2 + \lambda_2^2 + \lambda_3^2 - 3) \ . \tag{7.21}$$

We used the fact that according to assumption (6.4) the strain-energy function Ψ for the unit cube of rubber vanishes in the reference configuration (normalization condition). The term $\int_{t=0}^{t=T} R \mathrm{d}t$ in relation (7.21) represents the heat per unit reference volume within the closed time interval $t \in [0, T]$ (thermal work). All work done which appears as the strain energy is transformed to heat.

The total entropy change of the network $\Delta\eta$ induced by the thermodynamic process is, in accord with (7.17), given by

$$\Delta\eta = -\frac{\Psi}{\Theta} = -\frac{1}{2}Nk\frac{\overline{r^2}_{0\text{in}}}{r^2_{\text{out}}}(\lambda_1^2 + \lambda_2^2 + \lambda_3^2 - 3) \ . \tag{7.22}$$

By substituting the given values into (7.21) and (7.22) we find using Boltzmann's constant $k = 1.38 \cdot 10^{-23}\text{Nm/K}$ that $\int R \mathrm{d}t = -[3.0 \cdot 10^{21}\, 1.38 \cdot 10^{-23}\, 293.15\, (4 + 9 + 1/36 - 3)]/2 = -60.85\text{N/m}^2$. The negative sign means that energy in the form of heat is *destroyed* within the solid body. The entropy change gives $\Delta\eta = -60.85/293.15 = -0.208\text{N/m}^2\text{K}$ which shows clearly that entropy is decreasing as the rubber block is deformed. ∎

EXERCISES

1. A drunken man starts to walk on a flat field at a starting point O. He makes one step per second each step of length 0.5 m. The path of the walk of course meanders randomly (the man is drunken), which means that there is no correlation between the directions of successive steps.

   By applying relation (7.3) compute the average distance $\bar{r}$ from point O he has moved after three minutes.

2. A rubber band of initial cross-sectional area A_0 is applied to a mass m. At a certain temperature Θ the mass causes a 200% increase in length.

   Compute the number of chains N per unit volume for the assumption that the material is modeled as neo-Hookean and $\overline{r^2}_{0\text{in}} = \overline{r^2}_{\text{out}}$ (Boltzmann's constant $k = 1.38 \cdot 10^{-23}\text{Nm/K}$).

3. Two rubber bands, A and B, with identical material compositions and length l are tied together at their ends. Then the assembled band is stretched up to a total

length of $6l$ and fixed at this position. By assuming that the rubber band A is at temperature Θ_A, and the rubber band B at Θ_B, find the displacement of the knot at which the two bands are tied together.

7.3 Thermodynamic Potentials

To characterize continuous media within the context of thermodynamics we need to define *two* material functions, namely

(i) the **thermodynamic potential** characterizing *all* thermodynamic properties of a system, and

(ii) the **heat flux vector** describing heat transfer.

A thermodynamic potential is a function from which we may derive state variables characterizing a certain thermodynamic state of a system. In the following we define four common thermodynamic potentials. All of them are scalar-valued functions and assumed to be objective. In addition, the potentials are supposed to be at least twice differentiable with respect to all associated components.

For a supplementary account of the relevant topic see the classical work of TRUES-DELL and TOUPIN [1960]; see also the texts by, for example, MALVERN [1969], ZIEGLER [1983] and HAUPT [1993b].

Associated thermodynamic potentials.   One example of a thermodynamic potential is the uniquely defined *Helmholtz free-energy function* $\Psi = \Psi(\mathbf{F}, \Theta)$, measured per unit reference volume (in thermodynamics the Helmholtz free-energy function is frequently denoted by f or F). The value of the free energy is determined by the changes of two independent variables, i.e. the deformation gradient $\mathbf{F}$ and a non-mechanical variable given by the temperature Θ.

In the following we consider homogeneous materials, which means that the associated functions are independent of position in the medium. With the free energy which describes here non-isothermal thermoelastic processes, we may deduce directly physical expressions from the Clausius-Planck form of the second law of thermodynamics (4.153). For *all* admissible thermoelastic processes the identity $\mathcal{D}_{\text{int}} = \mathbf{P} : \dot{\mathbf{F}} - \dot{\Psi} - \eta\dot{\Theta} = 0$ holds, which means the internal dissipation $\mathcal{D}_{\text{int}}$ is zero. By applying the chain rule, time differentiation of the free energy $\Psi(\mathbf{F}, \Theta)$ gives the hypothetical change of the thermodynamic state. We obtain

$$\dot{\Psi} = \dot{\Psi}(\mathbf{F}, \Theta) = \mathbf{P} : \dot{\mathbf{F}} - \eta\dot{\Theta} = \left(\frac{\partial\Psi(\mathbf{F}, \Theta)}{\partial\mathbf{F}}\right)_\Theta : \dot{\mathbf{F}} + \left(\frac{\partial\Psi(\mathbf{F}, \Theta)}{\partial\Theta}\right)_\mathbf{F}\dot{\Theta} \ , \qquad (7.23)$$

which holds at every point of the continuum body and for all times. As usual, the subscripts in (7.23) indicate variables that are being held constant during the partial differentiation of Ψ. For convenience, in the following we will sometimes omit the subscripts. The coupled equation $(7.23)_3$ is known as the **Gibbs relation** for elastic solids (*Gibbs* postulated the equation only for the case of a fluid).

By comparing terms, we may evaluate physical expressions imposed on requirement (7.23) which must hold for any given $(\mathbf{F}, \Theta)$. Since $\dot{\mathbf{F}}$ and $\dot{\Theta}$ can be chosen arbitrarily

$$\mathbf{P} = \left(\frac{\partial \Psi(\mathbf{F}, \Theta)}{\partial \mathbf{F}} \right)_\Theta \qquad \text{and} \qquad \eta = - \left(\frac{\partial \Psi(\mathbf{F}, \Theta)}{\partial \Theta} \right)_\mathbf{F} , \qquad (7.24)$$

which are the general forms of constitutive equations for the first Piola-Kirchhoff stress $\mathbf{P}$ and the entropy η describing thermoelastic materials. Note that for the case of any *isothermal process* ($\Theta = $ const) the free-energy function Ψ is identified with the isothermal strain-energy function (compare with eq. (6.1)). Consequently, $\Psi(\mathbf{F}) = \Psi(\mathbf{F}, \Theta)|_{\Theta=\text{const}}$. By (7.24), the stress and the entropy are determined by the free energy Ψ, which has the status of a **potential** for the stress, the entropy and their respective conjugate thermodynamic variables.

From physical expressions (7.24) we deduce the *stress* and *entropy functions* depending on the deformation gradient $\mathbf{F}$ and the temperature Θ, i.e.

$$\mathbf{P} = \mathbf{P}(\mathbf{F}, \Theta) \quad , \qquad \eta = \eta(\mathbf{F}, \Theta) \quad . \qquad (7.25)$$

Eqs. (7.24) and (7.25), also known as **thermal equations of state**, are crucial in specifying material behavior and are completely determined once the free energy $\Psi = \Psi(\mathbf{F}, \Theta)$ is given.

Alternative constitutive equations for the stress and the entropy may be found by analogy with the treatment carried out in Section 6.1. The second Piola-Kirchhoff stress $\mathbf{S}$ follows from relation $(3.65)_2$, by means of $(7.24)_1$ and the analogue of (6.11), as

$$\mathbf{S} = \mathbf{F}^{-1}\mathbf{P} = \mathbf{F}^{-1} \left(\frac{\partial \Psi(\mathbf{F}, \Theta)}{\partial \mathbf{F}} \right)_\Theta = 2 \left(\frac{\partial \Psi(\mathbf{C}, \Theta)}{\partial \mathbf{C}} \right)_\Theta , \qquad (7.26)$$

where $\partial \Psi(\mathbf{C}, \Theta)/\partial \mathbf{C}$ is a symmetric tensor. From $(7.24)_2$, using the relation $(6.9)_1$, we obtain the alternative constitutive equation for the entropy, i.e.

$$\eta = - \left(\frac{\partial \Psi(\mathbf{C}, \Theta)}{\partial \Theta} \right)_\mathbf{C} . \qquad (7.27)$$

In order to describe the thermodynamic state of a system by an alternative thermodynamic potential we require that the entropy function $(7.25)_2$ is uniquely invertible with respect to Θ for each fixed $\mathbf{F}$, so that we have locally the condition $\partial \eta/\partial \Theta \neq 0$.

We assume that the inversion of eq. $(7.25)_2$ is given by $\Theta = \Theta(\mathbf{F}, \eta)$ and we postulate an associated thermodynamic potential e, which is the *internal-energy function* per unit reference volume introduced on p. 157. Knowing that the internal energy is related to the free energy Ψ through the Legendre transformation (4.152), we have the (canonical) representation

$$e = e(\mathbf{F}, \eta) = \Psi(\mathbf{F}, \Theta(\mathbf{F}, \eta)) + \Theta(\mathbf{F}, \eta)\eta \quad , \tag{7.28}$$

also known as the **caloric equation of state**. It is an equation that determines the internal energy as a function of the deformation gradient $\mathbf{F}$ and a non-mechanical variable, i.e. the entropy η.

With potential (7.28) we may deduce fundamental physical expressions from the entropy principle based on the Clausius-Planck inequality (4.141). For *all* admissible thermoelastic processes the second law of thermodynamics reduces to the identity $\mathcal{D}_{\text{int}} = \mathbf{P} : \dot{\mathbf{F}} - \dot{e} + \Theta\dot{\eta} = 0$. Hence, the *Gibbs relation* is obtained by determining the total rate of change of $e = e(\mathbf{F}, \eta)$ and by use of the chain rule. Thus,

$$\dot{e} = \dot{e}(\mathbf{F}, \eta) = \mathbf{P} : \dot{\mathbf{F}} + \Theta\dot{\eta} = \left(\frac{\partial e(\mathbf{F}, \eta)}{\partial \mathbf{F}}\right)_\eta : \dot{\mathbf{F}} + \left(\frac{\partial e(\mathbf{F}, \eta)}{\partial \eta}\right)_\mathbf{F}\dot{\eta} \quad , \tag{7.29}$$

whence, for arbitrary choices of $\dot{\mathbf{F}}$ and $\dot{\eta}$, we have the physical expressions

$$\mathbf{P} = \left(\frac{\partial e(\mathbf{F}, \eta)}{\partial \mathbf{F}}\right)_\eta \quad \text{and} \quad \Theta = \left(\frac{\partial e(\mathbf{F}, \eta)}{\partial \eta}\right)_\mathbf{F} \tag{7.30}$$

for the first Piola-Kirchhoff stress $\mathbf{P}$ and the temperature Θ. Note that for the case of an *isentropic process* ($\eta = \text{const}$) there exists a state function e, whose partial derivative with respect to $\mathbf{F}$ gives the corresponding first Piola-Kirchhoff stress $\mathbf{P}$.

Comparing the physical expression (6.1), i.e. $\mathbf{P} = \partial\Psi(\mathbf{F})/\partial\mathbf{F}$, with eqs. $(7.24)_1$ and $(7.30)_1$ we recognize that the strain energy $\Psi = \Psi(\mathbf{F})$ serves as the free energy $\Psi(\mathbf{F}) = \Psi(\mathbf{F}, \Theta)|_{\Theta=\text{const}}$ or as the internal energy $e(\mathbf{F}) = e(\mathbf{F}, \eta)|_{\eta=\text{const}}$, depending on the process considered, *isothermal* or *isentropic*.

Knowing that $e = e(\mathbf{F}, \eta)$, we deduce from (7.30) the *stress* and *temperature functions* which depend on $\mathbf{F}$ and η. We obtain the thermal equations of state in the form

$$\mathbf{P} = \mathbf{P}(\mathbf{F}, \eta) \quad , \quad \Theta = \Theta(\mathbf{F}, \eta) \quad . \tag{7.31}$$

The two thermodynamic potentials Ψ and e introduced are commonly applied in *solid mechanics*. They are suitable for modeling so-called **thermoelastic materials** (no 'memory effects' occur).

For the sake of completeness two additional potentials are reviewed briefly. These are the **Gibbs free energy** (or in the literature sometimes called the **Gibbs function** or **chemical potential**), denoted by g (or sometimes in the literature by G), and the

enthalpy, denoted by h (or sometimes by H). The two thermodynamic potentials g and h are used frequently in *fluid dynamics*.

We postulate that the *Gibbs free energy* $g = g(\mathbf{P}, \Theta)$ is a function of the first Piola-Kirchhoff stress $\mathbf{P}$ and the temperature Θ. Performing a Legendre transformation by analogy with (7.28) we may express the *enthalpy* h by means of the thermal equation of state $(7.31)_2$ as

$$h = h(\mathbf{P}, \eta) = g(\mathbf{P}, \Theta(\mathbf{F}, \eta)) + \Theta(\mathbf{F}, \eta)\eta \ , \tag{7.32}$$

which is a function of the first Piola-Kirchhoff stress $\mathbf{P}$ and the entropy η. Here, we have used two definitions of g and h which are associated with the free energy Ψ and the internal energy e by the transformations

$$g = g(\mathbf{P}, \Theta) = \Psi - \mathbf{P} : \mathbf{F} \ , \qquad h = h(\mathbf{P}, \eta) = e - \mathbf{P} : \mathbf{F} \ . \tag{7.33}$$

In order to find the *Gibbs relations* we determine the total rates of change of these two thermodynamic potentials by applying the chain rule, i.e.

$$\dot{g} = \dot{g}(\mathbf{P}, \Theta) = \left(\frac{\partial g(\mathbf{P}, \Theta)}{\partial \mathbf{P}}\right)_\Theta : \dot{\mathbf{P}} + \left(\frac{\partial g(\mathbf{P}, \Theta)}{\partial \Theta}\right)_\mathbf{P} \dot\Theta \ , \tag{7.34}$$

$$\dot{h} = \dot{h}(\mathbf{P}, \eta) = \left(\frac{\partial h(\mathbf{P}, \eta)}{\partial \mathbf{P}}\right)_\eta : \dot{\mathbf{P}} + \left(\frac{\partial h(\mathbf{P}, \eta)}{\partial \eta}\right)_\mathbf{P} \dot\eta \ . \tag{7.35}$$

Using the second law of thermodynamics in the forms of (4.153) and (4.141) (with $\mathcal{D}_{\mathrm{int}} = 0$) and the material time derivatives of transformations (7.33) we arrive simply at $\dot{g} = \dot{g}(\mathbf{P}, \Theta) = -\dot{\mathbf{P}} : \mathbf{F} - \eta\dot\Theta$ and $\dot{h} = \dot{h}(\mathbf{P}, \eta) = -\dot{\mathbf{P}} : \mathbf{F} + \Theta\dot\eta$. Hence, by comparing with Gibbs relations (7.34) and (7.35) we obtain expressions for the deformation gradient $\mathbf{F}$, the entropy η and the temperature Θ. Thus,

$$\mathbf{F} = -\left(\frac{\partial g(\mathbf{P}, \Theta)}{\partial \mathbf{P}}\right)_\Theta \quad \text{and} \quad \eta = -\left(\frac{\partial g(\mathbf{P}, \Theta)}{\partial \Theta}\right)_\mathbf{P} \ , \tag{7.36}$$

$$\mathbf{F} = -\left(\frac{\partial h(\mathbf{P}, \eta)}{\partial \mathbf{P}}\right)_\eta \quad \text{and} \quad \Theta = \left(\frac{\partial h(\mathbf{P}, \eta)}{\partial \eta}\right)_\mathbf{P} \ . \tag{7.37}$$

The Gibbs free energy g and the enthalpy h have the status of a potential from which we may derive $\mathbf{F}, \eta$ and $\mathbf{F}, \Theta$, respectively.

In order to characterize the properties of a thermoelastic material, the considered thermodynamic potential must be supplemented by a suitable constitutive equation for the Piola-Kirchhoff heat flux $\mathbf{Q}$, necessary to determine heat transfer. It may be introduced as a function of the deformation gradient, temperature and temperature gradient, i.e.

$$\mathbf{Q} = \mathbf{Q}(\mathbf{F}, \Theta, \mathrm{Grad}\Theta) \ , \tag{7.38}$$

satisfying the classical heat conduction inequality (see relation (4.140), i.e. the version in the material description). For a more specific constitutive assertion see, for example, the phenomenological Duhamel's law of heat conduction (4.144) on p. 169 (given in terms of material coordinates). A material for which the heat flux $\mathbf{Q} = \mathbf{o}$ and the heat source $R = 0$ vanish for any point and time is known as an **adiabatic material**.

EXERCISES

1. Using constitutive equations (7.24), (7.30) and (7.36), (7.37) obtain four relations combining $\mathbf{P}$, η, $\mathbf{F}$, Θ in the forms

$$\left(\frac{\partial \mathbf{P}}{\partial \Theta}\right)_{\mathbf{F}} = -\left(\frac{\partial \eta}{\partial \mathbf{F}}\right)_{\Theta} \quad \text{and} \quad \left(\frac{\partial \mathbf{P}}{\partial \eta}\right)_{\mathbf{F}} = \left(\frac{\partial \Theta}{\partial \mathbf{F}}\right)_{\eta} \quad , \quad (7.39)$$

$$\left(\frac{\partial \mathbf{F}}{\partial \eta}\right)_{\mathbf{P}} = -\left(\frac{\partial \Theta}{\partial \mathbf{P}}\right)_{\eta} \quad \text{and} \quad \left(\frac{\partial \mathbf{F}}{\partial \Theta}\right)_{\mathbf{P}} = \left(\frac{\partial \eta}{\partial \mathbf{F}}\right)_{\Theta} \quad . \quad (7.40)$$

These identities are known as the **thermodynamic Maxwell** (or **reciprocal**) **relations** and are very valuable in thermodynamic analysis.

2. Each of the four (most common) thermodynamic potentials introduced, i.e. Ψ, e, g, h, is related to any other by a Legendre transformation. Show that the identity

$$e - \Psi - h + g = 0$$

is satisfied.

7.4 Calorimetry

Two centuries ago **calorimetry** became a branch of experimental physics. In the present day calorimetry deals with both the *measurement* of the amount of heat generated (or destroyed) within a given body during a change of state, and with its *formulation* within the theory of continuum thermodynamics.

Specific heat capacity and latent heat. Firstly, we introduce the **specific heat capacity at constant deformation** ($\mathbf{F} = $ const) per unit reference volume, which is usually denoted by $c_{\mathbf{F}}$. It is the energy required to produce unit increase in the temperature of a unit volume of the body keeping the *deformation fixed*.

The specific heat capacity $c_{\mathbf{F}} = c_{\mathbf{F}}(\mathbf{F}, \Theta) > 0$ for all $(\mathbf{F}, \Theta)$ is, in general, defined to be a positive function of the form

$$c_{\mathbf{F}} = c_{\mathbf{F}}(\mathbf{F}, \Theta) = -\Theta \left(\frac{\partial^2 \Psi(\mathbf{F}, \Theta)}{\partial \Theta \partial \Theta}\right)_{\mathbf{F}} > 0 \quad . \quad (7.41)$$

It is proportional to the second derivative of the free energy Ψ. For a general compressible material $c_{\mathbf{F}}$ depends on the deformation gradient $\mathbf{F}$ as well as on the temperature Θ. The positiveness of $c_{\mathbf{F}}$ may be related to the stability of the material (see, for example, ŠILHAVÝ [1997, Section 17.3]).

Using $(7.24)_2$ and the Legendre transformation $\Psi = e - \eta\Theta$, the specific heat capacity may be represented by the alternative convenient form

$$c_{\mathbf{F}} = c_{\mathbf{F}}(\mathbf{F}, \Theta) = \Theta \left(\frac{\partial \eta(\mathbf{F}, \Theta)}{\partial \Theta} \right)_{\mathbf{F}}$$

$$= \Theta \frac{\partial \eta(\mathbf{F}, \Theta)}{\partial \Theta} + \frac{\partial \Psi(\mathbf{F}, \Theta)}{\partial \Theta} + \eta(\mathbf{F}, \Theta) = \left(\frac{\partial e(\mathbf{F}, \Theta)}{\partial \Theta} \right)_{\mathbf{F}} \quad . \tag{7.42}$$

Hence, the specific heat capacity at constant deformation $c_{\mathbf{F}}$ may also be expressed through the internal energy e which, in general, depends on $\mathbf{F}$ and Θ.

Secondly, we introduce the **latent heat** which is denoted by the *symmetric* tensor $\boldsymbol{v}$. It is a *spatial field* defined to be

$$\boldsymbol{v} = -\Theta \frac{\partial^2 \Psi(\mathbf{F}, \Theta)}{\partial \mathbf{F} \partial \Theta} \mathbf{F}^{\mathrm{T}} = -\Theta \mathbf{F} \left(\frac{\partial^2 \Psi(\mathbf{F}, \Theta)}{\partial \mathbf{F} \partial \Theta} \right)^{\mathrm{T}}$$

$$\text{or} \qquad v_{ab} = -\Theta F_{aA} \frac{\partial^2 \Psi}{\partial F_{bA} \partial \Theta} \quad . \tag{7.43}$$

Note that the latent heat $\boldsymbol{v}$ is proportional to the mixed second derivative of the free energy Ψ.

Structural thermoelastic heating (or cooling). We define the general relation for the **structural thermoelastic heating (or cooling)** $\mathcal{H}_e$ as the double contraction of the latent heat, as given in $(7.43)_1$, and the symmetric part of the spatial velocity gradient $\mathbf{l} = \dot{\mathbf{F}}\mathbf{F}^{-1}$, i.e. the rate of deformation tensor $\mathbf{d}$. Thus, with property (1.95)

$$\mathcal{H}_e = \boldsymbol{v} : \mathbf{d} = -\Theta \frac{\partial^2 \Psi(\mathbf{F}, \Theta)}{\partial \mathbf{F} \partial \Theta} \mathbf{F}^{\mathrm{T}} : \frac{1}{2} [\dot{\mathbf{F}}\mathbf{F}^{-1} + (\dot{\mathbf{F}}\mathbf{F}^{-1})^{\mathrm{T}}]$$

$$= -\Theta \frac{\partial^2 \Psi(\mathbf{F}, \Theta)}{\partial \mathbf{F} \partial \Theta} : \dot{\mathbf{F}} \quad . \tag{7.44}$$

The scalar quantity $\mathcal{H}_e$ represents the thermoelastic coupling effect. This so-called *Gough-Joule effect* occurs, for example, during an adiabatic stretching of a rubber band which typically changes its temperature, as pointed out in Section 7.1. In some problems the thermoelastic coupling effect is neglected due to the fact that this change of temperature is small.

For a thermoelastic process ($\mathcal{D}_{\text{int}} = 0$) the rate of change of the entropy, as derived in (4.142), may be written by means of $(7.25)_2$ and $(7.24)_2$ and the chain rule as

$$\Theta\dot{\eta}(\mathbf{F}, \Theta) = -\mathrm{Div}\mathbf{Q} + R = \Theta \frac{\partial \eta(\mathbf{F}, \Theta)}{\partial \mathbf{F}} : \dot{\mathbf{F}} + \Theta \frac{\partial \eta(\mathbf{F}, \Theta)}{\partial \Theta} \dot{\Theta}$$

$$= -\Theta \frac{\partial^2 \Psi(\mathbf{F}, \Theta)}{\partial \mathbf{F} \partial \Theta} : \dot{\mathbf{F}} - \Theta \frac{\partial^2 \Psi(\mathbf{F}, \Theta)}{\partial \Theta \partial \Theta} \dot{\Theta} \quad , \tag{7.45}$$

with $\mathrm{Div}\mathbf{Q}$ and R denoting the material divergence of the Piola-Kirchhoff heat flux $\mathbf{Q}$ and the heat source per unit time and per unit reference volume, respectively. Hence, from (7.45) we obtain finally the (coupled) energy balance equation in **temperature form** (the local evolution of the temperature Θ appears explicitly). Using definitions $(7.41)_2$ and $(7.44)_3$ we have

$$c_{\mathbf{F}}\dot{\Theta} = -\mathrm{Div}\mathbf{Q} - \mathcal{H}_e + R \qquad \text{or} \qquad c_{\mathbf{F}}\dot{\Theta} = -\frac{\partial Q_A}{\partial X_A} - \mathcal{H}_e + R \ . \qquad (7.46)$$

On comparison with the associated energy balance equation in *entropy form*, that is eq. (4.142), we recognize that the structural thermoelastic heating (or cooling) $\mathcal{H}_e$ appears only explicitly in the temperature form (7.46).

Consider the case of a process during which $\mathrm{Div}\mathbf{Q}$ vanishes. We deduce from (7.46) that the heat source R per unit time and per unit reference volume is given by

$$R = c_{\mathbf{F}}\dot{\Theta} + \mathcal{H}_e \ , \qquad (7.47)$$

which we may regard as defining the *theory of calorimetry*. For a historical study see TRUESDELL [1980, Chapter 2C].

Within the theory of *finite thermoelasticity*, in general, we observe three different types of thermomechanical coupling effects, namely

(i) the influence of a change in temperature on the stress (thermal stress),

(ii) structural thermoelastic heating (or cooling) – Gough-Joule effect, eq. (7.44), and

(iii) geometric coupling (influence of a change in deformation on heat conduction) (see, for example, eq. (4.144)).

EXERCISE

1. By means of physical expression $(7.36)_2$, show an alternative version of the energy balance equation in temperature form (7.46), i.e.

$$c_{\mathbf{P}}\dot{\Theta} = -\mathrm{Div}\mathbf{Q} - \mathcal{H}_{e\mathbf{P}} + R \qquad \text{or} \qquad c_{\mathbf{P}}\dot{\Theta} = -\frac{\partial Q_A}{\partial X_A} - \mathcal{H}_{e\mathbf{P}} + R \ ,$$

where $c_{\mathbf{P}} = c_{\mathbf{P}}(\mathbf{P}, \Theta) > 0$ denotes the **specific heat capacity at constant stress** $(\mathbf{P} = \mathrm{const})$, defined to be $c_{\mathbf{P}}(\mathbf{P}, \Theta) = -\Theta(\partial^2 g(\mathbf{P}, \Theta)/\partial\Theta\partial\Theta)|_{\mathbf{P}}$. In words: $c_{\mathbf{P}}$ is the energy required to produce unit increase in the temperature of a unit volume of the body keeping the *stress fixed*. Alternatively to eq. $(7.44)_3$ the term $\mathcal{H}_{e\mathbf{P}}$ represents the structural thermoelastic heating (or cooling) which is defined to be $\mathcal{H}_{e\mathbf{P}} = -\Theta(\partial^2 g(\mathbf{P}, \Theta)/\partial\mathbf{P}\partial\Theta) : \dot{\mathbf{P}}$.

7.5　Isothermal, Isentropic Elasticity Tensors

The following presentation of the isothermal and isentropic elasticity tensors is based on the concept introduced in Section 6.6. It is an extension of the purely mechanical framework to thermodynamics by one thermal variable, i.e. the *temperature* Θ or its conjugate quantity, the *entropy* η.

Isothermal elasticity tensor and stress-temperature tensor.　Suppose that a body admits the *right Cauchy-Green tensor* $\mathbf{C} = \mathbf{F}^{\mathrm{T}}\mathbf{F}$ and the *temperature* Θ as independent mechanical and thermal variables and suppose the existence of the Helmholtz free-energy function in the form of $\Psi = \Psi(\mathbf{C}, \Theta)$ (and equivalently $\Psi = \Psi(\mathbf{F}, \Theta)$).

Then, according to relation (7.26)$_3$, we can find the second Piola-Kirchhoff stress tensor $\mathbf{S}$ of a point at a certain time t, which may be seen as a nonlinear tensor-valued tensor function of the two variables $\mathbf{C}$ and Θ. By analogy with Section 6.6, we now compute the change in $\mathbf{S}$. According to considerations (1.247) and (1.248), we obtain the total differential

$$d\mathbf{S} = \mathbb{C} : \frac{1}{2}d\mathbf{C} + \mathbf{T}d\Theta \quad , \tag{7.48}$$

which gives expressions for a purely mechanical part, $\mathbb{C}$, and a (mixed) mechanical-thermal part, $\mathbf{T}$, in the material description. In the first term in (7.48) we have introduced the definition of the *fourth-order tensor* $\mathbb{C}$, which is proportional to the second partial derivative of Ψ with respect to $\mathbf{C}$. By analogy with eqs. (6.154) and (6.157) we write

$$\mathbb{C} = 2\frac{\partial \mathbf{S}(\mathbf{C}, \Theta)}{\partial \mathbf{C}} = 4\frac{\partial^2 \Psi(\mathbf{C}, \Theta)}{\partial \mathbf{C}\partial \mathbf{C}} \quad \text{or} \quad C_{ABCD} = 4\frac{\partial^2 \Psi}{\partial C_{AB}\partial C_{CD}} \quad , \tag{7.49}$$

evaluated at $(\mathbf{C}, \Theta)$, with the major symmetries $\mathbb{C} = \mathbb{C}^{\mathrm{T}}$ or $C_{ABCD} = C_{CDAB}$. Here we call $\mathbb{C}$ the **isothermal elasticity tensor** in the *material description* or the **referential tensor of isothermal elasticities**, which is defined by keeping the *temperature fixed* during the process. The **isothermal elasticity tensor** in the *spatial description* $\mathfrak{c} = J^{-1}\chi_*(\mathbb{C})$ (or the **spatial tensor of isothermal elasticities**) is defined as a push-forward (and Piola) transformation of $\mathbb{C}$ on each large index by analogy with relation (6.159).

The second term in eq. (7.48) is the **referential stress-temperature tensor** or the **referential thermal coefficient of stress**, denoted by $\mathbf{T}$, which is proportional to the mixed second partial derivative of Ψ with respect to $\mathbf{C}$ and Θ. It is a symmetric *second-order tensor* defined as

$$\mathbf{T} = \frac{\partial \mathbf{S}(\mathbf{C}, \Theta)}{\partial \Theta} = 2\frac{\partial^2 \Psi(\mathbf{C}, \Theta)}{\partial \mathbf{C}\partial \Theta} \quad \text{or} \quad T_{AB} = 2\frac{\partial^2 \Psi}{\partial C_{AB}\partial \Theta} \quad . \tag{7.50}$$

The spatial counterpart, denoted by $\mathbf{t}$, results via a standard push-forward (and Piola) transformation $\mathbf{t} = J^{-1}\chi_*(\mathbf{T}^{\sharp})$ of the (contravariant) referential stress-temperature tensor $\mathbf{T}^\sharp = \partial \mathbf{S}^\sharp / \partial \Theta$ by the motion χ. By analogy with relation (3.66) we find, by means of eq. $(7.50)_2$, that

$$\mathbf{t} = J^{-1}\mathbf{F}\mathbf{T}\mathbf{F}^\mathrm{T} = 2J^{-1}\mathbf{F}\frac{\partial^2 \Psi(\mathbf{C},\Theta)}{\partial \mathbf{C}\partial \Theta}\mathbf{F}^\mathrm{T}$$

(7.51)

or $\qquad t_{ab} = J^{-1}F_{aA}F_{bB}T_{AB} = J^{-1}F_{aA}F_{bB}\dfrac{\partial^2 \Psi}{\partial C_{AB}\partial \Theta}$.

Note that the symbols $\mathbf{T}$ and $\mathbf{t}$ have already been used and must not be confused with the traction vectors.

We now express the symmetric **spatial stress-temperature tensor t** in terms of the *latent heat* $\boldsymbol{v}$, as defined in eq. (7.43). Knowing the transformation $(\partial \Psi(\mathbf{F},\Theta)/\partial \mathbf{F})^\mathrm{T} = 2(\partial \Psi(\mathbf{C},\Theta)/\partial \mathbf{C})\mathbf{F}^\mathrm{T}$, which is in accord with eq. (6.11), we find from $(7.51)_2$ that

$$\mathbf{t} = J^{-1}\frac{\partial^2 \Psi(\mathbf{F},\Theta)}{\partial \mathbf{F}\partial \Theta}\mathbf{F}^\mathrm{T} = J^{-1}\mathbf{F}\left(\frac{\partial^2 \Psi(\mathbf{F},\Theta)}{\partial \mathbf{F}\partial \Theta}\right)^\mathrm{T} = -J^{-1}\frac{\boldsymbol{v}}{\Theta}$$

(7.52)

or $\qquad t_{ab} = J^{-1}F_{aA}\dfrac{\partial^2 \Psi}{\partial F_{bA}\partial \Theta} = -J^{-1}\dfrac{v_{ab}}{\Theta}$.

Both tensor quantities $\mathbf{T}$ and $\mathbf{t}$ measure the change of the stress in a process in which the temperature is raised by one unit keeping the *deformation fixed*.

Finally we compute the change in the function for the entropy $\eta = \eta(\mathbf{C},\Theta)$. With constitutive equation $(7.24)_2$ and the equivalence $\Psi(\mathbf{F},\Theta) = \Psi(\mathbf{C},\Theta)$, we obtain

$$d\eta = -2\frac{\partial^2 \Psi(\mathbf{C},\Theta)}{\partial \mathbf{C}\partial \Theta} : \frac{1}{2}d\mathbf{C} - \frac{\partial^2 \Psi(\mathbf{C},\Theta)}{\partial \Theta \partial \Theta}d\Theta \quad .$$

(7.53)

By applying the definitions of the referential stress-temperature tensor $\mathbf{T}$ and the specific heat capacity at constant deformation c_F, the entropy change may be expressed as

$$d\eta = -\mathbf{T} : \frac{1}{2}d\mathbf{C} + \frac{c_\mathrm{F}}{\Theta}d\Theta \quad ,$$

(7.54)

where the relations $(7.50)_2$ and $(7.41)_2$ are to be used.

Isentropic elasticity tensor and stress-entropy tensor.   Consider a body which admits the *right Cauchy-Green tensor* $\mathbf{C} = \mathbf{F}^\mathrm{T}\mathbf{F}$ and the *entropy* η as independent mechanical and thermal variables and consider the existence of the internal-energy function per unit reference volume in the form of $e = e(\mathbf{C},\eta)$ (and equivalently $e = e(\mathbf{F},\eta)$).

The **isentropic elasticity tensor** in the *material description* or the **referential tensor of isentropic elasticities**, denoted by $\mathbb{C}^\mathrm{ise}$, is derived from the internal energy e in the same way that the isothermal elasticity tensor is derived from the free energy Ψ. Hence, by analogy with the above, the change in $\mathbf{S}$ is given by

$$d\mathbf{S} = \mathbb{C}^\mathrm{ise} : \frac{1}{2}d\mathbf{C} + \mathbf{T}^\mathrm{ise}d\eta \quad ,$$

(7.55)

in which we have introduced the definition of the *fourth-order tensor*

$$\mathbb{C}^{\mathrm{ise}} = 2\frac{\partial \mathbf{S}(\mathbf{C}, \eta)}{\partial \mathbf{C}} = 4\frac{\partial^2 e(\mathbf{C}, \eta)}{\partial \mathbf{C} \partial \mathbf{C}} \qquad \text{or} \qquad C^{\mathrm{ise}}_{ABCD} = 4\frac{\partial^2 \Psi}{\partial C_{AB} \partial C_{CD}} \;, \qquad (7.56)$$

evaluated at $(\mathbf{C}, \eta)$. The isentropic elasticity tensor $\mathbb{C}^{\mathrm{ise}}$ is defined by keeping the *entropy fixed* during the process.

The second term in eq. (7.55) denotes the **referential stress-entropy tensor** $\mathbf{T}^{\mathrm{ise}}$, which is proportional to the mixed second partial derivative of e with respect to $\mathbf{C}$ and η. It is a symmetric *second-order tensor* defined as

$$\mathbf{T}^{\mathrm{ise}} = \frac{\partial \mathbf{S}(\mathbf{C}, \eta)}{\partial \eta} = 2\frac{\partial^2 e(\mathbf{C}, \eta)}{\partial \mathbf{C} \partial \eta} \qquad \text{or} \qquad T^{\mathrm{ise}}_{AB} = 2\frac{\partial^2 e}{\partial C_{AB} \partial \Theta} \;. \qquad (7.57)$$

The **isentropic elasticity tensor** in the *spatial description* $\mathsf{c}^{\mathrm{ise}} = J^{-1}\chi_*(\mathbb{C}^{\mathrm{ise}})$ (or the **spatial tensor of isentropic elasticities**) and the (contravariant) **spatial stress-entropy tensor** $\mathbf{t}^{\mathrm{ise}} = J^{-1}\chi_*(\mathbf{T}^{\mathrm{ise}\,\sharp})$ are derived from $\mathbb{C}^{\mathrm{ise}}$ and $\mathbf{T}^{\mathrm{ise}}$ by analogy with the above. In the above expressions, $\mathbb{C}^{\mathrm{ise}}$, $\mathbf{T}^{\mathrm{ise}}$ and $\mathbf{t}^{\mathrm{ise}}$ are the *isentropic* quantities analogous to the isothermal quantities $\mathbb{C}$, $\mathbf{T}$ and $\mathbf{t}$.

EXAMPLE 7.3   Obtain the fundamental relationship

$$\mathbb{C}^{\mathrm{ise}} = \mathbb{C} + \frac{\Theta}{c_{\mathrm{F}}}\mathbf{T} \otimes \mathbf{T} \qquad (7.58)$$

between the referential tensors of *isentropic* and *isothermal* elasticities, with the *specific heat capacity at constant deformation* c_{F} and the *referential stress-temperature tensor* $\mathbf{T}$, as defined in eqs. $(7.41)_2$ and $(7.50)_2$, respectively.

Solution.   If we postulate the existence of the Helmholtz free-energy function $\Psi = \Psi(\mathbf{C}, \Theta)$ we may derive the second Piola-Kirchhoff stress tensor $\mathbf{S}$ in the differential form according to (7.48). As can be seen from relation (7.48), changes in stress, $d\mathbf{S}$, are associated with changes in both the deformation and the temperature.

A thermodynamic process in which the entropy $\eta = \eta(\mathbf{C}, \Theta)$ is constant (fixed), $d\eta = 0$, necessarily implies a change in temperature. An explicit expression may be deduced from (7.54) in the form

$$d\Theta = \frac{\Theta}{c_{\mathrm{F}}}\mathbf{T} : \frac{1}{2}d\mathbf{C} \;. \qquad (7.59)$$

This result substituted back into eq. (7.48) leads to

$$d\mathbf{S} = \mathbb{C} : \frac{1}{2}d\mathbf{C} + \mathbf{T}d\Theta$$

$$= \underbrace{\left(\mathbb{C} + \frac{\Theta}{c_{\mathrm{F}}}\mathbf{T} \otimes \mathbf{T}\right)}_{\mathbb{C}^{\mathrm{ise}}} : \frac{1}{2}d\mathbf{C} \;, \qquad (7.60)$$

which furnishes the desired expression for the isentropic elasticity tensor in the material description for which the entropy is held constant during the thermodynamic process.

This important relation expresses the isentropic elasticity tensor $\mathbb{C}^{ise}$ in the material description as a function of the free energy $\Psi = \Psi(\mathbf{C}, \Theta)$ (compare with eqs. $(7.49)_2$, $(7.50)_2$). Note that in terms of the internal-energy function $e = e(\mathbf{C}, \eta)$, relation (7.60) reduces to $(7.56)_2$. ■

Some numerical aspects. The distinction between isothermal and isentropic elasticities plays a crucial role in the analysis of *nonlinear numerical stability*.

Coupled thermomechanical problems in solid mechanics may be solved numerically within one time step leading to simultaneous (monolithic) solutions of all the fields involved in the problem which have the feature of a good stability characteristic. However, this approach leads to large *non-symmetric* systems which are inefficient to solve and are associated with high computational cost. This type of fundamental numerical solution strategy goes back to NICKELL and SACKMAN [1968] and ODEN [1969, 1972].

Alternatively, the coupled system of nonlinear differential equations is often solved using the classical **staggered solution technique** (or also known as the **fractional-step method** or **staggered method**) (see, for example, YANENKO [1971], MARCHUK [1982, and references therein]). In this method, the key idea is to partition the monolithic system of equations into smaller (symmetric) sub-systems by making use of the physical meaning of the problem considered. Within the concept of a staggered solution technique the system can be solved sequentially with much lower computational cost. For each sub-system we can apply existing algorithms and solution strategies.

The classical (merely standard) staggered solution technique for a coupled thermomechanical problem in solid mechanics is based on the solution of a mechanical (isothermal) problem at a *fixed temperature* of the system (elastodynamic phase), which involves the *isothermal elasticity tensor* (7.49), followed by the solution of a heat conduction problem at a *fixed configuration* in the temperature form (7.46). This classical partition is referred to as the **isothermal operator split**, which was used within the context of coupled thermomechanical problems (see, for example, ARGYRIS et al. [1979, 1981, 1982], MIEHE [1988] and SIMO and MIEHE [1992]). However, this type of staggered solution technique is associated with the crucial restriction of conditional stability (see ARMERO and SIMO [1992, 1993]).

It emerges that an alternative partition of a strongly coupled thermomechanical problem leads to a so-called **unconditionally stable (time-stepping) solution technique**, characterized as independent from the chosen time step. This technique allows solutions in an efficient numerically accurate way. The analysis is based on the solution of a mechanical problem at a *fixed entropy* of the system (elastodynamic phase), which involves the *isentropic elasticity tensor* (7.56), followed by the solution of a heat conduction problem at a *fixed configuration* (thermal phase) in the entropy form (4.142).

This alternative methodology is referred to as the **isentropic operator split** when *dissipative materials* are involved (damage, viscous or plastic effects may occur). Likewise for *perfectly thermoelastic materials* the split is referred to as the **adiabatic operator split**. Within this solution technique it is possible to show that a defined *Lyapunov functional* for the coupled system of evolution equations – regarded as the canonical free-energy function for thermoelasticity, first introduced by DUHEM [1911, Vol.2, pp. 220-231], is decreasing along the flows for each of the two sub-problems involved. This approach was proposed by ARMERO and SIMO [1992] for linear and nonlinear thermoelasticity and by ARMERO and SIMO [1993] for finite thermoplasticity. For a successful application to rubber thermoelasticity see the papers by MIEHE [1995b] and HOLZAPFEL and SIMO [1996b].

It must be emphasized that this class of staggered solution technique can be applied not only to coupled thermomechanical problems in solid mechanics but also to coupled problems of, for example, fluid flow in a porous medium, magnetohydrodynamics in fluid mechanics or to stress-diffusion problems. All that must be done is to replace the *temperature* Θ and the *entropy* η by the associated field variables of the coupled problem considered.

EXERCISES

1. Recall the spatial stress-temperature tensor (7.51). By means of eq. (3.66) derive the alternative expression

$$\mathbf{t} = \frac{\partial \boldsymbol{\sigma}(\mathbf{C}, \Theta)}{\partial \Theta} \qquad \text{or} \qquad t_{ab} = \frac{\partial \sigma_{ab}}{\partial \Theta} \quad ,$$

   where $\boldsymbol{\sigma}$ denotes the symmetric Cauchy stress tensor. In addition, derive the spatial stress-entropy tensor which has the form $\mathbf{t}^{ise} = \partial \boldsymbol{\sigma}(\mathbf{C}, \eta)/\partial \eta$.

2. Suppose that a body admits the (spatial) *left Cauchy-Green tensor* $\mathbf{b} = \mathbf{F}\mathbf{F}^{\mathrm{T}}$ and the *temperature* Θ as independent variables and consider a free-energy function in the form of $\Psi = \Psi(\mathbf{b}, \Theta)$. By analogy with (6.38) we may find the associated constitutive equation in the form of $\boldsymbol{\tau} = 2\mathbf{b}(\partial \Psi(\mathbf{b}, \Theta)/\partial \mathbf{b})$, where $\boldsymbol{\tau}$ denotes the symmetric Kirchhoff stress tensor (note that this type of constitutive equation represents *isotropic* thermoelastic response only).

   Recall the kinematic relation (2.171) and obtain from the Oldroyd stress rate (5.59) of the Kirchhoff stress, using the chain rule,

$$\pounds_{\mathbf{v}}(\boldsymbol{\tau}^{\sharp}) = J(\mathbb{c} : \mathbf{d} + \mathbf{t}\dot{\Theta}) \quad , \tag{7.61}$$

with the definitions

$$J\mathbb{c} = 4\mathbf{b}\frac{\partial^2 \Psi(\mathbf{b}, \Theta)}{\partial \mathbf{b} \partial \mathbf{b}}\mathbf{b} \quad , \qquad J\mathbf{t} = 2\frac{\partial^2 \Psi(\mathbf{b}, \Theta)}{\partial \mathbf{b} \partial \Theta}\mathbf{b} = 2\mathbf{b}\frac{\partial^2 \Psi(\mathbf{b}, \Theta)}{\partial \mathbf{b} \partial \Theta} \qquad (7.62)$$

of the (fourth-order) isothermal elasticity tensor $\mathbb{c}$ in the spatial description and the (second-order) spatial stress-temperature tensor $\mathbf{t}$.

For explicit derivations of eqs. (7.61) and (7.62) see the work of MIEHE [1995b]. Therein, the definitions of the tensor variables $\mathbb{c}$ and $\mathbf{t}$ exclude the factor J.

7.6 Entropic Elastic Materials

We consider so-called **entropic elastic materials**, which have the property that the change in internal energy with deformation is small or even zero (recall Section 7.1). The underlying concept of **entropic elasticity** is used particularly for the thermo-mechanical description of rubber-like materials such as elastomers (see, for example, FLORY [1961], CHADWICK [1974], CHADWICK and CREASY [1984], MÜLLER [1985], KRAWITZ [1986], HADDOW and OGDEN [1990] and OGDEN [1992b]).

From experimental observations it is known that the bulk modulus for rubber-like materials considerably exceeds the shear modulus. For an *ideal* rubber the internal energy e is assumed to be a function of the temperature Θ alone, $e = e(\Theta)$, which is a typical characteristic of incompressible materials (see, for example, TRELOAR [2005, p. 34] and ANTHONY et al. [1942] for more details). This assumption leads to the **purely** (or **strictly**) **entropic theory** of rubber thermoelasticity (for a theoretical treatment see the work of CHADWICK [1974]). Consequently, the change in internal energy with deformation at a given *reference temperature* Θ_0 is constant. We assume that

$$e_0(\mathbf{F}) = 0 \quad . \qquad (7.63)$$

Here and elsewhere the subscript $(\bullet)_0$ characterizes quantities at a reference temperature Θ_0 so that, for example, $e_0(\mathbf{F}) = e(\mathbf{F}, \Theta_0)$.

Alternatively, a thermoelastic material obeys the **modified entropic theory** if its internal energy e is expressible as the sum of $e(\Theta)$, already known from the purely entropic theory, and the internal energy $e_0(J) = e(J, \Theta_0)$. The additional contribution $e_0(J)$ to the internal energy depends on the deformation only through the volume ratio J at a given reference temperature Θ_0 (for a theoretical treatment see the work of CHADWICK [1974] and CHADWICK and CREASY [1984]). Consequently, we conclude that

$$e_0(\mathbf{F}) = e_0(J) \quad . \qquad (7.64)$$

For notational simplicity we often use the same letter for different functions.

Change in temperature, internal energy, entropy.   For both the *purely* and the *modified* entropic theories of rubber thermoelasticity we may conclude from eq. $(7.42)_4$ that the specific heat capacity c_F is a function of the temperature Θ only. Thus, we write

$$c_F(\mathbf{F}, \Theta) = c(\Theta) \ . \tag{7.65}$$

Having this in mind we assume a **temperature change**

$$\vartheta = \Theta - \Theta_0 \tag{7.66}$$

between a selected reference state with reference temperature Θ_0 (the choice is quite arbitrary) and the current state with absolute temperature Θ. Hence, by means of (7.65) the change in internal energy for a process is determined by integrating eq. $(7.42)_4$ with respect to the temperature, i.e.

$$e(\mathbf{F}, \Theta) - e_0(\mathbf{F}) = \int_{\hat\Theta = \Theta_0}^{\hat\Theta = \Theta} c(\hat\Theta) \mathrm{d}\hat\Theta \ . \tag{7.67}$$

Recall from eqs. (7.64) and (7.63) that $e_0(\mathbf{F}) = e_0(J)$ for the modified entropic theory while $e_0(\mathbf{F})$ is assumed to be zero for the purely entropic theory.

Similarly, by means of (7.65), the entropy change simply results from $(7.42)_2$ by the integration

$$\eta(\mathbf{F}, \Theta) - \eta_0(\mathbf{F}) = \int_{\hat\Theta = \Theta_0}^{\hat\Theta = \Theta} c(\hat\Theta) \frac{\mathrm{d}\hat\Theta}{\hat\Theta} \ , \tag{7.68}$$

where $\eta_0(\mathbf{F}) = \eta(\mathbf{F}, \Theta_0)$ denotes the entropy at a reference temperature Θ_0.

General structure of the thermodynamic potential.   Using eqs. (7.67) and (7.68) and the Legendre transformation $\Psi(\mathbf{F}, \Theta) = e(\mathbf{F}, \Theta) - \Theta\eta(\mathbf{F}, \Theta)$, the thermodynamic potential in the form of the uniquely defined free energy Ψ may be expressed in terms of the internal energy e_0, the entropy η_0 and an additional function T for the purely thermal contribution. Thus,

$$\Psi(\mathbf{F}, \Theta) = e_0(\mathbf{F}) - \Theta\eta_0(\mathbf{F}) + T(\Theta) \ , \tag{7.69}$$

$$T(\Theta) = -\int_{\hat\Theta = \Theta_0}^{\hat\Theta = \Theta} c(\hat\Theta)(\Theta - \hat\Theta) \frac{\mathrm{d}\hat\Theta}{\hat\Theta} \ . \tag{7.70}$$

Another commonly used alternative form of the thermodynamic potential may be found by considering only state functions which are assumed to characterize the reference state at a given reference temperature Θ_0. These functions are interrelated by means of the Legendre transformation according to

$$\Psi_0(\mathbf{F}) = e_0(\mathbf{F}) - \Theta_0\eta_0(\mathbf{F}) \ . \tag{7.71}$$

Consider the thermodynamic potential (7.69) and substitute for the entropy η_0 (at a given temperature Θ_0) the expression which follows from transformation (7.71). Then, by means of the temperature change (7.66), we arrive, after some simple algebra, at the expression

$$\Psi(\mathbf{F}, \Theta) = \Psi_0(\mathbf{F}) \frac{\Theta}{\Theta_0} - e_0(\mathbf{F}) \frac{\vartheta}{\Theta_0} + T(\Theta) \quad , \tag{7.72}$$

which is due to CHADWICK [1974]. Instead of the entropy $\eta_0(\mathbf{F})$ this alternative form uses the isothermal free energy $\Psi_0(\mathbf{F})$, which is the change in strain energy for a deformation from the reference configuration to the current configuration at a fixed (constant) reference temperature Θ_0. The purely thermal contribution $T(\Theta)$ is given by eq. (7.70).

Note that for a thermoelastic material that obeys the modified entropic theory we require $e_0(\mathbf{F}) = e_0(J)$ in eqs. (7.69) and (7.72). However, within the purely entropic theory the characterization of a specific thermoelastic material is given basically by the entropy η_0 (or the isothermal free energy Ψ_0) only, since $e_0(\mathbf{F}) = 0$.

The structure of the free-energy function $\Psi(\mathbf{F}, \Theta)$ introduced in (7.69) and (7.72) is general in the sense that it may be used for the description of any entropic elastic material. A specification is accomplished with the choice of particular functions for the internal energy e_0 and the entropy η_0 (or the isothermal free energy Ψ_0) at a reference temperature Θ_0.

In order to perform the integration (7.70) we need an expression for the specific heat capacity as a function of the temperature. However, for some cases the specific heat capacity is assumed to be a positive *constant* over a given temperature range, and we write this as $c_0 > 0$. Then, the integrations in eqs. (7.67), (7.68) and (7.70) can be performed explicitly. The purely thermal contribution (7.70), for example, takes on the standard form

$$T(\Theta) = c_0 \left[\vartheta - \Theta \ln \left(\frac{\Theta}{\Theta_0} \right) \right] \quad , \tag{7.73}$$

which will enable us to determine the thermal contribution $T(\Theta)$ for incompressible materials with sufficient accuracy.

EXERCISES

1. An incompressible material with a certain volume, constant specific heat capacity c_0 and with temperature Θ is thrown into a very large lake with reference temperature Θ_0. After some time thermodynamic equilibrium is reached. Assume that the lake will absorb all the heat rejected by the material without any change in its temperature.

Determine the change in entropy for the material which changes its temperature by $\vartheta = \Theta - \Theta_0$.

2. Assume that a thermoelastic material obeys the purely entropic theory ($e_0(\mathbf{F}) = 0$). The change in the strain energy from the reference to the current configuration at a given (constant) temperature Θ_0 is given by $\Psi_0(\mathbf{F})$.

   (a) Recall eq. (7.44)$_3$ and show that for this type of thermoelastic material the structural thermoelastic heating (or cooling) $\mathcal{H}_e$ is governed by the relation

   $$\mathcal{H}_e = -\dot{\Psi}_0(\mathbf{F})\frac{\Theta}{\Theta_0} \quad . \tag{7.74}$$

   (b) By adopting the energy balance equation (7.46) and eq. (7.74) show that for every *adiabatic* process (in which the heat flux $Q_N = -\mathbf{Q} \cdot \mathbf{N}$ and the heat source R are zero for all points of the material and for all times) the temperature evolution from the reference to the current configuration is given explicitly by

   $$\Theta = \Theta_0 \exp\left(\frac{\Psi_0(\mathbf{F})}{\Theta_0 c_0}\right) \quad , \tag{7.75}$$

   where the strain energy vanishes in the reference configuration according to agreement (6.4). For convenience assume that the specific heat capacity is a constant c_0.

3. Consider a thermoelastic material with a given isothermal free energy $\Psi_0(\mathbf{F})$ and a constant specific heat capacity c_0. Assume that the material obeys the purely entropic theory.

   (a) Recall the physical expression (7.24)$_2$ and show that the evolution (change) of the entropy η is given by

   $$\dot{\eta}(\mathbf{F}, \Theta) = -\frac{\dot{\Psi}_0(\mathbf{F})}{\Theta_0} + c_0\frac{\dot{\Theta}}{\Theta} \quad .$$

   (b) Consider an *isentropic* process in which the entropy possessed by the given thermoelastic material remains constant. Deduce that

   $$c_0\dot{\Theta} = \dot{\Psi}_0(\mathbf{F})\frac{\Theta}{\Theta_0} \quad ,$$

   which gives the same temperature evolution as in (7.75). Interpret this result.

7.7 Thermodynamic Extension of Ogden's Material Model

In this section we particularize the general structure of the thermodynamic potential (7.69) (or its equivalent form (7.72)) introduced previously. Since the bulk modulus for rubber-like materials greatly exceeds the shear modulus it is most advantageous to employ the concept of decoupled (volumetric-isochoric) finite (hyper)elasticity, already introduced within the context of isothermal compressible hyperelasticity (see Section 6.4). This concept is based on a multiplicative split of the deformation gradient (or the corresponding right Cauchy-Green tensor) defined in eq. (6.79).

Our approach is purely phenomenological, providing a set of constitutive equations appropriate for numerical realization using the finite element method. The basic idea of the constitutive model presented for the isotropic thermoelastic behavior of elastomeric (rubber-like) materials incorporating large strains is due to CHADWICK [1974], while aspects for its computational implementation are addressed by MIEHE [1995b], HOLZAPFEL and SIMO [1996b] and REESE and GOVINDJEE [1998b].

Structure of the Helmholtz free-energy function.   A useful constitutive model for the isothermal and isotropic behavior of compressible (rubber-like) materials proposed by OGDEN [1972b] was presented in Section 6.5 on p. 244.

Very briefly we recall Ogden's strain-energy function expressed in terms of the volume ratio J, the modified principal stretches $\overline{\lambda}_a = J^{-1/3}\lambda_a$, $a = 1, 2, 3$, and a given (fixed) reference temperature Θ_0 (typically room temperature). The *decoupled* representation of the strain-energy function $\Psi_0 = \Psi(\lambda_1, \lambda_2, \lambda_3, \Theta_0)$ reads

$$\Psi_0 = \Psi_{\text{vol}0} + \Psi_{\text{iso}0} \quad, \tag{7.76}$$

where $\Psi_{\text{vol}0} = \Psi_{\text{vol}}(J, \Theta_0)$ and $\Psi_{\text{iso}0} = \Psi_{\text{iso}}(\overline{\lambda}_1, \overline{\lambda}_2, \overline{\lambda}_3, \Theta_0)$ are assumed to be objective scalar-valued functions characterizing the volumetric elastic response and the isochoric elastic response of the hyperelastic material.

By recalling (6.137)$_1$ and (6.139) we may specify the two response functions $\Psi_{\text{vol}0}$ and $\Psi_{\text{iso}0}$. Having in mind the notation introduced, we write

$$\Psi_{\text{vol}}(J, \Theta_0) = \kappa(\Theta_0)\mathcal{G}(J) \quad, \qquad \Psi_{\text{iso}}(\overline{\lambda}_1, \overline{\lambda}_2, \overline{\lambda}_3, \Theta_0) = \sum_{a=1}^{3} \overline{\omega}(\overline{\lambda}_a, \Theta_0) \quad, \tag{7.77}$$

$$\overline{\omega}(\overline{\lambda}_a, \Theta_0) = \sum_{p=1}^{N} \frac{\mu_p(\Theta_0)}{\alpha_p}(\overline{\lambda}_a^{\alpha_p} - 1) \quad, \qquad a = 1, 2, 3 \quad. \tag{7.78}$$

In addition, we must satisfy the (consistency) condition

$$2\mu_0 = \sum_{p=1}^{N} \mu_p(\Theta_0)\alpha_p \qquad \text{with} \qquad \mu_p(\Theta_0)\alpha_p > 0 \quad, \qquad p = 1, \ldots, N \quad, \tag{7.79}$$

where the parameter μ_0 denotes the shear modulus in the reference configuration at Θ_0.

The strain energy $\Psi_{\text{vol}}(J, \Theta_0)$ is associated with the volumetric elastic response and is, in general, decomposed into a bulk modulus $\kappa(\Theta_0)$ at a fixed reference temperature Θ_0 and a scalar-valued scalar function $\mathcal{G}(J)$. One example for $\mathcal{G}$ was introduced in eq. $(6.137)_2$. The strain energy $\Psi_{\text{iso}}(\bar{\lambda}_1, \bar{\lambda}_2, \bar{\lambda}_3, \Theta_0)$, however, is associated with the isochoric elastic behavior in the space of principal directions of $\overline{\mathbf{C}}$, i.e. the Valanis-Landel hypothesis (see VALANIS and LANDEL [1967]). The parameters $\mu_p(\Theta_0)$ denote the (constant) shear moduli at Θ_0 and α_p are dimensionless constants, $p = 1, \ldots, N$.

Next, we derive a relatively simple but very efficient thermodynamic potential for rubber-like materials which is basically a thermodynamic extension of Ogden's model. We employ the modified entropic theory with assumption (7.64). Applying the thermodynamic potential in the form of (7.72) and using the strain energy (7.76) together with relations (7.77) and (7.78), we obtain the non-isothermal free energy Ψ for isotropic thermoelastic material response, i.e.

$$\Psi(\lambda_1, \lambda_2, \lambda_3, \Theta) = \Psi_{\text{vol}}(J, \Theta) + \Psi_{\text{iso}}(\bar{\lambda}_1, \bar{\lambda}_2, \bar{\lambda}_3, \Theta) \ . \tag{7.80}$$

This decoupled structure is based on the definitions

$$\Psi_{\text{vol}} = \kappa(\Theta)\mathcal{G}(J) - e_0(J)\frac{\vartheta}{\Theta_0} + T(\Theta) \ , \qquad \Psi_{\text{iso}} = \sum_{a=1}^{3} \overline{\omega}(\bar{\lambda}_a, \Theta) \ , \tag{7.81}$$

$$\overline{\omega}(\bar{\lambda}_a, \Theta) = \sum_{p=1}^{N} \frac{\mu_p(\Theta)}{\alpha_p}(\bar{\lambda}_a^{\alpha_p} - 1) \ , \qquad a = 1, 2, 3 \ , \tag{7.82}$$

$$\kappa(\Theta) = \kappa(\Theta_0)\frac{\Theta}{\Theta_0} \ , \tag{7.83}$$

$$\mu_p(\Theta) = \mu_p(\Theta_0)\frac{\Theta}{\Theta_0} \ , \qquad p = 1, \ldots, N \ , \tag{7.84}$$

and on the condition (7.79). In $(7.81)_1$ the purely thermal contribution $T(\Theta)$ is given in eq. (7.70).

The first contribution $\Psi_{\text{vol}}(J, \Theta)$ to the thermoelastic response defined by $(7.81)_1$ is due to volume changes and purely thermal causes. The bulk modulus $\kappa(\Theta)$ in $(7.81)_1$ depends linearly on the absolute temperature Θ (see eq. (7.83)). Note that the *energetic contribution* $e_0(J)$ to the function $\Psi_{\text{vol}}(J, \Theta)$ occurs only within a modified entropic theory. An empirical expression was proposed by CHADWICK [1974] and has the form

$$\frac{e_0(J)}{\Theta_0} = 3\alpha_0\kappa(\Theta_0)\mathcal{G}(J) \qquad \text{with} \qquad \mathcal{G}(J) = \gamma^{-1}(J^\gamma - 1) \ . \tag{7.85}$$

This relationship is based upon experimental considerations, where $\gamma > 0$ is a positive non-dimensional parameter and the quantity $\alpha_0 = \alpha(\Theta_0)$ denotes the so-called

linear expansion coefficient relative to a selected reference state with reference temperature Θ_0. The empirical response function $\mathcal{G}(J)$, which is extensively studied by WOOD [1964] and CHADWICK [1974], is able to fit experimental data which are obtained from isothermal compression tests performed with different temperature values. For an instructive example on the basis of $\gamma = 1$ and the neo-Hookean model the reader is referred to the paper by OGDEN [1992b, Example 1] in which the symbol α_0 is used for the *volume* coefficient of thermal expansion. Observe that within a purely entropic theory the energetic contribution $e_0(J)$ vanishes (recall Section 7.6). Consequently, for this case the stress is proportional to the absolute temperature Θ, since $\Psi = \Psi_0(\Theta/\Theta_0) + T$.

The second contribution $\Psi_{\mathrm{iso}}(\overline{\lambda}_1, \overline{\lambda}_2, \overline{\lambda}_3, \Theta)$ to the thermoelastic response defined by $(7.81)_2$ is due to isochoric deformations. Here, we consider a representation of Ψ_{iso} in terms of the modified principal stretches $\overline{\lambda}_a$, $a = 1, 2, 3$. The shear moduli $\mu_p(\Theta)$, $p = 1, \ldots, N$, in (7.82) depend linearly on the absolute temperature Θ (see eq. (7.84)). A physical interpretation of this fact was presented within the context of Gaussian statistical theory of molecular networks which is valid for the region of small strains (compare with relation (7.19)).

The thermodynamic potential (7.80) describes the stress-strain-temperature behavior of rubber-like materials within the finite strain domain. As for the isothermal case, the thermodynamic extension of the *Mooney-Rivlin model* and the *neo-Hookean model* results from (7.82) by setting $N = 2$, $\alpha_1 = 2$, $\alpha_2 = -2$ and $N = 1$, $\alpha_1 = 2$, respectively. Observe that for an isothermal deformation process ($\Theta = \Theta_0$) the second and third term in $(7.81)_1$ vanish, the material parameters (7.83) and (7.84) change into constants, and consequently the free-energy function $\Psi = \Psi(\lambda_1, \lambda_2, \lambda_3, \Theta)$ changes into the strain-energy function $\Psi_0 = \Psi(\lambda_1, \lambda_2, \lambda_3, \Theta_0)$, as presented by eqs. (7.76), (7.77) and (7.78).

Consistent linearization.   Subsequently, we point out the consistent linearization process of the thermodynamic potential given by (7.80). In particular, as a first step, we compute the thermoelastic stress response, characterized by the *second Piola-Kirchhoff stress tensor* **S**, followed by a second step which determines the *isothermal elasticity tensor* $\mathbb{C}$ in the material description and the *referential stress-temperature tensor* **T**. The formulation is presented exclusively within the concept of spectral decomposition and is characterized by a geometric setting relative to the reference configuration.

In order to deduce the stress tensor $\mathbf{S} = 2\partial\Psi(\lambda_1, \lambda_2, \lambda_3, \Theta)/\partial\mathbf{C}$ we follow the procedure as shown in Example 6.7 on p. 245. By means of the decomposed structure (7.80) we may find the purely *volumetric* and purely *isochoric* stress contributions $\mathbf{S}(\lambda_1, \lambda_2, \lambda_3, \Theta) = \mathbf{S}_{\mathrm{vol}} + \mathbf{S}_{\mathrm{iso}}$, which are defined to be

$$\mathbf{S}_{\mathrm{vol}} = 2\frac{\partial\Psi_{\mathrm{vol}}(J, \Theta)}{\partial\mathbf{C}} = Jp\mathbf{C}^{-1} \ , \tag{7.86}$$

$$\mathbf{S}_{\text{iso}} = 2\frac{\partial \Psi_{\text{iso}}(\overline{\lambda}_1, \overline{\lambda}_2, \overline{\lambda}_3, \Theta)}{\partial \mathbf{C}} = \sum_{a=1}^{3} \underbrace{\frac{1}{\overline{\lambda}_a} \frac{\partial \Psi_{\text{iso}}}{\partial \overline{\lambda}_a} \hat{\mathbf{N}}_a \otimes \hat{\mathbf{N}}_a}_{S_{\text{iso}\,a}} , \tag{7.87}$$

for the general case $\lambda_1 \neq \lambda_2 \neq \lambda_3 \neq \lambda_1$ (see the analogues of eqs. (6.140)$_1$, (6.143)).

The constitutive equation for the hydrostatic pressure p, essential for relation (7.86)$_2$, may be specified in terms of the free energy (7.81)$_1$ as

$$p = \frac{\partial \Psi_{\text{vol}}(J, \Theta)}{\partial J} = \kappa(\Theta) \frac{d\mathcal{G}(J)}{dJ} - \frac{de_0(J)}{dJ} \frac{\vartheta}{\Theta_0} , \tag{7.88}$$

where the term $d\mathcal{G}(J)/dJ$ was particularized in eq. (6.141)$_1$ or (6.141)$_2$ depending on whether the scalar-valued function (6.137)$_2$ or (6.138) is used.

In addition, with free energy (7.81)$_2$ (and (7.82)) we may compute the three principal isochoric stress functions $S_{\text{iso}\,a}$, $a = 1, 2, 3$, in the form

$$\begin{aligned}
S_{\text{iso}\,a} &= \frac{1}{\overline{\lambda}_a} \frac{\partial \Psi_{\text{iso}}(\overline{\lambda}_1, \overline{\lambda}_2, \overline{\lambda}_3, \Theta)}{\partial \overline{\lambda}_a} = \frac{1}{\overline{\lambda}_a^2} \left(\overline{\lambda}_a \frac{\partial \Psi_{\text{iso}}}{\partial \overline{\lambda}_a} - \frac{1}{3} \sum_{b=1}^{3} \overline{\lambda}_b \frac{\partial \Psi_{\text{iso}}}{\partial \overline{\lambda}_b} \right) \\
&= \frac{1}{\overline{\lambda}_a^2} \sum_{p=1}^{N} \mu_p(\Theta) \left(\overline{\lambda}_a^{\alpha_p} - \frac{1}{3} \sum_{b=1}^{3} \overline{\lambda}_b^{\alpha_p} \right) , \qquad a = 1, 2, 3 , \tag{7.89}
\end{aligned}$$

which are needed for (7.87)$_2$ (compare with the derivation which led to eqs. (6.144)$_2$ and (6.145)).

The derived set of expressions (7.86)–(7.89) completely defines the constitutive model for rubber-like materials, allowing thermoelastic deformations with strain changes unrestricted in magnitude. It is a straightforward thermodynamic extension of Ogden's model known from the isothermal regime, i.e. (6.140), (6.141) and (6.143)–(6.145).

Alternative stress measures follow directly from eqs. (7.86) and (7.87) by means of suitable transformations. For example, the stress response expressed by the Cauchy stress tensor $\boldsymbol{\sigma}$ simply results from a push-forward (and Piola) transformation $\boldsymbol{\sigma} = J^{-1}\chi_*(\mathbf{S}^\sharp) = J^{-1}\mathbf{FSF}^{\mathrm{T}}$ of $\mathbf{S}$.

As a second step in the consistent linearization process, we compute the change in the stress tensor $\mathbf{S}$, i.e. $d\mathbf{S} = \mathbb{C} : d\mathbf{C}/2 + \mathbf{T}d\Theta$, with the definitions of the isothermal elasticity tensor $\mathbb{C}$ in the material description, i.e. (7.49), and the referential stress-temperature tensor $\mathbf{T}$, i.e. (7.50). Based on the decomposed structure of the derived stress response (7.86) and (7.87), we may obtain the decoupled representation

$$\mathbb{C} = \mathbb{C}_{\text{vol}} + \mathbb{C}_{\text{iso}} \qquad \text{and} \qquad \mathbf{T} = \mathbf{T}_{\text{vol}} + \mathbf{T}_{\text{iso}} \tag{7.90}$$

for the tensors $\mathbb{C}$ and $\mathbf{T}$, where the first expression represents the familiar additive split of the fourth-order (isothermal) elasticity tensor $\mathbb{C}$ (compare with the considerations on p. 254 and subsequently). The second expression consists of second-order tensors only. Analogously to $(7.90)_1$, it is composed of a purely volumetric contribution $\mathbf{T}_{\text{vol}}$ and a purely isochoric contribution $\mathbf{T}_{\text{iso}}$.

The explicit forms of the isothermal elasticity tensors $\mathbb{C}_{\text{vol}}$ and $\mathbb{C}_{\text{iso}}$ are adopted from isothermal finite elasticity and are based on eqs. $(6.166)_4$ and (6.196), respectively. We bear in mind that the underlying free energies Ψ_{vol} and Ψ_{iso} depend on both the three principal stretches $\lambda_a = J^{1/3}\overline{\lambda}_a$, $a = 1, 2, 3$, and the temperature Θ.

The isothermal elasticity tensor $\mathbb{C}_{\text{vol}}$, as given by eq. $(6.166)_4$, requires the constitutive equation for the hydrostatic pressure p (which is given in (7.88)) and the scalar function $\tilde{p}$. With reference to specification (7.88), we obtain finally from (6.167) the explicit form

$$\tilde{p} = \kappa(\Theta)\left(\frac{\mathrm{d}\mathcal{G}(J)}{\mathrm{d}J} + J\frac{\mathrm{d}^2\mathcal{G}(J)}{\mathrm{d}J\mathrm{d}J}\right) - \left(\frac{\mathrm{d}e_0(J)}{\mathrm{d}J} + J\frac{\mathrm{d}^2e_0(J)}{\mathrm{d}J\mathrm{d}J}\right)\frac{\vartheta}{\Theta_0} \quad . \tag{7.91}$$

Considering the isothermal elasticity tensor $\mathbb{C}_{\text{iso}}$ in the spectral form, as given by eq. (6.196), we have just to take care of the coefficients

$$\frac{1}{\lambda_b}\frac{\partial S_{\text{iso}\,a}}{\partial \lambda_b} \quad \text{and} \quad \frac{S_{\text{iso}\,b} - S_{\text{iso}\,a}}{\lambda_b^2 - \lambda_a^2} \quad , \tag{7.92}$$

which depend on the material model in question. The three values $S_{\text{iso}\,a}$, $a = 1, 2, 3$, denote the principal values of the second Piola-Kirchhoff stress tensor $\mathbf{S}_{\text{iso}}$ and are given by eq. $(7.89)_2$. Hence, the second term in (7.92) is already determined. In order to determine the first term recall the closed form solution (6.197) and just consider that the shear moduli μ_p, $p = 1, \ldots, N$, are temperature dependent according to relation (7.84).

Finally we compute the referential stress-temperature tensor $\mathbf{T}$ defined by $(7.50)_1$. By recalling the constitutive equations $(7.86)_2$ and $(7.87)_2$ we may compute the purely volumetric and purely isochoric contributions to the referential stress-temperature tensor. They are defined in the sense that

$$\mathbf{T}_{\text{vol}} = \frac{\partial \mathbf{S}_{\text{vol}}}{\partial \Theta} = Jp'\mathbf{C}^{-1} \quad , \tag{7.93}$$

$$\mathbf{T}_{\text{iso}} = \frac{\partial \mathbf{S}_{\text{iso}}}{\partial \Theta} = \sum_{a=1}^{3} S'_{\text{iso}\,a}\hat{\mathbf{N}}_a \otimes \hat{\mathbf{N}}_a \tag{7.94}$$

for $\lambda_1 \neq \lambda_2 \neq \lambda_3 \neq \lambda_1$, where the values p' and $S'_{\text{iso}\,a}$, $a = 1, 2, 3$, depend on the material model in question. They are the derivatives of the hydrostatic pressure p and the three principal isochoric stress functions $S_{\text{iso}\,a}$, $a = 1, 2, 3$, with respect to the temperature Θ.

From $(7.88)_2$ and $(7.89)_3$ we find, using the relations for the bulk modulus (7.83) and the shear moduli (7.84), which are temperature dependent quantities, that

$$p' = \frac{\partial p}{\partial \Theta} = \frac{\kappa(\Theta_0)}{\Theta_0} \frac{\mathrm{d}\mathcal{G}(J)}{\mathrm{d}J} - \frac{\mathrm{d}e_0(J)}{\mathrm{d}J} \frac{1}{\Theta_0} \quad , \tag{7.95}$$

$$S'_{\mathrm{iso}\,a} = \frac{\partial S_{\mathrm{iso}\,a}}{\partial \Theta} = \frac{S_{\mathrm{iso}\,a}}{\Theta} \quad , \qquad a = 1, 2, 3 \quad . \tag{7.96}$$

Observe the similar structure of the referential stress-temperature tensors (7.93) and (7.94) to the volumetric-isochoric stress response, as defined in relations (7.86) and (7.87). The *spatial* counterparts $\mathbb{c}$ and $\mathbf{t}$ of the tensors defined in (7.90) result via a standard push-forward (and Piola) transformation.

Heat conduction.   The considered thermoelastic problem, for which $(\mathcal{D}_{\mathrm{int}} = 0)$, is governed essentially by *Cauchy's first equation of motion* (see, for example, the local forms (4.53) or (4.63)) and by the *balance of (mechanical and thermal) energy* in entropy or temperature form (see eq. $(7.45)_1$ or (7.46)). Hence, in regard to the energy balance equation we need an additional constitutive equation for the heat flux vector governing heat transfer. One example which satisfies the heat conduction inequality is Duhamel's law of heat conduction (see the considerations on p. 170). For the class of thermally isotropic materials we may express the constitutive equation as

$$\mathbf{Q}(\mathbf{C}, \Theta, \mathrm{Grad}\Theta) = -k_0(\Theta)\mathbf{C}^{-1}\mathrm{Grad}\Theta \tag{7.97}$$

(i.e. Fourier's law of heat conduction, which should be compared with eq. (4.148)), where $\mathbf{Q}$ is the Piola-Kirchhoff heat flux and $k_0 \geq 0$ denotes the coefficient of thermal conductivity associated with the reference configuration. Note that this coefficient is, in general, not a constant. In fact, for vulcanized elastomers, k_0 decreases linearly with increasing temperature (see SIRCAR and WELLS [1981]) according to

$$k_0(\Theta) = k_0(\Theta_0)[1 - \xi(\Theta - \Theta_0)] \quad , \tag{7.98}$$

where $k_0(\Theta_0)$ denotes the coefficient of thermal conductivity at the reference temperature Θ_0 and ξ is a **softening parameter**.

The solution of the coupled thermomechanical problem may be performed by adopting the staggered solution technique. Within a time step this technique leads to a decomposition of the coupled problem (compare with the statements on p. 331). As a result we must solve two smaller, in general, symmetric decoupled sub-problems on a staggered basis. For algorithmic aspects of the entropic theory of rubber thermoelasticity see, for example, MIEHE [1995b] and HOLZAPFEL and SIMO [1996b].

EXERCISES

1. Consider the scalar-valued function $\mathcal{G}(J) = \beta^{-2}(\beta\ln J + J^{-\beta} - 1)$ according to eq. (6.137)$_2$ (due to *Ogden*) and the energetic contribution $e_0(J)/\Theta_0 = 3\alpha_0\kappa(\Theta_0)\gamma^{-1}(J^\gamma - 1)$, with $\gamma > 0$, according to eq. (7.85) (due to *Chadwick*).

   Use (7.88)$_2$, (7.91) and the relation for the bulk modulus (7.83) in order to obtain the constitutive equations

$$Jp = \kappa(\Theta_0)\left(\beta^{-1}(1 - J^{-\beta})\frac{\Theta}{\Theta_0} - 3\alpha_0 J^\gamma(\Theta - \Theta_0)\right) \quad,$$

$$J\tilde{p} = \kappa(\Theta_0)\left(J^{-\beta}\frac{\Theta}{\Theta_0} - 3\alpha_0\gamma J^\gamma(\Theta - \Theta_0)\right) \quad,$$

   which completely determine the isothermal elasticity tensor $\mathbb{C}_{\text{vol}}$ in the material description.

2. Consider the extension of the isothermal Ogden material to the non-isothermal domain (7.80)–(7.84) (including the quantity $e_0(J)$ which represents an energetic (volumetric) contribution to the free energy). Recall the definition of the structural thermoelastic heating (or cooling) $\mathcal{H}_e$, i.e. eq. (7.44)$_3$, and show that $\mathcal{H}_e$ may be written in the decoupled structure of the form $\mathcal{H}_e = \mathcal{H}_{e\,\text{vol}} + \mathcal{H}_{e\,\text{iso}}$, with the definitions

$$\mathcal{H}_{e\,\text{vol}} = (\dot{e}_0(J) - \dot{\Psi}_{\text{vol}\,0})\frac{\Theta}{\Theta_0} \quad, \qquad \mathcal{H}_{e\,\text{iso}} = -\dot{\Psi}_{\text{iso}\,0}\frac{\Theta}{\Theta_0} \quad.$$

   The response functions $\Psi_{\text{vol}\,0} = \Psi_{\text{vol}}(J, \Theta_0)$, $\Psi_{\text{iso}\,0} = \Psi_{\text{iso}}(\bar\lambda_1, \bar\lambda_2, \bar\lambda_3, \Theta_0)$ are given by (7.77) with (7.78).

   The analogue of the decoupled structure of $\mathcal{H}_e$ was derived by MIEHE [1995b]. In his work the structural thermoelastic heating $\mathcal{H}_e$ is, however, based on the multiplicative split of the (spatial) left Cauchy-Green tensor $\mathbf{b} = \mathbf{F}\mathbf{F}^T$ and is defined with the opposite sign.

7.8   Simple Tension of Entropic Elastic Materials

The aim of this section is to illustrate the ability and performance of Ogden's model for the non-isothermal domain as outlined in the last section. We set up the basic equations required to describe the realistic physical stress-strain-temperature response of rubber-like materials. In particular, here we consider the simple tension of an entropic elastic material, a class of material defined in Section 7.6. A representative example concerned with the adiabatic stretching of a rubber band will contribute to a deeper insight in the coupled thermomechanical phenomena.

Thermoelastic deformation.   Before we start our studies with the simple tension of entropic elastic materials it is most beneficial to point out briefly some aspects of the **thermoelastic deformation** of a continuum body.

Consider a fixed reference configuration of a body with the geometrical region Ω_0 corresponding to a fixed reference time $t = 0$. The position of a typical point may be identified by the position vector $\mathbf{X}$ (with material coordinates (X_1, X_2, X_3)) relative to a fixed origin O (see Figure 7.5). The reference configuration is assumed to be *stress-free* and possesses a homogeneous (uniform) reference temperature value $\Theta_0 (> 0)$. A map of the reference configuration Ω_0 to a current configuration (with the new region) Ω is characterized by the macroscopic motion $\mathbf{x} = \chi(\mathbf{X}, t)$ for all $\mathbf{X} \in \Omega_0$ and for all times t. The motion carries a typical point $\mathbf{X} \in \Omega_0$ to a point $\mathbf{x} \in \Omega$ which is characterized by the spatial coordinates (x_1, x_2, x_3).

As a measure of the thermoelastic deformation we use the deformation gradient $\mathbf{F}(\mathbf{X}, t)$ and the volume ratio $J(\mathbf{X}, t) = \det\mathbf{F}(\mathbf{X}, t) > 0$. Very often it is convenient to decompose the (local) motion $\chi(\mathbf{X}, t) = \chi_M[\chi_\Theta(\mathbf{X}, t)]$ into two successive motions χ_M and χ_Θ according to

$$\mathbf{F} = \frac{\partial\chi(\mathbf{X}, t)}{\partial\mathbf{X}} = \mathbf{F}_M\mathbf{F}_\Theta \qquad \text{and} \qquad J = J_M J_\Theta > 0 \ , \tag{7.99}$$

with the definitions

$$\mathbf{F}_M = \frac{\partial\chi_M(\mathbf{X}_\Theta, t)}{\partial\mathbf{X}_\Theta} \ , \qquad J_M = \det\mathbf{F}_M > 0 \ , \tag{7.100}$$

$$\mathbf{F}_\Theta = \frac{\partial\chi_\Theta(\mathbf{X}, t)}{\partial\mathbf{X}} \ , \qquad J_\Theta = \det\mathbf{F}_\Theta > 0 \ . \tag{7.101}$$

The multiplicative decomposition (7.99) separates the total thermoelastic deformation into a purely *mechanical* contribution $\mathbf{F}_M$, J_M and a purely *thermal* contribution $\mathbf{F}_\Theta$, J_Θ, which represents the *Duhamel-Neumann hypothesis* for the nonlinear theory (see, for example, CARLSON [1972, p. 310]).

The two successive motions establish a new **intermediate (imagined) configuration** with geometrical region Ω_Θ, as illustrated in Figure 7.5. The new configuration is assumed to be *isolated* from the body so that a *thermal stress-free* deformation may occur. Hence, a relative temperature field $\vartheta = \Theta - \Theta_0$ causes a (free) thermal expansion (or contraction) about the reference configuration Ω_0 characterized by the associated variables $\mathbf{F}_\Theta$ and J_Θ. The intermediate configuration with the region Ω_Θ is given by the macroscopic motion $\mathbf{X}_\Theta = \chi_\Theta(\mathbf{X}, t)$ which carries points $\mathbf{X}$ located at Ω_0 to points

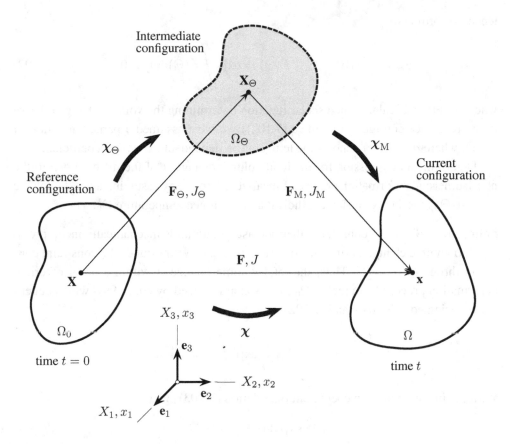

Figure 7.5 Multiplicative decomposition of the thermoelastic deformation into a purely *mechanical* contribution $\mathbf{F}_M$, $J_M = \det\mathbf{F}_M$ and a purely *thermal* contribution $\mathbf{F}_\Theta$, $J_\Theta = \det\mathbf{F}_\Theta$.

$\mathbf{X}_\Theta$ in the intermediate configuration Ω_Θ. A typical point $\mathbf{X}_\Theta$ is characterized by the coordinates $(X_{\Theta 1}, X_{\Theta 2}, X_{\Theta 3})$.

According to the multiplicative split (7.99) we have defined, additionally, a macroscopic motion $\mathbf{x} = \boldsymbol{\chi}_M(\mathbf{X}_\Theta, t)$ at a constant (fixed) temperature Θ along with the *stress-producing* deformation gradient $\mathbf{F}_M$ and the volume change J_M. A so-called **mechanically incompressible material** for which $J_M = \det\mathbf{F}_M = 1$, keeps the volume constant during a motion $\boldsymbol{\chi}_M$.

EXAMPLE 7.4 Consider a mechanically incompressible and thermoelastic material under a non-isothermal deformation process. The thermoelastic material is assumed to be thermally isotropic so that the deformation gradient $\mathbf{F}_\Theta$ may be given by an isotropic tensor according to

$$\mathbf{F}_\Theta = F(\Theta)\mathbf{I} \ , \qquad F(\Theta) = \exp[\int_{\Theta_0}^{\Theta} \alpha(\hat\Theta)\mathrm{d}\hat\Theta] > 0 \ , \tag{7.102}$$

where $F(\Theta)$ is a scalar-valued scalar function determining the volume change relative to the reference configuration. In eq.(7.102)$_2$ we have assumed a particularization of $F(\Theta)$, where $\alpha = \alpha(\Theta)$ denotes the temperature dependent expansion coefficient.

Determine an expression for the total volume change J of the material due to the non-isothermal deformation process. Linearize the result by assuming a constant value $\alpha_0 = \alpha(\Theta_0)$ for the expansion coefficient at a reference temperature Θ_0.

Solution.   Since the considered thermoelastic material is mechanically incompressible (no volume change during an isothermal process) we introduce the constraint condition through $J_\mathrm{M} = 1$. Thus, the total volume change J within a non-isothermal deformation process from region Ω_0 to Ω is characterized by eq. (7.99)$_2$ which degenerates, using eqs. (7.101)$_2$ and (7.102), to

$$J = J_\Theta(\Theta) = \det\mathbf{F}_\Theta = \exp[\int_{\Theta_0}^{\Theta} 3\alpha(\hat\Theta)\mathrm{d}\hat\Theta] > 0 \ . \tag{7.103}$$

With the linear expansion coefficient α_0, relation (7.103)$_3$ gives

$$J = \exp[3\alpha_0(\Theta - \Theta_0)] \ , \tag{7.104}$$

and linearization leads to the approximate solution

$$J \approx 1 + 3\alpha_0(\Theta - \Theta_0) \ . \tag{7.105}$$

It defines the total volume change J of a mechanically incompressible and thermally isotropic material within the infinitesimal strain theory. The approximate solution (7.105) is a well-known relation in linear continuum mechanics. It may be viewed as the volume of a unit cube at temperature Θ with values α_0 and Θ_0 which correspond to a reference state.   ∎

Entropic elasticity for a stretched piece of rubber.   In the following, attention will be confined to the thermoelastic description of isotropic and entropic elastic materials (such as elastomers) at finite strains. The stress-deformation-temperature response of a piece of rubber under simple tension, in particular, is examined. We assume that the material under consideration obeys the modified entropic theory of rubber thermoelasticity. Furthermore, the material is assumed to be mechanically incompressible, which motivates the use of a multiplicative split of the thermoelastic deformation, as introduced above. Consequently, the total volume change is given in $(7.103)_1$, i.e. $J = J_\Theta(\Theta) \, (\det \mathbf{F}_M = 1)$.

Now we consider a thin sheet of rubber stretched in one direction from the reference (undeformed) state to $\lambda_1 = \lambda$ (simple tension). Then, obeying condition $J = J_\Theta(\Theta) = \lambda_1 \lambda_2 \lambda_3$, we deduce by symmetry that $\lambda_2 = \lambda_3 = (J/\lambda)^{1/2}$. Hence, the Helmholtz free-energy function per unit reference volume

$$\Psi(\lambda, \Theta) = \Psi(\lambda, (J/\lambda)^{1/2}, \Theta) \tag{7.106}$$

is given in terms of one independent mechanical variable and one thermal variable, i.e. the stretch ratio λ and the temperature Θ. For notational convenience we use the same Greek letter Ψ for different free-energy functions.

Since the material is isotropic it is appropriate to use the thermodynamic extension of Ogden's model, as discussed in Section 7.7 (see eqs. (7.80)–(7.84)). Recall that, in general, one contribution to the free energy is due to volume changes and purely thermal causes (compare with relation $(7.81)_1$). Since we study a mechanically incompressible material we need only to consider the purely thermal contribution. We use, without loss of generality, the standard form (7.73), i.e. $T(\Theta) = c_0[\vartheta - \Theta \ln(\Theta/\Theta_0)]$, where the specific heat capacity $c_0 > 0$ is a positive constant.

Hence, the free energy relative to the reference configuration, which is stress-free and free of thermal expansion (or contraction), results from eqs. (7.80)–(7.82) and (7.84), using the temperature change $\vartheta = \Theta - \Theta_0$ and the specified kinematic relations, in the form

$$\Psi(\lambda, \Theta) = \frac{\Theta}{\Theta_0} \sum_{p=1}^{N} \frac{\mu_p(\Theta_0)}{\alpha_p} \left[\lambda^{\alpha_p} + 2 \left(\frac{J}{\lambda} \right)^{\alpha_p/2} - 3 \right]$$

$$+ c_0 \left[(\Theta - \Theta_0) - \Theta \ln \left(\frac{\Theta}{\Theta_0} \right) \right] , \tag{7.107}$$

where the additional condition (7.79) must be enforced. Note that the volume change due to thermal expansion governed by $J = J_\Theta(\Theta)$ is of considerable importance, which will be pointed out in more detail within the Example 7.5 below.

Next, we derive the associated thermal equations of state, namely the *stress* and *entropy functions*, as given in eqs. (7.24) and (7.25). For simple tension we may write

$$P = \frac{\partial \Psi(\lambda, \Theta)}{\partial \lambda} \qquad \text{and} \qquad \eta = -\frac{\partial \Psi(\lambda, \Theta)}{\partial \Theta} , \tag{7.108}$$

with given free energy Ψ substituted from (7.107),

$$P = \frac{\Theta}{\Theta_0} \sum_{p=1}^{N} \frac{\mu_p(\Theta_0)}{\lambda} \left[\lambda^{\alpha_p} - \left(\frac{J}{\lambda} \right)^{\alpha_p/2} \right] , \tag{7.109}$$

$$\eta = -\frac{1}{\Theta_0} \sum_{p=1}^{N} \frac{\mu_p(\Theta_0)}{\alpha_p} \left[\lambda^{\alpha_p} + (2 + 3\alpha_0\Theta\alpha_p) \left(\frac{J}{\lambda}\right)^{\alpha_p/2} - 3 \right] + c_0 \ln\left(\frac{\Theta}{\Theta_0}\right) \ . \quad (7.110)$$

Relations (7.109) and (7.110) specify the (non-zero) nominal stress $P_1 = P$ (also called the first Piola-Kirchhoff stress) $P_2 = P_3 = 0$, and the entropy η which are generated by the stretch ratio λ and the temperature Θ.

In order to describe the coupled thermomechanical problem of the piece of stretched rubber completely we must add a constitutive equation for the heat flux vector governing heat transfer. For the class of thermally isotropic materials we may adopt Fourier's law of heat conduction and refer to eq. (7.97). Having in mind the specified kinematic relations, the inverse of the right Cauchy-Green tensor, $\mathbf{C}^{-1}$, as needed in eq. (7.97), may be given in the form of its matrix representation

$$[\mathbf{C}^{-1}] = \begin{bmatrix} \lambda^{-2} & 0 & 0 \\ 0 & \lambda/J & 0 \\ 0 & 0 & \lambda/J \end{bmatrix} , \quad (7.111)$$

where the diagonal elements are the eigenvalues of $\mathbf{C}^{-1}$.

EXAMPLE 7.5　　Consider the stretching of a mechanically incompressible piece of rubber, for example a rubber band, obeying the modified entropic theory of rubber thermoelasticity. The rubber band is elongated rapidly so that no time remains for isothermal removal of heat. Hence, the homogeneous deformation process is viewed as an adiabatic process for which the heat flux on the boundary surface is zero and, in addition, heat sources are zero (thermal energy cannot be generated or destroyed within the material). The non-isothermal deformation process is assumed to be reversible.

In the present example attention is paid to the effects of *structural thermoelastic heating (or cooling)* and to the *stress-strain-temperature response* of the rubber band. In particular, show how the nominal stress P depends on the temperature change $\vartheta = \Theta - \Theta_0$ at a fixed elongation, i.e. a fixed stretch λ, and derive the temperature evolution of the rubber band with stretch. Finally, discuss the results of this classical example of rubber thermoelasticity, which demonstrates one of the great differences between *rubber* and *'hard'* solids, namely the distinctive effects of temperature.

As known from Section 6.5 the Ogden model excellently replicates the finite extensibility domain of rubber-like materials. Hence, as a basis for the constitutive model take the thermodynamic extension of Ogden's model with three pairs of constants ($N = 3$) characterized by the Helmholtz free-energy function (7.107). The constants α_p, $\mu_p(\Theta_0)$, $p = 1, 2, 3$, at the reference temperature Θ_0 are those given by OGDEN [1972a], listed in eq. (6.121) of this text. The volume change due to thermal expansion (or contraction) $J = J_\Theta(\Theta)$ is assumed to be governed by relation (7.104). The specific heat capacity and the linear expansion coefficient are given by $c_0 = 1.83 \cdot 10^3 \text{Nm/kgK}$ and $\alpha_0 = 22.333 \cdot 10^{-5} \text{K}^{-1}$ (see CYR [1988]), respectively. We assume that for the reference state the specimen and its environment constitute a system, which is in thermodynamic equilibrium. The homogeneous temperature field is given by the value $\Theta_0 = 293.15 \text{K} \ (= 20°\text{C})$.

Solution. The relationship between the nominal stress P and the temperature change $\vartheta = \Theta - \Theta_0$ is given explicitly by the theoretical solution (7.109), with volume change (7.104), and by experimental results due to ANTHONY et al. [1942]. The respective diagram for various fixed values of λ is depicted in Figure 7.6.

The figure is supplemented by a curve indicating the evolution of the temperature change ϑ with respect to the stretch ratio λ. Since the adiabatic process is assumed to be reversible the energy balance equation (4.142) reduces to $\Theta\dot{\eta} = 0$. Consequently, the entropy ($\eta = \text{const}$) possessed by the material remains constant, which indicates that the deformation process is also isentropic. The condition $\eta = \text{const}$ gives the relation between the temperature change ϑ and the stretch λ, which is based upon the constitutive equation for the entropy, as derived in eq. (7.110), with eq. (7.104). The analytical solution is completed by experimental data due to JOULE [1859].

The qualitative fit to the experimental data given by JOULE [1859] and ANTHONY et al. [1942] is satisfying; nevertheless, the physical properties of the vulcanized rubber strip used in *Joule's* and *Anthony's* experiments are only partly documented and may differ from those given above.

One of the very first works reporting simple observations regarding thermal effects due to deflections of vulcanized rubber bands was published by GOUGH [1805]. The first work which explored experimentally the crucial physical properties of elastic india-rubber was presented by JOULE [1859]. The phenomena of structural thermoelastic heating (or cooling) of stressed rubber-like materials is called the *Gough-Joule effect*. A wide range of publications followed over more than one century, which mostly deal with theoretical formulations of more or less approximate theories. The history of these investigations is discussed in the classical book by FLORY [1953, Section XI.1a], an overview is given by TRELOAR [2005, Chapters 2, 5], CHADWICK [1974] and PRICE [1976, and references therein]. For a detailed account of the relevant results see also OGDEN [1992b].

Numerical analyses for this type of adiabatic (isentropic) simple tension test within the finite element context were given by MIEHE [1995b] and HOLZAPFEL and SIMO [1996b]. The algorithmic solution procedure reduces merely to *one* step, which is concerned with solving a mechanical problem with *isentropic elasticities* as indicated explicitly through expression (7.58). During the solution process the entropy constraint condition $\Theta\dot{\eta} = 0$ must be enforced, which means that the entropy at each time-step

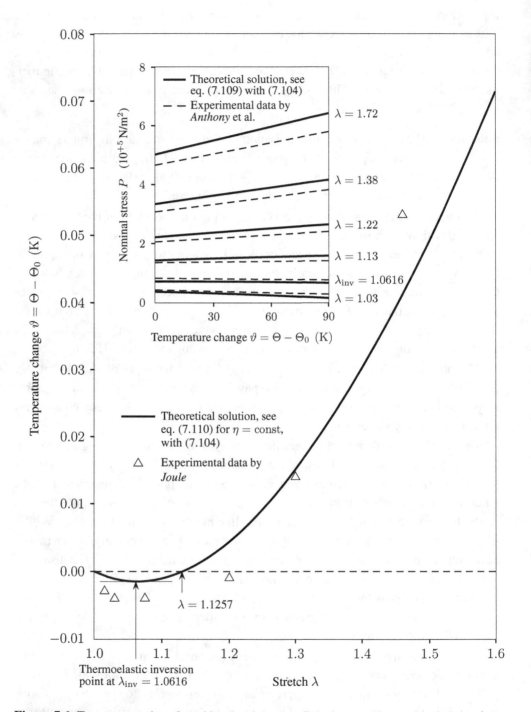

Figure 7.6 Temperature rise of a rubber band due to adiabatic stretching – showing the *thermoelastic inversion point* – and nominal stress P versus temperature change $\vartheta = \Theta - \Theta_0$ for various stretch ratios λ for a three-term Ogden elastic material.

is unchanged. As a direct result the temperature and the displacement field follow according to Figure 7.6.

Thermoelastic inversion point. For a small stretch ratio λ the rubber band indicates an initial *cooling effect* (see Figure 7.6), which increases first with deformation and changes to a *heating effect* at a certain minimum point, which entered into the literature as the so-called **thermoelastic inversion point**. The associated extension, labeled as λ_{inv}, characterizes a configuration of the sample in which the acting force is independent of the temperature.

To gain more insight in the interesting thermoelastic inversion phenomena the following observations are emphasized:

(i) An analytical investigation of the reversible adiabatic (isentropic) process – characterized by constitutive relation (7.110), with $\eta = \text{const}$ – leads, within an adequate linearization process, to an explicit expression for the stretch at the inversion point, denoted by λ_{inv}. Specifically, from $\partial\Theta/\partial\lambda = 0$ we obtain the approximate solution

$$\lambda_{inv}^3 \approx 1 + 3\alpha_0\Theta_0 \tag{7.112}$$

(see JAMES and GUTH [1943] among many others). With given data thermoelastic inversion takes place for an extension of about $\lambda_{inv} = 1.0616$ at which the maximum temperature drop occurs (see Figure 7.6).

A second point of interest is at $\vartheta = 0$, where a heating process takes place. Note that this behavior is contrary to that for a metallic spring, which cools continuously on stretching within the elastic domain. An approximate solution for λ at $\vartheta = 0$ may be derived from eq. (7.110) using an adequate linearization process. We obtain a quadratic equation in λ, i.e. $\lambda^2 + \lambda - 2\lambda_{inv}^3 \approx 0$ (see also JAMES and GUTH [1943]). Remarkably, the stretch ratio at $\vartheta = 0$ is only influenced by the location of the thermoelastic inversion point; with the given data the numerical value is $\lambda = 1.1257$ (see Figure 7.6).

(ii) In view of eq. (7.109) it is crucial to remark that for a fixed λ the stress is a nonlinear function with respect to the temperature. The nonlinearity is clearly caused by the volume thermal expansion term $J = J_\Theta(\Theta)$ characterized by eq. (7.104).

As a direct consequence the term J_Θ governs the change in the slope $\partial P/\partial\Theta$ of the stress-temperature curves, as seen in Figure 7.6. If the elongation is small the slope is negative and the deformation behavior is dominated by thermal expansion (energetic contribution). For larger strain ranges the slope $\partial P/\partial\Theta$ becomes positive. It emerges that the reversal in the slope of the weak nonlinear stress-temperature plot is indicated at $\lambda = \lambda_{inv}$, which can be shown using eq. (7.109) along with condition $\partial P/\partial\Theta = 0$.

At the thermoelastic inversion point thermal expansion and entropy contraction balance. Hence, thermoelastic inversion is governed clearly by *thermal expansion*, which is expressed by relation (7.104), i.e. $J = \exp[3\alpha_0(\Theta - \Theta_0)]$. As can be seen from (7.112), thermoelastic inversion depends basically on the linear expansion coefficient $\alpha_0 \neq 0$. Alternatively, it is interesting to note that the inversion point also occurs at the same value of extension by fixing the load of a strip of rubber and increasing the temperature. A connection of the mentioned observations may be found within the context of the thermodynamic Maxwell relations (7.39) and (7.40), which motivates the inclusion of the two diagrams within one figure (Figure 7.6) (see also the theoretical study by OGDEN [1992b]).

By setting $\alpha_0 = 0$ (which gives $e_0(J) = 0$) we may recover the purely entropic theory of rubber thermoelasticity as a special case. With the assumed constant specific heat capacity c_0 the temperature evolution is then given by an explicit function, i.e. $\Theta = \Theta_0 \exp[\Psi_0/(\Theta_0 c_0)]$ (compare with Exercise 2 on p. 336). Here, $\Psi_0 = \Psi(\lambda, \Theta_0)$ denotes the strain energy for the deformation at a fixed reference temperature Θ_0, i.e. eq. (7.107) with $\Theta = \Theta_0$. ∎

One-dimensional finite thermoelasticity.   We consider here the one-dimensional case of finite thermoelasticity. A rod (with uniform cross-section) is imagined as being stretched to λ in the direction of its (x-)axis, with the associated kinematic relation $x = \lambda X$ (uniform non-isothermal deformation). The rod capable of supporting finite thermoelastic deformations admits the *stretch ratio* λ and the *temperature* Θ as independent mechanical and thermal variables.

From the Helmholtz free-energy function $\Psi = \Psi(\lambda, \Theta)$ (measured per unit reference volume), which is assumed to exist, we are able to deduce the stress and entropy functions $P = P(\lambda, \Theta)$ and $\eta = \eta(\lambda, \Theta)$, respectively. They are nonlinear scalar-valued functions which depend on the two scalar variables λ and Θ and have the same form as (7.108). Note that the constitutive relation (7.108)$_1$ determines the nominal stress P by keeping the temperature Θ fixed while (7.108)$_2$ determines the entropy η at a fixed stretch λ.

For a more profound understanding of the constitutive relations it is beneficial to note the differential mathematical relationship between the nominal stress P and the entropy η. For the continuous function Ψ the well-known property

$$\frac{\partial}{\partial \Theta}\left(\frac{\partial \Psi}{\partial \lambda}\right)_{\Theta} = \frac{\partial}{\partial \lambda}\left(\frac{\partial \Psi}{\partial \Theta}\right)_{\lambda} \tag{7.113}$$

holds. Using eqs. (7.108)$_1$ and (7.108)$_2$, (7.113) can be rewritten in an expression which relates the stress and the entropy function according to

$$\left(\frac{\partial P}{\partial \Theta}\right)_\lambda = -\left(\frac{\partial \eta}{\partial \lambda}\right)_\Theta . \tag{7.114}$$

This identity shows that the change in stress P with the temperature Θ of, for example, a rod at fixed stretch λ is equal to the negative value of the change in entropy η with the stretch ratio λ of that rod at fixed temperature Θ. Relation (7.114) characterizes the **thermodynamic Maxwell relation** for the one-dimensional case, fundamental in rubber thermoelasticity (see WALL [1965, p. 314], TRELOAR [2005, p. 30] and CARLSON [1972, p. 304]). Compare also the identities (7.39) and (7.40), valid for the three-dimensional case.

Furthermore, it is important to note that for polymers in the 'rubbery' state the term on the left-hand side of eq. (7.114) is found to be positive at large extensions λ, and negative at small extensions, as seen clearly in Figure 7.6 (recall Example 7.5 on p. 348).

All that remains is the computation of the isothermal elasticity tensor in the material description (consistent linearized tangent moduli) and the referential stress-temperature tensor. With this aim in view we derive the change in the second Piola-Kirchhoff stress $S = S(\lambda, \Theta)$, defined as $S = \lambda^{-1}P = \lambda^{-1}(\partial\Psi(\lambda, \Theta)/\partial\lambda)$ (compare with relations (6.50), with (6.47), valid for three dimensions). Knowing that the Green-Lagrange strain, denoted by E, is $(\lambda^2 - 1)/2$ (compare with relation (2.69), valid for three dimensions), we may find

$$dS = \left(\frac{\partial S}{\partial \lambda}\right)_\Theta d\lambda + \left(\frac{\partial S}{\partial \Theta}\right)_\lambda d\Theta$$

$$= C dE + T d\Theta , \tag{7.115}$$

where $dE = \lambda d\lambda$. Here we have introduced the definitions

$$C = \lambda^{-1}\frac{\partial S(\lambda, \Theta)}{\partial \lambda} \qquad \text{and} \qquad T = \frac{\partial S(\lambda, \Theta)}{\partial \Theta} \tag{7.116}$$

of the *isothermal elasticity tensor* C in the material description (fixed temperature during the process) and the *referential stress-temperature tensor* T.

The scalar-valued functions $(7.115)_2$ and (7.116) are the one-dimensional counterparts of the tensor-valued functions (7.48) and $(7.49)_1$, $(7.50)_1$, respectively. Observe that by means of $S = \lambda^{-1}P$ and the thermodynamic Maxwell relation (7.114) we may find an equivalent of the referential stress-temperature tensor $(7.116)_2$ in the form

$$\lambda T = \left(\frac{\partial P}{\partial \Theta}\right)_\lambda = -\left(\frac{\partial \eta}{\partial \lambda}\right)_\Theta . \tag{7.117}$$

In order to complete our presentation of one-dimensional finite thermoelasticity we have to add a constitutive equation for the heat flux. For one dimension, Fourier's law of heat conduction in a coupled thermomechanical regime reads

$$Q(\lambda, \Theta, \Theta') = -k_0(\Theta)\lambda^{-2}\Theta' \ . \tag{7.118}$$

The temperature gradient along the axis of the considered rod is denoted by Θ' while Q denotes the Piola-Kirchhoff heat flux, which is, in the one-dimensional case, a scalar-valued function.

EXERCISES

1. Consider the Helmholtz free-energy function $\Psi = \Psi(\lambda, \Theta)$ in terms of the stretch ratio λ and the temperature Θ characterizing a *one-dimenional* constitutive problem of finite thermoelasticity. According to relations (7.108) we may derive constitutive equations for the nominal stress $P = P(\lambda, \Theta)$ and the entropy $\eta = \eta(\lambda, \Theta)$. Assume that the specific heat capacity is a positive constant $c_0 > 0$.

   (a) Using the chain rule, show that the change in entropy may be expressed as

   $$d\eta = -T dE + \frac{c_0}{\Theta} d\Theta \ , \tag{7.119}$$

   where $dE = \lambda d\lambda$. The referential stress-temperature tensor T is given in eq. $(7.116)_2$ (or by the equivalent of eq. (7.117)).

   (b) Using eqs. $(7.115)_2$ and (7.119) show that the isentropic elasticity tensor C^{ise} in the material description (for a fixed entropy η during a process) is governed by the relationship

   $$C^{\mathrm{ise}} = C + \frac{\Theta}{c_0}T^2 \ . \tag{7.120}$$

   Note that relations (7.119), (7.120) are the one-dimenional counterparts of relations (7.54), (7.58).

2. Suppose that a rod (considered as a one-dimensional structure) admits the stretch ratio λ and the entropy η as independent variables and consider the existence of the internal-energy function per unit reference volume in the form of $e = e(\lambda, \eta)$.

   Derive the second Piola-Kirchhoff stress $S = S(\lambda, \eta)$ and show that its change is $dS = C^{\mathrm{ise}} dE + T^{\mathrm{ise}} d\eta$, with the definitions

   $$C^{\mathrm{ise}} = \lambda^{-1}\frac{\partial S(\lambda, \eta)}{\partial \lambda} \ , \qquad T^{\mathrm{ise}} = \frac{\partial S(\lambda, \eta)}{\partial \eta}$$

   of the *isentropic elasticity tensor* C^{ise} in the material description and the *referential stress-entropy tensor* T^{ise}, evaluated at (λ, η) (compare with Section 7.5 for the three-dimensional case, in particular, with relations $(7.56)_1$ and $(7.57)_1$).

3. Consider an adiabatic (isentropic) stretching of a mechanically incompressible rubber band and study the homogeneous deformation process in the large strain domain up to a stretch ratio $\lambda = 8$. Take the material properties and assumptions according to Example 7.5 (see p. 348). As a basis for the constitutive model take the thermodynamic extension of Ogden's model, with $N = 3$, and compare with the coupled thermomechanical Mooney-Rivlin model, by setting $N = 2$, and the neo-Hookean and Varga models, by setting $N = 1$.

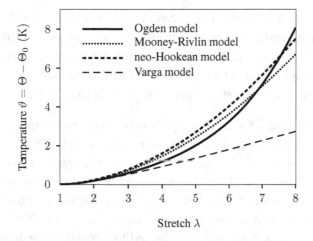

Figure 7.7 Temperature rise ϑ of a rubber band due to adiabatic stretching for the large strain domain. Comparison between four different thermoelastic models.

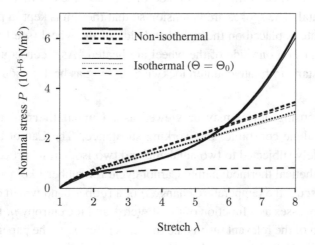

Figure 7.8 Nominal stress P versus stretch ratio λ. Comparison between a coupled thermomechanical (non-isothermal) and a decoupled (isothermal) deformation process.

In particular, for $N = 2$, assume $\alpha_1 = 2$, $\alpha_2 = -2$ and $\mu_1(\Theta_0) = 0.875\mu_0$, $\mu_2(\Theta_0) = -0.125\mu_0$, so that $\mu_1(\Theta_0)/\mu_2(\Theta_0) = -7$ (see ANAND [1986]), with the shear modulus $\mu_0 = 4.225 \cdot 10^5 \mathrm{N/m}^2$ in the reference configuration. For the neo-Hookean model ($N = 1$) assume $\alpha_1 = 2$ and $\mu_1(\Theta_0) = \mu_0$ and for the Varga model ($N = 1$), $\alpha_1 = 1$ and $\mu_1(\Theta_0) = 2\mu_0$.

(a) Based on the Ogden, Mooney-Rivlin, neo-Hookean and Varga material models derive the temperature evolution $\vartheta = \Theta - \Theta_0$ of the rubber band due to the stretch ratio λ.

(b) Derive a relationship between the nominal stress P and the stretch ratio λ for a non-isothermal deformation process. Show the difference compared with the (classical) solution, which is based on the isothermal theory (set $\Theta = \Theta_0$). Note that for an isothermal deformation process the free energy Ψ reduces to the strain energy, as given in (6.119) and eqs. (6.127)–(6.129).

Figures 7.7 and 7.8 show a comparison between the thermoelastic Ogden, Mooney-Rivlin, neo-Hookean and Varga material models. Observe the sharp upturn in the stress at high elongations ($\lambda \geq 6$) for the Ogden material model as indicated in Figure 7.8. This fact may partly be explained physically within more advanced theories obeying Langevin distribution function (non-Gaussian statistical theory), as discussed by, for example, TRELOAR [2005, Chapter 6]. The observed rise of stress is mainly caused by limited extensibility of the polymer chains themselves and by strain-induced crystallization (see FLORY [1976] for further insight).

4. A bicycle wheel with spokes made of rubber bands is mounted with the axle horizontal. The spokes are in tension so that the rim is kept in place. An electric heat plate is placed on the right-hand side of the wheel (see Figure 7.9) so that the spokes on one side of the wheel are heated. As a consequence, the bicycle wheel starts to rotate counterclockwise, as long as heat is induced. Explain this effect.

This amusing device may be viewed as a **Carnot thermal engine**, in which rubber alone constitutes the working substance. The Carnot thermal engine is alternately subjected to two adiabatic and two isothermal processes. Investigate a hypothetical thermodynamic Carnot cycle for rubber-like materials. In particular, discuss the temperature change ϑ of a rubber band which occurs during the four processes as a function of the stretch λ and the entropy η. For a detailed exposition of the relevant results the reader is referred to the paper by HOLZAPFEL and SIMO [1996b].

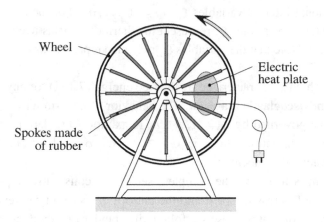

Figure 7.9 A wheel with rubber spokes starts to rotate counterclockwise when heated on one side.

7.9   Thermodynamics with Internal Variables

In this section we link together *finite elasticity* and *non-equilibrium thermodynamics*. We consider the thermodynamics of continuous media within the large strain regime and apply the theory of **finite thermoviscoelasticity**.

We use a thermodynamic approach with internal variables which leads to a very general description of materials involving irreversible (dissipative) effects, such as damage, relaxation and/or creep and plastic deformations. It generalizes finite thermoelasticity, as outlined in Section 7.3, in the sense that additional thermodynamic variables (known as internal variables) are incorporated with the aim of representing the irreversible mechanism of the (inelastic) structural material behavior. A general discussion of constitutive models with internal variables was emphasized in Section 6.9.

A fully coupled three-dimensional thermomechanical model for viscous materials is examined. It is particularly suited for the thermoviscoelastic behavior of dissipative elastomeric (rubber-like) materials under varying temperatures at finite strains. Constitutive equations for the stress, the entropy and the internal variables are specified.

Finite thermoviscoelasticity.   We define a Helmholtz free-energy function (measured per unit reference volume) as

$$\Psi = \Psi(\mathbf{F}, \Theta, \boldsymbol{\xi}_1, \dots, \boldsymbol{\xi}_m) \ . \tag{7.121}$$

The thermodynamic state is completely characterized through the set of independent variables $(\mathbf{F}, \Theta, \boldsymbol{\xi}_1, \dots, \boldsymbol{\xi}_m)$, i.e. the deformation gradient $\mathbf{F}$, the absolute temperature

Θ and m additional internal variables $\boldsymbol{\xi}_\alpha$, $\alpha = 1, \ldots, m$. For the case of thermovis-coelasticity the tensor variables $\boldsymbol{\xi}_\alpha$ represent the thermoviscoelastic contribution to the material response. Note that the variables $\boldsymbol{\xi}_\alpha$ may also represent *damage* and/or *plastic* mechanisms.

A material which is characterized by the free energy (7.121) for any point and time we call a **thermoviscoelastic material**. The behavior of a thermoviscoelastic material is assumed to be governed by $\alpha = 1, \ldots, m$ relaxation (retardation) processes with given relaxation (retardation) times $\tau_\alpha(\Theta) \in (0, \infty)$, $\alpha = 1, \ldots, m$, which are, in general, temperature dependent.

The next aim is to derive the complete set of the constitutive equations for the first Piola-Kirchhoff stress tensor $\mathbf{P}$ and the entropy η (per unit reference volume) in the general form. For that purpose we follow the standard methods presented in Sections 6.9, 7.3 and use the Clausius-Planck form of the second law of thermodynamics (4.153), i.e. the internal dissipation inequality $\mathcal{D}_{\text{int}} = \mathbf{P} : \dot{\mathbf{F}} - \dot{\Psi} - \eta\dot{\Theta} \geq 0$ (the Clausius-Planck form assumes non-negative entropy production due to conduction of heat, i.e. $-(1/\Theta)\mathbf{Q} \cdot \text{Grad}\Theta \geq 0$ (compare with Section 4.6)). By means of the chain rule, time differentiation of the free energy $\Psi(\mathbf{F}, \Theta, \boldsymbol{\xi}_1, \ldots, \boldsymbol{\xi}_m)$ gives the hypothetical change in the thermodynamic state, that is the **Gibbs relation** for thermoviscoelastic materials, namely

$$\dot{\Psi} = \dot{\Psi}(\mathbf{F}, \Theta, \boldsymbol{\xi}_1, \ldots, \boldsymbol{\xi}_m) = \mathbf{P} : \dot{\mathbf{F}} - \eta\dot{\Theta} - \mathcal{D}_{\text{int}}$$

$$= \left(\frac{\partial\Psi(\mathbf{F}, \Theta, \boldsymbol{\xi}_1, \ldots, \boldsymbol{\xi}_m)}{\partial\mathbf{F}}\right)_{\Theta, \boldsymbol{\xi}_\alpha} : \dot{\mathbf{F}} + \left(\frac{\partial\Psi(\mathbf{F}, \Theta, \boldsymbol{\xi}_1, \ldots, \boldsymbol{\xi}_m)}{\partial\Theta}\right)_{\mathbf{F}, \boldsymbol{\xi}_\alpha} \dot{\Theta}$$

$$+ \sum_{\alpha=1}^{m}\left(\frac{\partial\Psi(\mathbf{F}, \Theta, \boldsymbol{\xi}_1, \ldots, \boldsymbol{\xi}_m)}{\partial\boldsymbol{\xi}_\alpha}\right)_{\mathbf{F}, \Theta} : \dot{\boldsymbol{\xi}}_\alpha \ , \tag{7.122}$$

where $\dot{\boldsymbol{\xi}}_\alpha$, $\alpha = 1, \ldots, m$, denote the internal strain rates.

Since the rates $\dot{\mathbf{F}}$ and $\dot{\Theta}$ can be chosen arbitrarily we find the constitutive equations

$$\mathbf{P} = \left(\frac{\partial\Psi(\mathbf{F}, \Theta, \boldsymbol{\xi}_1, \ldots, \boldsymbol{\xi}_m)}{\partial\mathbf{F}}\right)_{\Theta, \boldsymbol{\xi}_\alpha}, \qquad \eta = -\left(\frac{\partial\Psi(\mathbf{F}, \Theta, \boldsymbol{\xi}_1, \ldots, \boldsymbol{\xi}_m)}{\partial\Theta}\right)_{\mathbf{F}, \boldsymbol{\xi}_\alpha} \tag{7.123}$$

for the first Piola-Kirchhoff stress $\mathbf{P}$ and the entropy η, and we deduce a remainder inequality

$$\mathcal{D}_{\text{int}} = \sum_{\alpha=1}^{m}\boldsymbol{\Xi}_\alpha : \dot{\boldsymbol{\xi}}_\alpha \geq 0 \ , \qquad \boldsymbol{\Xi}_\alpha = -\left(\frac{\partial\Psi(\mathbf{F}, \Theta, \boldsymbol{\xi}_1, \ldots, \boldsymbol{\xi}_m)}{\partial\boldsymbol{\xi}_\alpha}\right)_{\mathbf{F}, \Theta}. \tag{7.124}$$

The inequality $\mathcal{D}_{\text{int}} \geq 0$ characterizes the internal dissipation in the viscous material which generates heat in an irreversible manner.

The defined tensor variables $\boldsymbol{\Xi}_\alpha$, $\alpha = 1, \ldots, m$, correspond to the internal tensor variables $\boldsymbol{\xi}_\alpha$ according to the internal constitutive equations $(7.124)_2$, which is the thermodynamic extension of relation (6.233) (or (6.244)). By analogy with a linear solid, Ξ_α, $\alpha = 1, \ldots, m$, are to be interpreted as (internal) non-equilibrium stresses (compare with Section 6.10).

From physical expressions (7.123) and $(7.124)_2$ we deduce functions for the first Piola-Kirchhoff stress $\mathbf{P}$, the entropy η and the internal (stress-like) variables Ξ_α which depend on $\mathbf{F}, \Theta, \boldsymbol{\xi}_1, \ldots, \boldsymbol{\xi}_m$. Thus,

$$\mathbf{P} = \mathbf{P}(\mathbf{F}, \Theta, \boldsymbol{\xi}_1, \ldots, \boldsymbol{\xi}_m) \quad , \qquad \eta = \eta(\mathbf{F}, \Theta, \boldsymbol{\xi}_1, \ldots, \boldsymbol{\xi}_m) \quad , \qquad (7.125)$$

$$\Xi_\alpha = \Xi_\alpha(\mathbf{F}, \Theta, \boldsymbol{\xi}_1, \ldots, \boldsymbol{\xi}_m) \quad , \qquad \alpha = 1, \ldots, m \quad . \qquad (7.126)$$

The fundamental inequality $(7.124)_1$, which characterizes internal dissipation, must be satisfied by a suitable set of evolution equations for the internal strain rates $\dot{\boldsymbol{\xi}}_\alpha$, $\alpha = 1, \ldots, m$, described generally in the form

$$\dot{\boldsymbol{\xi}}_\alpha(t) = \mathcal{A}_\alpha(\mathbf{F}, \Theta, \boldsymbol{\xi}_1, \ldots, \boldsymbol{\xi}_m) \quad , \qquad \alpha = 1, \ldots, m \quad . \qquad (7.127)$$

The equations of evolution (rate equations) (7.127) describe the way in which an irreversible process evolves.

Note that the non-equilibrium stresses characterize the current 'distance from equilibrium' and vanish at the state of *thermodynamic equilibrium*. In view of (7.127) we may write the equations $\mathcal{A}_\alpha(\mathbf{F}, \Theta, \boldsymbol{\xi}_1, \ldots, \boldsymbol{\xi}_m) = \mathbf{O}$, $\alpha = 1, \ldots, m$, as time t goes to infinity; further, $\Xi_\alpha = -\partial\Psi/\partial\boldsymbol{\xi}_\alpha|_{t\to\infty} \equiv \mathbf{O}$, $\alpha = 1, \ldots, m$. This implies that with reference to $(7.124)_1$ the internal dissipation vanishes ($\mathcal{D}_{\text{int}} = 0$). Then equilibrium is reached and the values for stress and entropy remain constant. They are governed by the potential relations as derived in Section 7.3 (see eqs. (7.24)).

The limiting case of thermodynamic equilibrium states that the thermodynamic process is reversible and the continuum responds *fully thermoelastically*. All associated thermodynamic potentials, as outlined in Section 7.3 for finite thermoelasticity, are approximated asymptotically.

In summary: the response of a thermoviscoelastic material is defined through the constitutive equations (7.123) and $(7.124)_2$ (or (7.125) and (7.126)), the internal dissipation $(7.124)_1$ and the evolution equations as outlined generally in (7.127). In addition, these equations are supplemented by a suitable constitutive equation for the heat flux vector, necessary to determine heat transfer. The Piola-Kirchhoff heat flux $\mathbf{Q}$ may be introduced as a function of the deformation gradient, temperature, temperature gradient and internal variables, i.e.

$$\mathbf{Q} = \mathbf{Q}(\mathbf{F}, \Theta, \text{Grad}\Theta, \boldsymbol{\xi}_1, \ldots, \boldsymbol{\xi}_m) \quad , \qquad (7.128)$$

and must satisfy the inequality $\mathbf{Q} \cdot \text{Grad}\Theta \leq 0$.

Structural thermoviscoelastic heating (or cooling).   The specific heat capacity c_F at constant deformation per unit reference volume was introduced and discussed in Section 7.4. Within the theory of thermodynamics with internal variables we define the specific heat capacity to have a positive value which depends on the deformation gradient $\mathbf{F}$, the temperature field Θ and, additionally, on the internal variables $\boldsymbol{\xi}_\alpha$, $\alpha = 1, \dots, m$. Thus, we write

$$c_F = c_F(\mathbf{F}, \Theta, \boldsymbol{\xi}_1, \dots, \boldsymbol{\xi}_m) = -\Theta \left(\frac{\partial^2 \Psi(\mathbf{F}, \Theta, \boldsymbol{\xi}_1, \dots, \boldsymbol{\xi}_m)}{\partial \Theta \partial \Theta} \right)_{\mathbf{F}, \boldsymbol{\xi}_\alpha} > 0 \qquad (7.129)$$

for all $(\mathbf{F}, \Theta, \boldsymbol{\xi}_1, \dots, \boldsymbol{\xi}_m)$. Here the specific heat capacity is the energy required to produce unit increase in the temperature of a unit volume of the body keeping the *deformation* and the *internal variables fixed*. For notational convenience we shall use the same symbol c_F for the specific heat capacity introduced in Section 7.4.

Recall that within the theory of finite thermoelasticity we observe, in general, three different thermomechanical coupling effects (see p. 327). In addition, finite thermoviscoelasticity incorporates *viscous dissipation* according to $(7.124)_1$ and **structural thermoviscoelastic heating (or cooling)**, denoted $\mathcal{H}_{in}$ and defined as

$$\mathcal{H}_{in} = -\Theta \sum_{\alpha=1}^{m} \frac{\partial^2 \Psi(\mathbf{F}, \Theta, \boldsymbol{\xi}_1, \dots, \boldsymbol{\xi}_m)}{\partial \boldsymbol{\xi}_\alpha \partial \Theta} : \dot{\boldsymbol{\xi}}_\alpha \quad . \qquad (7.130)$$

For the case in which the quantities $\boldsymbol{\xi}_\alpha$, $\alpha = 1, \dots, m$, represent plastic contributions, eq. (7.130) defines **structural inelastic (plastic) heating (or cooling)**.

By analogy with the derivation which led to the energy balance equation in temperature form (7.46) we proceed now by determining the change in entropy. With the equations of state $(7.125)_2$ and $(7.123)_2$ and by means of the chain rule we deduce that

$$\dot{\eta}(\mathbf{F}, \Theta, \boldsymbol{\xi}_1, \dots, \boldsymbol{\xi}_m) = \frac{\partial \eta}{\partial \mathbf{F}} : \dot{\mathbf{F}} + \frac{\partial \eta}{\partial \Theta} \dot{\Theta} + \sum_{\alpha=1}^{m} \frac{\partial \eta}{\partial \boldsymbol{\xi}_\alpha} : \dot{\boldsymbol{\xi}}_\alpha$$

$$= -\frac{\partial^2 \Psi}{\partial \mathbf{F} \partial \Theta} : \dot{\mathbf{F}} - \frac{\partial^2 \Psi}{\partial \Theta \partial \Theta} \dot{\Theta} - \sum_{\alpha=1}^{m} \frac{\partial^2 \Psi}{\partial \boldsymbol{\xi}_\alpha \partial \Theta} : \dot{\boldsymbol{\xi}}_\alpha \qquad (7.131)$$

(the arguments of the functions have been omitted for simplicity). By multiplying this equation with the temperature Θ and using eqs. $(7.44)_3$, $(7.129)_2$ and (7.130), we find that

$$\Theta \dot{\eta} = \mathcal{H}_e + c_F \dot{\Theta} + \mathcal{H}_{in} \quad . \qquad (7.132)$$

On comparison with relation (4.142) we obtain finally the energy balance equation in *temperature form*, i.e.

$$c_F \dot{\Theta} = -\text{Div} \mathbf{Q} + \mathcal{D}_{int} - \mathcal{H}_e - \mathcal{H}_{in} + R \quad . \qquad (7.133)$$

From the energy balance equation (7.46) we know that, within the theory of finite thermoelasticity, the evolution of the temperature Θ is influenced by the material divergence $\text{Div}\mathbf{Q}$ of the Piola-Kirchhoff heat flux $\mathbf{Q}$, the structural thermo*elastic* heating $\mathcal{H}_e$ and the heat source R. However, relation (7.133) indicates that due to viscous effects the quantity $c_F \dot{\Theta}$ depends additionally on the internal dissipation $\mathcal{D}_{int}$ and the structural thermo*viscoelastic* heating $\mathcal{H}_{in}$, as defined in eqs. (7.124) and (7.130), and this is particularly important to the thermomechanical behavior of viscous materials.

A constitutive model for finite thermoviscoelasticity. Many materials which behave elastically at ordinary (room) temperatures display pronounced inelastic characteristics at elevated temperatures, solid polymers being important examples. The molecular network of vulcanized rubber exhibits (nearly) no stress relaxation in the low temperature range of the 'rubbery' state. However, in the temperature range of 100 to 150^0C stress relaxation-experiments of vulcanized rubber at constant deformation (see TOBOLSKY et al. [1944]) indicate a rapid stress-decay which may be explained by chemical rupture of the three-dimensional network. The phenomenon of stress relaxation of rubber is strongly influenced by temperature changes but is independent of both the deformed state of the network and the absence or presence of carbon-black fillers in the rubber (see TOBOLSKY et al. [1944], TOBOLSKY [1960, Chapter 5] and LEE et al. [1966]).

In the remaining part of this chapter we consider a three-dimensional constitutive model for dissipative continuous media capable of accommodating thermoviscoelastic changes at finite strains. The model is considered to be an extension of the three-dimensional viscoelastic model (proposed in Section 6.10) to the thermodynamic regime satisfying the entropy inequality principle. The mathematical structure of the thermoviscoelastic material model is based on the concept of internal state variables and is well-suited for numerical realization using the finite element method. The material model is based on a common assumption for viscous materials, namely that the evolution equations for the internal variables are linear. This simplification enables us to characterize states close to thermodynamic equilibrium. For thermoviscoelastic models that account for finite perturbations away from thermodynamic equilibrium the reader is referred to LION [1997b] and REESE and GOVINDJEE [1998b].

We choose a geometric setting relative to the reference configuration and postulate for non-isothermal processes a Helmholtz-free energy function Ψ (measured per unit reference volume) in the form

$$\Psi(\mathbf{C}, \Theta, \mathbf{\Gamma}_1, \ldots, \mathbf{\Gamma}_m) = \Psi_\infty(\mathbf{C}, \Theta) + \sum_{\alpha=1}^{m} \Upsilon_\alpha(\mathbf{C}, \Theta, \mathbf{\Gamma}_\alpha) \ , \qquad (7.134)$$

valid for some closed time interval $t \in [0, T]$ of interest. We require that

$$\Psi_\infty(\mathbf{I}, \Theta_0) = 0 \ , \qquad \sum_{\alpha=1}^{m} \Upsilon_\alpha(\mathbf{I}, \Theta_0, \mathbf{I}) = 0 \ , \qquad (7.135)$$

where $\Theta_0(> 0)$ is a given homogeneous reference temperature relative to a selected stress-free reference configuration. The set of independent variables $(\mathbf{C}, \Theta, \boldsymbol{\Gamma}_1, \ldots, \boldsymbol{\Gamma}_m)$, i.e. the (symmetric) right Cauchy-Green tensor $\mathbf{C}$, the absolute temperature Θ and the (symmetric) internal variables $\boldsymbol{\Gamma}_\alpha$, $\alpha = 1, \ldots, m$ (not accessible to direct observation), completely characterizes the thermodynamic state. The internal variables $\boldsymbol{\Gamma}_\alpha$ are considered as inelastic (viscous) strains akin to the strain measure $\mathbf{C}$.

The first term $\Psi_\infty(\mathbf{C}, \Theta)$ in (7.134) characterizes the *equilibrium state* of the solid. We employ the subscript $(\bullet)_\infty$ to designate functions which represent the hyperelastic behavior of sufficiently slow processes. An efficient free energy Ψ_∞ describing the stress-strain-temperature response of rubber-like materials at finite strains, which is based on the concept of entropic elasticity, may be adopted from Section 7.7 (see eq. (7.80)).

The second term $\sum_{\alpha=1}^{m} \Upsilon_\alpha(\mathbf{C}, \Theta, \boldsymbol{\Gamma}_\alpha)$ in (7.134) represents the configurational free energy ('dissipative' potential) and characterizes the *non-equilibrium state* of the solid (relaxation and/or creep behavior). The potential Υ_α has to satisfy the thermodynamic restrictions imposed on the second law of thermodynamics (namely the non-negativeness of the internal dissipation) for any thermodynamic process. For a general form of Υ_α and a detailed discussion of this issue the reader is referred to HOLZAPFEL and SIMO [1996c].

Following arguments analogous to those which led from (7.121) to relations (7.123) and (7.124) we find, using the property (6.11) and $\mathbf{S} = \mathbf{F}^{-1}\mathbf{P}$, physical expressions for the (symmetric) second Piola-Kirchhoff stress tensor $\mathbf{S}$ and the entropy η (per unit reference volume) in the forms

$$\mathbf{S} = \mathbf{S}_\infty(\mathbf{C}, \Theta) + \sum_{\alpha=1}^{m} \mathbf{Q}_\alpha(\mathbf{C}, \Theta, \boldsymbol{\Gamma}_\alpha) \ , \tag{7.136}$$

$$\eta = \eta_\infty(\mathbf{C}, \Theta) + \sum_{\alpha=1}^{m} \eta_\alpha(\mathbf{C}, \Theta, \boldsymbol{\Gamma}_\alpha) \ , \tag{7.137}$$

and a remainder inequality

$$\mathcal{D}_{\text{int}} = -\sum_{\alpha=1}^{m} 2\frac{\partial \Upsilon_\alpha(\mathbf{C}, \Theta, \boldsymbol{\Gamma}_\alpha)}{\partial \boldsymbol{\Gamma}_\alpha} : \frac{1}{2}\dot{\boldsymbol{\Gamma}}_\alpha \geq 0 \ , \tag{7.138}$$

where $\dot{\boldsymbol{\Gamma}}_\alpha$, $\alpha = 1, \ldots, m$, denote the internal strain rates. The inequality governs the non-negativeness of the internal dissipation $\mathcal{D}_{\text{int}}$ in the thermoviscoelastic material.

We have introduced the definitions

$$\mathbf{S}_\infty = 2\frac{\partial \Psi_\infty(\mathbf{C}, \Theta)}{\partial \mathbf{C}} \ , \qquad \eta_\infty = -\frac{\partial \Psi_\infty(\mathbf{C}, \Theta)}{\partial \Theta} \ , \tag{7.139}$$

$$\mathbf{Q}_\alpha = 2\frac{\partial \Upsilon_\alpha(\mathbf{C}, \Theta, \boldsymbol{\Gamma}_\alpha)}{\partial \mathbf{C}} \quad , \qquad \eta_\alpha = -\frac{\partial \Upsilon_\alpha(\mathbf{C}, \Theta, \boldsymbol{\Gamma}_\alpha)}{\partial \Theta} \qquad (7.140)$$

of the contributions to the stress and the entropy, with $\alpha = 1, \ldots, m$.

As a result of the mathematical structure given in (7.134) the second Piola-Kirchhoff stress $\mathbf{S}$ and the entropy η are decomposed into *equilibrium* and *non-equilibrium parts* according to eqs. (7.136) and (7.137), respectively. The stress contribution $\mathbf{S}_\infty$ and the entropy contribution η_∞ are associated with the fully thermoelastic response which we describe within the framework of finite thermolasticity introduced in Section 7.3. In particular, we may adopt the relations (7.26)$_3$ and (7.27) by using Ψ_∞ instead of Ψ. According to definitions (7.140) the second terms in eqs. (7.136) and (7.137), i.e. $\sum_{\alpha=1}^{m} \mathbf{Q}_\alpha$ and $\sum_{\alpha=1}^{m} \eta_\alpha$, contribute to the stresses and the entropy and are responsible for the viscous response of the material.

The variables $\mathbf{Q}_\alpha$, $\alpha = 1, \ldots, m$, denote non-equilibrium stresses and are related (conjugate) to the (right Cauchy-Green) strain-like variables $\boldsymbol{\Gamma}_\alpha$. Hence, $\mathbf{Q}_\alpha$ are internal tensor variables with the internal constitutive equations

$$\mathbf{Q}_\alpha = -2\frac{\partial \Upsilon_\alpha(\mathbf{C}, \Theta, \boldsymbol{\Gamma}_\alpha)}{\partial \boldsymbol{\Gamma}_\alpha} \quad , \qquad \alpha = 1, \ldots, m \quad , \qquad (7.141)$$

which restrict the configurational free energy Υ_α in view of eq. (7.140)$_1$ (compare also with the discussion in Section 6.10, in particular, eq. (6.244)). Hence, considering (7.138), the internal dissipation takes on the form $\mathcal{D}_{\text{int}} = \sum_{\alpha=1}^{m} \mathbf{Q}_\alpha : \dot{\boldsymbol{\Gamma}}_\alpha / 2 \geq 0$, and vanishes at the state of thermodynamic equilibrium, since $\mathbf{Q}_\alpha = -2\partial\Upsilon_\alpha/\partial\boldsymbol{\Gamma}_\alpha|_{t\to\infty} \equiv \mathbf{O}$, $\alpha = 1, \ldots, m$.

In order to describe thermoviscoelastic processes the proposed constitutive model must be complemented by suitable *equations of evolution* (rate equations). In particular, we want to specify the evolution of the non-equilibrium stresses $\mathbf{Q}_\alpha$, $\alpha = 1, \ldots, m$, involved. A simple set of linear evolution equations for $\mathbf{Q}_\alpha$ is assumed to have the form

$$\dot{\mathbf{Q}}_\alpha + \frac{\mathbf{Q}_\alpha}{\tau_\alpha} = \dot{\mathbf{S}}_\alpha - \mathbf{Q}_{\alpha_{\text{cpl}}} \quad , \qquad \alpha = 1, \ldots, m \quad , \qquad (7.142)$$

where (7.142) is valid for some semi-open time interval $t \in (0, T]$ and for small perturbations away from the equilibrium state (for small strain rates). As usual, we start from a stress-free reference configuration which requires the values $\mathbf{Q}_\alpha|_{t=0} = \mathbf{O}$ for the internal variables to be zero at initial time $t = 0$. The first-order differential equations (7.142) require additional data in the form of initial conditions $\mathbf{Q}_{\alpha\,0+}$ at time $t = 0^+$.

The term $\mathbf{Q}_{\alpha_{\text{cpl}}}$, $\alpha = 1, \ldots, m$, represents thermomechanical coupling effects which come from temperature dependent material parameters. Note that this term vanishes for purely mechanically based theories (see eq. (6.252)). The second-order tensors

$\mathbf{Q}_{\alpha_{\mathrm{cpl}}}$ have to be determined such that evolution equations (7.142) are dissipative and compatible with the internal constitutive equations (7.141).

The second Piola-Kirchhoff stresses $\mathbf{S}_\alpha$, $\alpha = 1, \ldots, m$, in evolution equations (7.142) correspond to the free energies $\Psi_\alpha(\mathbf{C}, \Theta)$ (with $\Psi_\alpha(\mathbf{I}, \Theta_0) = 0$ at the reference configuration). They are responsible for the viscoelastic contribution and are related to the α-relaxation (retardation) process. We characterize the material variables $\mathbf{S}_\alpha$ by the constitutive equations

$$\mathbf{S}_\alpha = 2\frac{\partial \Psi_\alpha(\mathbf{C}, \Theta)}{\partial \mathbf{C}} \quad , \qquad \alpha = 1, \ldots, m \quad . \tag{7.143}$$

The stresses $\mathbf{S}_\alpha$ only depend on the *external variables* $\mathbf{C}$ and Θ. For the case of solid polymers that are composed of identical polymer chains we may replace Ψ_α by the free energy Ψ_∞ and adopt a relation similar to that given by eq. (6.256), namely $\Psi_\alpha(\mathbf{C}, \Theta) = \beta_\alpha^\infty \Psi_\infty(\mathbf{C}, \Theta)$, where $\beta_\alpha^\infty \in [0, \infty)$ are given non-dimensional free-energy factors. Note that Ψ_α must be related to the configurational free energy Υ_α in such a way that the second law of thermodynamics is satisfied for any thermodynamic process (see HOLZAPFEL and SIMO [1996c]).

The α-relaxation process is associated with the relaxation time $\tau_\alpha \in (0, \infty)$, which, in general, depends on the absolute temperature Θ (see TOBOLSKY et al. [1944]). The temperature dependence on the relaxation time may be related to the **activation energy** E_a of the relaxation process and expressed according to the **Arrhenius equation**,

$$\tau = A\exp\left(\frac{E_a}{\mathcal{R}\Theta}\right) \quad , \tag{7.144}$$

where A represents a constant for the reacting substance and $\mathcal{R} = 8.31\mathrm{Nm/Kmol}$ denotes the **gas constant** (or **universal gas constant**). For an explicit derivation of the empirical exponential function (7.144) the reader is referred to SPERLING [1992, Appendix 10.1]. The form of the functional dependence of the relaxation time on temperature predicts the physical observation that viscoelastic effects occur faster as temperature increases. The Arrhenius equation in the form of (7.144) is representative of most polymer relaxations (LEE et al. [1966] and TOBOLSKY et al. [1944]).

EXAMPLE 7.6   The purpose of this example is to illustrate the introduced phenomenological constitutive model for thermoviscoelastic materials by means of simple particularizations. The free-energy functions Υ_α and Ψ_α are considered to be of the quadratic forms

$$\Upsilon_\alpha = \mu_\alpha(\Theta)\mathrm{tr}[(\mathbf{E} - \mathbf{\Lambda}_\alpha)^2] \quad , \qquad \Psi_\alpha = \mu_\alpha(\Theta)\mathrm{tr}(\mathbf{E}^2) \quad , \tag{7.145}$$

(with $\alpha = 1, \ldots, m$), where $\mu_\alpha > 0$ is a *temperature dependent* (Lamé-type) shear modulus characterizing the thermoviscoelastic behavior of the α-relaxation process with given relaxation time $\tau_\alpha > 0$. The functions (7.145) are of *Saint-Venant Kirchhoff-type* (compare with eq. (6.151)), in which, for simplicity, only the shear moduli are attached to the functions. The elastic strains are described by the (symmetric) Green-Lagrange strain tensor $\mathbf{E}$ defined by (2.69), while

$$\Lambda_\alpha = \frac{1}{2}(\Gamma_\alpha - \mathbf{I}) \ , \qquad \alpha = 1, \ldots, m \tag{7.146}$$

denote the inelastic (viscous) strain measures expressed by the (symmetric) second-order tensors Λ_α.

Consider a coupling term $\mathbf{Q}_{\alpha_{cpl}}$, $\alpha = 1, \ldots, m$, and a viscous dissipation $\mathcal{D}_{int}$ of the forms

$$\mathbf{Q}_{\alpha_{cpl}} = 2\dot{\mu}_\alpha(\Theta)\Lambda_\alpha \ , \qquad \mathcal{D}_{int} = \sum_{\alpha=1}^{m} \frac{1}{\hat{\eta}_\alpha}|\mathbf{Q}_\alpha|^2 \geq 0 \ , \tag{7.147}$$

with the relationship $\hat{\eta}_\alpha = 2\mu_\alpha\tau_\alpha$, motivated by the linear theory of viscoelasticity. The parameter $\hat{\eta}_\alpha > 0$ characterizes the viscosity of the α-relaxation process (usually denoted in the literature by the symbol η, but to avoid confusion with the entropy η we place an accent over the symbol), while $|\mathbf{Q}_\alpha| = (\mathbf{Q}_\alpha : \mathbf{Q}_\alpha)^{1/2}$ denotes the *norm* of the tensor $\mathbf{Q}_\alpha$, which is a non-negative real number. Hence, the non-equilibrium stresses generate a non-negative dissipation such that the inequality is satisfied.

Obtain all the relevant thermodynamic relations, in particular the constitutive equations for the stress and the entropy. Specify the evolution equations introduced in (7.142) and discuss the state of thermodynamic equilibrium. For a nice rheological interpretation of this type of thermoviscoelastic model, given by the thermomechanical description (7.145)–(7.147), the reader is referred to the exercises below in this section.

Solution.   By adopting (7.140)₁ and particularization (7.145)₁ we may derive the non-equilibrium stresses $\mathbf{Q}_\alpha$, which characterize the current 'distance from equilibrium'. By means of property (1.252)₂, relation (2.69) and the chain rule we obtain

$$\mathbf{Q}_\alpha = 2\frac{\partial \Upsilon_\alpha(\mathbf{C}, \Theta, \Gamma_\alpha)}{\partial \mathbf{C}} = 2\mu_\alpha(\Theta)\frac{\partial \mathrm{tr}[(\mathbf{E} - \Lambda_\alpha)^2]}{\partial(\mathbf{E} - \Lambda_\alpha)} : \frac{\partial(\mathbf{E} - \Lambda_\alpha)}{\partial \mathbf{C}}$$

$$= 2\mu_\alpha(\Theta)(\mathbf{E} - \Lambda_\alpha) \ , \qquad \alpha = 1, \ldots, m \ . \tag{7.148}$$

It is important to note that a straightforward differentiation of Υ_α with respect to Γ_α gives the same result as for $\mathbf{Q}_\alpha$, as can be seen by recalling (7.141) and using property (1.252)₂, the relation (7.146) and the chain rule.

In order to compute the entropy generated by the relaxation process, we use (7.140)₂ and particularization (7.145)₁. Thus, we obtain

$$\eta_\alpha = -\frac{\partial \Upsilon_\alpha(\mathbf{C}, \Theta, \Gamma_\alpha)}{\partial \Theta} = -\frac{\partial \mu_\alpha(\Theta)[\mathrm{tr}(\mathbf{E} - \Lambda_\alpha)^2]}{\partial \Theta}$$

$$= -\mu'_\alpha(\Theta)[\mathrm{tr}(\mathbf{E} - \Lambda_\alpha)^2] \ , \qquad \alpha = 1, \ldots, m \ , \tag{7.149}$$

with the common notation $\mu'_\alpha(\Theta) = d\mu_\alpha(\Theta)/d\Theta$.

Hence, the stress and entropy response at time t follow from eqs. (7.136) and (7.137) as

$$\mathbf{S} = \mathbf{S}_\infty + \sum_{\alpha=1}^{m} 2\mu_\alpha(\Theta)(\mathbf{E} - \mathbf{\Lambda}_\alpha) \ , \tag{7.150}$$

$$\eta = \eta_\infty - \sum_{\alpha=1}^{m} \mu'_\alpha(\Theta)\text{tr}[(\mathbf{E} - \mathbf{\Lambda}_\alpha)^2] \ . \tag{7.151}$$

Before proceeding to examine the evolution equations it is first necessary to determine the second Piola-Kirchhoff stresses $\mathbf{S}_\alpha$. From (7.143) and (7.145)$_2$ we find, by means of property (1.252)$_2$, relation (2.69) and the chain rule, that

$$\mathbf{S}_\alpha = 2\frac{\partial \Psi_\alpha(\mathbf{C}, \Theta)}{\partial \mathbf{C}} = 2\mu_\alpha(\Theta)\frac{\partial \text{tr}(\mathbf{E}^2)}{\partial \mathbf{E}} : \frac{\partial \mathbf{E}}{\partial \mathbf{C}} = 2\mu_\alpha(\Theta)\mathbf{E} \ , \tag{7.152}$$

with $\alpha = 1, \dots, m$. Hence, from (7.142) and with help of the product rule and assumption (7.147)$_1$ we obtain finally the evolution equations for the non-equilibrium stresses, namely

$$\dot{\mathbf{Q}}_\alpha + \frac{\mathbf{Q}_\alpha}{\tau_\alpha} = 2\mu_\alpha(\Theta)\dot{\mathbf{E}} + 2\dot{\mu}_\alpha(\Theta)(\mathbf{E} - \mathbf{\Lambda}_\alpha) \ , \qquad \alpha = 1, \dots, m \ , \tag{7.153}$$

which are valid for some semi-open time interval $t \in (0, T]$.

On comparing the given viscous dissipation (7.147)$_2$ with (7.138) (by means of (7.141)), we conclude that $\mathbf{Q}_\alpha : (\mathbf{Q}_\alpha/\hat{\eta}_\alpha - \dot{\mathbf{\Gamma}}_\alpha/2) = 0$. Hence, for the case in which $\mathbf{Q}_\alpha$ is different from zero we find that the expressions in parentheses must vanish. Thus, using (7.146), we obtain

$$\mathbf{Q}_\alpha = \hat{\eta}_\alpha \dot{\mathbf{\Lambda}}_\alpha \ , \qquad \alpha = 1, \dots, m \ , \tag{7.154}$$

which is viewed as the classical Newtonian constitutive equation for the shear stress applied to simple shear (see eq. (5.100)). In eq. (7.154), $\mathbf{Q}_\alpha$ may be interpreted as the non-equilibrium shear stress, $\dot{\mathbf{\Lambda}}_\alpha$ as the shear rate (of a dashpot) and $\hat{\eta}_\alpha$ has the characteristic of the Newtonian shear viscosity.

The state of thermodynamic equilibrium requires that $\mathbf{Q}_\alpha = \mathbf{O}$ for $t \to \infty$, and that the internal dissipation $\mathcal{D}_{\text{int}}$ vanishes (see eq. (7.147)$_2$). In view of (7.154) this implies that $\dot{\mathbf{\Lambda}}_\alpha|_{t\to\infty} = \mathbf{O}$. For this limiting case the thermodynamic process is reversible and the material response is fully thermoelastic. ■

Note that the vast majority of polymers exhibit the well-known **Newtonian shear thinning phenomenon** (pseudoplasticity) which means that with respect to Newtonian characteristics the shear rate $\dot{c}$ increases faster than the shear stress σ_{12} increases (see, for example, BARNES et al. [1989, Chapter 2]). A model for both Newtonian and non-Newtonian materials is the extensively used **power law model** (see, for example, ROSEN [1979, and references therein]). This empirical model is frequently written in the form

$$\sigma_{12} = \sigma_{21} = m\dot{c}^n \ , \qquad (7.155)$$

which, in this form, is only valid for simple shear. Here, n is the **power law factor** (or the **flow behavior index**) and m is a (temperature dependent) parameter called the **viscosity index**. Shear thinning occurs if $n < 1$.

This model is used for a large number of engineering applications because it can be fitted to experimental results for various materials and reduces to a Newtonian fluid for n equal to 1, in which case m is known as the *viscosity* of the Newtonian fluid (compare with eqs. (7.154) or (5.100)). For typical parameter values see BARNES et al. [1989, p. 22]. For an overview of different types of rheological models the reader is referred to, for example, ROSEN [1979] and SCHOFF [1988, p. 455].

EXERCISES

1. The rheological model as illustrated in Figure 7.10 (referred to briefly as a *thermomechanical device*) is a suitable spring-and-dashpot model representing quantitatively the mechanical behavior of real thermoviscoelastic materials.

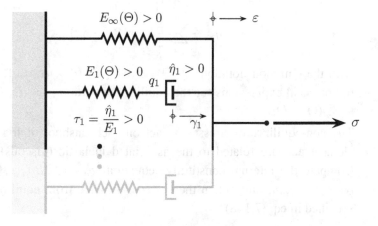

Figure 7.10 Rheological model with temperature dependent moduli.

For simplicity the thermomechanical device is assumed to have unit area and unit length so that stresses and strains are to be interpreted as forces and extensions (contractions), respectively.

It is considered to be a one-dimensional *generalized Maxwell model* with springs of *Hookean* type and dashpots of *Newtonian* type. The temperature dependent Young's moduli and the Newtonian shear viscosities are given by $E_\infty(\Theta) > 0$, $E_\alpha(\Theta) > 0$ and $\hat{\eta}_\alpha > 0$, $\alpha = 1, \ldots, m$, respectively. We now define the free energy $\psi(\varepsilon, \Theta, \gamma_1, \ldots, \gamma_m) = \psi_\infty(\varepsilon, \Theta) + \sum_{\alpha=1}^m v_\alpha(\varepsilon, \Theta, \gamma_\alpha)$, with the quadratic forms

$$\psi_\infty(\varepsilon, \Theta) = \frac{1}{2}E_\infty(\Theta)\varepsilon^2 \ , \tag{7.156}$$

$$v_\alpha(\varepsilon, \Theta, \gamma_\alpha) = \frac{1}{2}E_\alpha(\Theta)(\varepsilon - \gamma_\alpha)^2 \ , \qquad \alpha = 1, \ldots, m \ , \tag{7.157}$$

and the requirements $\psi_\infty(0, \Theta_0) = 0$ and $v_\alpha(0, \Theta_0, 0) = 0$, $\alpha = 1, \ldots, m$. The energy function ψ depends on the external variable ε (measuring the total linear strain), the absolute temperature Θ and the inelastic (viscous) strains γ_α, $\alpha = 1, \ldots, m$. The free energy v_α is responsible for the α-relaxation process of the α-Maxwell element with given relaxation time $\tau_\alpha > 0$.

(a) Based on assumptions (7.156) and (7.157) obtain the explicit constitutive equations for the total linear stress $\sigma = \partial\psi/\partial\varepsilon$ and the total entropy $\eta = -\partial\psi/\partial\Theta$ in the forms

$$\sigma = \sigma_\infty + \sum_{\alpha=1}^m \underbrace{E_\alpha(\Theta)(\varepsilon - \gamma_\alpha)}_{q_\alpha} \ , \tag{7.158}$$

$$\eta = \eta_\infty - \sum_{\alpha=1}^m \frac{1}{2}E'_\alpha(\Theta)(\varepsilon - \gamma_\alpha)^2 \ , \tag{7.159}$$

with the common notation $E'_\alpha(\Theta) = \mathrm{d}E_\alpha(\Theta)/\mathrm{d}\Theta$, $\alpha = 1, \ldots, m$, and the physical expressions for the equilibrium parts $\sigma_\infty = E_\infty(\Theta)\varepsilon$, $\eta_\infty = -E'_\infty(\Theta)\varepsilon^2/2$.

The non-equilibrium stresses q_α act on each dashpot of the α-Maxwell element and are related to the associated inelastic (viscous) strains γ_α. Compute the internal constitutive equations $q_\alpha = -\partial v_\alpha(\varepsilon, \Theta, \gamma_\alpha)/\partial\gamma_\alpha$, $\alpha = 1, \ldots, m$, and obtain the current 'distance from equilibrium' q_α as specified in eq. (7.158).

(b) From Figure 7.10 using equilibrium derive the total linear stress σ and establish eq. (7.158). Interpret the result as a superposition of the equilibrium stress σ_∞ and the non-equilibrium stresses q_α, $\alpha = 1, \ldots, m$.

The considered thermoviscoelastic model presents a thermodynamic extension of the viscoelastic constitutive model introduced in Example 6.10 (see p. 286). Remarkably, the equilibrium equation (7.158) and the constitutive equation for the entropy (7.159) constitute the one-dimensional linear counterparts of eqs. (7.150) and (7.151), respectively. Hence, the purely phenomenological thermoviscoelastic model presented in Example 7.6 can be viewed as a nonlinear multi-dimensional generalization of the linear rheological model, as illustrated in Figure 7.10.

2. Recall the proposed one-dimensional thermoviscoelastic model from the previous exercise. The dashpots in the rheological model (Figure 7.10) characterize the dissipation mechanism. According to a Newtonian viscous fluid we may relate q_α to the strain rates $\dot\gamma_\alpha$ by the linear constitutive equations $q_\alpha = \hat\eta_\alpha \dot\gamma_\alpha$, $\alpha = 1, \ldots, m$.

   (a) Consider the time derivative of the non-equilibrium stresses $q_\alpha = E_\alpha(\Theta)$ $(\varepsilon - \gamma_\alpha)$ and obtain the physically based evolution equations for the internal variables, namely

$$\dot q_\alpha + \frac{q_\alpha}{\tau_\alpha} = E_\alpha(\Theta)\dot\varepsilon + \dot E_\alpha(\Theta)(\varepsilon - \gamma_\alpha) \ , \qquad \alpha = 1, \ldots, m \ , \quad (7.160)$$

   where the relations $\tau_\alpha = \hat\eta_\alpha / E_\alpha$, $\alpha = 1, \ldots, m$, should be used.

   (b) Knowing that q_α and $\dot\gamma_\alpha$ are the stresses and the strain rates acting on each dashpot, derive the rate of work dissipated within the considered thermomechanical device and derive the non-negative expression

$$\mathcal{D}_{\text{int}} = \sum_{\alpha=1}^{m} \frac{q_\alpha^2}{\hat\eta_\alpha} \geq 0 \ . \qquad (7.161)$$

   Discuss the thermostatic limit and, in particular, specify in which parts of the device the stresses and the entropy remain.

Note that the evolution equations (7.160) and the internal dissipation (7.161) constitute the one-dimensional linear counterparts of eqs. (7.153) and (7.147)$_2$.

8 Variational Principles

The last chapter deals with the formulation of the field equations in the form of variational principles and methods.

The variational approach in various forms is often taken as the *cornerstone* for the development of discretization techniques such as the well established finite element methodology. The finite element method is today becoming widely used in industrial applications because of its predictive capability and general effectiveness in providing approximate solutions for the underlying initial boundary-value problems. On the finite element method, which is one of the most powerful numerical techniques, a large amount of literature is available. See, for example, the books by ODEN [1972], STRANG and FIX [1988b], ZIENKIEWICZ and TAYLOR [1989, 1991], BREZZI and FORTIN [1991], REDDY [1993], BATHE [1996], CRISFIELD [1991, 1997], BELYTSCHKO et al. [2000], HUGHES [2000], and WRIGGERS [2001]. For a description of a finite element program solving problems of continua in the *nonlinear* context (with available computer software) the reader is referred to, for example, ZIENKIEWICZ and TAYLOR [1991], CRISFIELD [1997] and BONET and WOOD [1997]. It is pointed out, however, that this reference list is by no means complete on this subject.

Variational principles are particularly powerful tools for the evaluation of continuous bodies and belong to the fundamental principles in mathematics and mechanics. It is important to note that the finite element method need not necessarily depend upon the existence of a variational principle. However, good approximate solutions are often related to the weak forms of field equations, which are consequences of the stationarity condition of a functional.

In this chapter we discuss (and compare) the most important variational principles leading to the finite element method. We focus attention solely on isothermal processes, for which the temperature remains constant. Coupling between mechanical and thermal quantities is not considered; hence, only the mechanical balance principles enter the variational formulations.

We start by explaining the notion of virtual displacements and variations and continue with the principle of virtual work, which is fundamental for a large number of efficient finite element formulations. The powerful concept of linearization is reviewed briefly and the principle of virtual work in both the material and spatial descriptions is linearized explicitly. We present some of the basic ideas of two and three-field variational principles particularly designed to capture kinematic constraints such as incompressibility.

The reader who wishes for additional information on the rich area of variational principles should consult the books by TRUESDELL and TOUPIN [1960], VAINBERG [1964], DUVAUT and LIONS [1972], ODEN and REDDY [1976] and WASHIZU [1982].

8.1   Virtual Displacements, Variations

Consider a continuum body $\mathcal{B}$ with a typical particle $\mathsf{P} \in \mathcal{B}$ at a given instant of time t. As usual, points $\mathsf{X} \in \Omega_0$ and $\mathsf{x} \in \Omega$ characterize the positions $\mathbf{X}$ and $\mathbf{x}$ of that particle in the reference configuration Ω_0 at time $t = 0$ and the current configuration Ω at a subsequent time $t > 0$. In the following we indicate the displacement vector field of P as $\mathbf{u}$, pointing from the reference configuration of the continuum body into the current configuration, i.e. from $\mathbf{X}$ to $\mathbf{x}$ (see Figure 8.1).

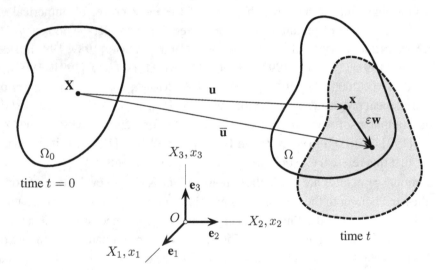

Figure 8.1 Virtual configuration in the neighborhood of $\mathbf{u}$, given by $\bar{\mathbf{u}} = \mathbf{u} + \varepsilon\mathbf{w}$.

Next, consider some arbitrary and entirely new vector field $\mathbf{w}$ at point x which yields a **virtual**, slightly modified deformed configuration in the neighborhood of $\mathbf{u}$. The virtual configuration is characterized by the modified displacement vector field $\bar{\mathbf{u}}$ according to

$$\bar{\mathbf{u}} = \mathbf{u} + \varepsilon\mathbf{w} \quad , \tag{8.1}$$

where ε is a scalar parameter.

The displacement vector field is regarded as a continuous and differentiable function of space and time. It may be written in the spatial or material form, i.e. $\mathbf{u}(\mathbf{x}, t)$, $\mathbf{U}(\mathbf{X}, t)$, as introduced in Section 2.2. In order to keep the notation as simple as possible, we agree not to use this distinction any longer, we write subsequently $\mathbf{u}(\mathbf{x}, t) = \mathbf{u}(\mathbf{X}, t)$. It will be clear from the text if the displacement field actually depends on spatial or material coordinates. In addition, within this chapter, the position and time arguments will often be omitted, for convenience.

Virtual displacement field.   Following *Lagrange* we know that the difference between two neighboring displacement fields, i.e. $\bar{\mathbf{u}}$ and $\mathbf{u}$, is called the **(first) variation** of the displacement field $\mathbf{u}$, denoted by $\delta\mathbf{u}$. We write

$$\delta\mathbf{u} = \bar{\mathbf{u}} - \mathbf{u} = \varepsilon\mathbf{w} \quad . \tag{8.2}$$

In mechanics $\delta\mathbf{u}$ is also known as the **virtual displacement field**. The variation of $\mathbf{u}$ is assumed to be an *arbitrary, infinitesimal* (since $\varepsilon \to 0$) and a **virtual change**, i.e. an imaginary (not a 'real') change. Note that $d\mathbf{u}$ also characterizes an *infinitesimal change* of $\mathbf{u}$. However, $d\mathbf{u}$ refers to an **actual change**. The variation of the time-dependent displacement vector field $\mathbf{u}$ is performed at *fixed* instant of time.

The virtual displacement field $\delta\mathbf{u}$ is totally independent of the actual displacement field $\mathbf{u}$ and may be expressed in terms of spatial coordinates or material coordinates. Omitting the time argument t, we have

$$\delta\mathbf{u}(\mathbf{x}) = \delta\mathbf{u}(\chi(\mathbf{X})) = \delta\mathbf{U}(\mathbf{X}) \quad . \tag{8.3}$$

For simplicity, we agree, by analogy with the relevant notation introduced above, to write $\delta\mathbf{u}(\mathbf{x}) = \delta\mathbf{u}(\mathbf{X})$ for the virtual displacement field $\delta\mathbf{u}$.

We discuss briefly the two fundamental commutative properties of the δ-process. For the gradient of $\delta\mathbf{u}$ we find by means of relation $(8.2)_1$ that

$$\mathrm{grad}(\delta\mathbf{u}) = \mathrm{grad}\bar{\mathbf{u}} - \mathrm{grad}\mathbf{u} \quad , \tag{8.4}$$

while, on the other hand, the variation of $\mathrm{grad}\mathbf{u} = \partial\mathbf{u}/\partial\mathbf{x}$ yields by analogy with $(8.2)_1$

$$\delta(\mathrm{grad}\mathbf{u}) = \mathrm{grad}\bar{\mathbf{u}} - \mathrm{grad}\mathbf{u} \quad . \tag{8.5}$$

On comparing eq. (8.4) with (8.5) we find finally the commutative property

$$\delta(\mathrm{grad}\mathbf{u}) = \mathrm{grad}(\delta\mathbf{u}) \quad , \tag{8.6}$$

which shows that the variation of the derivative of a function ($\bullet$) is equal to the derivative of the variation of that function ($\bullet$). In an analogous manner we may derive the second characteristic commutative property of the δ-process, namely, that the order of variation and definite integral is interchangeable.

By analogy with the transformation (2.48) we may relate the gradient with respect to the current position of a particle to the material gradient, defined on region Ω_0. For subsequent use we note the relation

$$\mathrm{grad}\delta\mathbf{u} = \mathrm{Grad}\delta\mathbf{u}\,\mathbf{F}^{-1} \qquad \text{or} \qquad \frac{\partial \delta u_a}{\partial x_b} = \frac{\partial \delta u_a}{\partial X_A} F_{Ab}^{-1} \quad . \tag{8.7}$$

For more details on the variation see COURANT and HILBERT [1968a, 1968b]. A clearly arranged summary of the calculus of variations may also be found in the book by FUNG [1965, Chapter 10].

First variation of a function in material description.   In the following let $\mathcal{F} = \mathcal{F}(\mathbf{u})$ be a smooth (possibly time-dependent) vector function. The single argument of $\mathcal{F}$ is the displacement vector variable $\mathbf{u}$ given in the *material* description. We agree that the value of $\mathcal{F}$, which characterizes some physical quantity, is either a scalar, vector or tensor. (Note the abuse of notation in regard to Section 2.3 where $\mathcal{F} = \mathcal{F}(\mathbf{X}, t)$ characterizes a smooth material field).

In order to obtain the *first variation* of the vector function $\mathcal{F}$ we must evaluate simply the directional derivative (or Gâteaux derivative) of $\mathcal{F}(\mathbf{u})$ at any fixed $\mathbf{u}$ in the direction of $\delta\mathbf{u}$, which we denote as $D_{\delta\mathbf{u}}\mathcal{F}(\mathbf{u})$ (recall the concept introduced in Section 1.8). We may consider the definition

$$\delta\mathcal{F}(\mathbf{u}, \delta\mathbf{u}) = D_{\delta\mathbf{u}}\mathcal{F}(\mathbf{u}) = \frac{\mathrm{d}}{\mathrm{d}\varepsilon}\mathcal{F}(\mathbf{u} + \varepsilon\delta\mathbf{u})|_{\varepsilon=0} \tag{8.8}$$

and say that $\delta\mathcal{F}(\mathbf{u}, \delta\mathbf{u})$ is the first variation of the function $\mathcal{F}(\mathbf{u})$ in the direction of the virtual displacement field $\delta\mathbf{u}$. It is the ordinary differentiation of $\mathcal{F}(\mathbf{u} + \varepsilon\delta\mathbf{u})$ with respect to the scalar parameter ε.

Note that the variational operator $\delta(\bullet)$ and the Gâteaux operator $D(\bullet)$ are *linear*. The usual properties of differentiation are valid, i.e. the chain rule, product rule and so forth.

EXAMPLE 8.1   Show that the first variation $\delta\mathbf{F}$ of the deformation gradient $\mathbf{F}$ may be expressed as

$$\delta\mathbf{F} = \mathrm{Grad}\delta\mathbf{u} \qquad \text{or} \qquad \delta F_{aA} = \frac{\partial \delta u_a}{\partial X_A} = \mathrm{Grad}_{X_A}\delta u_a \quad . \tag{8.9}$$

In addition, verify that the first variation $\delta\mathbf{F}^{-1}$ of the inverse of the deformation gradient $\mathbf{F}^{-1}$ is given by

$$\delta\mathbf{F}^{-1} = -\mathbf{F}^{-1}\mathrm{grad}\delta\mathbf{u} \qquad \text{or} \qquad \delta F_{Aa}^{-1} = -F_{Ab}^{-1}\frac{\partial \delta u_b}{\partial x_a} = -F_{Ab}^{-1}\mathrm{grad}_{x_a}\delta u_b \quad . \tag{8.10}$$

Solution.   With the definition (2.39) of the deformation gradient and rule (8.8), we may compute the directional derivative of $\mathbf{F}$ in the direction of the virtual displacement field $\delta\mathbf{u}$ at the position $\mathbf{u}(\mathbf{x})$ of the current configuration, i.e.

$$\delta\mathbf{F} = D_{\delta\mathbf{u}}\mathbf{F} = \frac{\mathrm{d}}{\mathrm{d}\varepsilon}\mathbf{F}(\mathbf{u} + \varepsilon\delta\mathbf{u})|_{\varepsilon=0}$$

$$= \frac{\mathrm{d}}{\mathrm{d}\varepsilon}(\mathbf{F} + \varepsilon\mathrm{Grad}\delta\mathbf{u})|_{\varepsilon=0} = \mathrm{Grad}\delta\mathbf{u} \quad . \tag{8.11}$$

Alternatively, knowing that the operators $\delta(\bullet)$ and $\partial(\bullet)$ commute we obtain simply result (8.9) by thinking of $\delta(\bullet)$ as a *linear* operator. Applying relation $(2.45)_2$ we conclude that

$$\delta\mathbf{F} = \delta(\mathrm{Grad}\mathbf{u} + \mathbf{I}) = \delta(\mathrm{Grad}\mathbf{u}) = \mathrm{Grad}\delta\mathbf{u} \quad . \tag{8.12}$$

In order to show eq. (8.10) we start with the variation of the identity $\mathbf{F}^{-1}\mathbf{F} = \mathbf{I}$ which gives $\delta\mathbf{F}^{-1} = -\mathbf{F}^{-1}\delta\mathbf{F}\mathbf{F}^{-1}$ (compare with eq. $(2.145)_1$). Substituting (8.9) and using transformation (8.7) we find that

$$\delta\mathbf{F}^{-1} = -\mathbf{F}^{-1}(\mathrm{Grad}\delta\mathbf{u}\,\mathbf{F}^{-1}) = -\mathbf{F}^{-1}\mathrm{grad}\delta\mathbf{u} \quad , \tag{8.13}$$

which is the desired expression (8.10).   ■

Finally we establish the variation of the Green-Lagrange strain tensor $\mathbf{E}$. By recalling the definition (2.69) of $\mathbf{E}$, we obtain with the product rule that $\delta\mathbf{E} = [\delta(\mathbf{F}^{\mathrm{T}})\mathbf{F} + \mathbf{F}^{\mathrm{T}}\delta\mathbf{F}]/2$. With relation (8.9) and property (1.84) we arrive at

$$\delta\mathbf{E} = \frac{1}{2}[(\mathbf{F}^{\mathrm{T}}\mathrm{Grad}\delta\mathbf{u})^{\mathrm{T}} + \mathbf{F}^{\mathrm{T}}\mathrm{Grad}\delta\mathbf{u}] = \mathrm{sym}(\mathbf{F}^{\mathrm{T}}\mathrm{Grad}\delta\mathbf{u}) \quad , \tag{8.14}$$

or, in index notation,

$$\delta E_{AB} = \frac{1}{2}\left(F_{aB}\frac{\partial\delta u_a}{\partial X_A} + F_{aA}\frac{\partial\delta u_a}{\partial X_B}\right) \quad , \tag{8.15}$$

which is an important relation used in subsequent studies. The notation $\mathrm{sym}(\bullet)$ is used to indicate the symmetric part of a tensor (compare with eq. $(1.112)_1$). Since $\mathbf{E} = (\mathbf{C} - \mathbf{I})/2$, we find additionally that $\delta\mathbf{E} = (\delta\mathbf{C})/2$.

First variation of a function in spatial description.   Let $f = f(\mathbf{u})$ be a smooth (possibly time-dependent) vector function in the *spatial* description. Note that the value of the function $f(\mathbf{u}) = \chi_*(\mathcal{F}(\mathbf{u}))$, which we consider as the push-forward of $\mathcal{F}$, is either a scalar, vector or tensor.

In order to obtain the first variation of f we formally apply the important concept of *pull-back* and *push-forward* operations introduced in Section 2.5. The variation of f is obtained by the following three steps:

(i) compute the *pull-back* of f to the reference configuration, which results in the associated function $\mathcal{F}(\mathbf{u}) = \chi_*^{-1}(f(\mathbf{u}))$;

(ii) apply the concept of variation to $\mathcal{F}$, as introduced in eq. (8.8), and

(iii) carry out the *push-forward* of the result to the current configuration.

This concept is actually the same as for the computation of the *Lie time derivative* introduced in Section 2.8. Instead of the direction $\mathbf{v}$ used for the Lie time derivative we take here the virtual displacement field $\delta\mathbf{u}$.

Consequently, for the *first variation* of a vector function f given in the spatial description we merely write, with reference to eq. (2.189),

$$\delta f(\mathbf{u}, \delta\mathbf{u}) = \chi_*(D_{\delta\mathbf{u}}\chi_*^{-1}(f)) = \chi_*(D_{\delta\mathbf{u}}\mathcal{F}) \quad . \tag{8.16}$$

Since $D_{\delta\mathbf{u}}\mathcal{F}(\mathbf{u}) = \delta\mathcal{F}(\mathbf{u}, \delta\mathbf{u})$ according to (8.8)$_1$, we obtain

$$\delta f(\mathbf{u}, \delta\mathbf{u}) = \chi_*(\delta\mathcal{F}(\mathbf{u}, \delta\mathbf{u})) \quad . \tag{8.17}$$

Therefore, the first variation of function $f = f(\mathbf{u})$ is the push-forward of the first variation of the associated function $\mathcal{F}(\mathbf{u}) = \chi_*^{-1}(f(\mathbf{u}))$ in the direction of the virtual displacement field $\delta\mathbf{u}$. If $f = f(\mathbf{u})$ is a scalar-valued function, then $f = f(\mathbf{u}) = \mathcal{F}(\mathbf{u})$ and the variation of f coincides with the variation of its associated function $\mathcal{F}$; thus, $\delta f = \delta\mathcal{F}$.

Note that in our terminology the introduced operator δ is used for the variation of a function in both the material and spatial descriptions.

EXAMPLE 8.2   Show that the first variation $\delta\mathbf{e}$ of the Euler-Almansi strain tensor $\mathbf{e}$ may be expressed as

$$\delta\mathbf{e} = \frac{1}{2}(\mathrm{grad}^T\delta\mathbf{u} + \mathrm{grad}\delta\mathbf{u}) = \mathrm{sym}(\mathrm{grad}\delta\mathbf{u}) \quad , \tag{8.18}$$

or, in index notation, as

$$\delta e_{ab} = \frac{1}{2}\left(\frac{\partial\delta u_b}{\partial x_a} + \frac{\partial\delta u_a}{\partial x_b}\right) \quad . \tag{8.19}$$

Solution.   Compare Example 2.15 on p. 107. From rule (8.16) we obtain the variation of the spatial tensor $\mathbf{e}$, i.e.

$$\delta\mathbf{e} = \mathbf{F}^{-T}(D_{\delta\mathbf{u}}(\mathbf{F}^T\mathbf{e}\mathbf{F}))\mathbf{F}^{-1} = \mathbf{F}^{-T}(D_{\delta\mathbf{u}}\mathbf{E})\mathbf{F}^{-1} \quad , \tag{8.20}$$

which is the push-forward of the directional derivative of the associated Green-Lagrange strain tensor $\mathbf{E} = \chi_*^{-1}(\mathbf{e})$ in the direction of $\delta\mathbf{u}$.

By means of rule $(8.8)_1$, relation $(8.14)_1$ and transformation (8.7) we find from $(8.20)_2$ that

$$\delta\mathbf{e} = \mathbf{F}^{-T}\delta\mathbf{E}\,\mathbf{F}^{-1} = \frac{1}{2}(\mathbf{F}^{-T}\mathrm{Grad}^T\delta\mathbf{u} + \mathrm{Grad}\delta\mathbf{u}\,\mathbf{F}^{-1})$$

$$= \frac{1}{2}(\mathrm{grad}^T\delta\mathbf{u} + \mathrm{grad}\delta\mathbf{u}) \quad . \tag{8.21}$$

According to eq. (1.112) the variation of the spatial tensor $\mathbf{e}$ is the symmetric part of the tensor $\mathrm{grad}\delta\mathbf{u}$, which gives the desired result $(8.18)_2$. ∎

EXERCISES

1. Show that the first variations of the volume ratio J and the inverse right Cauchy-Green tensor $\mathbf{C}^{-1}$ are

$$\delta J = J\mathrm{div}\,\delta\mathbf{u} \quad , \qquad \delta\mathbf{C}^{-1} = \mathbf{F}^{-1}(\mathrm{grad}^T\delta\mathbf{u} + \mathrm{grad}\delta\mathbf{u})\,\mathbf{F}^{-T}$$

2. Show that the first variations of the spatial line, surface and volume elements are

$$\delta(d\mathbf{x}) = \mathrm{grad}\delta\mathbf{u}d\mathbf{x} \quad ,$$

$$\delta(d\mathbf{s}) = (\mathrm{div}\delta\mathbf{u}\,\mathbf{I} - \mathrm{grad}^T\delta\mathbf{u})d\mathbf{s} \quad ,$$

$$\delta(dv) = \mathrm{div}\delta\mathbf{u}dv \quad ,$$

   where $\mathbf{I}$ denotes the second-order unit tensor. Compare eqs. $(2.175)_3$, $(2.180)_3$ and $(2.182)_2$.

3. Show the commutative property (8.6) by means of the rule (8.16) and (8.7).

8.2 Principle of Virtual Work

In the following two sections we study variational principles with only one field of unknowns, called **single-field variational principles**. In particular, we introduce work and stationary principles in which the displacement vector $\mathbf{u}$ is the only unknown field.

These principles are fundamental and will become essential in establishing finite element formulations.

Initial boundary-value problem. The finite element method requires the formulation of the balance laws in the form of variational principles.

As one of the most fundamental balance laws we recall Cauchy's first equation of motion (i.e. balance of mechanical energy) discussed in Section 4.3. Knowing that the spatial velocity field $\mathbf{v}$ may be expressed as the time rate of change of the displacement field $\mathbf{u}$, we may write Cauchy's first equation of motion, i.e. (4.53), as

$$\text{div}\boldsymbol{\sigma} + \mathbf{b} = \rho\ddot{\mathbf{u}} \quad . \tag{8.22}$$

From the fundamental standpoint adopted in Section 4.3, the Cauchy stress tensor is governed by the symmetry condition $\boldsymbol{\sigma} = \boldsymbol{\sigma}^{\mathrm{T}}$ deriving from the *balance of angular momentum*. The spatial mass density of the material is $\rho = J^{-1}\rho_0$, which describes *continuity of mass*. The body force $\mathbf{b}$ per unit current volume which acts on a particle in region Ω is considered to be a prescribed (given) force while the term $\rho\ddot{\mathbf{u}}$ characterizes the inertia force per unit current volume. Note that when we write eq. (8.22) we mean $\text{div}\boldsymbol{\sigma}(\mathbf{x}, t) + \mathbf{b}(\mathbf{x}, t) = \rho(\mathbf{x}, t)\ddot{\mathbf{u}}(\mathbf{x}, t)$ at every point $\mathbf{x} \in \Omega$ and for all times t.

In the following we consider boundary conditions and initial conditions for the motion $\mathbf{x} = \boldsymbol{\chi}(\mathbf{X}, t)$ required to satisfy the second-order differential equation (8.22). We assume subsequently that the boundary surface $\partial\Omega$ of a continuum body $\mathcal{B}$ occupying region Ω is decomposed into disjoint parts so that

$$\partial\Omega = \partial\Omega_{\mathrm{u}} \cup \partial\Omega_{\sigma} \qquad \text{with} \qquad \partial\Omega_{\mathrm{u}} \cap \partial\Omega_{\sigma} = \emptyset \quad . \tag{8.23}$$

Figure 8.2 illustrates the decomposition of the boundary surface $\partial\Omega$ in a two-dimensional space at time t.

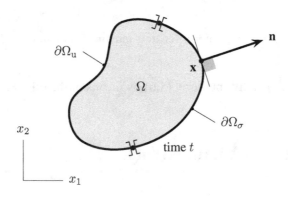

Figure 8.2 Partition of a boundary surface $\partial\Omega$.

We distinguish two classes of boundary conditions, namely the **Dirichlet boundary conditions**, which correspond to a displacement field $\mathbf{u} = \mathbf{u}(\mathbf{x}, t)$, and the **von Neumann boundary conditions**, which are identified physically with the surface traction $\mathbf{t} = \mathbf{t}(\mathbf{x}, t, \mathbf{n})$.

We write

$$\mathbf{u} = \bar{\mathbf{u}} \quad \text{on} \quad \partial\Omega_u \;, \qquad \mathbf{t} = \boldsymbol{\sigma}\mathbf{n} = \bar{\mathbf{t}} \quad \text{on} \quad \partial\Omega_\sigma \;, \qquad (8.24)$$

where the overbars ($\bar{\bullet}$) denote prescribed (given) functions on the boundaries $\partial\Omega_{(\bullet)} \subset \partial\Omega$ of a continuum body occupying the region Ω. The unit exterior vector normal to the boundary surface $\partial\Omega_\sigma$ is characterized by $\mathbf{n}$. The prescribed displacement field $\bar{\mathbf{u}}$ and the prescribed Cauchy traction vector $\bar{\mathbf{t}}$ (force measured per unit current surface area) are specified on a portion $\partial\Omega_u \subset \partial\Omega$ and on the remainder $\partial\Omega_\sigma$, respectively. Note that in the previous section the symbol $\bar{\mathbf{u}}$ also stands for the modified displacement field.

We call the prescribed body force $\mathbf{b}$ and surface traction $\bar{\mathbf{t}}$ **loads**. We say that the continuum body is subjected to **holonomic external constraints** if $\mathbf{u} = \bar{\mathbf{u}}$ on the boundary surface $\partial\Omega_u$. External constraints are **nonholonomic** if they are given by an inequality.

The second-order differential equations (8.22) themselves require additional data in the form of **initial conditions**. The displacement field $\mathbf{u}|_{t=0}$ and the velocity field $\dot{\mathbf{u}}|_{t=0}$ at initial time $t = 0$ are specified as

$$\mathbf{u}(\mathbf{x}, t)|_{t=0} = \mathbf{u}_0(\mathbf{X}) \;, \qquad \dot{\mathbf{u}}(\mathbf{x}, t)|_{t=0} = \dot{\mathbf{u}}_0(\mathbf{X}) \;, \qquad (8.25)$$

where $(\bullet)_0$ denotes a prescribed function in Ω_0. Since we agreed to consider a stress-free reference configuration at $t = 0$, the initial values $(\bullet)_0$ are assumed to be zero in our case. However, in dynamics the configuration at $t = 0$ is sometimes not chosen as a reference configuration.

In order to achieve *compatibility* of the boundary and initial conditions we require additionally on $\partial\Omega_u$ that

$$\bar{\mathbf{u}}(\mathbf{x}, 0) = \mathbf{u}_0(\mathbf{X}) \;, \qquad \dot{\bar{\mathbf{u}}}(\mathbf{x}, 0) = \dot{\mathbf{u}}_0(\mathbf{X}) \;. \qquad (8.26)$$

Now, the problem is to find a motion that satisfies eq. (8.22) with the prescribed boundary and initial conditions (8.24), (8.25) and compatibility conditions (8.26).

This leads to the formulation in the **strong form** (or **classical form**) of the **initial boundary-value problem (IBVP)**. Given the body force, and the boundary and initial conditions, find the displacement field $\mathbf{u}$ so that (considering only mechanical variables)

$$\left. \begin{aligned} \operatorname{div}\boldsymbol{\sigma} + \mathbf{b} &= \rho\ddot{\mathbf{u}} \;, \\ \mathbf{u} &= \bar{\mathbf{u}} \quad &\text{on} \quad \partial\Omega_u \;, \\ \mathbf{t} = \boldsymbol{\sigma}\mathbf{n} &= \bar{\mathbf{t}} \quad &\text{on} \quad \partial\Omega_\sigma \;, \\ \mathbf{u}(\mathbf{x}, t)|_{t=0} &= \mathbf{u}_0(\mathbf{X}) \;, \\ \dot{\mathbf{u}}(\mathbf{x}, t)|_{t=0} &= \dot{\mathbf{u}}_0(\mathbf{X}) \;. \end{aligned} \right\} \qquad (8.27)$$

Note that the set (8.27) of equations generally defines a *nonlinear* initial boundary-value problem for the unknown displacement field **u**. In addition, we need a constitutive equation for the stress σ which is, in general, a nonlinear function of the displacement field **u**.

If data depend on time and the acceleration is assumed to vanish, i.e. $\ddot{\mathbf{u}} = \mathbf{o}$, the considered problem is called **quasi-static**. For this case the Cauchy's equation of equilibrium is subjected to the conditions (8.24), (8.25)$_1$, and the requirement that eq. (8.26)$_1$ holds for compatibility.

If the data are independent of time the problem is referred to as **static**. For this case we consider a body in static equilibrium for which the set (8.27) of equations reduces to the associated nonlinear **boundary-value problem** (BVP) of elastostatics, i.e.

$$\left.\begin{array}{lll} \operatorname{div}\boldsymbol{\sigma} + \mathbf{b} = \mathbf{o} & , & \\[2mm] \mathbf{u} = \bar{\mathbf{u}} & \text{on} & \partial\Omega_{\mathrm{u}} \quad , \\[2mm] \mathbf{t} = \boldsymbol{\sigma}\mathbf{n} = \bar{\mathbf{t}} & \text{on} & \partial\Omega_{\sigma} \quad . \end{array}\right\} \tag{8.28}$$

Thus, the solution of a static problem at a point of a continuum body depends only on the data of the boundary and not on initial conditions (there is no need for initial conditions).

Principle of virtual work in spatial description.   An analytical solution of the nonlinear initial boundary-value problem described is only possible for some special cases. Therefore, on the basis of variational principles, solution strategies such as the finite element method are often used in order to achieve approximate solutions.

In order to develop the principle of virtual work we start with Cauchy's first equation of motion (4.53) which we multiply with an arbitrary vector-valued function $\eta = \eta(\mathbf{x})$, defined on the current configuration Ω of the body. Integration over the region Ω of the body yields the scalar-valued function

$$f(\mathbf{u}, \eta) = \int_{\Omega}(-\operatorname{div}\boldsymbol{\sigma} - \mathbf{b} + \rho\ddot{\mathbf{u}}) \cdot \eta \mathrm{d}v = 0 \quad . \tag{8.29}$$

For the first argument of function f we conveniently introduce the displacement vector field **u** rather then the motion χ for a given time t. The second argument of f is a function $\eta = \eta(\mathbf{x}) = \eta(\chi(\mathbf{X}, t))$ (at a *fixed* instant of time t), called a **test function** (or **weighting function**). It is a smooth function with $\eta = \mathbf{o}$ on the boundary surface $\partial\Omega_{\mathrm{u}}$. Eq. (8.29) is known as the **weak form** of the equation of motion with respect to the spatial configuration. Equations in the weak form often remain valid for discontinuous problems such as *shocks* where most of the variables undergo a discontinuous variation. For this type of problem differential equations are not necessarily appropriate.

Since η is arbitrary, the vector equation $\mathrm{div}\boldsymbol{\sigma} + \mathbf{b} = \rho\ddot{\mathbf{u}}$ on Ω is *equivalent* to the weak form (8.29). The method used to prove this important property goes along with the **fundamental lemma** of the calculus of variations. The solution of the problem in the strong form is identical to the solution in the weak form. For further details see the books by HUGHES [2000] and MARSDEN and HUGHES [1994].

Subsequently, applying the product rule (1.290) to the term $\mathrm{div}\boldsymbol{\sigma} \cdot \boldsymbol{\eta}$, i.e.

$$\mathrm{div}\boldsymbol{\sigma} \cdot \boldsymbol{\eta} = \mathrm{div}(\boldsymbol{\sigma}\boldsymbol{\eta}) - \boldsymbol{\sigma} : \mathrm{grad}\boldsymbol{\eta} \quad , \tag{8.30}$$

and using the divergence theorem in the form of (1.301), eq. (8.29) may be written as

$$f(\mathbf{u}, \boldsymbol{\eta}) = \int_{\Omega} [\boldsymbol{\sigma} : \mathrm{grad}\boldsymbol{\eta} - (\mathbf{b} - \rho\ddot{\mathbf{u}}) \cdot \boldsymbol{\eta}] dv - \int_{\partial\Omega} \boldsymbol{\sigma}\boldsymbol{\eta} \cdot \mathbf{n} ds = 0 \quad . \tag{8.31}$$

Since η vanishes on the part of the boundary surface $\partial\Omega_{\mathrm{u}}$ where $\bar{\mathbf{u}}$ is prescribed, the surface integral only needs to be integrated over the portion $\partial\Omega_{\sigma} \subset \partial\Omega$. By use of boundary conditions (8.24) and by formulating the initial conditions (8.25) in the weak form, we obtain the following set of scalar equations known as the **variational problem**:

$$\left.\begin{array}{c} f(\mathbf{u}, \boldsymbol{\eta}) = \displaystyle\int_{\Omega} [\boldsymbol{\sigma} : \mathrm{grad}\boldsymbol{\eta} - (\mathbf{b} - \rho\ddot{\mathbf{u}}) \cdot \boldsymbol{\eta}] dv - \int_{\partial\Omega_{\sigma}} \bar{\mathbf{t}} \cdot \boldsymbol{\eta} ds = 0 \quad , \\[4mm] \displaystyle\int_{\Omega} \mathbf{u}(\mathbf{x}, t)|_{t=0} \cdot \boldsymbol{\eta} dv = \int_{\Omega} \mathbf{u}_0(\mathbf{X}) \cdot \boldsymbol{\eta} dv \quad , \\[4mm] \displaystyle\int_{\Omega} \dot{\mathbf{u}}(\mathbf{x}, t)|_{t=0} \cdot \boldsymbol{\eta} dv = \int_{\Omega} \dot{\mathbf{u}}_0(\mathbf{X}) \cdot \boldsymbol{\eta} dv \quad . \end{array}\right\} \tag{8.32}$$

This set of equations characterizes the **weak form** (or **variational form**) of the **initial boundary-value problem**. It is the equivalent counterpart in the strong form (8.27) which is satisfied when (8.32) is satisfied. Note that the stress boundary conditions on the portion $\partial\Omega_{\sigma}$ are part in the weak form $(8.32)_1$, so they are often referred to as **natural** boundary conditions. However, the conditions $\mathbf{u} = \bar{\mathbf{u}}$ which are prescribed over the boundary surface $\partial\Omega_{\mathrm{u}}$ are called **essential** boundary conditions of the variational problem.

Hence, variational problems are related to initial boundary-value problems which are described through differential equations and initial and boundary conditions. The differential equation is usually called the **Euler-Lagrange equation** in the weak formulation which in our case is Cauchy's first equation of motion $(8.27)_1$. Formulations in the weak form are mathematically helpful for investigations of existence, uniqueness or stability of solutions (see, for example, MARSDEN and HUGHES [1994, Sections 6.1-6.5]).

Note that the test function η in (8.32) is *arbitrary*. If we look upon η as the virtual displacement field $\delta\mathbf{u}$, defined on the *current configuration*, then the formulation in the weak form of the initial boundary-value problem (8.32) leads to the fundamental **principle of virtual work** (or **principle of virtual displacement**). Considering the symmetry of σ and the variation of the Euler-Almansi strain tensor $\delta\mathbf{e}$, as derived in eq. (8.18)$_2$, we arrive at the principle of virtual work in the *spatial description* expressed in terms of the virtual displacement, i.e.

$$f(\mathbf{u}, \delta\mathbf{u}) = \int_{\Omega} [\sigma : \delta\mathbf{e} - (\mathbf{b} - \rho\ddot{\mathbf{u}}) \cdot \delta\mathbf{u}]dv - \int_{\partial\Omega_\sigma} \bar{\mathbf{t}} \cdot \delta\mathbf{u}ds = 0 \quad , \tag{8.33}$$

with the additional initial conditions $\int_\Omega \mathbf{u}(\mathbf{x}, t)|_{t=0} \cdot \delta\mathbf{u}dv = \int_\Omega \mathbf{u}_0(\mathbf{X}) \cdot \delta\mathbf{u}dv$ and $\int_\Omega \dot{\mathbf{u}}(\mathbf{x}, t)|_{t=0} \cdot \delta\mathbf{u}dv = \int_\Omega \dot{\mathbf{u}}_0(\mathbf{X}) \cdot \delta\mathbf{u}dv$. An equation of type (8.33) is typically called a **variational equation**.

The smooth virtual displacement field $\delta\mathbf{u}$ is arbitrary over the region Ω and over the boundary surface $\partial\Omega_\sigma$ where the traction vector $\bar{\mathbf{t}}$ is prescribed. We require that $\delta\mathbf{u}$ vanishes on $\partial\Omega_u$, where the displacement field $\bar{\mathbf{u}}$ is prescribed (see the boundary conditions (8.24)$_1$). The virtual displacement field is assumed to be *infinitesimal*, which is not a requirement for an arbitrary test function. It is an imaginary (not a 'real') change of the continuum which is subjected to the loadings.

The principle of virtual work is the simplest variational principle and it states: the virtual stress work $\sigma : \delta\mathbf{e}$ at fixed σ is equal to the work done by the body force $\mathbf{b}$ and inertia force $\rho\ddot{\mathbf{u}}$ per unit current volume and the surface traction $\bar{\mathbf{t}}$ per unit current surface along $\delta\mathbf{u}$ removed from the current configuration.

The functions

$$\delta W_{\text{int}}(\mathbf{u}, \delta\mathbf{u}) = \int_\Omega \sigma : \delta\mathbf{e}dv \quad , \tag{8.34}$$

$$\delta W_{\text{ext}}(\mathbf{u}, \delta\mathbf{u}) = \int_\Omega \mathbf{b} \cdot \delta\mathbf{u}dv + \int_{\partial\Omega_\sigma} \bar{\mathbf{t}} \cdot \delta\mathbf{u}ds \quad , \tag{8.35}$$

are known as **internal (mechanical) virtual work** δW_{int} and **external (mechanical) virtual work** δW_{ext}.

In the first case the stress σ does internal work along the virtual strains $\delta\mathbf{e}$. In the second case external work is done by the loads, which are the body force $\mathbf{b}$ and the surface traction $\bar{\mathbf{t}}$, along the virtual displacement $\delta\mathbf{u}$ about region Ω and its boundary surface $\partial\Omega$, respectively. For vanishing accelerations $\ddot{\mathbf{u}}$, the internal virtual work equals the external virtual work, i.e. $\delta W_{\text{int}} = \delta W_{\text{ext}}$.

It is important to emphasize that the principle of virtual work does not necessitate the existence of a potential. No statement in regard to a particular material is invoked. Therefore, the principle of virtual work is general in the sense that it is applicable to any material including inelastic materials.

Pressure boundary loading. In the following we are concerned with an important load case, the pressure boundary loading, which is caused by liquids or gases, for example, water or wind. Pressure loads are deformation dependent and of crucial interest for finite deformation problems.

We consider a *pressure boundary condition* on the current boundary surface $\partial\Omega_\sigma \subset \partial\Omega$. In particular, we consider a traction vector $\bar{\mathbf{t}} = \sigma\mathbf{n} = p\mathbf{n}$ per unit current surface acting in the direction of the (pointwise) outward unit vector $\mathbf{n} = \mathbf{n}(\mathbf{x})$. The unit normal vector is perpendicular to the pressure loaded surface $\partial\Omega_\sigma$ of the body with region Ω. Further, we assume that the normal pressure p is a given *constant*. An example in which a pressure boundary condition typically exists is inflation of a balloon.

The external virtual work done by the constant pressure p along the virtual displacement $\delta\mathbf{u}$ is then defined by

$$\delta W_{\text{ext}}(\mathbf{u}, \delta\mathbf{u}) = p \int_{\partial\Omega_\sigma} \mathbf{n} \cdot \delta\mathbf{u} \, ds \quad, \tag{8.36}$$

where ds denotes an infinitesimal surface element in the current configuration.

The external virtual work of the pressure boundary condition is discussed in more detail by SCHWEIZERHOF and RAMM [1984], BUFLER [1984], BONET and WOOD [1997, Section 6.5] and SIMO et al. [1991b] describing applications to axisymmetric problems.

In the following we introduce briefly a parametrization of the current boundary surface $\partial\Omega_\sigma$ which is very useful for implementation in a finite element program. The parameter plane with region Ω_ξ is characterized by ξ_1 and ξ_2 (see Figure 8.3). The parametrization of the surface on which p is prescribed is given in the form $\mathbf{x} = \gamma(\xi_1, \xi_2, t) \subset \partial\Omega_\sigma$ ($x_a = \gamma_a(\xi_1, \xi_2, t)$) at fixed time t. Hence, the outward *unit* vector $\mathbf{n}$ may be expressed as the cross product of the displacement dependent vectors $\partial\gamma/\partial\xi_1$ and $\partial\gamma/\partial\xi_2$. The infinitesimal surface element ds follows from eq. (1.32) and the use of the chain rule. We write

$$\mathbf{n} = \frac{\dfrac{\partial\gamma}{\partial\xi_1} \times \dfrac{\partial\gamma}{\partial\xi_2}}{\left|\dfrac{\partial\gamma}{\partial\xi_1} \times \dfrac{\partial\gamma}{\partial\xi_2}\right|} \quad, \qquad ds = \left|\frac{\partial\gamma}{\partial\xi_1} \times \frac{\partial\gamma}{\partial\xi_2}\right| d\xi_1 d\xi_2 \quad. \tag{8.37}$$

These relations enable the external virtual work (8.36) to be expressed as

$$\delta W_{\text{ext}}(\mathbf{u}, \delta\mathbf{u}) = p \int_{\Omega_\xi} \left(\frac{\partial\gamma}{\partial\xi_1} \times \frac{\partial\gamma}{\partial\xi_2}\right) \cdot \delta\mathbf{u} \, d\xi_1 d\xi_2 \quad, \tag{8.38}$$

which is appropriate for finite element approximations.

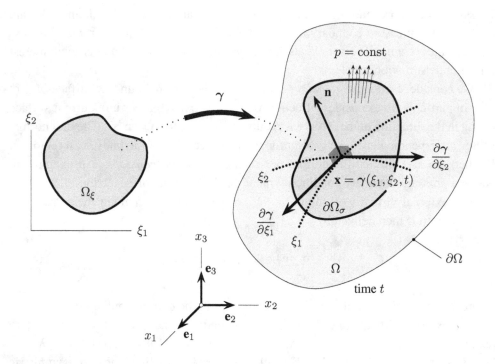

Figure 8.3 Constant pressure boundary loading and parametrization of the pressure loaded surface $\partial\Omega_\sigma$.

Principle of virtual work in material description. Now we are in a position to express the principle of virtual work in terms of material variables.

We assume a region Ω_0 of the continuum body which is bounded by a reference boundary surface $\partial\Omega_0$. This boundary surface is partitioned into disjoint parts (compare with the associated partition (8.23)) so that

$$\partial\Omega_0 = \partial\Omega_{0\,u} \cup \partial\Omega_{0\,\sigma} \qquad \text{with} \qquad \partial\Omega_{0\,u} \cap \partial\Omega_{0\,\sigma} = \emptyset \ . \tag{8.39}$$

As a point of departure we recall the equation of motion in the material description (4.63). We use the form

$$\mathrm{Div}\mathbf{P} + \mathbf{B} = \rho_0\ddot{\mathbf{u}} \ , \tag{8.40}$$

corresponding to (8.22). Here, $\mathbf{P}$, $\mathbf{B}$ and $\rho_0\ddot{\mathbf{u}}$ denote the first Piola-Kirchhoff stress tensor, the reference body force and the inertia force per unit reference volume, respectively.

For the Dirichlet and von Neumann boundary conditions, i.e. $\mathbf{u} = \mathbf{u}(\mathbf{X}, t)$ and $\mathbf{T} = \mathbf{T}(\mathbf{X}, t, \mathbf{N})$, we write, by analogy with (8.24),

$$\mathbf{u} = \bar{\mathbf{u}} \qquad \text{on} \qquad \partial\Omega_{0\,u} \ , \qquad \mathbf{T} = \mathbf{PN} = \bar{\mathbf{T}} \qquad \text{on} \qquad \partial\Omega_{0\,\sigma} \ , \tag{8.41}$$

where the unit exterior vector normal to the boundary surface $\partial\Omega_{0\sigma}$ is characterized by **N**. The prescribed displacement field $\bar{\mathbf{u}}$ and the prescribed first Piola-Kirchhoff traction vector $\bar{\mathbf{T}}$ (force measured per unit reference surface area) are specified on the disjoint parts $\partial\Omega_{0u}$ and $\partial\Omega_{0\sigma}$, respectively. The second-order differential equation (8.40) must be supplemented by initial conditions for the displacement field and the velocity field at the instant of time $t = 0$ (see eq. (8.25)).

Using the above concept, we may show the principle of virtual work in the *material description* expressed in terms of the virtual displacement, i.e.

$$\mathcal{F}(\mathbf{u}, \delta\mathbf{u}) = \int_{\Omega_0} [\mathbf{P} : \text{Grad}\delta\mathbf{u} - (\mathbf{B} - \rho_0\ddot{\mathbf{u}}) \cdot \delta\mathbf{u}]dV - \int_{\partial\Omega_{0\sigma}} \bar{\mathbf{T}} \cdot \delta\mathbf{u}dS = 0 \quad, \tag{8.42}$$

with the virtual displacement field $\delta\mathbf{u}$ (here defined on the *reference configuration*) satisfying the condition $\delta\mathbf{u} = \mathbf{o}$ on the part of the boundary surface $\partial\Omega_{0u}$ where the displacement field $\bar{\mathbf{u}}$ is prescribed. The surface traction $\bar{\mathbf{T}}$ acts on the portion $\partial\Omega_{0\sigma} \subset \partial\Omega_0$. According to relation (3.1), **T** has the same direction as **t**, but $\mathbf{T} \neq \mathbf{t}$. It is important to note that the description of the variational equation (8.42) is **equivalent** to that of (8.33).

EXAMPLE 8.3 Show that the material form of the principle of virtual work, as given in (8.42), can be obtained alternatively by a *pull-back* operation of relation (8.31) to the reference configuration.

Solution. In order to show (8.42) we must express the internal and external virtual work δW_{int} and δW_{ext} in eqs. (8.34) and (8.35) and the contribution $\int_\Omega \rho\ddot{\mathbf{u}} \cdot \delta\mathbf{u}dv$ in terms of material variables.

We begin by considering the internal virtual work (8.34). With the help of identities $(8.18)_2$, (2.51) and transformation (8.7) we obtain

$$\int_\Omega \boldsymbol{\sigma} : \delta\mathbf{e}dv = \int_\Omega \boldsymbol{\sigma} : \text{grad}\delta\mathbf{u}dv = \int_{\Omega_0} J\boldsymbol{\sigma} : \text{Grad}\delta\mathbf{u}\,\mathbf{F}^{-1}dV \quad, \tag{8.43}$$

where the symmetry of the Cauchy stress tensor $\boldsymbol{\sigma}$ is to be used.

Applying property (1.95) and Piola transformation (3.8) we obtain the canonical representation of the material version, i.e.

$$\delta W_{\text{int}}(\mathbf{u}, \delta\mathbf{u}) = \int_\Omega \boldsymbol{\sigma} : \delta\mathbf{e}dv = \int_{\Omega_0} \mathbf{P} : \text{Grad}\delta\mathbf{u}dV \quad. \tag{8.44}$$

The external virtual work δW_{ext} in the form of eq. (8.35) may be transformed by means of the relation for the body force **b**, i.e. $\mathbf{b} = J^{-1}\mathbf{B}$, and the change in volume which is given by $dv = JdV$. In addition, we must show the equivalence of the prescribed traction vectors. With relation (3.1) and boundary conditions $(8.24)_2$ and $(8.41)_2$ we deduce that $\bar{t}ds = \bar{T}dS$. Hence,

$$\delta W_{\text{ext}}(\mathbf{u}, \delta\mathbf{u}) = \int_{\Omega} \mathbf{b} \cdot \delta\mathbf{u}dv + \int_{\partial\Omega_\sigma} \bar{\mathbf{t}} \cdot \delta\mathbf{u}ds = \int_{\Omega_0} \mathbf{B} \cdot \delta\mathbf{u}dV + \int_{\partial\Omega_{0\,\sigma}} \bar{\mathbf{T}} \cdot \delta\mathbf{u}dS \ . \quad (8.45)$$

The remaining term in eq. (8.42), i.e. the inertia force $\rho_0\ddot{\mathbf{u}}$ per unit reference volume over the region Ω_0, may simply be established from the third term in the associated eq. (8.31) by means of $\rho_0 dV = \rho dv$, i.e. conservation of mass in the local form (4.6). This result together with (8.44) and (8.45) leads to the desired relation (8.42). ∎

EXERCISE

1. Starting at eq. $(8.44)_2$, show that the internal virtual work δW_{int} may also be expressed as the contraction of the symmetric second Piola-Kirchhoff stress tensor $\mathbf{S}$ and the virtual Green-Lagrange strain tensor $\delta\mathbf{E}$ defined in eq. (8.14).

   In addition, show that δW_{int} may also be given in terms of the Mandel stress tensor $\mathbf{\Sigma} = \mathbf{CS}$ and the variation of the symmetric Green-Lagrange strain tensor, i.e. $\delta\mathbf{C} = 2\delta\mathbf{E}$, by using eq. (3.67) as

$$\delta W_{\text{int}} = \int_{\Omega_0} \mathbf{P} : \text{Grad}\delta\mathbf{u}dV = \int_{\Omega_0} \mathbf{S} : \delta\mathbf{E}dV = \int_{\Omega_0} \mathbf{\Sigma} : \frac{1}{2}\mathbf{C}^{-1}\delta\mathbf{C}dV \ . \quad (8.46)$$

8.3 Principle of Stationary Potential Energy

In the principle of virtual work, as derived in the last section, the stresses are considered without their connections to the strains. We have not taken into account a particular material.

In this section we assume a conservative mechanical system (compare with p. 159) requiring the existence of an energy functional Π for both the stresses and the loads. The assumption of the existence of Π is common in many fields in solid mechanics. The loads may depend on the motion, but they must emanate from a functional. A formulation based on energy functionals is very useful, for example, for the development of robust numerical algorithms that are based on optimization techniques.

In the following we introduce a stationary energy principle in which the displacement vector field $\mathbf{u}$ is taken to be the only fundamental unknown.

Stationary energy principle.   From now on we assume that the loads do not depend on the motion of the body, which is usually the case, for example, for body forces. It means that the directions of the loads remain parallel and their values unchanged throughout the deformation of the body. We say that such loads are **'dead'**.

Instead of a vibrating body we consider a body in static equilibrium under the action of specified *'dead'* loadings and boundary conditions on $\partial\Omega_{0\,\mathrm{u}}$ and $\partial\Omega_{0\,\sigma}$, according to eq. (8.41). Then the total potential energy Π of the system is given as the sum of the internal and external potential energy, Π_{int} and Π_{ext}, i.e.

$$\Pi(\mathbf{u}) = \Pi_{\mathrm{int}}(\mathbf{u}) + \Pi_{\mathrm{ext}}(\mathbf{u}) \quad , \tag{8.47}$$

$$\Pi_{\mathrm{int}}(\mathbf{u}) = \int_{\Omega_0} \Psi(\mathbf{F}(\mathbf{u}))\mathrm{d}V \quad , \qquad \Pi_{\mathrm{ext}}(\mathbf{u}) = -\int_{\Omega_0} \mathbf{B} \cdot \mathbf{u}\mathrm{d}V - \int_{\partial\Omega_{0\,\sigma}} \overline{\mathbf{T}} \cdot \mathbf{u}\mathrm{d}S \quad , \tag{8.48}$$

where $\Psi = \Psi(\mathbf{F})$ denotes the strain-energy function per unit reference volume, as introduced in Section 6.1.

Since the deformation gradient $\mathbf{F}$ depends on the displacement vector field $\mathbf{u}$ by the relation according to eq. $(2.45)_2$, i.e. $\mathbf{F} = \mathrm{Grad}\mathbf{u} + \mathbf{I}$, we indicate explicitly the dependence of $\mathbf{F}$ on $\mathbf{u}$ and write $\mathbf{F} = \mathbf{F}(\mathbf{u})$. For Π_{int} and Π_{ext} we will also indicate subsequently the dependence on $\mathbf{u}$. Note that for a rigid body the term Π_{int} is zero. The *'dead'* loadings, given by the external forces $\mathbf{B}$ and $\overline{\mathbf{T}}$, are distributed over the volume of the continuum body and its von Neumann boundary, respectively.

One main objective of common engineering interests is to find the state of equilibrium (the deformed configuration) for which the potential is **stationary**. The stationary position of the total potential energy Π is obtained by requiring the directional derivative with respect to the displacements $\mathbf{u}$ to vanish in all directions $\delta\mathbf{u}$. Compute

$$\delta\Pi(\mathbf{u}, \delta\mathbf{u}) = D_{\delta\mathbf{u}}\Pi(\mathbf{u}) = \frac{\mathrm{d}}{\mathrm{d}\varepsilon}\Pi(\mathbf{u} + \varepsilon\delta\mathbf{u})|_{\varepsilon=0} = 0 \quad , \tag{8.49}$$

which is known as the **principle of stationary potential energy**, another fundamental variational principle in mechanics. In other words, we require that the first variation of the total potential energy, denoted $\delta\Pi$, vanishes. The variation of Π clearly is a function of both $\mathbf{u}$ and $\delta\mathbf{u}$. The arbitrary vector field $\delta\mathbf{u}$ is consistent with the conditions imposed on the continuum body. Thus, $\delta\mathbf{u} = \mathbf{o}$ over $\partial\Omega_{\mathrm{u}}$, where boundary displacements are prescribed.

In order to decide if the solution corresponds to a *maximum*, a *minimum* or a *saddle point* we must determine the **second variation** of the total potential energy Π, denoted by $\delta^2\Pi(\mathbf{u}, \delta\mathbf{u}, \Delta\mathbf{u}) = D^2_{\delta\mathbf{u},\Delta\mathbf{u}}\Pi(\mathbf{u})$. Here, $\Delta\mathbf{u}$ is the increment of the displacement field $\mathbf{u}$ which will be discussed later in Section 8.4. The quantity $D^2_{\delta\mathbf{u},\Delta\mathbf{u}}\Pi(\mathbf{u})$ is obtained from the directional derivative of variational equation (8.49) with respect to $\mathbf{u}$ in the direction $\Delta\mathbf{u}$ (i.e. the second directional derivative of Π with respect to $\mathbf{u}$), which is either a maximum $(D^2_{\delta\mathbf{u},\Delta\mathbf{u}}\Pi(\mathbf{u}) < 0)$, a minimum $(D^2_{\delta\mathbf{u},\Delta\mathbf{u}}\Pi(\mathbf{u}) > 0)$, or a saddle point $(D^2_{\delta\mathbf{u},\Delta\mathbf{u}}\Pi(\mathbf{u}) = 0)$.

EXAMPLE 8.4   Show that the directional derivative of the total potential energy Π, as given by (8.47), (8.48), with respect to $\mathbf{u}$ in the arbitrary direction $\delta\mathbf{u}$ leads to

$$D_{\delta\mathbf{u}}\Pi(\mathbf{u}) = D_{\delta\mathbf{u}}\Pi_{\text{int}}(\mathbf{u}) + D_{\delta\mathbf{u}}\Pi_{\text{ext}}(\mathbf{u}) \;, \tag{8.50}$$

$$D_{\delta\mathbf{u}}\Pi_{\text{int}}(\mathbf{u}) = \delta W_{\text{int}} \qquad \text{and} \qquad D_{\delta\mathbf{u}}\Pi_{\text{ext}}(\mathbf{u}) = -\delta W_{\text{ext}} \;. \tag{8.51}$$

The internal virtual work δW_{int} and the external virtual work δW_{ext} are given through eqs. (8.44) and (8.45), respectively.

Furthermore, show that the stationary position of the total potential energy Π gives the principle of virtual work for a body in static equilibrium, as given in eq. (8.42).

Solution.   Since the loads $\mathbf{B}$ and $\overline{\mathbf{T}}$ do not depend on the deformation of the body, they do not contribute to the directional derivative. Hence, from $(8.49)_1$, we find, with use of expressions (8.47) and (8.48), that

$$
\begin{aligned}
D_{\delta\mathbf{u}}\Pi(\mathbf{u}) &= \frac{\mathrm{d}}{\mathrm{d}\varepsilon}\Pi(\mathbf{u} + \varepsilon\delta\mathbf{u})|_{\varepsilon=0} \\
&= \frac{\mathrm{d}}{\mathrm{d}\varepsilon}[\int_{\Omega_0}\Psi(\mathbf{F}(\mathbf{u} + \varepsilon\delta\mathbf{u}))\mathrm{d}V - \int_{\Omega_0}\mathbf{B}\cdot(\mathbf{u} + \varepsilon\delta\mathbf{u})\mathrm{d}V \\
&\quad - \int_{\partial\Omega_{0\,\sigma}}\overline{\mathbf{T}}\cdot(\mathbf{u} + \varepsilon\delta\mathbf{u})\mathrm{d}S]|_{\varepsilon=0} \;.
\end{aligned}
\tag{8.52}
$$

Interchanging differentiation and integration and applying the chain rule, we obtain

$$D_{\delta\mathbf{u}}\Pi(\mathbf{u}) = \int_{\Omega_0}\frac{\partial\Psi(\mathbf{F}(\mathbf{u}))}{\partial\mathbf{F}} : D_{\delta\mathbf{u}}\mathbf{F}(\mathbf{u})\mathrm{d}V - \int_{\Omega_0}\mathbf{B}\cdot\delta\mathbf{u}\mathrm{d}V - \int_{\partial\Omega_{0\,\sigma}}\overline{\mathbf{T}}\cdot\delta\mathbf{u}\mathrm{d}S \;, \tag{8.53}$$

where $D_{\delta\mathbf{u}}\mathbf{F}$ denotes the directional derivative of the deformation gradient $\mathbf{F}$ at $\mathbf{u}$ in the direction of $\delta\mathbf{u}$, derived in eq. (8.11).

In order to specify the first integral in eq. (8.53) we recall the physical expression $(6.1)_1$ and use result $(8.11)_4$ to obtain

$$
\begin{aligned}
\int_{\Omega_0}\frac{\partial\Psi(\mathbf{F}(\mathbf{u}))}{\partial\mathbf{F}} : D_{\delta\mathbf{u}}\mathbf{F}(\mathbf{u})\mathrm{d}V &= \int_{\Omega_0}\frac{\partial\Psi(\mathbf{F}(\mathbf{u}))}{\partial\mathbf{F}} : \frac{\mathrm{d}}{\mathrm{d}\varepsilon}\mathbf{F}(\mathbf{u} + \varepsilon\delta\mathbf{u})|_{\varepsilon=0}\mathrm{d}V \\
&= \int_{\Omega_0}\mathbf{P} : \mathrm{Grad}\delta\mathbf{u}\mathrm{d}V \;.
\end{aligned}
\tag{8.54}
$$

By recalling definitions (8.44)$_2$ and (8.45)$_2$ and combining (8.53) with (8.54)$_2$ we find the desired results (8.50) and (8.51). Hence, stationary condition (8.49) yields precisely the principle of virtual work (8.42) for a configuration in static equilibrium which renders stationary the functional Π.

We conclude that the total potential energy Π is stationary for arbitrary variations $\delta\mathbf{u}$, which means evaluating $\delta\Pi(\mathbf{u}, \delta\mathbf{u}) = 0$ with respect to the displacements, *if and only if* the nonlinear variational equation (8.42) (for $\ddot{\mathbf{u}} = \mathbf{o}$) is satisfied (equilibrium state).

Finally, note that for the purpose of computing the stationary position of Π the magnitude of the virtual displacements need not be small, as is sometimes stated in the literature. However, in order to achieve a first-order approximation the magnitude of the virtual displacements must be small. ∎

Penalty method for incompressibility.   The principle of virtual work is not the appropriate variational approach to invoke kinematic constraint conditions such as incompressibility, contact boundary conditions or Kirchhoff-Love (kinematic) conditions on plates and shells often occurring in engineering applications.

A numerical analysis of nearly incompressible and incompressible materials necessitates so-called **multi-field variational principles** in which additional variables are incorporated. Multi-field variational principles, dealt with in more detail in Sections 8.5 and 8.6, lead to *mixed* or *hybrid methods* for finite elements.

Nevertheless, a single-field variational approach with the displacement $\mathbf{u}$ as the only field variable is very often used in order to approximate, for example, incompressible materials. This leads to the so-called **penalty method**, which is based on the simple (physical) idea of modeling an incompressible material as slightly compressible by using a large value of the bulk modulus. Of course, following this idea, an incompressible material can be obtained by taking the limit infinity for the bulk modulus. However, the result of this idea from the numerical point of view is that we always work with a slightly compressible material since the incompressible limit can never be achieved.

To be more precise, rather than employing the strain-energy function in the form of $\Psi = \Psi(\mathbf{F})$, it is standard to use the unique decoupled representation of the strain-energy function in the form

$$\Psi(\mathbf{C}) = \Psi_{\text{vol}}(J) + \Psi_{\text{iso}}(\overline{\mathbf{C}}) \qquad \text{with} \qquad \Psi_{\text{vol}}(J) = \kappa \mathcal{G}(J) \ , \qquad (8.55)$$

so that energy functional (8.47) takes on the **penalty form**

$$\Pi_{\text{p}}(\mathbf{u}) = \int_{\Omega_0} [\Psi_{\text{vol}}(J(\mathbf{u})) + \Psi_{\text{iso}}(\overline{\mathbf{C}}(\mathbf{u}))]\mathrm{d}V + \Pi_{\text{ext}}(\mathbf{u}) \ , \qquad (8.56)$$

with the external potential energy Π_{ext} given by eq. (8.48)$_2$.

Here, $J = J(\mathbf{u}) = (\det \mathbf{C})^{1/2}$ defines the volume ratio and $\overline{\mathbf{C}} = \overline{\mathbf{C}}(\mathbf{u}) = J^{-2/3}\mathbf{C}$ the corresponding modified right Cauchy-Green tensor, as introduced in eq. $(6.79)_2$. The *strictly convex* function Ψ_{vol} describes the *volumetric elastic response* while Ψ_{iso} is associated with the *isochoric elastic response* of the hyperelastic material. We require $\Psi_{\mathrm{vol}}(J)$ and $\Psi_{\mathrm{iso}}(\overline{\mathbf{C}})$ to be zero *if and only if* $J = 1$ and $\overline{\mathbf{C}} = \mathbf{I}$, ensuring that the reference configuration is stress-free.

According to eq. $(8.55)_2$ the volumetric contribution Ψ_{vol} is characterized by a (positive) **penalty parameter** $\kappa > 0$ which is independent of the deformation. The parameter κ may be viewed as the bulk modulus. The function $\mathcal{G}$ is motivated mathematically. It is known as the **penalty function** and may adopt the simple form

$$\mathcal{G}(J) = \frac{1}{2}(J(\mathbf{u}) - 1)^2 \ , \tag{8.57}$$

often used in numerical computations. Consequently, the meaning of the function Ψ_{vol} as used in eq. $(8.55)_1$ differs significantly from its meaning in eq. (6.85), in which Ψ_{vol} is of physical relevance.

We now derive the stationary position of Π_{p} with respect to the displacement field, which is basically a procedure according to Example 8.4. Starting with the fundamental condition (8.49) and following the steps which have led to eq. (8.54) we find, using decomposition (8.55), that

$$D_{\delta\mathbf{u}}\Pi_{\mathrm{p}}(\mathbf{u}) = \int_{\Omega_0} \left(\frac{\partial\Psi_{\mathrm{vol}}(J(\mathbf{u}))}{\partial\mathbf{C}} + \frac{\partial\Psi_{\mathrm{iso}}(\overline{\mathbf{C}}(\mathbf{u}))}{\partial\mathbf{C}} \right) : D_{\delta\mathbf{u}}\mathbf{C}(\mathbf{u})\mathrm{d}V$$

$$+ D_{\delta\mathbf{u}}\Pi_{\mathrm{ext}}(\mathbf{u}) = 0 \ , \tag{8.58}$$

where $D_{\delta\mathbf{u}}\mathbf{C}$ denotes the directional derivative of the right Cauchy-Green tensor $\mathbf{C}$ at $\mathbf{u}$ in the direction of $\delta\mathbf{u}$, which is $2\delta\mathbf{E}$ (see Section 8.1).

A specification of the integral in eq. (8.58) implies, by means of the chain rule and relation $(6.82)_1$, that

$$\int_{\Omega_0} \left(\frac{\partial\Psi_{\mathrm{vol}}}{\partial\mathbf{C}} + \frac{\partial\Psi_{\mathrm{iso}}}{\partial\mathbf{C}} \right) : D_{\delta\mathbf{u}}\mathbf{C}\mathrm{d}V = \int_{\Omega_0} \left(\frac{\mathrm{d}\Psi_{\mathrm{vol}}}{\mathrm{d}J}\frac{\partial J}{\partial\mathbf{C}} + \frac{\partial\Psi_{\mathrm{iso}}}{\partial\mathbf{C}} \right) : D_{\delta\mathbf{u}}\mathbf{C}\mathrm{d}V$$

$$= \int_{\Omega_0} \left(J\frac{\mathrm{d}\Psi_{\mathrm{vol}}}{\mathrm{d}J}\mathbf{C}^{-1} + 2\frac{\partial\Psi_{\mathrm{iso}}}{\partial\mathbf{C}} \right) : \delta\mathbf{E}\mathrm{d}V \tag{8.59}$$

(the arguments of the functions have been omitted for simplicity).

With reference to eq. $(6.91)_1$ the term $\mathrm{d}\Psi_{\mathrm{vol}}/\mathrm{d}J$ defines the hydrostatic pressure p. Hence, recalling definitions $(6.89)_2$ and $(6.90)_1$, we conclude that the terms in parentheses of $(8.59)_2$ are associated with the volumetric and isochoric stress contributions $\mathbf{S}_{\mathrm{vol}}$ and $\mathbf{S}_{\mathrm{iso}}$, respectively.

Using the second Piola-Kirchhoff stress tensor $\mathbf{S}$, which is based on the additive decomposition $(6.88)_2$, we achieve finally the principle of virtual work (for a configuration in static equilibrium) in the form

$$D_{\delta\mathbf{u}}\Pi_{\mathrm{p}}(\mathbf{u}) = \int_{\Omega_0} \mathbf{S} : \delta\mathbf{E}\mathrm{d}V + D_{\delta\mathbf{u}}\Pi_{\mathrm{ext}}(\mathbf{u}) = 0 \quad . \tag{8.60}$$

Note that the integral in eq. (8.60), i.e. the directional derivative of the internal potential energy Π_{int} with respect to $\mathbf{u}$ in the arbitrary direction $\delta\mathbf{u}$, precisely gives the internal virtual work δW_{int} (see eq. $(8.46)_2$). The second term in eq. (8.60) is related to the external virtual work according to $(8.51)_2$.

A parametrization of the hydrostatic pressure p is simply obtained from eq. $(8.55)_2$ by means of assumption (8.57), i.e.

$$p = \frac{\mathrm{d}\Psi_{\mathrm{vol}}(J(\mathbf{u}))}{\mathrm{d}J} = \kappa\frac{\mathrm{d}\mathcal{G}(J(\mathbf{u}))}{\mathrm{d}J} = \kappa(J(\mathbf{u}) - 1) \quad . \tag{8.61}$$

In contrast to eqs. $(6.140)_2$ and (6.141), this is an artificial constitutive equation for p designed to prevent a significant volumetric response, as already pointed out.

The user-specified penalty parameter κ is merely an adjustable numerical parameter which is often chosen through numerical experiments. Clearly, with increasing κ the violation of the constraint is reduced. If we take the restriction on the value $\kappa \to \infty$, the constraint condition is exactly enforced, and then eq. (8.56) represents a functional for an incompressible material with $J = 1$.

Unfortunately, for an approximation technique such as the finite element method, the stiffness matrix becomes increasingly *ill-conditioned* for increasing κ (see, for example, BERTSEKAS [1982] and LUENBERGER [1984] for detailed studies). For that case the **reduced integration method**, and the later proposed **selective-reduced integration method** (which are equivalent to certain types of mixed finite element methods, as discussed in MALKUS and HUGHES [1978] and HUGHES [2000]) is often used to underintegrate (weaken) the penalty function.

However, penalty methods are attractive because they are based on a simple variational principle with all its computational advantages and are very effective to implement in a finite element program.

EXERCISES

1. The classical **Hamilton's variational principle** represents a generalization of the principle of stationary potential energy (8.49) to continuum dynamics. It is presented by the stationary condition

$$\delta \int_{t_0}^{t_1} L(\mathbf{u}, \dot{\mathbf{u}})\mathrm{d}t = D_{\delta\mathbf{u}} \int_{t_0}^{t_1} L(\mathbf{u}, \dot{\mathbf{u}})\mathrm{d}t = 0 \qquad \text{where} \qquad L(\mathbf{u}, \dot{\mathbf{u}}) = \Pi(\mathbf{u}) - \mathcal{K}(\dot{\mathbf{u}}) \quad ,$$

with $\delta\mathbf{u}$ denoting the variation of the displacement vector field, which is a function of *position* and *time*. The functional L (in the literature sometimes introduced as $-L$) is integrated with respect to time t over the closed time interval $t \in [t_0, t_1]$ (t_0 and t_1 are two arbitrary instants of time). We now assume the restriction that at the times t_0 and t_1, the variation $\delta\mathbf{u}$ of the displacement vector field vanishes at all points of the continuum body, i.e.

$$\delta\mathbf{u}|_{t=t_0} = \delta\mathbf{u}|_{t=t_1} = \mathbf{o} \quad . \tag{8.62}$$

The scalar-valued functionals Π and $\mathcal{K}$ denote the total potential energy (4.114) and the kinetic energy (4.83) of the moving body, as usual. The potential energy of the loads exists and is given by eq. $(8.48)_2$.

Show that the vanishing variation of the functional L with the imposed restrictions (8.62) gives the principle of virtual work (8.42) for all $\delta\mathbf{u}$ which are zero on $\partial\Omega_{\mathrm{u}}$ throughout the entire closed time interval $t \in [t_0, t_1]$.

2. Consider a constant normal pressure p applied to a boundary surface enclosing a certain region. Show that there exists the associated potential

$$\Pi_{\mathrm{ext}}(\mathbf{u}) = p \int_\Omega \mathrm{d}v = p \int_{\Omega_0} J(\mathbf{X}, t)\mathrm{d}V \quad ,$$

whose variation gives the external virtual work δW_{ext}, as defined in eq. (8.36), i.e. $D_{\delta\mathbf{u}}\Pi_{\mathrm{ext}}(\mathbf{u}) = \delta W_{\mathrm{ext}}(\mathbf{u}, \delta\mathbf{u}) = p \int_{\partial\Omega_\sigma} \mathbf{n} \cdot \delta\mathbf{u}\,\mathrm{d}s$.

8.4   Linearization of the Principle of Virtual Work

Variational principles such as the principle of virtual work in the forms of (8.33) or (8.42) are generally nonlinear in the unknown displacement vector field $\mathbf{u}$. Typically, the nonlinearities are due to *geometric* and *material* contributions, i.e. the kinematics of the body and the constitutive equation of the material, respectively.

As mentioned above, one main objective of engineering analysis is to find the unknown field $\mathbf{u}$ which is the solution of the associated nonlinear boundary-value problem. Usually, (exact) closed-form mathematical solutions of a set of nonlinear partial differential equations are only available for some special engineering problems; they are rather complicated and often unusable.

In order to keep the complexities of engineering problems intact, approximate numerical solutions, based on, for example, the finite element method, are required. A very common and simple numerical technique to solve nonlinear equations is to employ the reliable *incremental/iterative solution technique* of *Newton's type*. It is an efficient method with the feature of a quadratic convergence rate near the solution point. This technique requires a *consistent linearization* of all the quantities associated with the considered nonlinear problem generating efficient *recurrence update formulas*. The nonlinear problem then is replaced by a sequence of linear problems which are easy to solve at each iteration.

Linearization is a systematic process which is based on the concept of directional derivatives, see the pioneering work of HUGHES and PISTER [1978]; for the more generalized concept see the book by MARSDEN and HUGHES [1994, Chapter 4], and for an application to rods and plates the work of WRIGGERS [1988] among others. For the concepts of linearization and directional derivative and their applications in nonlinear continuum mechanics see also the textbook by BONET and WOOD [1997].

Concept of linearization.   The following part of this section deals with the linearization of a nonlinear and smooth (possibly time-dependent) function $\mathcal{F} = \mathcal{F}(\mathbf{u})$ in the *material* description which is either scalar-valued, vector-valued or tensor-valued. The single argument of $\mathcal{F}$ is the displacement vector variable $\mathbf{u}$.

Consider $\mathbf{u}$, then the fundamental relationship for the linearization of the nonlinear function $\mathcal{F}$ is based on the first-order (Taylor's) expansion, which is expressed as

$$\mathcal{F}(\mathbf{u}, \Delta\mathbf{u}) = \mathcal{F}(\mathbf{u}) + \Delta\mathcal{F}(\mathbf{u}, \Delta\mathbf{u}) + o(\Delta\mathbf{u}) \quad , \tag{8.63}$$

where $\Delta(\bullet)$ denotes the linearization operator similar to $\delta(\bullet)$. The operator $\Delta(\bullet)$ is also *linear* and the usual properties of differentiation are valid. The quantity $\Delta\mathbf{u}$ denotes the **increment** of the displacement field $\mathbf{u}$, here expressed in the reference configuration.

The remainder $o(\Delta\mathbf{u})$, characterized by the *Landau order* symbol $o(\bullet)$, is a small error that tends to zero faster than $\Delta\mathbf{u} \rightarrow \mathbf{o}$, i.e. $\lim_{\Delta\mathbf{u}\rightarrow\mathbf{o}} o(\Delta\mathbf{u})/|\Delta\mathbf{u}| = \mathbf{o}$.

Within the classical solution technique of *Newton's method*, Taylor's expansion is truncated after the first derivative of $\mathcal{F}$. Hence, the first term in (8.63) is a constant part, i.e. an approximate solution for a given state $\mathbf{u}$. The second term $\Delta\mathcal{F}$ is the linear change in $\mathcal{F}$ due to $\Delta\mathbf{u}$ at $\mathbf{u}$. It is the directional derivative of $\mathcal{F}$ at given $\mathbf{u}$ (fixed) in the direction of the incremental displacement field $\Delta\mathbf{u}$, i.e.

$$\Delta\mathcal{F}(\mathbf{u}, \Delta\mathbf{u}) = D_{\Delta\mathbf{u}}\mathcal{F}(\mathbf{u}) = \frac{\mathrm{d}}{\mathrm{d}\varepsilon}\mathcal{F}(\mathbf{u} + \varepsilon\Delta\mathbf{u})|_{\varepsilon=0} \quad , \tag{8.64}$$

where the linear Gâteaux operator $D(\bullet)$ is with respect to the incremental displacement field $\Delta\mathbf{u}$. We say that $\Delta\mathcal{F}(\mathbf{u}, \Delta\mathbf{u})$ is the **linearization** (or **linear approximation**) of $\mathcal{F}$ at $\mathbf{u}$.

Note that in regard to eq. (8.8) the *first variation* $D_{\delta\mathbf{u}}\mathcal{F}(\mathbf{u})$ of a vector function $\mathcal{F}(\mathbf{u})$ and the *linearization* $D_{\Delta\mathbf{u}}\mathcal{F}(\mathbf{u})$ of that vector function are based on the same concept of directional derivatives. By taking notice of this equivalence of variation and linearization, all relations derived in the previous Section 8.1 can be adopted here; we just use the symbol $\Delta(\bullet)$ instead of $\delta(\bullet)$.

For example, relation (8.6) and transformation (8.7) read

$$\Delta(\text{grad}\mathbf{u}) = \text{grad}(\Delta\mathbf{u}) \quad , \tag{8.65}$$

$$\text{grad}\Delta\mathbf{u} = \text{Grad}\Delta\mathbf{u}\,\mathbf{F}^{-1} \qquad \text{or} \qquad \frac{\partial\Delta u_a}{\partial x_b} = \frac{\partial\Delta u_a}{\partial X_A} F_{Ab}^{-1} \quad . \tag{8.66}$$

In addition, the linearizations of tensors $\mathbf{F}$, $\mathbf{F}^{-1}$, $\mathbf{E}$ are

$$\Delta\mathbf{F} = D_{\Delta\mathbf{u}}\mathbf{F} = \text{Grad}\Delta\mathbf{u} \quad , \tag{8.67}$$

$$\Delta\mathbf{F}^{-1} = D_{\Delta\mathbf{u}}\mathbf{F}^{-1} = -\mathbf{F}^{-1}\text{grad}\Delta\mathbf{u} \quad , \tag{8.68}$$

$$\Delta\mathbf{E} = D_{\Delta\mathbf{u}}\mathbf{E} = \frac{1}{2}[(\mathbf{F}^{\mathrm{T}}\text{Grad}\Delta\mathbf{u})^{\mathrm{T}} + \mathbf{F}^{\mathrm{T}}\text{Grad}\Delta\mathbf{u}] = \text{sym}(\mathbf{F}^{\mathrm{T}}\text{Grad}\Delta\mathbf{u}) \quad , \tag{8.69}$$

which are analogous to eqs. (8.9), (8.10), (8.14).

EXAMPLE 8.5   Show that the linearization $\Delta\delta\mathbf{E}$ of the virtual Green-Lagrange strain tensor $\delta\mathbf{E} = \text{sym}(\mathbf{F}^{\mathrm{T}}\text{Grad}\delta\mathbf{u})$, as derived in eq. $(8.14)_2$, may be expressed as

$$\Delta\delta\mathbf{E} = \text{sym}(\text{Grad}^{\mathrm{T}}\Delta\mathbf{u}\,\text{Grad}\delta\mathbf{u}) \quad . \tag{8.70}$$

Solution.   According to the rule (8.64), we compute the directional derivative of $\delta\mathbf{E}$ in the direction of $\Delta\mathbf{u}$ at $\mathbf{u}$, i.e.

$$\Delta\delta\mathbf{E} = D_{\Delta\mathbf{u}}\delta\mathbf{E} = \frac{\mathrm{d}}{\mathrm{d}\varepsilon}\delta\mathbf{E}(\mathbf{u} + \varepsilon\Delta\mathbf{u})|_{\varepsilon=0}$$

$$= \frac{\mathrm{d}}{\mathrm{d}\varepsilon}\text{sym}[(\mathbf{F}(\mathbf{u} + \varepsilon\Delta\mathbf{u}))^{\mathrm{T}}\text{Grad}\delta\mathbf{u}]|_{\varepsilon=0} \quad . \tag{8.71}$$

Since the virtual displacement field $\delta\mathbf{u}$ is independent of the displacement, the term $\text{Grad}\delta\mathbf{u}$ is not affected by the linearization. Knowing that $\mathrm{d}/\mathrm{d}\varepsilon\,\mathbf{F}(\mathbf{u} + \varepsilon\Delta\mathbf{u})|_{\varepsilon=0} = \text{Grad}\Delta\mathbf{u}$ (see eq. (8.67)), we find the desired result (8.70).   ∎

In order to linearize a nonlinear smooth vector function $f = f(\mathbf{u})$ in the *spatial* description we adopt the concept for the first variation of f introduced on p. 375. By analogy with relation (8.16), we may write

$$\Delta f(\mathbf{u}, \Delta\mathbf{u}) = \chi_*(D_{\Delta\mathbf{u}}\chi_*^{-1}(f)) = \chi_*(D_{\Delta\mathbf{u}}\mathcal{F}) \tag{8.72}$$

for the **linearization** (or **linear approximation**) of f. Since $D_{\Delta\mathbf{u}}\mathcal{F}(\mathbf{u}) = \Delta\mathcal{F}(\mathbf{u}, \Delta\mathbf{u})$ according to (8.64) we obtain

$$\Delta f(\mathbf{u}, \Delta\mathbf{u}) = \chi_*(\Delta\mathcal{F}(\mathbf{u}, \Delta\mathbf{u})) \quad , \tag{8.73}$$

which is analogous to eq.(8.17). For notational simplicity, the linearization operator Δ

is not particularly marked when applied to a function in the spatial description, as for the δ-process.

Note that the operators required for the *Lie time derivatives*, the *variations* and *linearizations* of spatial tensor variables are formally the same. They are based on the concept of directional derivative. For the Lie time derivative the considered direction of the derivative is **v**, while for the variation and linearization it is the virtual displacement field $\delta\mathbf{u}$ and the incremental displacement field $\Delta\mathbf{u}$, respectively. Compare relations (2.189), (8.16) and (8.72).

EXAMPLE 8.6   Show that the linearization $\Delta\delta\mathbf{e}$ of the virtual Euler-Almansi strain tensor $\delta\mathbf{e}$, which is a spatial tensor field according to eq. (8.18), may be expressed as

$$\Delta\delta\mathbf{e} = \text{sym}(\text{grad}^\text{T}\Delta\mathbf{u}\,\text{grad}\delta\mathbf{u}) \quad . \tag{8.74}$$

Solution.   Since we apply the systematic technique of linearization to material quantities, as a first step we pull-back the variation of the Euler-Almansi strain tensor $\delta\mathbf{e}$, that is the inversion of eq. $(8.20)_2$, which yields the variation $\delta\mathbf{E}$ of the associated Green-Lagrange strain tensor $\mathbf{E}$. The linearization of $\delta\mathbf{E}$ is carried out by analogy with Example 8.5. In the last step the push-forward operation on $\Delta\delta\mathbf{E}$, as given in (8.70), is

$$\chi_*(\Delta\delta\mathbf{E}) = \mathbf{F}^{-\text{T}}\text{sym}(\text{Grad}^\text{T}\Delta\mathbf{u}\,\text{Grad}\delta\mathbf{u})\mathbf{F}^{-1}$$
$$= \text{sym}(\text{grad}^\text{T}\Delta\mathbf{u}\,\text{grad}\delta\mathbf{u}) \quad , \tag{8.75}$$

where the relations (8.66) and (8.7) should be used.   ∎

Linearization of the principle of virtual work in material description.   In order to linearize the principle of virtual work in the material description we recall the nonlinear variational equation (8.42). For simplicity we consider a purely static problem, so that $\ddot{\mathbf{u}} = \mathbf{o}$. In addition, we assume the loads **B** and $\overline{\mathbf{T}}$ to be 'dead' (independent of the deformation of the body), so that the corresponding linearization of the external virtual work (8.45) vanishes, i.e. $D_{\Delta\mathbf{u}}\delta W_{\text{ext}}(\mathbf{u}, \delta\mathbf{u}) = 0$. This is certainly not the case for some other types of loads, like the pressure loads discussed on p. 383 and subsequently.

Hence, the linearization of the variational equation (8.42) only affects the internal virtual work δW_{int}, on which we will focus subsequently. For our purpose we take the material (or Lagrangian) form $(8.46)_2$, i.e.

$$\delta W_{\text{int}}(\mathbf{u}, \delta\mathbf{u}) = \int_{\Omega_0} \mathbf{S}(\mathbf{E}(\mathbf{u})) : \delta\mathbf{E}(\mathbf{u})\text{d}V \quad . \tag{8.76}$$

Note that the Green-Lagrange strain tensor **E** depends on the displacement field **u** through the relationship (2.91).

Now we may adopt rule (8.64) in order to compute the linearization of (8.76), i.e.

$$D_{\Delta u}\delta W_{int}(\mathbf{u}, \delta\mathbf{u}) = \frac{d}{d\varepsilon}\delta W_{int}(\mathbf{u} + \varepsilon\Delta\mathbf{u})|_{\varepsilon=0}$$

$$= \frac{d}{d\varepsilon}[\int_{\Omega_0} \mathbf{S}(\mathbf{E}(\mathbf{u} + \varepsilon\Delta\mathbf{u})) : \delta\mathbf{E}(\mathbf{u} + \varepsilon\Delta\mathbf{u})dV]|_{\varepsilon=0} \ . \qquad (8.77)$$

Interchanging differentiation and integration and using the product rule results in

$$D_{\Delta u}\delta W_{int}(\mathbf{u}, \delta\mathbf{u}) = \int_{\Omega_0} [\mathbf{S}(\mathbf{E}(\mathbf{u})) : D_{\Delta u}\delta\mathbf{E}(\mathbf{u}) + \delta\mathbf{E}(\mathbf{u}) : D_{\Delta u}\mathbf{S}(\mathbf{E}(\mathbf{u}))]dV \ , \qquad (8.78)$$

where $D_{\Delta u}\delta\mathbf{E}$ characterizes the directional derivative of $\delta\mathbf{E}$ at $\mathbf{u}$ in the direction of $\Delta\mathbf{u}$, i.e. the linearization of $\delta\mathbf{E}$ according to eq. (8.70).

In order to specify the linearization $D_{\Delta u}\mathbf{S}$ of the (symmetric) second Piola-Kirchhoff stress tensor $\mathbf{S}$ in eq. (8.78), use the chain rule to obtain

$$D_{\Delta u}\mathbf{S}(\mathbf{E}(\mathbf{u})) = \frac{\partial\mathbf{S}(\mathbf{E}(\mathbf{u}))}{\partial\mathbf{E}} : D_{\Delta u}\mathbf{E}(\mathbf{u}) = \mathbb{C}(\mathbf{u}) : D_{\Delta u}\mathbf{E}(\mathbf{u}) \ , \qquad (8.79)$$

with $D_{\Delta u}\mathbf{E}$ denoting the linearization of the Green-Lagrange strain tensor $\mathbf{E}$ (see relations (8.69)). It is important to emphasize that the term $\partial\mathbf{S}(\mathbf{E})/\partial\mathbf{E}$ is precisely the elasticity tensor $\mathbb{C}$ in the material description, as defined in eq. (6.155). It is a fourth-order tensor which possesses the *minor symmetries* $C_{ABCD} = C_{BACD} = C_{ABDC}$.

Hence, eq. (8.78) may be re-expressed as

$$D_{\Delta u}\delta W_{int}(\mathbf{u}, \delta\mathbf{u}) = \int_{\Omega_0} [\mathbf{S}(\mathbf{E}(\mathbf{u})) : D_{\Delta u}\delta\mathbf{E}(\mathbf{u}) + \delta\mathbf{E}(\mathbf{u}) : \mathbb{C}(\mathbf{u}) : D_{\Delta u}\mathbf{E}(\mathbf{u})]dV \ . \qquad (8.80)$$

Finally, we use the explicit expression (8.70) and property (1.95) for the first term of the integral in eq. (8.80) and relations (8.14) and (8.69) for the second term. Since the stress tensor $\mathbf{S}$ is symmetric and the elasticity tensor $\mathbb{C}$ has minor symmetries, the linearization of the internal virtual work in the material description leads to the set of linear increments

$$D_{\Delta u}\delta W_{int}(\mathbf{u}, \delta\mathbf{u}) = \int_{\Omega_0} (\text{Grad}\delta\mathbf{u} : \text{Grad}\Delta\mathbf{u}\,\mathbf{S}$$

$$+ \mathbf{F}^T\text{Grad}\delta\mathbf{u} : \mathbb{C} : \mathbf{F}^T\text{Grad}\Delta\mathbf{u})dV \ , \qquad (8.81)$$

or, in index notation,

$$D_{\Delta u}\delta W_{int}(\mathbf{u}, \delta\mathbf{u}) = \int_{\Omega_0} \left(\frac{\partial\delta u_a}{\partial X_B}\frac{\partial\Delta u_a}{\partial X_D}S_{BD} + F_{aA}\frac{\partial\delta u_a}{\partial X_B}C_{ABCD}F_{bC}\frac{\partial\Delta u_b}{\partial X_D}\right)dV$$

$$= \int_{\Omega_0} \frac{\partial\delta u_a}{\partial X_B}(\delta_{ab}S_{BD} + F_{aA}F_{bC}C_{ABCD})\frac{\partial\Delta u_b}{\partial X_D}dV \ , \qquad (8.82)$$

which describes the fully nonlinear (finite) deformation case. The terms $\delta_{ab}S_{BD}$ and $F_{aA}F_{bC}C_{ABCD}$ represent the effective elasticity tensor, which has the nature of the *(tangent) stiffness matrix.*

Relations (8.81) and (8.82) are linear with respect to $\delta\mathbf{u}$ and $\Delta\mathbf{u}$ depending on $\mathbf{X}$. These relations show a clear mathematical structure in the sense that $\delta\mathbf{u}$ and $\Delta\mathbf{u}$ can be interchanged without altering the result of the integral; thus $D_{\Delta\mathbf{u}}\delta W_{\text{int}}(\mathbf{u}, \delta\mathbf{u}) = D_{\delta\mathbf{u}}\delta W_{\text{int}}(\mathbf{u}, \Delta\mathbf{u})$. Relations (8.81) and (8.82) lead to a *symmetric* (tangent) stiffness matrix upon discretization. Note that, for example, the set of nonlinear equations associated with nonlinear heat conduction results in a different mathematical structure leading to a *non-symmetric* stiffness matrix.

The first term in eq. (8.81) comes from the current state of stress and represents the so-called **geometrical stress contribution** (in the literature sometimes called the **initial stress contribution**) to the linearization. Since S_{AB} is not the initial stress (it is in fact the current stress), the terminology is misleading. Within an *incremental/iterative solution technique* we can think of S_{AB} as the initial stress at every increment, so the term initial stress contribution has some meaning. The second term in eq. (8.81) represents the so-called **material contribution** to the linearization.

The linearized principle of virtual work (8.81) constitutes the starting point for approximation techniques such as the finite element method, typically leading to the **geometrical** (or **initial stress**) **stiffness matrix** and to the **material stiffness matrix**.

Note that for some cases it is more convenient to discretize the nonlinear variational equation as a first step and to linearize the result with respect to the positions of the nodal points as a second step.

Linearization of the principle of virtual work in spatial description.   In order to linearize the principle of virtual work in the spatial description we recall the nonlinear variational equation (8.33).

As above we consider the static case ($\ddot{\mathbf{u}} = \mathbf{o}$) and assume the loads $\mathbf{b}$ and $\bar{\mathbf{t}}$ to be independent of the motion of the body. Only the linearization of the internal virtual work δW_{int} in the spatial description remains. We adopt δW_{int} in the spatial (or Eulerian) form (8.34). The idea is first to pull-back the spatial quantities to the reference configuration, so they correspond with the internal virtual work in the material description. Then they are linearized, as above, and as a last step it is necessary to push-forward the linearized terms.

Starting with the equivalence

$$\delta W_{\text{int}}(\mathbf{u}, \delta\mathbf{u}) = \int_{\Omega} \boldsymbol{\sigma}(\mathbf{e}(\mathbf{u})) : \delta\mathbf{e}(\mathbf{u}) dv = \int_{\Omega_0} \mathbf{S}(\mathbf{E}(\mathbf{u})) : \delta\mathbf{E}(\mathbf{u}) dV \quad , \quad (8.83)$$

we consider the linerization of the internal virtual work in the material description which we have derived in (8.78), i.e.

$$D_{\Delta \mathbf{u}}\delta W_{\text{int}} = \int_{\Omega_0} (\mathbf{S} : D_{\Delta \mathbf{u}}\delta \mathbf{E} + \delta \mathbf{E} : D_{\Delta \mathbf{u}}\mathbf{S})\mathrm{d}V \tag{8.84}$$

(the arguments have been omitted).

Hence, the push-forward operation on the second Piola-Kirchhoff stress tensor $\mathbf{S}$ yields, according to (3.64), the Kirchhoff-stress tensor τ, which is related to the Cauchy stress tensor by $\tau = J\sigma$. Pushing forward the linearized variation of the Green-Lagrange strain tensor, $\Delta \delta \mathbf{E}$, yields the linearized variation of the Euler-Almansi strain tensor, $\Delta \delta \mathbf{e}$, as discussed in Example 8.6. Computing the push-forward of $\delta \mathbf{E}$ results in $\delta \mathbf{e}$, as introduced in Section 8.1.

Finally, we derive the push-forward of the linearized second Piola-Kirchhoff stress tensor, i.e. the last term in (8.84), which will yield the linearized Kirchhoff stress tensor $\Delta \tau$. We write

$$\Delta \tau = \chi_*(D_{\Delta \mathbf{u}}\mathbf{S}) = \mathbf{F}(D_{\Delta \mathbf{u}}\mathbf{S})\mathbf{F}^{\mathrm{T}} \quad , \tag{8.85}$$

with $D_{\Delta \mathbf{u}}\mathbf{S}$ given by (8.79), i.e. $D_{\Delta \mathbf{u}}\mathbf{S} = \mathbb{C} : D_{\Delta \mathbf{u}}\mathbf{E}$. By use of (8.69) and (8.66), the term $\mathbf{F}(D_{\Delta \mathbf{u}}\mathbf{S})\mathbf{F}^{\mathrm{T}}$ in eq. $(8.85)_2$ may be written as

$$\mathbf{F}(\mathbb{C} : \mathbf{F}^{\mathrm{T}}\mathrm{Grad}\Delta \mathbf{u})\mathbf{F}^{\mathrm{T}} = \mathbf{F}(\mathbb{C} : \mathbf{F}^{\mathrm{T}}\mathrm{grad}\Delta \mathbf{u}\,\mathbf{F})\mathbf{F}^{\mathrm{T}} \quad , \tag{8.86}$$

where we have also employed the minor symmetries of $\mathbb{C}$. In order to proceed it is more instructive to employ index notation. With the definition (6.159) of the spatial elasticity tensor $(\mathbb{c})_{abcd} = c_{abcd}$, eq. (8.86) is equivalent to

$$F_{aA}C_{ABCD}F_{cC}\frac{\partial \Delta u_c}{\partial x_d}F_{dD}F_{bB} = F_{aA}F_{bB}F_{cC}F_{dD}C_{ABCD}\frac{\partial \Delta u_c}{\partial x_d}$$

$$= Jc_{abcd}\frac{\partial \Delta u_c}{\partial x_d} \quad . \tag{8.87}$$

Hence, the linearization of the spatial Kirchhoff stress tensor, i.e. (8.85), gives the useful relation

$$\Delta \tau = J\mathbb{c} : \mathrm{grad}\Delta \mathbf{u} \quad . \tag{8.88}$$

Note that the increment $\Delta \tau$ denotes the linearized tensor-valued function τ according to the concept of directional derivative introduced in (8.72). Replacing the associated direction $\Delta \mathbf{u}$, used in the directional derivative, by the velocity vector $\mathbf{v}$, $\Delta \tau$ and $\mathrm{grad}\Delta \mathbf{u}$ result in the Lie time derivative $\pounds_v(\tau)$ of τ and the spatial velocity gradient $\mathbf{l}$ (defined by $(2.141)_4$), respectively. By using the symmetries of $\mathbb{c}$ relation (8.88) reads $\pounds_v(\tau) = J\mathbb{c} : \mathbf{d}$, which proves (6.161).

Considering the push-forward operations derived we obtain finally, from (8.84) with relation $dv = JdV$ and some rearranging, the linearized internal virtual work in the spatial description, i.e.

$$D_{\Delta u}\delta W_{int}(\mathbf{u}, \delta\mathbf{u}) = \int_{\Omega} (\text{grad}\delta\mathbf{u} : \text{grad}\Delta\mathbf{u}\,\boldsymbol{\sigma} + \text{grad}\delta\mathbf{u} : \mathbb{c} : \text{grad}\Delta\mathbf{u})dv \quad, \qquad (8.89)$$

or, in index notation,

$$D_{\Delta u}\delta W_{int}(\mathbf{u}, \delta\mathbf{u}) = \int_{\Omega} \left(\frac{\partial\delta u_a}{\partial x_b}\frac{\partial\Delta u_a}{\partial x_d}\sigma_{bd} + \frac{\partial\delta u_a}{\partial x_b}c_{abcd}\frac{\partial\Delta u_c}{\partial x_d} \right) dv$$

$$= \int_{\Omega} \frac{\partial\delta u_a}{\partial x_b} (\delta_{ac}\sigma_{bd} + c_{abcd}) \frac{\partial\Delta u_c}{\partial x_d} dv \quad, \qquad (8.90)$$

where $\delta_{ac}\sigma_{bd} + c_{abcd}$ represents the effective elasticity tensor in the spatial description. Relations (8.89) and (8.90) are linear with respect to the terms $\delta\mathbf{u}$ and $\Delta\mathbf{u}$. They describe the fully nonlinear (finite) deformation case and have a similar symmetric structure to the linearized eqs. (8.81) and (8.82).

Formulations according to (8.81) and (8.89) are in the literature sometimes called **total-Lagrangian** and **updated-Lagrangian**, respectively. This really means that integrals are calculated over the respective regions of the reference and the current configuration. However, it is important to emphasize that the derived material representations (8.81) and (8.82) of the linearized virtual internal work are **equivalent** to the spatial versions (8.89) and (8.90). The two representations are based on the use of change of variables, and the results are the same in both cases.

In order to recapture the small deformation (but nonlinear elastic) case (in the literature often called the materially nonlinear case) we fix the geometry, i.e. we do not distinguish between initial and current geometry. Further, we do not account for the initial stress contribution $\delta_{ac}\sigma_{bd}$ and ignore the quadratic terms in the Green-Lagrange strain tensor. In addition, for the fully linear case, the coefficients of the elasticity tensor are given and are not functions of strains anymore.

EXAMPLE 8.7   An alternative approach to derive the linearized internal virtual work (8.89) in the spatial description is to use the formal equivalence of the material time derivative of scalar-valued functions in the spatial description with the directional derivative of these functions in the direction $\Delta\mathbf{u}$ (compare with Section 2.8).

Carry out the material time derivative of the internal virtual work δW_{int} in the spatial description and use this property to obtain its linearization. This approach circumvents the extensive pull-back and push-forward operations.

Solution.   We again start with δW_{int} in the spatial form (8.34) and change the domain of integration with $dv = JdV$ so that

$$\delta W_{\text{int}}(\mathbf{u}, \delta \mathbf{u}) = \int_{\Omega} \boldsymbol{\sigma}(\mathbf{u}) : \delta \mathbf{e}(\mathbf{u}) dv = \int_{\Omega_0} J\boldsymbol{\sigma}(\mathbf{u}) : \delta \mathbf{e}(\mathbf{u}) dV \quad , \tag{8.91}$$

where the Euler-Almansi strain tensor $\mathbf{e}$ depends on the displacement field $\mathbf{u}$ through the relationship (2.92). The Cauchy stress tensor $\boldsymbol{\sigma}$ also depends on $\mathbf{u}$, here a function of the current position $\mathbf{x}$.

Employing material time derivatives and the product rule we find from $(8.91)_2$ that

$$\overline{\delta W_{\text{int}}(\mathbf{u}, \delta \mathbf{u})} = \int_{\Omega_0} [\boldsymbol{\tau}(\mathbf{u}) : \overline{\delta \mathbf{e}(\mathbf{u})} + \delta \mathbf{e}(\mathbf{u}) : \overline{\boldsymbol{\tau}(\mathbf{u})}] dV \quad , \tag{8.92}$$

where additionally the relationship between the symmetric Kirchhoff-stress tensor $\boldsymbol{\tau}$ and $\boldsymbol{\sigma}$, i.e. $\boldsymbol{\tau} = J\boldsymbol{\sigma}$, is to be used. Recall that the superposed dot denotes the material time derivative, as usual.

Firstly, we derive the material time derivative of the virtual Euler-Almansi strain tensor $\delta \mathbf{e}$. By means of (8.7) and $(2.145)_2$, we find from $(8.18)_2$ that

$$\overline{\delta \mathbf{e}} = \overline{\text{sym}(\text{grad}\delta \mathbf{u})} = \overline{\text{sym}(\text{Grad}\delta \mathbf{u}\, \mathbf{F}^{-1})}$$

$$= \text{sym}(\text{Grad}\delta \mathbf{u}\overline{\mathbf{F}^{-1}}) = \text{sym}[\text{Grad}\delta \mathbf{u}(-\mathbf{F}^{-1}\mathbf{l})]$$

$$= -\text{sym}(\text{grad}\delta \mathbf{u}\, \mathbf{l}) \quad , \tag{8.93}$$

with the spatial velocity gradient $\mathbf{l} = \text{grad}\mathbf{v}$ (recall definition (2.137)).

Secondly, we focus attention on the crucial material time derivative of the Kirchhoff stress tensor $\boldsymbol{\tau}$. We start with the relation for the (objective) *Oldroyd stress rate* of the spatial stress field $\boldsymbol{\tau}$. Recall the *Oldroyd stress rate* of $\boldsymbol{\tau}$ which is identical to the Lie time derivative of $\boldsymbol{\tau}$, and with reference to eq. (5.59) expressed as $\mathcal{L}_{\mathbf{v}}(\boldsymbol{\tau}) = \dot{\boldsymbol{\tau}} - \mathbf{l}\boldsymbol{\tau} - \boldsymbol{\tau}\mathbf{l}^{\text{T}}$. By means of $\mathcal{L}_{\mathbf{v}}(\boldsymbol{\tau}) = J\mathbb{c} : \mathbf{d}$, i.e. eq. (6.161), we conclude that

$$\dot{\boldsymbol{\tau}} = J\mathbb{c} : \mathbf{d} + \mathbf{l}\boldsymbol{\tau} + \boldsymbol{\tau}\mathbf{l}^{\text{T}} \quad , \tag{8.94}$$

where $\mathbb{c}$ and $\mathbf{d} = \text{sym}(\mathbf{l})$ are the spatial elasticity tensor and the rate of deformation tensor (compare with definitions (6.159) and (2.148), respectively).

Substituting relations $(8.93)_5$, (8.18) and (8.94) into (8.92) and using the symmetry of $\boldsymbol{\tau}$ and the minor symmetries of $\mathbb{c}$, i.e. $c_{abcd} = c_{bacd} = c_{abdc}$, we obtain

$$\overline{\delta W_{\text{int}}(\mathbf{u}, \delta \mathbf{u})} = \int_{\Omega_0} [\boldsymbol{\tau} : (-\text{grad}\delta \mathbf{u}\, \mathbf{l}) + \text{grad}\delta \mathbf{u} : J\mathbb{c} : \mathbf{d}$$

$$+ \text{grad}\delta \mathbf{u} : \mathbf{l}\boldsymbol{\tau} + \text{grad}\delta \mathbf{u} : \boldsymbol{\tau}\mathbf{l}^{\text{T}}] dV \quad . \tag{8.95}$$

By use of property (1.95) the sum of the first and last terms in eq. (8.95) vanishes and we obtain

$$\overline{\delta W_{\text{int}}(\mathbf{u}, \delta\mathbf{u})} = \int_{\Omega_0} (\text{grad}\delta\mathbf{u} : J\mathbb{c} : \mathbf{d} + \text{grad}\delta\mathbf{u} : \mathbf{l}\tau)dV \quad . \tag{8.96}$$

Applying now the formal equivalence of the material time derivatives and the directional derivatives, and replacing the spatial velocity field $\mathbf{v}$ by the linear increment $\Delta\mathbf{u}$, we may rewrite (8.96) as

$$D_{\Delta\mathbf{u}}\delta W_{\text{int}}(\mathbf{u}, \delta\mathbf{u}) = \int_{\Omega_0} (\text{grad}\delta\mathbf{u} : J\mathbb{c} : \text{grad}\Delta\mathbf{u} + \text{grad}\delta\mathbf{u} : \text{grad}\Delta\mathbf{u}\,\tau)dV \quad . \tag{8.97}$$

Changing the domain of integration back and using the stress transformation $\tau = J\sigma$ again we arrive finally at the linearized internal virtual work in the spatial description, i.e. (8.89). ∎

<hr>

EXERCISES

1. Take the part $\delta_{ac}\sigma_{bd}$ of the effective elasticity tensor due to the current stresses in the spatial description.

   (a) Show that this term possesses the *major symmetry*, that means that nothing changes by the manipulation

   $$\delta_{ac}\sigma_{bd} = \delta_{ca}\sigma_{db} \quad .$$

   (b) Further, show that it does not possess the *minor symmetry*, i.e.

   $$\delta_{ac}\sigma_{bd} \neq \delta_{ab}\sigma_{cd} \quad , \qquad \delta_{ac}\sigma_{bd} \neq \delta_{dc}\sigma_{ba} \quad .$$

   However, for the case in which the term $\delta_{ac}\sigma_{bd}$ is absent, under the assumption of a hyperelastic material the minor *and* the major symmetry of $\mathbb{c}_{abcd}$ hold. Discuss the consequences in regard to finite element discretizations.

2. Show that the linearization of the *internal virtual work* δW_{int} may also be written as

   $$D_{\Delta\mathbf{u}}\delta W_{\text{int}}(\mathbf{u}, \delta\mathbf{u}) = \int_{\Omega_0} \text{Grad}\delta\mathbf{u} : \mathbb{A} : \text{Grad}\Delta\mathbf{u}dV \quad ,$$

   where $\mathbb{A}$ is a (mixed) fourth-order tensor useful in numerical implementations. It is known as the **(first) elasticity tensor** with the definition

   $$\mathbb{A} = \frac{\partial\mathbf{P}(\mathbf{F})}{\partial\mathbf{F}} \qquad \text{or} \qquad A_{aAbB} = \frac{\partial P_{aA}}{\partial F_{bB}} \quad ,$$

where $\mathbf{P}$ denotes the first Piola-Kirchhoff stress tensor depending on the deformation gradient $\mathbf{F}$.

Note that the linearization of δW_{int} contains one term only (compare with relation (8.81)).

3. Carry out the linearization of the *external virtual work* δW_{ext}, assuming pressure boundary loading, as derived in eq. (8.38). Consider a constant pressure load p and show by means of the product rule and relation (2.5) that

$$D_{\Delta\mathbf{u}}\delta W_{ext}(\mathbf{u}, \delta\mathbf{u}) = p\int_{\Omega_\xi} \left(\frac{\partial\Delta\mathbf{u}}{\partial\xi_1} \times \frac{\partial\gamma}{\partial\xi_2} - \frac{\partial\Delta\mathbf{u}}{\partial\xi_2} \times \frac{\partial\gamma}{\partial\xi_1} \right) \cdot \delta\mathbf{u}\,d\xi_1 d\xi_2 \ . \quad (8.98)$$

Note that the two terms in eq. (8.98) are not symmetric in $\delta\mathbf{u}$ and $\Delta\mathbf{u}$. In finite element discretizations the associated tangent stiffness matrix is also not, in general, symmetric.

8.5   Two-field Variational Principles

So far we have considered single-field variational principles such as the principle of virtual work. It is not always the best principle to choose, particularly when constraint conditions are imposed on the deformation. In finite element analyses of problems which are associated with constraint conditions, significant numerical difficulties must be expected within the context of a **Galerkin method**, i.e. a standard *displacement-based method* in which only the displacement field is discretized. This method exhibits rather poor numerical performances such as penalty sensitivity and *ill-conditioning* of the stiffness matrix, as is well-known from

(i) the numerical analysis of rubber, which is frequently modeled as a nearly incompressible or incompressible material,

(ii) bending dominated (plate and shell) problems,

(iii) elastoplastic problems that are based on J_2-flow theory (the plastic flow is isochoric), and

(iv) Stokes' flow, which mathematically yields a problem identical to that for isotropic incompressible elasticity.

In the computational literature these devastating numerical difficulties are referred to as **locking phenomena**. Essentially, these locking difficulties arise from the over-stiffening of the system and are associated with a significant loss of accuracy observed, in particular, with low-order finite elements (for fundamental studies and for more references see the books by ZIENKIEWICZ and TAYLOR [1989, 1991] and HUGHES [2000]).

To eliminate these difficulties inherent in the conventional single-field variational approach a great deal of research effort by engineers and mathematicians has been devoted to the developments of efficient so-called **mixed finite element methods** (see HUGHES [2000, and references therein]). For a more mathematically oriented presentation see the work of BREZZI and FORTIN [1991].

For these types of methods the constraints imposed on the deformation are dealt with within a variational sense resulting in effective *multi-field variational principles*. This approach has attracted considerable attention in the computational mechanics literature. Besides the usual displacement field, a mixed (finite element) method incorporates one or more additional fields (typically the internal pressure field, the volume ratio field ...) which are treated as independent variables. The basic idea within the mixed finite element method is to discretize these additional variables independently with the aim of achieving *nonlocking* and *stable* numerical solutions in the incompressible limit.

Lagrange-multiplier method. In the subsequent part we focus attention on a suitable variational approach which captures nearly incompressible and incompressible hyperelastic material response.

Rubber or rubber-like materials may show a very high resistance to volumetric changes compared with that to isochoric changes. Typically, the ratio of the bulk modulus to the shear modulus is roughly of four orders of magnitude (exhibiting an almost incompressible response), and a very careful numerical treatment is needed. A standard displacement-based method cannot be applied directly to these types of problems, because locking or instabilities will occur. Hence, almost all hyperelastic materials which show a nearly incompressible or incompressible deformation behavior are treated with a mixed finite element formulation.

The **Lagrange-multiplier method**, in which a constraint is introduced using a scalar parameter called the **Lagrange multiplier**, is often used to prevent volumetric locking. Utilizing p as the Lagrange multiplier to enforce the incompressibility constraint $J = 1$, we may formulate a functional Π_L in the decoupled representation

$$\Pi_L(\mathbf{u}, p) = \Pi_{int}(\mathbf{u}, p) + \Pi_{ext}(\mathbf{u}) \ . \tag{8.99}$$

$$\Pi_{int}(\mathbf{u}, p) = \int_{\Omega_0} [p(J(\mathbf{u}) - 1) + \Psi_{iso}(\overline{\mathbf{C}}(\mathbf{u}))] dV \ , \tag{8.100}$$

where the term $p(J(\mathbf{u}) - 1)$ denotes the Lagrange-multiplier term, with the volume ratio $J = J(\mathbf{u}) = (\det \mathbf{C})^{1/2}$. The function $\Psi_{iso} = \Psi_{iso}(\overline{\mathbf{C}})$ characterizes the isochoric elastic response of the hyperelastic material with the corresponding modified right Cauchy-Green tensor $\overline{\mathbf{C}} = \overline{\mathbf{C}}(\mathbf{u}) = J^{-2/3}\mathbf{C}$. The loads do not depend on the motion of the body, so that the external potential energy Π_{ext} is given by the standard relation $(8.48)_2$.

Note that the Lagrange-multiplier term vanishes for the case of incompressible finite elasticity. In addition, it is important to emphasize that the solution of the Lagrange-multiplier method can be recovered from the penalty method just by taking the restriction on the value $\kappa \to \infty$ (compare with the outline given on p. 389). Hence, the penalty method may also be seen as an approximation to the Lagrange-multiplier method.

The Lagrange multiplier p plays the role of the (physical) *hydrostatic pressure*. The described formulation gives rise to a two-field mixed finite element implementation involving $\mathbf{u}$ and p as the independent field variables. Since we consider extra numbers of unknowns this technique requires additional computational effort.

The main objective is now to derive a two-field variational principle for finite elasticity by finding the stationary point of functional Π_{L}. In other words, we must compute and set to zero the directional derivatives of Π_{L} with respect to both the displacement field $\mathbf{u}$ and the hydrostatic pressure field p.

In addition to the virtual displacement field $\delta\mathbf{u}$ we need an arbitrary smooth scalar function $\delta p(\mathbf{x}) = \delta p(\boldsymbol{\chi}(\mathbf{X})) = \delta p(\mathbf{X})$, which is interpreted as the **virtual pressure field**, here defined in the reference configuration. We find the stationary conditions with respect to $\mathbf{u}$ and p as

$$D_{\delta\mathbf{u}}\Pi_{\text{L}}(\mathbf{u}, p) = 0 \qquad \text{and} \qquad D_{\delta p}\Pi_{\text{L}}(\mathbf{u}, p) = 0 \tag{8.101}$$

for all $\delta\mathbf{u}$ satisfying $\delta\mathbf{u} = \mathbf{o}$ on the boundary surface $\partial\Omega_{0\,\text{u}}$ and all δp.

Firstly, we compute the directional derivative of (8.99) in the direction of an arbitrary virtual displacement $\delta\mathbf{u}$. By means of the chain rule and the relations derived in Section 8.1, i.e. $D_{\delta\mathbf{u}}J = \delta J = J\text{div}\delta\mathbf{u}$ and $D_{\delta\mathbf{u}}\mathbf{C} = \delta\mathbf{C} = 2\delta\mathbf{E}$, we obtain, with reference to $(8.101)_1$, the weak form

$$D_{\delta\mathbf{u}}\Pi_{\text{L}} = \int_{\Omega_0} \left(pD_{\delta\mathbf{u}}J + 2\frac{\partial\Psi_{\text{iso}}}{\partial\mathbf{C}} : \frac{1}{2}D_{\delta\mathbf{u}}\mathbf{C} \right) dV + D_{\delta\mathbf{u}}\Pi_{\text{ext}}$$

$$= \int_{\Omega_0} \left(Jp\text{div}\delta\mathbf{u} + 2\frac{\partial\Psi_{\text{iso}}}{\partial\mathbf{C}} : \delta\mathbf{E} \right) dV + D_{\delta\mathbf{u}}\Pi_{\text{ext}} = 0 \tag{8.102}$$

(the arguments have been omitted). The contribution due to the loads is given by eq. $(8.51)_2$ as $D_{\delta\mathbf{u}}\Pi_{\text{ext}} = -\delta W_{\text{ext}}$, with the external virtual work δW_{ext}, i.e. (8.45).

In order to rewrite the term $Jp\text{div}\delta\mathbf{u}$ of $(8.102)_2$ we use the analogue of eq. $(1.279)_2$ and invoke relation (8.7) and property (1.95) to obtain

$$\text{div}\delta\mathbf{u} = \mathbf{I} : \text{grad}\delta\mathbf{u} = \mathbf{I} : \text{Grad}\delta\mathbf{u}\,\mathbf{F}^{-1}$$

$$= \mathbf{F}^{-\text{T}} : \text{Grad}\delta\mathbf{u} = \mathbf{F}^{-1}\mathbf{F}^{-\text{T}} : \mathbf{F}^{\text{T}}\text{Grad}\delta\mathbf{u}$$

$$= \mathbf{C}^{-1} : \mathbf{F}^{\text{T}}\text{Grad}\delta\mathbf{u} \quad, \tag{8.103}$$

with the inverse right Cauchy-Green tensor $\mathbf{C}^{-1} = \mathbf{F}^{-1}\mathbf{F}^{-T}$. Since $\mathbf{C}^{-1}$ is symmetric we find finally, by analogy with property (1.115) and with $(8.14)_2$, that

$$\mathrm{div}\delta\mathbf{u} = \mathbf{C}^{-1} : \delta\mathbf{E} \ . \tag{8.104}$$

This result substituted back into $(8.102)_2$ gives

$$D_{\delta\mathbf{u}}\Pi_{\mathrm{L}}(\mathbf{u}, p) = \int_{\Omega_0} \left(J(\mathbf{u})p\mathbf{C}^{-1}(\mathbf{u}) + 2\frac{\partial\Psi_{\mathrm{iso}}(\overline{\mathbf{C}}(\mathbf{u}))}{\partial\mathbf{C}} \right) : \delta\mathbf{E}(\mathbf{u})\mathrm{d}V$$
$$+ D_{\delta\mathbf{u}}\Pi_{\mathrm{ext}}(\mathbf{u}) = 0 \ . \tag{8.105}$$

By recalling the unique additive decomposition for the stress, i.e. eq. $(6.88)_2$ with definitions $(6.89)_2$ and $(6.90)_1$, we recognize that the two terms in parentheses of the variational equation (8.105) give precisely the second Piola-Kirchhoff stress tensor $\mathbf{S} = \mathbf{S}(\mathbf{E}(\mathbf{u}))$. Hence, the integral in (8.105) characterizes the volumetric and isochoric contributions to the internal virtual work δW_{int} (see eq. $(8.46)_2$).

Relation (8.105) is identified as the standard principle of virtual work expressed in the reference configuration (for a configuration in static equilibrium), i.e. $\delta W_{\mathrm{int}} - \delta W_{\mathrm{ext}} = 0$. The associated Euler-Lagrange equation is Cauchy's equation of equilibrium.

Secondly, we compute the directional derivative of (8.99) in the direction of an arbitrary virtual pressure δp. With reference to $(8.101)_2$, we obtain the weak form

$$D_{\delta p}\Pi_{\mathrm{L}}(\mathbf{u}, p) = \int_{\Omega_0} (J(\mathbf{u}) - 1)\delta p\mathrm{d}V = 0 \ . \tag{8.106}$$

As can be seen, the pressure variable p maintains the incompressibility constraint and we find the associated Euler-Lagrange equation to be $J = 1$ (see also the study by LE TALLEC [1994]).

The variational equations (8.105) and (8.106) provide the fundamental basis for a finite element implementation.

Linearization of the Lagrange-multiplier method. In order to solve the nonlinear equations (8.105) and (8.106) for the two independent variables $\mathbf{u}$ and p on an incremental/iterative basis, a *Newton type* method is usually employed (recall Section 8.4). In preparation for an incremental/iterative solution technique a systematic linearization of (8.105) and (8.106) with respect to $\mathbf{u}$ and p, essentially the second variation of (8.99), is required.

In order to linearize variational equation (8.106) in the directions of the increments $\Delta\mathbf{u}$ and Δp we recall that $D_{\Delta\mathbf{u}}J = J\mathrm{div}\Delta\mathbf{u}$. Employing the concept of directional derivative we obtain

$$D^2_{\delta p,\Delta\mathbf{u}}\Pi_{\mathrm{L}}(\mathbf{u}, p) = \int_{\Omega_0} J(\mathbf{u})\mathrm{div}\Delta\mathbf{u}\,\delta p\mathrm{d}V \ , \qquad D^2_{\delta p,\Delta p}\Pi_{\mathrm{L}}(\mathbf{u}, p) = 0 \ . \tag{8.107}$$

The linearization of the principle of virtual work (8.105) in the direction of the increment Δp gives

$$D^2_{\delta\mathbf{u},\Delta p}\Pi_{\mathrm{L}}(\mathbf{u}, p) = \int_{\Omega_0} J(\mathbf{u})\Delta p\mathbf{C}^{-1}(\mathbf{u}) : \delta\mathbf{E}(\mathbf{u})dV = \int_{\Omega_0} J(\mathbf{u})\Delta p\operatorname{div}\delta\mathbf{u}dV \quad, \quad (8.108)$$

while a linearization process in the direction of $\Delta\mathbf{u}$ was already carried out in detail within the last section, leading to

$$D^2_{\delta\mathbf{u},\Delta\mathbf{u}}\Pi_{\mathrm{L}}(\mathbf{u}, p) = \int_{\Omega_0} (\operatorname{Grad}\delta\mathbf{u} : \operatorname{Grad}\Delta\mathbf{u}\,\mathbf{S}$$

$$+\mathbf{F}^{\mathrm{T}}\operatorname{Grad}\delta\mathbf{u} : (\mathbb{C}_{\mathrm{vol}} + \mathbb{C}_{\mathrm{iso}}) : \mathbf{F}^{\mathrm{T}}\operatorname{Grad}\Delta\mathbf{u})dV \quad . \quad (8.109)$$

Since (8.105) is based on the additive decomposition of the stresses $\mathbf{S}$, we obtain the decoupled representation of the elasticity tensor $\mathbb{C} = \mathbb{C}_{\mathrm{vol}} + \mathbb{C}_{\mathrm{iso}}$, with the definitions

$$\mathbb{C}_{\mathrm{vol}} = 2\frac{\partial(J(\mathbf{u})p\mathbf{C}^{-1}(\mathbf{u}))}{\partial\mathbf{C}} \quad \text{and} \quad \mathbb{C}_{\mathrm{iso}} = 4\frac{\partial^2\Psi_{\mathrm{iso}}(\overline{\mathbf{C}}(\mathbf{u}))}{\partial\mathbf{C}\partial\mathbf{C}} \quad . \quad (8.110)$$

For an explicit treatment of these expressions recall Section 6.6, in particular relations $(6.166)_4$ and (6.168). Note that for the considered case the scalar quantity $\tilde{p}$ must be replaced by p in eq. $(6.166)_4$.

Since the structure of the linearized principle of virtual work is symmetric and because of the symmetry between eqs. $(8.107)_1$ and (8.108) a finite element implementation of this set of equations will lead to a symmetric (tangent) stiffness matrix.

In order to use these equations within a finite element regime so-called **interpolation functions** must be invoked separately for the displacement field $\mathbf{u}$, the pressure field p and their variations $\delta\mathbf{u}$ and δp, respectively (see SUSSMAN and BATHE [1987], ZIENKIEWICZ and TAYLOR [1989] among others). A well-considered choice of these functions is a crucial task in order to alleviate volumetric locking. It was observed that a discontinuous (constant) pressure and a continuous displacement interpolation over a typical finite element domain is computationally more efficient than with the choice of functions of the same order for $\mathbf{u}$ and p.

Perturbed Lagrange-multiplier method.   The Lagrange-multiplier method results in a stiffness matrix which is not positive definite for incompressible materials. In order to overcome the numerical difficulties associated with this fact and to avoid ill-conditioning of the stiffness matrix associated with the penalty approach, regularization procedures such as the so-called **perturbed Lagrange-multiplier method** have been introduced successfully (see, for example, GLOWINSKI and LE TALLEC [1982, 1984]).

It may be viewed as a two-field variational principle in which the functional (8.99) is *perturbed* by a penalty term. Thus,

$$\Pi_{\mathrm{PL}}(\mathbf{u}, p) = \int_{\Omega_0} [p(J(\mathbf{u}) - 1) + \Psi_{\mathrm{iso}}(\overline{\mathbf{C}}(\mathbf{u}))] dV$$

$$- \frac{1}{2} \int_{\Omega_0} \frac{1}{\kappa} p^2 dV + \Pi_{\mathrm{ext}}(\mathbf{u}) \tag{8.111}$$

(see the work of CHANG et al. [1991]), where the third (penalty) term in functional (8.111) regularizes (relaxes) the incompressibility constraint $J = 1$ involved in the first term of the integral. The Lagrange multiplier p which enforces the constraint no longer has the meaning of a pressure, in contrast to the Lagrange-multiplier method.

The positive penalty parameter κ may be viewed as a (constant) bulk modulus. Incompressible materials can be treated by replacing $1/\kappa$ with zero, so the first and the third terms in (8.111) vanish. For this incompressible limit the penalty method, the Lagrange-multiplier method and the perturbed Lagrange-multiplier method lead to identical equations.

By taking for the term $(J(\mathbf{u}) - 1)$ a more general, sufficiently smooth and strictly convex function $\mathcal{G}(J(\mathbf{u}))$ (so that $\mathcal{G}(J(\mathbf{u})) = 0$ *if and only if* $J = 1$) the functional (8.111) is identical to that proposed by BRINK and STEIN [1996]. Functional (8.111) may also be identified with a special form of the two-field variational principle given by ATLURI and REISSNER [1989]. This work deals with a general framework for incorporating volume constraints into multi-field variational principles. In addition, note that the formulation (8.111) also reduces from a mixed (finite element) formulation proposed by SUSSMAN and BATHE [1987].

EXERCISES

1. Consider the decoupled strain energy formulation proposed in (8.55), i.e. $\Psi(\mathbf{C}) = \Psi_{\mathrm{vol}}(J) + \Psi_{\mathrm{iso}}(\overline{\mathbf{C}})$, with $\Psi_{\mathrm{vol}}(J) = \kappa \mathcal{G}(J)$, and treat the displacement $\mathbf{u}$ and the hydrostatic pressure p as independent field variables (permitted to be varied). Require that $\mathrm{d}\mathcal{G}(J)/\mathrm{d}J = 0$ *if and only if* $J = 1$.

   (a) Derive the stationary condition with respect to $\mathbf{u}$ and incorporate the definition of the volumetric stress contribution, i.e. (6.89), with the constitutive equation for the hydrostatic pressure p according to (6.91)$_1$. Show that the resulting variational equation is, in accordance with the principle of virtual work, in the form (8.105).

   (b) Obtain the additional variational equation in the form

$$\int_{\Omega_0} \left(\frac{\mathrm{d}\mathcal{G}(J(\mathbf{u}))}{\mathrm{d}J} - \frac{1}{\kappa} p \right) \delta p dV = 0 \ , \tag{8.112}$$

which is relation (8.61) enforced in a weak sense. Interpret the result for the case $\kappa \to \infty$.

Note that variational equation (8.112) cannot simply be obtained by taking the first variation of the energy functional.

This type of two-field variational principle was proposed by DE BORST et al. [1988] and VAN DEN BOGERT et al. [1991].

2. Consider the augmented functional (8.111).

   (a) Derive the stationary conditions with respect to $\mathbf{u}$ and p, i.e. $D_{\delta\mathbf{u}}\Pi_{PL}(\mathbf{u}, p) = 0$ and $D_{\delta p}\Pi_{PL}(\mathbf{u}, p) = 0$, for arbitrary variations $\delta\mathbf{u}$ and δp, respectively. Show that the Euler-Lagrange equations are Cauchy's equation of equilibrium and the artificial constitutive equation $p = \kappa(J(\mathbf{u}) - 1)$.

   (b) Show that the linearization of the Euler-Lagrange equations in the weak forms gives

   $$D^2_{\delta p, \Delta p}\Pi_{PL}(\mathbf{u}, p) = -\int_{\Omega_0} \frac{1}{\kappa}\Delta p\, \delta p\, \mathrm{d}V \quad,$$

   and three further equations which are in accord with $(8.107)_1$, (8.108) and the linearized principle of virtual work, which has basically the form of eq. (8.109). Note that, thereby, the elasticity tensor $\mathbb{C} = \mathbb{C}_{\mathrm{vol}} + \mathbb{C}_{\mathrm{iso}}$ is given explicitly by eqs. $(6.166)_4$ and (6.168).

3. Study a two-field variational principle which involves *displacements* and *stresses* as independent field variables. A formulation of this type leads to the classical **Hellinger-Reissner variational principle**, widely used in, for example, nonlinear theories of plates and shells (see HELLINGER [1914] and REISSNER [1950]).

   Assume that the constitutive relation $\mathbf{P}(\mathbf{F}) = \partial\Psi(\mathbf{F})/\partial\mathbf{F}$, as introduced in $(6.1)_1$, is invertible, which is not a valid assumption, in general. (It is important to emphasize that, in general, there does not exist a unique deformation gradient $\mathbf{F}$ corresponding to a given first Piola-Kirchhoff stress tensor $\mathbf{P}$ (see OGDEN [1977, 1997, Section 6.2.2]). Define a **complementary strain-energy function** $\Psi_c(\mathbf{P})$ so that a *Legendre transformation* gives

   $$\Psi_c(\mathbf{P}) = \mathbf{P} : \mathbf{F} - \Psi(\mathbf{F}) \quad,$$

   where $\mathbf{P}$ and $\mathbf{F}$ are the first Piola-Kirchhoff stress tensor and the deformation gradient, respectively. Hence, the functional

   $$\Pi_{HR}(\mathbf{u}, \mathbf{P}) = \int_{\Omega_0} [\mathbf{P} : \mathbf{F} - \Psi_c(\mathbf{P})]\mathrm{d}V - \int_{\Omega_0} \mathbf{B} \cdot \mathbf{u}\,\mathrm{d}V$$

$$-\int_{\partial\Omega_{0\,\sigma}} \overline{\mathbf{T}}\cdot\mathbf{u}\mathrm{d}S - \int_{\partial\Omega_{0\,u}} \mathbf{T}\cdot(\mathbf{u}-\overline{\mathbf{u}})\mathrm{d}S \quad,$$

which is valid for large strains, is referred to as the **Hellinger-Reissner functional**. Here, the prescribed loads are $\mathbf{B}$ on Ω_0 and $\overline{\mathbf{T}}$ on $\partial\Omega_{0\,\sigma}$ and are assumed to be independent of the motion of the body. The third quantity prescribed is the displacement field $\overline{\mathbf{u}}$ acting on the boundary surface $\partial\Omega_{0\,u}$. Note the relation $\mathbf{T} = \mathbf{PN}$ introduced in $(3.3)_2$.

(a) Invoke the stationarity of Π_{HR} and determine the weak form of the elastic equilibrium equation. Since the principle is based on treating the displacement $\mathbf{u}$ and the stress $\mathbf{P}$ as independent field variables (permitted to be varied), evaluate separately

$$D_{\delta\mathbf{u}}\Pi_{\mathrm{HR}}(\mathbf{u},\mathbf{P}) = 0 \quad, \qquad D_{\delta\mathbf{P}}\Pi_{\mathrm{HR}}(\mathbf{u},\mathbf{P}) = 0 \quad.$$

The first variations of $\mathbf{u}$ and $\mathbf{P}$ are arbitrary vector-valued and tensor-valued functions for which the restriction $\delta\mathbf{u} = \mathbf{o}$ over the boundary surface $\partial\Omega_{0\,u}$ holds.

(b) Show that the associated Euler-Lagrange equations for the volume of the body and the (Dirichlet and von Neumann) boundary conditions are

$$\mathrm{Div}\mathbf{P} + \mathbf{B} = \mathbf{o} \quad, \qquad \mathbf{F}(\mathbf{P}) = \frac{\partial\Psi_{\mathrm{c}}(\mathbf{P})}{\partial\mathbf{P}} \quad,$$

$$\mathbf{u} = \overline{\mathbf{u}} \quad \text{on} \quad \partial\Omega_{0\,u} \quad, \qquad \mathbf{T} = \mathbf{PN} = \overline{\mathbf{T}} \quad \text{on} \quad \partial\Omega_{0\,\sigma} \quad.$$

The first relation represents Cauchy's equation of equilibrium in the material description, while the second relation denotes the inverse form of the constitutive equation for a hyperelastic material (in some cases not available).

(c) Alternatively to the functional stated above find Π_{HR} for which displacements and strains are the independent variables (instead of $\mathbf{u}$ and $\mathbf{P}$).

8.6 Three-field Variational Principles

We are interested in a suitable variational approach in order to capture nearly incompressible and incompressible materials. It is recognized that a constant pressure interpolation over the finite element within the framework of two-field variational principles leads to unpleasant pressure oscillation. There are some remedies for this problem in the computational literature, for example, the pressure smoothing technique by HUGHES et al. [1979]. However, a certain improvement is based on the idea of introducing additional independent field variables such as the volume ratio, leading to a more efficient three-field variational principle.

Simo-Taylor-Pister variational principle.   A very efficient variational principle that takes account of nearly incompressible response was originally proposed by SIMO et al. [1985] and is known as the mixed **Jacobian-pressure formulation** (for relevant applications to elastomers see SIMO [1987] and SIMO and TAYLOR [1991a]). It emanates from a three-field variational principle of HU [1955] and WASHIZU [1955]. Thereby, besides the displacement and pressure fields **u** and p, a third additional kinematic field variable, which we denote by $\tilde{J}$, is treated independently within finite element discretizations. The principle is decomposed into volumetric, isochoric and external parts and is defined by the expression

$$\Pi_{\mathrm{STP}}(\mathbf{u}, p, \tilde{J}) = \int_{\Omega_0} [\Psi_{\mathrm{vol}}(\tilde{J}) + p(J(\mathbf{u}) - \tilde{J}) + \Psi_{\mathrm{iso}}(\overline{\mathbf{C}}(\mathbf{u}))]\mathrm{d}V + \Pi_{\mathrm{ext}}(\mathbf{u}) \ . \quad (8.113)$$

Following *Simo-Taylor-Pister* the first two terms in the three-field variational principle are responsible for the nearly incompressible behavior of the material. They describe volume-changing (dilational) deformations and are expressed by J, p and the new variable $\tilde{J}$. The kinematic variable $\tilde{J}$ enters the functional as a constraint which is enforced by the Lagrange multiplier p. The Lagrange multiplier is an independent field variable which may be identified as the hydrostatic pressure.

In addition to the virtual displacement and pressure fields $\delta\mathbf{u}$ and δp, we introduce an arbitrary smooth (vector) function $\delta\tilde{J}(\mathbf{x}) = \delta\tilde{J}(\chi(\mathbf{X})) = \delta\tilde{J}(\mathbf{X})$ for the constraint, which we call the **virtual volume change** (here defined on the reference configuration). In equilibrium, functional (8.113) must be stationary. The necessary conditions for the stationarity of functional Π_{STP} with respect to the three field variables $(\mathbf{u}, p, \tilde{J})$ are evaluated separately. We require

$$\left. \begin{array}{cc} D_{\delta\mathbf{u}}\Pi_{\mathrm{STP}}(\mathbf{u}, p, \tilde{J}) = 0 \ , & D_{\delta p}\Pi_{\mathrm{STP}}(\mathbf{u}, p, \tilde{J}) = 0 \ , \\ D_{\delta\tilde{J}}\Pi_{\mathrm{STP}}(\mathbf{u}, p, \tilde{J}) = 0 \end{array} \right\} \quad (8.114)$$

for all $\delta\mathbf{u}$ satisfying $\delta\mathbf{u} = \mathbf{o}$ on the part of the boundary surface $\partial\Omega_{0\,\mathrm{u}}$ where displacements $\bar{\mathbf{u}}$ are prescribed and all δp, $\delta\tilde{J}$.

Differentiating functional Π_{STP} with respect to changes in **u** gives the weak form of the elastic equilibrium, i.e. the principle of virtual work in the form of (8.105). For an explicit derivation recall the manipulations of the last section.

A straightforward differentiation of Π_{STP} with respect to changes in the field variables $p, \tilde{J}$ gives the weak enforcement of the equivalence between J and $\tilde{J}$, and the constitutive equation for the volumetric changes, i.e.

$$D_{\delta p}\Pi_{\mathrm{STP}}(\mathbf{u}, p, \tilde{J}) = \int_{\Omega_0} (J(\mathbf{u}) - \tilde{J})\delta p\, dV = 0 \; , \qquad (8.115)$$

$$D_{\delta \tilde{J}}\Pi_{\mathrm{STP}}(\mathbf{u}, p, \tilde{J}) = \int_{\Omega_0} \left(\frac{d\Psi_{\mathrm{vol}}(\tilde{J})}{d\tilde{J}} - p\right) \delta \tilde{J}\, dV = 0 \; . \qquad (8.116)$$

For arbitrary δp, the variational equation (8.115) results in the Euler-Lagrange equation $J - \tilde{J} = 0$. It implies that the additional independent variable $\tilde{J}$ equals $J = (\det \mathbf{C}(\mathbf{u}))^{1/2}$, i.e. the kinematic constraint associated with the volumetric behavior. For arbitrary $\delta \tilde{J}$, eq. (8.116) results in the second constraint condition in the local form, that is the Euler-Lagrange equation $d\Psi_{\mathrm{vol}}/d\tilde{J} - p = 0$. This is the standard constitutive equation implying the volumetric stresses to be equal to the hydrostatic pressure.

A finite element procedure in which the dilatation $\tilde{J}$ and the pressure variables p are discretized by the same local interpolations as for the displacement field $\mathbf{u}$ would not give any advantage. To prevent volumetric locking an appropriate choice of the interpolation functions for the volumetric variables $p, \tilde{J}$ and their variations $\delta p, \delta \tilde{J}$ is crucial. A simple formulation arises by discretizing the dilatation and pressure variables over a typical finite element domain with the same *discontinuous* (constant) function which need not be continuous across the finite element boundaries. This approach is known as the **mean dilatation method** and is proposed in the notable work of NAGTEGAAL et al. [1974] who recognized the effect of volumetric locking in elastoplastic J_2-flow theory.

Since the interpolation functions are discontinuous, the volumetric variables $p, \tilde{J}$ can be eliminated on the finite element level, a process known as **static condensation** in the computational mechanics literature. Therefore, the variational equations (8.115) and (8.116) need not be solved on the global level leading back to a reduced *displacement-based method*.

The work of BRINK and STEIN [1996] is a comparative study of various multi-field variational principles. It emerges that under certain conditions the above three-field variational principle and some two-field principles yield the same discrete result in each step of the Newton method.

EXERCISES

1. Consider the functional (8.113) with the three independent field variables $(\mathbf{u}, p, \tilde{J})$ and the associated variation equations (8.115), (8.116), (8.105), and show that for each step of the Newton type method the problem is completely described by the

set of linearized equations

$$D^2_{\delta p, \Delta \mathbf{u}} \Pi_{\text{STP}}(\mathbf{u}, p, \tilde{J}) = \int_{\Omega_0} J(\mathbf{u}) \text{div} \Delta \mathbf{u} \, \delta p \, \mathrm{d}V \quad ,$$

$$D^2_{\delta p, \Delta \tilde{J}} \Pi_{\text{STP}}(\mathbf{u}, p, \tilde{J}) = -\int_{\Omega_0} \Delta \tilde{J} \delta p \, \mathrm{d}V \quad ,$$

$$D^2_{\delta \tilde{J}, \Delta p} \Pi_{\text{STP}}(\mathbf{u}, p, \tilde{J}) = -\int_{\Omega_0} \Delta p \, \delta \tilde{J} \, \mathrm{d}V \quad ,$$

$$D^2_{\delta \tilde{J}, \Delta \tilde{J}} \Pi_{\text{STP}}(\mathbf{u}, p, \tilde{J}) = \int_{\Omega_0} \frac{\mathrm{d}^2 \Psi_{\text{vol}}(\tilde{J})}{\mathrm{d}\tilde{J} \mathrm{d}\tilde{J}} \Delta \tilde{J} \, \delta \tilde{J} \, \mathrm{d}V \quad ,$$

and the linearized principle of virtual work, which has the form of eqs. (8.109) and (8.108).

2. The described functional $\Pi_{\text{STP}}(\mathbf{u}, p, \tilde{J})$ takes into account only the volumetric strain and stress components. Study a more general and very powerful type of a **Hu-Washizu variational principle** fundamental for various finite element methods, i.e.

$$\Pi_{\text{HW}}(\mathbf{u}, \mathbf{F}, \mathbf{P}) = \int_{\Omega_0} (\Psi(\mathbf{F}) - \mathbf{P} : \mathbf{F} - \mathbf{B} \cdot \mathbf{u} - \text{Div} \mathbf{P} \cdot \mathbf{u}) \mathrm{d}V$$

$$+ \int_{\partial \Omega_{0\,\sigma}} \mathbf{u} \cdot (\mathbf{T} - \overline{\mathbf{T}}) \mathrm{d}S - \int_{\partial \Omega_{0\,\text{u}}} \mathbf{T} \cdot (\mathbf{u} - \overline{\mathbf{u}}) \mathrm{d}S \quad ,$$

with the three independent variables $\mathbf{u}$, $\mathbf{F}$, $\mathbf{P}$ and the prescribed quantities $\mathbf{B}$ on Ω_0, $\overline{\mathbf{T}}$ on $\partial \Omega_{0\,\sigma}$ and $\overline{\mathbf{u}}$ on $\partial \Omega_{0\,\text{u}}$. The loads $\mathbf{B}$ and $\overline{\mathbf{T}}$ are assumed to be conservative and the first Piola-Kirchhoff traction vector $\mathbf{T}$ is given in eq. $(3.3)_2$.

(a) With identity (1.290) show that the Hu-Washizu variational principle can be posed as a generalization of the principle of virtual work, i.e.

$$\Pi_{\text{HW}}(\mathbf{u}, \mathbf{F}, \mathbf{P}) = \Pi - \int_{\Omega_0} \mathbf{P} : (\mathbf{F} - \text{Grad}\mathbf{u}) \mathrm{d}V - \int_{\partial \Omega_{0\,\text{u}}} \mathbf{T} \cdot (\mathbf{u} - \overline{\mathbf{u}}) \mathrm{d}S \quad ,$$

where the total potential energy Π is given in (8.47) and (8.48).

(b) Invoke the stationarity of Π_{HW} with respect to $\mathbf{u}, \mathbf{F}$ and $\mathbf{P}$. The vector-valued and tensor-valued functions $\delta \mathbf{u}$ and $\delta \mathbf{F}, \delta \mathbf{P}$ are arbitrary with the conditions $\delta \mathbf{u} = \mathbf{o}$ over the boundary surface $\partial \Omega_{0\,\text{u}}$ and $\delta \mathbf{P} = \mathbf{O}$ on $\partial \Omega_{0\,\sigma}$. Show that the associated Euler-Lagrange equations for the functional Π_{HW}

are

$$\text{Div}\mathbf{P} + \mathbf{B} = \mathbf{o} \ , \qquad \mathbf{P} = \frac{\partial\Psi(\mathbf{F})}{\partial\mathbf{F}} \ , \qquad \mathbf{F} = \text{Grad}\mathbf{u} + \mathbf{I} \ ,$$

with the (Dirichlet and von Neumann) boundary conditions

$$\mathbf{u} = \bar{\mathbf{u}} \qquad \text{on} \qquad \partial\Omega_{0\,\mathrm{u}} \ , \qquad \mathbf{T} = \mathbf{P}\mathbf{N} = \bar{\mathbf{T}} \qquad \text{on} \qquad \partial\Omega_{0\,\sigma}$$

for the body under consideration.

Bibliography

Note: Numbers in parentheses following the reference indicate the chapters in which it is cited.

Abè, H., Hayashi, K., and Sato, M., eds. [1996], *Data Book on Mechanical Properties of Living Cells, Tissues, and Organs*, Springer-Verlag, New York. (6)

Abraham, R., and Marsden, J.E. [1978], *Foundations of Mechanics*, 2nd edn., The Benjamin/Cummings Publishing Company, Reading, Massachusetts. (4)

Abraham, R., Marsden, J.E., and Ratiu, T. [1988], *Manifolds, Tensor Analysis, and Applications*, 2nd edn., Springer-Verlag, New York. (1)

Adams, L.H., and Gibson, R.E. [1930], The compressibility of rubber, *Journal of the Washington Academy of Sciences* **20**, 213–223. (6)

Alexander, H. [1971], Tensile instability of initially spherical balloons, *International Journal of Engineering Science* **9**, 151–162. (6)

Anand, L. [1986], Moderate deformations in extension-torsion of incompressible isotropic elastic materials, *Journal of the Mechanics and Physics of Solids* **34**, 293–304. (6,7)

Anand, L. [1996], A constitutive model for compressible elastomeric solids, *Computational Mechanics* **18**, 339–355. (6)

Anthony, R.L., Caston, R.H., and Guth, E. [1942], Equations of state for natural and synthetic rubber-like materials. I, *The Journal of Physical Chemistry* **46**, 826–840. (7)

Argyris, J.H., and Doltsinis, J.St. [1979], On the large strain inelatic analysis in natural formulation. Part I: Quasistatic problems, *Computer Methods in Applied Mechanics and Engineering* **20**, 213–251. (7)

Argyris, J.H., and Doltsinis, J.St. [1981], On the natural formulation and analysis of large deformation coupled thermomechanical problems, *Computer Methods in Applied Mechanics and Engineering* **25**, 195–253. (7)

415

416

Argyris, J.H., Doltsinis, J.St., Pimenta, P.M., and Wüstenberg, H. [1982], Thermomechanical response of solids at high strains – natural approach, *Computer Methods in Applied Mechanics and Engineering* **32**, 3–57. (7)

Armero, F., and Simo, J.C. [1992], A new unconditionally stable fractional step method for nonlinear coupled thermomechanical problems, *International Journal for Numerical Methods in Engineering* **35**, 737–766. (7)

Armero, F., and Simo, J.C. [1993], A priori stability estimates and unconditionally stable product formula algorithms for nonlinear coupled thermoplasticity, *International Journal of Plasticity* **9**, 749–782. (7)

Arruda, E.M., and Boyce, M.C. [1993], A three-dimensional constitutive model for the large stretch behavior of rubber elastic materials, *Journal of the Mechanics and Physics of Solids* **41**, 389–412. (6)

Atluri, S.N. [1984], Alternate stress and conjugate strain measures, and mixed variational formulations involving rigid rotations, for computational analyses of finitely deformed solids, with application to plates and shells – I, *Computers and Structures* **18**, 93-116. (4)

Atluri, S.N., and Reissner, E. [1989], On the formulation of variational theorems involving volume constraints, *Computational Mechanics* **5**, 337–344. (8)

Ball, J.M. [1977], Convexity conditions and existence theorems in nonlinear elasticity, *Archive for Rational Mechanics and Analysis* **63**, 337–403. (6)

Barenblatt, G.I., and Joseph, D.D., eds. [1997], *Collected papers of R.S. Rivlin*, Volume 1,2, Springer-Verlag, New York. (6)

Barnes, H.A., Hutton, J.F., and Walters, K. [1989], *An Introduction to Rheology*, Rheology series Volume 3, Elsevier, New York. (2,7)

Bathe, K.-J. [1996], *Finite Element Procedures*, Prentice-Hall, Englewood Cliffs, New Jersey. (8)

Beatty, M.F. [1987], Topics in finite elasticity: Hyperelasticity of rubber, elastomers, and biological tissues – with examples, *Applied Mechanics Reviews* **40**, 1699–1734. (6)

Beatty, M.F., and Stalnaker, D.O. [1986], The Poisson function of finite elasticity, *Journal of Applied Mechanics* **53**, 807–813. (6)

Belytschko, T., Liu, W.K., and Moran, B. [2000], *Nonlinear Finite Elements for Continua and Structures*, John Wiley & Sons, Chichester. (8)

Bergström, J.S., and Boyce, M.C. [1998], Constitutive modeling of the large strain time-dependent behavior of elastomers, *Journal of the Mechanics and Physics of Solids* **46**, 931–954. (6)

Bertsekas, D.P. [1982], *Constrained Optimization and Lagrange Multiplier Methods*, Academic Press, New York. (8)

Betten, J. [1987a], *Tensorrechnung für Ingenieure*, B.G. Teubner, Stuttgart. (6)

Betten, J. [1987b], Formulation of anisotropic constitutive equations, in: J.P. Boehler, ed., *Applications of Tensor Functions in Solid Mechancis*, CISM Courses and Lectures No. 292, International Centre for Mechanical Sciences, Springer-Verlag, Wien, 227–250. (6)

Biot, M.A. [1965], *Mechanics of Incremental Deformations*, Wiley, New York. (3)

Blatz, P.J. [1971], On the thermostatic behavior of elastomers, in: *Polymer Networks, Structure and mechanical Properties*, Plenum Press, New York, 23–45. (6)

Blatz, P.J., and Ko, W.L. [1962], Application of finite elasticity theory to the deformation of rubbery materials, *Transactions of the Society of Rheology* **6**, 223–251. (6)

Bonet, J., and Burton, A.J. [1998], A simple orthotropic, transversely isotropic hyperelastic constitutive equation for large strain computations, *Computer Methods in Applied Mechanics and Engineering* **162**, 151–164. (6)

Bonet, J., and Wood, R.D. [1997], *Nonlinear Continuum Mechanics for Finite Element Analysis*, Cambridge University Press, Cambridge. (6,8)

de Borst, R., van den Bogert, P.A.J., and Zeilmaker, J. [1988], Modelling and analysis of rubberlike materials, *Heron* **33**, 1–57. (8)

Bowen, R.M. [1976a], Theory of mixtures, in: A.C. Eringen, ed., *Continuum Physics*, Volume III, Academic Press, New York. (6)

Bowen, R.M., and Wang, C.-C. [1976b], *Introduction to Vectors and Tensors*, Volume 1,2, Plenum Press, New York. (1,2)

Brezzi, F., and Fortin, M. [1991], *Mixed and Hybrid Finite Element Methods*, Springer-Verlag, New York. (8)

Bridgman, P.W. [1945], The compression of 61 substances to 25.000 kg/cm^2 determined by a new rapid method, *Proceedings of the American Academy of Arts and Sciences* **76**, 9–24. (6)

Brink, U., and Stein, E. [1996], On some mixed finite element methods for incompressible and nearly incompressible finite elasticity, *Computational Mechanics* **19**, 105–119. (8)

Bueche, F. [1960], Molecular basis of the Mullins effect, *Journal of Applied Polymer Science* **4**, 107–114. (6)

Bueche, F. [1961], Mullins effect and rubber-filler interaction, *Journal of Applied Polymer Science* **5**, 271–281. (6)

Bufler, H. [1984], Pressure loaded structures under large deformations, *Zeitschrift für Angewandte Mathematik und Mechanik* **64**, 287–295. (8)

Callen, H.B. [1985], *Thermodynamics and an Introduction to Thermostatistics*, 2nd edn., John Wiley & Sons, New York. (4,7)

Carlson, D.E. [1972], Linear thermoelasticity, in: S. Flügge, ed., *Encyclopedia of Physics*, Volume VIa/2, Springer-Verlag, Berlin, 297–346. (7)

Chadwick, P. [1974], Thermo-mechanics of rubberlike materials, *Philosophical Transactions of the Royal Society of London* **A276**, 371–403. (7)

Chadwick, P. [1975], Applications of an energy-momentum tensor in non-linear elastostatics, *Journal of Elasticity* **5**, 249–258. (6)

Chadwick, P. [1999], *Continuum Mechanics, Concise Theory and Problems*, Dover, New York. (1,2)

Chadwick, P., and Creasy, C.F.M. [1984], Modified entropic elasticity of rubberlike materials, *Journal of the Mechanics and Physics of Solids* **32**, 337–357. (7)

Chadwick, P., and Ogden, R.W. [1971a], On the definition of elastic moduli, *Archive for Rational Mechanics and Analysis* **44**, 41–53. (6)

Chadwick, P., and Ogden, R.W. [1971b], A theorem of tensor calculus and its application to isotropic elasticity, *Archive for Rational Mechanics and Analysis* **44**, 54–68. (6)

Chang, T.Y.P., Saleeb, A.F., and Li, G. [1991], Large strain analysis of rubber-like materials based on a perturbed Lagrangian variational principle, *Computational Mechanics* **8**, 221–233. (8)

Christensen, R.M. [1982], *Theory of Viscoelasticity. An Introduction*, 2nd edn., Academic Press, New York. (6)

Ciarlet, P.G. [1988], *Mathematical Elasticity. Volume I: Three-Dimensional Elasticity, Studies in Mathematics and its Applications*, North-Holland, Amsterdam. (2,6)

Ciarlet, P.G., and Geymonat, G. [1982], Sur les lois de comportement en élasticité non linéaire compressible, *Comptes Rendus Hebdomadaires des Séances de l'Académie des Sciences, Série II* **295**, 423–426. (6)

Coleman, B.D., and Gurtin, M.E. [1967], Thermodynamics with internal state variables, *Journal of Chemistry and Physics* **47**, 597–613. (6)

Coleman, B.D., and Noll, W. [1963], The thermodynamics of elastic materials with heat conduction and viscosity, *Archive for Rational Mechanics and Analysis* **13**, 167–178. (4,6)

Courant, R., and Hilbert, D. [1968a], *Methoden der mathematischen Physik*, Volume 1, 3rd edn., Springer-Verlag, Berlin. Heidelberger Taschenbücher Volume 30. (8)

Courant, R., and Hilbert, D. [1968b], *Methoden der mathematischen Physik*, Volume 2, 2nd edn., Springer-Verlag, Berlin. Heidelberger Taschenbücher Volume 31. (8)

Crisfield, M.A. [1991], *Non-linear Finite Element Analysis of Solids and Structures, Essentials*, Volume 1, John Wiley & Sons, Chichester. (8)

Crisfield, M.A. [1997], *Non-linear Finite Element Analysis of Solids and Structures, Advanced Topics*, Volume 2, John Wiley & Sons, Chichester. (8)

Curnier, A. [1994], *Computational Methods in Solid Mechanics*, Kluwer Academic Publishers, Dordrecht, The Netherlands. (6)

Cyr, D.R.St. [1988], Rubber natural, in: J.I. Kroschwitz, ed., *Encyclopedia of Polymer Science and Engineering*, Volume 14, John Wiley & Sons, New York, 687–716. (7)

Daniel, I.M., and Ishai, O. [1994], *Engineering Mechanics of Composite Materials*, Oxford University Press, Oxford. (6)

Danielson, D.A. [1997], *Vectors and Tensors in Engineering and Physics*, 2nd edn., Addison-Wesley Publishing Company, Reading, Massachusetts. (1)

Dorfmann, A., and Muhr, A., eds. [1999], *Constitutive Models for Rubber*, Balkema, Rotterdam. (6)

Duffett, G., and Reddy, B.D. [1983], The analysis of incompressible hyperelastic bodies by the finite element method, *Computer Methods in Applied Mechanics and Engineering* **41**, 105–120. (6)

Duhem, P. [1911], *Traité d'Énergétique ou de Thermodynamique Générale*, Gauthier-Villars, Paris. (7)

Duvaut, G., and Lions, J.L. [1972], *Les Inéquations en Mécanique et en Physique*, Dunod, Paris. (8)

Ericksen, J.L. [1977], Special topics in elastostatics, in: *Advances in Applied Mechanics*, Volume 17, Academic Press, New York, 189–244. (7)

Ericksen, J.L. [1998], *Introduction to the Thermodynamics of Solids*, revised edn., Springer-Verlag, New York. (6)

Eshelby, J.D. [1975], The elastic energy-momentum tensor, *Journal of Elasticity* **5**, 321–335. (6)

Flory, P.J. [1953], *Principles of Polymer Chemistry*, Cornell University Press, Ithaca. (7)

Flory, P.J. [1956], Theory of elastic mechanisms in fibrous proteins, *Journal of the American Chemical Society* **78**, 5222–5235. (7)

Flory, P.J. [1961], Thermodynamic relations for highly elastic materials, *Transactions of the Faraday Society* **57**, 829–838. (6,7)

Flory, P.J. [1969], *Statistical Mechanics of Chain Molecules*, Wiley – Interscience, New York. (7)

Flory, P.J. [1976], Statistical thermodynamics of random networks, *Proceedings of the Royal Society of London* **A351**, 351–380. (7)

Flory, P.J., and Erman, B. [1982], Theory of elasticity of polymer networks, *Macromolecules* **15**, 800–806. (6)

Fung, Y.C. [1965], *Foundation of Solid Mechanics*, Prentice-Hall, Englewood Cliffs, New Jersey. (8)

Fung, Y.C. [1990], *Biomechanics. Motion, Flow, Stress, and Growth*, Springer-Verlag, New York. (6)

Fung, Y.C. [1993], *Biomechanics. Mechanical Properties of Living Tissues*, 2nd edn., Springer-Verlag, New York. (6)

Fung, Y.C. [1997], *Biomechanics. Circulation*, 2nd edn., Springer-Verlag, New York. (6)

Fung, Y.C., Fronek, K., and Patitucci, P. [1979], Pseudoelasticity of arteries and the choice of its mathematical expression, *American Physiological Society* **237**, H620–H631. (6)

Gent, A.N. [1962], Relaxation processes in vulcanized rubber. I. Relation among stress relaxation, creep, recovery and hysteresis, *Journal of Applied Polymer Science* **6**, 433–441. (6)

Glowinski, R., and Le Tallec, P. [1984], Finite element analysis in nonlinear incompressible elasticity, in: J.T. Oden, and G.F. Carey, eds., *Finite elements, Special Problems in Solid Mechanics*, Volume V, Prentice-Hall, Englewood Cliffs, New Jersey. (8)

Glowinski, R., and Le Tallec, P. [1989], *Augmented Lagrangian and Operator Splitting Methods in Nonlinear Mechanicss*, SIAM, Philadelphia. (8)

Gough, J. [1805], A description of a property of Caoutchouc or indian rubber; with some reflections on the case of the elasticity of this substance, *Memoirs of the Literary and Philosophical Society of Manchester* **1**, 288–295. (7)

Govindjee, S., and Simo, J.C. [1991], A micro-mechanically based continuum damage model for carbon black-filled rubbers incorporating the Mullins' effect, *Journal of the Mechanics and Physics of Solids* **39**, 87–112. (6)

Govindjee, S., and Simo, J.C. [1992a], Transition from micro-mechanics to computationally efficient phenomenology: carbon black filled rubbers incorporating Mullins' effect, *Journal of the Mechanics and Physics of Solids* **40**, 213–233. (6)

Govindjee, S., and Simo, J.C. [1992b], Mullins' effect and the strain amplitude dependence of the storage modulus, *International Journal of Solids and Structures* **29**, 1737–1751. (6)

Govindjee, S., and Simo, J.C. [1993], Coupled stress-diffusion: case II, *Journal of the Mechanics and Physics of Solids* **41**, 863–867. (6)

Green, A.E., and Adkins, J.E. [1970], *Large Elastic Deformations*, 2nd edn., Oxford University Press, Oxford. (6)

Green, M.S., and Tobolsky, A.V. [1946], A new approach to the theory of relaxing polymeric media, *The Journal of Physical Chemistry* **14**, 80–92. (6)

Gurtin, M.E. [1981a], *An Introduction to Continuum Mechanics*, Academic Press, Boston. (1,2,5,6)

Gurtin, M.E., and Francis, E.C. [1981b], Simple rate-independent model for damage, *AIAA Journal of Spacecraft* **18**, 285–288. (6)

Guth, E. [1966], Statistical mechanics of polymers, *Journal of Polymer Science* **C12**, 89–109. (7)

Guth, E., and Mark, H. [1935], Zur innermolekularen Statistik, insbesondere bei Kettenmolekülen I, *Monatshefte für Chemie und verwandte Teile anderer Wissenschaften* **65**, 93–121. (7)

Haddow, J.B., and Ogden, R.W. [1990], Thermoelasticity of rubber-like solids at small strains, in: G. Eason, and R.W. Ogden, eds., *Elasticity, Mathematical Methods and Applications, the Ian N. Sneddon 70th Birthday Volume*, Ellis Horwood, Chichester, 165–179. (7)

Halmos, P.R. [1958], *Finite-Dimensional Vector Spaces*, 2nd edn., Van Nostrand-Reinhold, New York. (1)

Harwood, J.A.C., and Payne, A.R. [1966a], Stress softening in natural rubber vulcanizates. Part III. Carbon black-filled vulcanizates, *Journal of Applied Polymer Science* **10**, 315–324. (6)

Harwood, J.A.C., and Payne, A.R. [1966b], Stress softening in natural rubber vulcanizates. Part IV. Unfilled vulcanizates, *Journal of Applied Polymer Science* **10**, 1203–1211. (6)

Harwood, J.A.C., Mullins, L., and Payne, A.R. [1965], Stress softening in natural rubber vulcanizates. Part II. Stress softening effects in pure gum and filler loaded rubbers, *Journal of Applied Polymer Science* **9**, 3011–3021. (6)

Haughton, D.M. [1980], Post-bifurcation of perfect and imperfect spherical elastic membranes, *International Journal of Solids and Structures* **16**, 1123–1133. (6)

Haughton, D.M. [1987], Inflation and bifurcation of thick-walled compressible elastic spherical shells, *IMA Journal of Applied Mathematics* **39**, 259–272. (6)

Haughton, D.M., and Ogden, R.W. [1978], On the incremental equations in non-linear elasticity – II. Bifurcation of pressurized spherical shells, *Journal of the Mechanics and Physics
   of Solids* **26**, 111–138. (6)

Haupt, P. [1993a], On the mathematical modelling of material behavior in continuum mechanics, *Acta Mechanica* **100**, 129–154. (6)

Haupt, P. [1993b], Thermodynamics of solids, in: W. Muschik, ed., *Non-Equilibrium Thermodynamics with Applications to Solids*, CISM Courses and Lectures No. 336, International
   Centre for Mechanical Sciences, Springer-Verlag, Wien, 65–138. (6,7)

Hayashi, K. [1993], Experimental approaches on measuring the mechanical properties and constitutive laws of arterial walls, *ASME Journal of Biomechanical Engineering* **115**, 481–
   488. (6)

Hellinger, E. [1914], Die allgemeinen Ansätze der Mechanik der Kontinua, in: F. Klein, and
   C. Müller, eds., *Enzyklopädie der Mathematischen Wissenschaften*, Volume IV, Pt. 4,
   Teubner Verlag, Leipzig, 601–694. (8)

Herakovich, C.T. [1998], *Mechanics of Fibrous Composites*, John Wiley & Sons, New York. (6)

Hill, R. [1968], On constitutive inequalities for simple materials, *Journal of the Mechanics and
   Physics of Solids* **16**, 229–242. (2)

Hill, R. [1970], Constitutive inequalities for isotropic elastic solids under finite strain, *Proceedings of the Royal Society of London* **A314**, 457–472. (2)

Hill, R. [1975], On the elasticity and stability of perfect crystals at finite strain, *Mathematical
   Proceedings of the Cambridge Philosophical Society* **77**, 225–240. (7)

Hill, R. [1978], Aspects of invariance in solid mechanics, in: *Advances in Applied Mechanics*,
   Volume 18, Academic Press, New York, 1–75. (2)

Hill, R. [1981], Invariance relations in thermoelasticity with generalized variables, *Mathematical Proceedings of the Cambridge Philosophical Society* **90**, 373–384. (6)

Hoger, A. [1987], The stress conjugate to logarithmic strain, *International Journal of Solids
   and Structures* **23**, 1645–1656. (2)

Hoger, A., and Carlson, D.E. [1984], Determination of the stretch and rotation in the polar
   decomposition of the deformation gradient, *Quarterly Applied Mathematics* **42**, 113–
   117. (2)

Holzapfel, G.A. [1996a], On large strain viscoelasticity: Continuum formulation and finite element applications to elastomeric structures, *International Journal for Numerical Methods
   in Engineering* **39**, 3903–3926. (6)

Holzapfel, G.A. [2001], Biomechanics of soft tissue, in: J. Lemaitre, ed., *The Handbook of Materials Behavior Models*, Volume III, Multiphysics Behaviors, Chapter 10, Composite Media, Biomaterials, Academic Press, Boston, 1049–1063. (6)

Holzapfel, G.A., and Simo, J.C. [1996b], Entropy elasticity of isotropic rubber-like solids at finite strains, *Computer Methods in Applied Mechanics and Engineering* **132**, 17–44. (7)

Holzapfel, G.A., and Simo, J.C. [1996c], A new viscoelastic constitutive model for continuous media at finite thermomechanical changes, *International Journal of Solids and Structures* **33**, 3019–3034. (7)

Holzapfel, G.A., and Weizsäcker, H.W. [1998], Biomechanical behavior of the arterial wall and its numerical characterization, *Computers in Biology and Medicine* **28**, 377–392. (6)

Holzapfel, G.A., and Gasser, T.C. [2001], A viscoelastic model for fiber-reinforced composites at finite strains: Continuum basis, computational aspects and applications, *Computer Methods in Applied Mechanics and Engineering* **190**, 4379–4430. (6)

Holzapfel, G.A., Stadler, M., and Ogden, R.W. [1999], Aspects of stress softening in filled rubbers incorporating residual strains, in: A. Dorfmann, and A. Muhr, eds., *Constitutive Models for Rubber*, Balkema, Rotterdam, 189–193. (6)

Holzapfel, G.A., Gasser, T.C., and Ogden, R.W. [2000], A new constitutive framework for arterial wall mechanics and a comparative study of material models, *Journal of Elasticity* **61**, 1–48. (6)

Holzapfel, G.A., Eberlein, R., Wriggers, P., and Weizsäcker, H.W. [1996d], Large strain analysis of soft biological membranes: Formulation and finite element analysis, *Computer Methods in Applied Mechanics and Engineering* **132**, 45–61. (6)

Holzapfel, G.A., Eberlein, R., Wriggers, P., and Weizsäcker, H.W. [1996e], A new axisymmetrical membrane element for anisotropic, finite strain analysis of arteries, *Communications in Numerical Methods in Engineering* **12**, 507–517. (6)

Hu, H.-C. [1955], On some variational principles in the theory of elasticity and the theory of plasticity, *Scientia Sinica* **4**, 33–54. (8)

Hughes, T.J.R. [2000], *The Finite Element Method: Linear Static and Dynamic Finite Element Analysis*, Dover, New York. (8)

Hughes, T.J.R., and Pister, K.S. [1978], Consistent linearization in mechanics of solids and structures, *Computers and Structures* **8**, 391–397. (6,8)

Hughes, T.J.R., and Winget, J. [1980], Finite rotation effects in numerical integration of rate constitutive equations arising in large-deformation analysis, *International Journal for Numerical Methods in Engineering* **15**, 1413–1418. (6)

Hughes, T.J.R., Liu, W.K., and Brooks, A. [1979], Review of finite element analysis of incompressible viscous flows by the penalty function formulation, *Journal of Computational Physics* **30**, 1–60. (8)

Humphrey, J.D. [1995], Mechanics of the arterial wall: Review and directions, *Critical Reviews in Biomedical Engineering* **23**, 1–162. (6)

Humphrey, J.D. [1998], Computer methods in membrane biomechanics, *Computer Methods in Biomechanics and Biomedical Engineering* **1**, 171–210. (6)

Hutter, K. [1977], The foundations of thermodynamics, its basic postulates and implications. A review of modern thermodynamics, *Acta Mechanica* **27**, 1–54. (4)

James, H.M., and Guth, E. [1943], Theory of the elastic properties of rubber, *Journal of Chemical Physics* **11**, 455–481. (7)

James, H.M., and Guth, E. [1949], Simple representation of network theory of rubber, with a discussion of other theories, *Journal of Polymer Science* **4**, 153–182. (7)

Johnson, M.A., and Beatty, M.F. [1993a], The Mullins effect in uniaxial extension and its influence on the transverse vibration of a rubber string, *Continuum Mechanics and Thermodynamics* **5**, 83–115. (6)

Johnson, M.A., and Beatty, M.F. [1993b], A constitutive equation for the Mullins effect in stress controlled uniaxial extension experiments, *Continuum Mechanics and Thermodynamics* **5**, 301–318. (6)

Jones, R.M. [1999], *Mechanics of Composite Materials*, 2nd edn., Taylor & Francis, Philadelphia. (6)

Jones, D.F., and Treloar, L.R.G. [1975], The properties of rubber in pure homogeneous strain, *Journal of Physics D: Applied Physics* **8**, 1285–1304. (6)

Joule, J.P. [1859], On some thermo-dynamic properties of solids, *Philosophical Transactions of the Royal Society of London* **A149**, 91–131. (7)

Kachanov, L.M. [1958], Time of the rupture process under creep conditions, *Izvestija Akademii Nauk Sojuza Sovetskich Socialisticeskich Respubliki (SSSR) Otdelenie Techniceskich Nauk (Moskra)* **8**, 26–31. (6)

Kachanov, L.M. [1986], *Introduction to Continuum Damage Mechanics*, Martinus Nijhoff Publishers, Dordrecht, The Netherlands. (6)

Kaliske, M., and Rothert, H. [1997], Formulation and implementation of three-dimensional viscoelasticity at small and finite strains, *Computational Mechanics* **19**, 228–239. (6)

Kawabata, S., and Kawai, H. [1977], Strain energy density functions of rubber vulcanizations from biaxial extension, in: H.-J. Cantow et al., eds., *Advances in Polymer Science*, Volume 24, Springer-Verlag, Berlin, 90–124. (6)

Kestin, J. [1979], *A Course in Thermodynamics*, Volume I,II, McGraw-Hill, New York. (4)

Knauss, W., and Emri, I. [1981], Non-linear viscoelasticity based on free volume considerations, *Computers and Structures* **13**, 123–128. (6)

Koh, S.L., and Eringen, A.C. [1963], On the foundations of non-linear thermo-viscoelasticity, *International Journal of Engineering Science* **1**, 199–229. (6)

Krajcinovic, D. [1996], *Damage Mechanics*, North-Holland, Amsterdam. (6)

Krawitz, A. [1986], *Materialtheorie. Mathematische Beschreibung des Phänomenologischen Thermomechanischen Verhaltens*, Springer-Verlag, Berlin. (7)

Kuhn, W. [1938], Die Bedeutung der Nebenvalenzkräfte für die elastischen Eigenschaften hochmolekularer Stoffe, *Angewandte Chemie* **51**, 640–647. (7)

Kuhn, W. [1946], Dependence of the average transversal on the longitudinal dimensions of statistical coils formed by chain molecules, *Journal of Polymer Science* **1**, 380–388. (7)

Kuhn, W., and Grün, F. [1942], Beziehungen zwischen elastischen Konstanten und Dehnungs-doppelbrechung hochelastischer Stoffe, *Kolloid-Zeitschrift* **101**, 248–271. (7)

Lee, E.H. [1969], Elastic-plastic deformation at finite strains, *Journal of Applied Mechanics* **36**, 1–6. (6)

Lee, S.M., ed. [1990], *International Encyclopedia of Composites*, Volume 1,2,3, VCH Publishers, New York. (6)

Lee, S.M., ed. [1991], *International Encyclopedia of Composites*, Volume 4,5, VCH Publishers, New York. (6)

Lee, T.C.P., Sperling, L.H., and Tobolsky, A.V. [1966], Thermal stability of elastomeric networks at high temperatures, *Journal of Applied Polymer Science* **10**, 1831–1836. (7)

Lemaitre, J. [1996], *A Course on Damage Mechanics*, 2nd revised and enlarged edn., Springer-Verlag, Berlin. (6)

Lemaitre, J., and Chaboche, J.-L. [1990], *Mechanics of Solid Materials*, Cambridge University Press, Cambridge. (6)

Le Tallec, P. [1994], Numerical methods for nonlinear three-dimensional elasticity, in: P.G. Ciarlet, and J.L. Lions, eds., *Handbook of Numerical Analysis*, Volume III, North-Holland, Elsevier, 465–622. (6,8)

Lion, A. [1996], A constitutive model for carbon black filled rubber: experimental investigations and mathematical representation, *Continuum Mechanics and Thermodynamics* **6**, 153–169. (6)

Lion, A. [1997a], On the large deformation behavior of reinforced rubber at different temperatures, *Journal of the Mechanics and Physics of Solids* **45**, 1805–1834. (6)

Lion, A. [1997b], A physically based method to represent the thermo-mechanical behaviour of elastomers, *Acta Mechanica* **123**, 1–25. (7)

Lubliner, J. [1985], A model of rubber viscoelasticity, *Mechanics Research Communications* **12**, 93–99. (6)

Luenberger, D.G. [1984], *Linear and Nonlinear Programming*, Addison-Wesley Publishing Company, Reading, Massachusetts. (8)

Malkus, D.S., and Hughes, T.J.R. [1978], Mixed finite element methods – reduced and selective integration techniques:A unification of concept, *Computer Methods in Applied Mechanics and Engineering* **15**, 63–81. (8)

Malvern, L.E. [1969], *Introduction to the Mechanics of a Continuous Medium*, Prentice-Hall, Englewood Cliffs, New Jersey. (2,3,4,6,7)

Man, C.-S., and Guo, Z.-H. [1993], A basis-free formula for time rate of Hill's strain tensors, *International Journal of Solids and Structures* **30**, 2819–2842. (2)

Marchuk, G.I. [1982], *Methods of Numerical Mathematics*, 2nd edn., Springer-Verlag, New York. (7)

Mark, J.E., and Erman, B. [1988], *Rubberlike Elasticity a Molecular Primer*, John Wiley & Sons, New York. (6,7)

Marsden, J.E., and Hughes, T.J.R. [1994], *Mathematical Foundations of Elasticity*, Dover, New York. (1,2,6,8)

McCrum, N.G., Buckley, C.P., and Bucknall, C.B. [1997], *Principles of Polymer Engineering*, 2nd edn., Oxford University Press, Oxford. (6,7)

Miehe, C. [1988], Zur numerischen Behandlung thermomechanischer Prozesse, Technischer Bericht F 88/6, Forschungs- und Seminarberichte aus dem Bereich der Mechanik der Universität Hannover. (7)

Miehe, C. [1994], Aspects of the formulation and finite element implementation of large strain isotropic elasticity, *International Journal for Numerical Methods in Engineering* **37**, 1981–2004. (2,6)

Miehe, C. [1995a], Discontinuous and continuous damage evolution in Ogden-type large-strain elastic materials, *European Journal of Mechanics, A/Solids* **14**, 697–720. (6)

Miehe, C. [1995b], Entropic thermoelasticity at finite strains. Aspects of the formulation and numerical implementation, *Computer Methods in Applied Mechanics and Engineering* **120**, 243–269. (7)

Miehe, C. [1996], Numerical computation of algorithmic (consistent) tangent moduli in large-strain computational inelasticity, *Computer Methods in Applied Mechanics and Engineering* **134**, 223–240. (6)

Miehe, C., and Keck, J. [2000], Superimposed finite elastic-viscoelastic-plastoelastic stress response with damage in filled rubbery polymers. Experiments, modelling and algorithmic implementation, *Journal of the Mechanics and Physics of Solids*, to appear. (6)

Miehe, C., and Stein, E. [1992], A canonical model of multiplicative elasto-plasticity. Formulation and aspects of the numerical implementation, *European Journal of Mechanics, A/Solids* **11**, 25–43. (6)

Mooney, M. [1940], A theory of large elastic deformation, *Journal of Applied Physics* **11**, 582–592. (6)

Morman, Jr., K.N. [1986], The generalized strain measure with application to nonhomogeneous deformations in rubber-like solids, *Journal of Applied Mechanics* **53**, 726–728. (2)

Müller, I. [1985], *Thermodynamics*, Pitman Advanced Publishing Program, Boston. (7)

Mullins, L. [1947], Effect of stretching on the properties of rubber, *Journal of Rubber Research* **16**, 275–289. (6)

Mullins, L. [1969], Softening of rubber by deformation, *Rubber Chemistry and Technology* **42**, 339–362. (6)

Mullins, L., and Thomas, A.G. [1960], Determination of degree of crosslinking in natural rubber vulcanizates. Part V. Effect of network flaws due to free chain ends, *Journal of Polymer Science* **43**, 13–21. (7)

Mullins, L., and Tobin, N.R. [1957], Theoretical model for the elastic behavior of filler-reinforced vulcanized rubbers, *Rubber Chemistry and Technology* **30**, 551–571. (6)

Mullins, L., and Tobin, N.R. [1965], Stress softening in rubber vulcanizates. Part I. Use of a strain amplification factor to describe the elastic behavior of filler-reinforced vulcanized rubber, *Journal of Applied Polymer Science* **9**, 2993–3009. (6)

Naghdi, P.M., and Trapp, J.A. [1975], The significance of formulating plasticity theory with reference to loading surfaces in strain space, *International Journal of Engineering Science* **13**, 785–797. (6)

Nagtegaal, J.C., Parks, D.M., and Rice, J.R. [1974], On numerically accurate finite element solutions in the fully plastic range, *Computer Methods in Applied Mechanics and Engineering* **4**, 153–177. (8)

Needleman, A. [1977], Inflation of spherical rubber balloons, *International Journal of Solids and Structures* **13**, 409–421. (6)

Needleman, A., Rabinowitz, S.A., Bogen, D.K., and McMahon, T.A. [1983], A finite element model of the infarcted left ventricle, *Journal of Biomechanics* **16**, 45–58. (6)

Nickell, R.E., and Sackman, J.L. [1968], Approximate solutions in linear, coupled thermoelasticity, *Journal of Applied Mechanics* **35**, 255–266. (7)

Oden, J.T. [1969], Finite element analysis of nonlinear problems in the dynamical theory of coupled thermoelasticity, *Nuclear Engineering and Design* **10**, 465–475. (7)

Oden, J.T. [1972], *Finite Elements of Nonlinear Continua*, McGraw-Hill, New York. (7,8)

Oden, J.T., and Reddy, J.N. [1976], *Variational Methods in Theoretical Mechanics*, Springer-Verlag, Heidelberg. (8)

Ogden, R.W. [1972a], Large deformation isotropic elasticity – on the correlation of theory and experiment for incompressible rubberlike solids, *Proceedings of the Royal Society of London* **A326**, 565–584. (6,7)

Ogden, R.W. [1972b], Large deformation isotropic elasticity: on the correlation of theory and experiment for compressible rubberlike solids, *Proceedings of the Royal Society of London* **A328**, 567–583. (6,7)

Ogden, R.W. [1977], Inequalities associated with the inversion of elastic stress-deformation relation and their implications, *Mathematical Proceedings of the Cambridge Philosophical Society* **81**, 313–324. (8)

Ogden, R.W. [1982], Elastic deformations of rubberlike solids, in: H.G. Hopkins, and M.J. Sewell, eds., *Mechanics of Solids, the Rodney Hill 60th Anniversary Volume*, Pergamon Press, Oxford, 499–537. (6)

Ogden, R.W. [1986], Recent advances in the phenomenological theory of rubber elasticity, *Rubber Chemistry and Technology* **59**, 261–383. (6)

Ogden, R.W. [1987], Aspects of the phenomenological theory of rubber thermoelasticity, *Polymer* **28**, 379–385. (6)

Ogden, R.W. [1992a], Nonlinear elasticity: Incremental equations and bifurcation phenomena, *Nonlinear Equations in the Applied Sciences* **2**, 437–468. (6)

Ogden, R.W. [1992b], On the thermoelastic modeling of rubberlike solids, *Journal of Thermal Stresses* **15**, 533–557. (7)

Ogden, R.W. [1997], *Non-linear Elastic Deformations*, Dover, New York. (1,2,5,6,8)

Ogden, R.W., and Roxburgh, D.G. [1999a], A pseudo-elastic model for the Mullins effect in filled rubber, *Proceedings of the Royal Society of London* **A455**, 2861–2877. (6)

Ogden, R.W., and Roxburgh, D.G. [1999b], An energy-based model of the Mullins effect, in: A. Dorfmann, and A. Muhr, eds., *Constitutive Models for Rubber*, Balkema, Rotterdam, 23–28. (6)

Ortiz, M. [1999], Nanomechanics of defects in solids, in: *Advances in Applied Mechanics*, Volume 36, Academic Press, New York, 1–79. (2)

Price, C. [1976], Thermodynamics of rubber elasticity, *Proceedings of the Royal Society of London* **A351**, 331–350. (7)

Raoult, A. [1986], Non-polyconvexity of the stored energy function of a Saint Venant-Kirchhoff material, *Aplikace Mathematiky* **6**, 417–419. (6)

Reddy, J.N. [1993], *An Introduction to the Finite Element Method*, 2nd edn., McGraw-Hill, Boston. (8)

Reese, S., and Govindjee, S. [1998a], A theory of finite viscoelasticity and numerical aspects, *International Journal of Solids and Structures* **35**, 3455–3482. (6)

Reese, S., and Govindjee, S. [1998b], Theoretical and numerical aspects in the thermoviscoelastic material behavior of rubber-like polymers, *Mechanics of time-dependent materials* **1**, 357–396. (7)

Reissner, E. [1950], On a variational theorem in elasticity, *Journal of Mathematics and Physics* **29**, 90–95. (8)

Rhodin, J.A.G. [1980], Architecture of the vessel wall, in: D.F. Bohr, A.D. Somlyo, and H.V. Sparks, Jr., eds., *Handbook of Physiology, The Cardiovascular System*, Section 2, Volume 2, American Physiologial Society, Bethesda, Maryland, 1–31. (6)

Rivlin, R.S. [1948], Large elastic deformations of isotropic materials. IV. Further developments of the general theory, *Philosophical Transactions of the Royal Society of London* **A241**, 379–397. (6)

Rivlin, R.S. [1949a], Large elastic deformations of isotropic materials. V. The problem of flexure, *Proceedings of the Royal Society of London* **A195**, 463–473. (6)

Rivlin, R.S. [1949b], Large elastic deformations of isotropic materials. VI. Further results in the theory of torsion, shear and flexure, *Philosophical Transactions of the Royal Society of London* **A242**, 173–195. (6)

Rivlin, R.S. [1970], An introduction to non-linear continuum mechanics, in: R.S. Rivlin, ed., *Non-linear Continuum Theories in Mechanics and Physics and their Applications*, Edizioni Cremonese, Rome, 151–309. (6)

Rivlin, R.S., and Ericksen, J.L. [1955], Stress-deformation relations for isotropic materials, *Journal of Rational Mechanics and Analysis* **4**, 323–425. Reprinted in Rational Mechanics of Materials. International Science Review Series, New York: Gordon & Breach [1965]. (5)

Rivlin, R.S., and Saunders, D.W. [1951], Large elastic deformations of isotropic materials. VII. Experiments on the deformation of rubber, *Philosophical Transactions of the Royal Society of London* **A243**, 251–288. (2)

Rosen, M.R. [1979], Characterization of non-Newtonian flow, *Polymer Plastics Technology and Engineering* **12**, 1–42. (7)

Roy, C.S. [1880–1882], The elastic properties of the arterial wall, *The Journal of Physiology* **3**, 125–159. (7)

Saleeb, A.F., Chang, T.Y.P., and Arnold, S.M. [1992], On the development of explicit robust schemes for implementation of a class of hyperelastic models in large-strain analysis of rubbers, *International Journal for Numerical Methods in Engineering* **33**, 1237–1249. (2)

Scanlan, J. [1960], The effect of network flaws on the elastic properties of vulcanizates, *Journal of Polymer Science* **43**, 501–508. (7)

Schoff, C.K. [1988], Rheological measurements, in: J.I. Kroschwitz, ed., *Encyclopedia of Polymer Science and Engineering*, Volume 14, John Wiley & Sons, New York, 454–541. (7)

Schröder, J. [1996], Theoretische und algorithmische Konzepte zur phänomenologischen Beschreibung anisotropen Materialverhaltens, Technischer Bericht F 96/3, Forschungs- und Seminarberichte aus dem Bereich der Mechanik der Universität Hannover. (6)

Schur, I. [1968], Vorlesungen über Invariantentheorie, in: H. Grunsky, ed., *Die Grundlehren der mathematischen Wissenschaften*, Volume 143, Springer-Verlag, Berlin. (6)

Schweizerhof, K.H., and Ramm, E. [1984], Displacement dependent pressure loads in nonlinear finite element analysis, *Computers and Structures* **18**, 1099–1114. (8)

Seki, W., Fukahori, Y., Iseda, Y., and Matsunaga, T. [1987], A large-deformation finite-element analysis for multilayer elastomeric bearings, *Rubber Chemistry and Technology* **60**, 856–869. (6)

Seth, B.R. [1964], Generalized strain measure with applications to physical problems, in: M. Reiner, and D. Abir, eds., *Second-Order Effects in Elasticity, Plasticity, and Fluid Dynamics*, Pergamon Press, Oxford, 162–172. (2)

Shen, M., and Croucher, M. [1975], Contribution of internal energy to the elasticity of rubberlike materials, *Journal of Macromolecular Science C – Reviews in Macromolecular Chemistry* **12**, 287–329. (7)

Sidoroff, F. [1974], Un modèle viscoélastique non linéaire avec configuration intermédiaire, *Journal de Mécanique* **13**, 679–713. (6)

Šilhavý, M. [1997], *The Mechanics and Thermodynamics of Continuous Media*, Springer-Verlag, New York. (4,6,7)

Simmonds, J.G. [1994], *A Brief on Tensor Analysis*, 2nd edn., Springer-Verlag, New York. (1)

Simo, J.C. [1987], On a fully three-dimensional finite-strain viscoelastic damage model: Formulation and computational aspects, *Computer Methods in Applied Mechanics and Engineering* **60**, 153–173. (6,8)

Simo, J.C., and Hughes, T.J.R. [1998], *Computational Inelasticity*, Springer-Verlag, New York. (6)

Simo, J.C., and Miehe, C. [1992], Associative coupled thermoplasticity at finite strains: Formulation, numerical analysis and implementation, *Computer Methods in Applied Mechanics and Engineering* **98**, 41–104. (6,7)

Simo, J.C., and Taylor, R.L. [1991a], Quasi-incompressible finite elasticity in principal stretches. Continuum basis and numerical algorithms, *Computer Methods in Applied Mechanics and Engineering* **85**, 273–310. (2,6,8)

Simo, J.C., Taylor, R.L., and Pister, K.S. [1985], Variational and projection methods for the volume constraint in finite deformation elasto-plasticity, *Computer Methods in Applied Mechanics and Engineering* **51**, 177–208. (6,8)

Simo, J.C., Taylor, R.L., and Wriggers, P. [1991b], A note on finite-element implementation of pressure boundary loading, *Communications in Applied Numerical Methods* **7**, 513–525. (8)

Simon, B.R., Kaufmann, M.V., McAfee, M.A., and Baldwin, A.L. [1993], Finite element models for arterial wall mechanics, *ASME Journal of Biomechanical Engineering* **115**, 489–496. (6)

Sircar, A.K., and Wells, J.L. [1981], Thermal conductivity of elastomer vulcanizates by differential scanning calorimetry, *Rubber Chemistry and Technology* **55**, 191–207. (7)

So, H., and Chen, U.D. [1991], A nonlinear mechanical model for solid-filled rubbers, *Polymer Engineering and Science* **31**, 410–416. (6)

de Souza Neto, E.A., Perić, D., and Owen, D.R.J. [1994], A phenomenological three-dimensional rate-independent continuum damage model for highly filled polymers: Formulation and computational aspects, *Journal of the Mechanics and Physics of Solids* **42**, 1533–1550. (6)

de Souza Neto, E.A., Perić, D., and Owen, D.R.J. [1998], Continuum modelling and numerical simulation of material damage at finite strains, *Archives of Computational Methods in Engineering* **5**, 311–384. (6)

Spencer, A.J.M. [1971], Theory of invariants, in: A.C. Eringen, ed., *Continuum Physics*, Volume I, Academic Press, New York. (6)

Spencer, A.J.M. [1980], *Continuum Mechanics*, Longman, London. (5)

Spencer, A.J.M. [1984], Constitutive theory for strongly anisotropic solids, in: A.J.M. Spencer, ed., *Continuum Theory of the Mechanics of Fibre-Reinforced Composites*, CISM Courses and Lectures No. 282, International Centre for Mechanical Sciences, Springer-Verlag, Wien, 1–32. (6)

Sperling, L.H. [1992], *Introduction to Physical Polymer Science*, 2nd edn., John Wiley & Sons, New York. (6,7)

Stern, H.J. [1967], *Rubber: Natural and Synthetic*, MacLaren, London. (6)

Strang, G. [1988a], *Linear Algebra and its Applications*, 3rd edn., Saunders Harcourt Brace Jovanovich, San Diego. (1)

Strang, G., and Fix, G.J. [1988b], *An Analysis of the Finite Element Method*, Wellesley-Cambridge Press, Wellesley. (8)

Sullivan, J.L. [1986], The relaxation and deformational properties of a carbon-black filled elastomer in biaxial tension, *Journal of Applied Polymer Science* **24**, 161–173. (6)

Sussman, T., and Bathe, K.-J. [1987], A finite element formulation for nonlinear incompressible elastic and inelastic analysis, *Computers and Structures* **26**, 357–409. (6,8)

Taylor, R.L., Pister, K.S., and Goudreau, G.L. [1970], Thermomechanical analysis of viscoelastic solids, *International Journal for Numerical Methods in Engineering* **2**, 45–59. (6)

Ting, T.T. [1985], Determination of $\mathbf{C}^{1/2}$, $\mathbf{C}^{-1/2}$ and more general isotropic tensor functions of $\mathbf{C}$, *Journal of Elasticity* **15**, 319–323. (2)

Tobolsky, A.V. [1960], *Properties and Structure of Polymers*, John Wiley & Sons, New York. (7)

Tobolsky, A.V., Prettyman, I.B., and Dillon, J.H. [1944], Stress relaxation of natural and synthetic rubber stocks, *Journal of Applied Physics* **15**, 380–395. (7)

Treloar, L.R.G. [1943a], The elasticity of a network of long-chain molecules – I, *Transactions of the Faraday Society* **39**, 36–41. (6,7)

Treloar, L.R.G. [1943b], The elasticity of a network of long-chain molecules – II, *Transactions of the Faraday Society* **39**, 241–246. (6,7)

Treloar, L.R.G. [1944], Stress-strain data for vulcanized rubber under various types of deformation, *Transactions of the Faraday Society* **40**, 59–70. (6)

Treloar, L.R.G. [1954], The photoelastic properties of short-chain molecular networks, *Transactions of the Faraday Society* **50**, 881–896. (6)

Treloar, L.R.G. [1976], The mechanics of rubber elasticity, *Proceedings of the Royal Society of London* **A351**, 301–330. (6)

Treloar, L.R.G. [2005], *The Physics of Rubber Elasticity*, 3rd edn., Oxford University Press, Oxford (reprint edition). (2,6,7)

Truesdell, C. [1977], *A First Course in Rational Continuum Mechanics*, Volume I, Academic Press, New York. (2)

Truesdell, C. [1980], *The Tragicomical History of Thermodynamics 1822-1854, Studies in the History of Mathematics and Physical Sciences 4*, Springer-Verlag, New York. (7)

Truesdell, C. [1984], *Rational Thermodynamics*, 2nd edn., Springer-Verlag, New York. (6)

Truesdell, C., and Noll, W. [2004], *The Non-Linear Field Theories of Mechanics*, 3rd edn., Springer-Verlag, Berlin. (1,2,3,5,6)

Truesdell, C., and Toupin, R.A. [1960], The classical field theories, in: S. Flügge, ed., *Encyclopedia of Physics*, Volume III/1, Springer-Verlag, Berlin, 226–793. (4,6,7,8)

Tsai, S.W., and Hahn, H.T. [1980], *Introduction to Composite Materials*, Technomic Publishing Company, Lancaster. (6)

Twizell, E.H., and Ogden, R.W. [1983], Non-linear optimization of the material constants in Ogden's stress-deformation function for incompressible isotropic elastic materials, *Journal of the Australian Mathematical Society* **B24**, 424–434. (6)

Vainberg, M.M. [1964], *Variational Methods for the Study of Nonlinear Operators*, Holden-Day, San Francisco. (8)

Valanis, K.C. [1972], *Irreversible Thermodynamics of Continuous Media, Internal Variable Theory*, CISM Courses and Lectures No. 77, International Centre for Mechanical Sciences, Springer-Verlag, Wien. (6)

Valanis, K.C., and Landel, R.F. [1967], The strain-energy function of a hyperelastic material in terms of the extension ratios, *Journal of Applied Physics* **38**, 2997–3002. (6,7)

van den Bogert, P.A.J., de Borst, R., Luiten, G.T., and Zeilmaker, J. [1991], Robust finite elements for 3D-analysis of rubber-like materials, *Engineering Computations* **8**, 3–17. (8)

Varga, O.H. [1966], *Stress-strain behavior of elastic materials, Selected problems of large deformations*, Wiley – Interscience, New York. (6)

Wall, F.T. [1965], *Chemical Thermodynamics*, 2nd edn., Freeman, San Francisco. (7)

Wang, C.-C., and Truesdell, C. [1973], *Introduction to Rational Elasticity*, Noordhoff, Leyden. (2)

Ward, I.M., and Hadley, D.W. [1993], *An Introduction to the Mechanical Properties of Solid Polymers*, John Wiley & Sons, New York. (6,7)

Washizu, K. [1955], On the variational principles of elasticity and plasticity, Technical Report No. 25-18, Aeroelastic and Structures Research Laboratory, MIT, Cambridge, Massachusetts. (8)

Washizu, K. [1982], *Variational Methods in Elasticity and Plasticity*, 3rd edn., Pergamon Press, Oxford. (8)

Weiner, J.H. [1983], *Statistical Mechanics of Elasticity*, John Wiley & Sons, New York. (7)

Weiss, J.A., Maker, B.N., and Govindjee, S. [1996], Finite element implementation of incompressible, transversely isotropic hyperelasticity, *Computer Methods in Applied Mechanics and Engineering* **135**, 107–128. (6)

Wilmański, K. [1998], *Thermomechanics of Continua*, Springer-Verlag, Berlin. (4)

Wood, L.A., and Martin, G.M. [1964], Compressibility of natural rubber at pressures below 500 kg/cm^2, *Journal of Research of the National Bureau of Standards* **68A**, 259–268. (7)

Wriggers, P. [1988], Konsistente Linearisierung in der Kontinuumsmechanik und ihre Anwendung auf die Finite-Element-Methode, Technischer Bericht F 88/4, Forschungs- und Seminarberichte aus dem Bereich der Mechanik der Universität Hannover. (2,8)

Wriggers, P. [2001], *Nichtlineare Finite-Element-Methoden*, Springer-Verlag, Berlin. (8)

Yanenko, N.N. [1971], *The Method of Fractional Steps*, Springer-Verlag, New York. English translation edited by M. Holt. (7)

Yeoh, O.H. [1990], Characterization of elastic properties of carbon-black-filled rubber vulcanizates, *Rubber Chemistry and Technology* **63**, 792–805. (6)

Zheng, Q.-S. [1994], Theory of representations for tensor functions – a unified invariant approach to constitutive equations, *Applied Mechanics Reviews* **47**, 545–587. (6)

Ziegler, H. [1983], *An Introduction to Thermomechanics*, North-Holland, Amsterdam. (7)

Zienkiewicz, O.C., and Taylor, R.L. [1989], *The Finite Element Method. Basic Formulation and Linear Problems*, Volume 1, 4th edn., McGraw-Hill, London. (8)

Zienkiewicz, O.C., and Taylor, R.L. [1991], *The Finite Element Method. Solid and Fluid Mechanics, Dynamics and Nonlinearity*, Volume 2, 4th edn., McGraw-Hill, London. (8)

Index

evolution equation, 281, 282, 286–290, 302, 359, 363, 366, 369
expansion coefficient, linear, 339, 346, 349, 352
extension
  strip-biaxial, 93
  uniform (uniaxial), 92, 124
external constraints
  holonomic, 379
  nonholonomic, 379
external variables, 278, 282, 287, 289, 364, 368

fiber, 77–79, 267, 273
  direction, 84, 96, 266–269, 273–276,
  extensible, 270
  inextensible, 270, 275–277
fiber-reinforced composite, 265, 266, 272, 274
fibers, families of, 265, 273–276
field
  harmonic, 50, 52, 68, 69, 137
  material, 64, 65
  mechanical, 306
  scalar; *see* scalar field
  spatial, 64–67
  tensor; *see* tensor field
  thermal, 306
  vector; *see* vector field
Finger deformation tensor, 81
finite element method
  displacement-based, 402, 403, 411
  hybrid, 389
  Jacobian-pressure formulation, 410
  mixed, 389, 391, 403, 404
first law of thermodynamics, 164–166, 175, 319
first Piola-Kirchhoff stress tensor, 111–114, 127, 128, 199, 207
  Euclidean transformation of, 190, 191
first Piola-Kirchhoff traction vector, 111, 113, 114, 144
  in material description, 165
  in spatial description, 163
first variation of a function
  in material description, 374, 375
  in spatial description, 375–377
flow
  irrotational, 149
  steady, 149
flow behavior index, 367

fluid, 205
  elastic, 125, 126, 204
  Reiner-Rivlin, 202, 204
  viscous, 202, 203
    Newtonian, 203, 204, 286, 287, 367, 369
fluid mechanics, 205
flux 52, 136, 139, 176, 150
  entropy; *see* entropy flux
  heat; *see* heat flux
force
  body; *see* body force
  contact, 111
  external, 110, 387
  inertia, 142, 378, 382, 384, 386
  internal, 110
  resultant, 110,128, 142
  retractive, 310, 311
  thermodynamic; *see* thermodynamic force
forces, system of, 147–149, 153
Fourier's law of heat conduction, 171, 342, 348, 354
fourth-order tensor, 22–24
  transpose of, 23
fractional-step method, 331
frame-indifferent spatial fields, 182, 185, 186; *see also* objectivity
free energy, 173, 206, 267–270, 273–275, 280, 298, 321–326, 328–335, 347, 357, 361, 364; *see also* Helmholtz free-energy function
  configurational, 284, 285, 304, 362, 364
  Gibbs, 323, 324
free-energy factor, 364
free index, 4, 18, 34, 43
free vibration, 154
freely jointed chain, 312–315, 319
friction, 160, 166, 311
function
  convex, strictly, 229, 244, 303, 390, 407
  linear, 32, 41, 162
  nonlinear, 41, 351, 380, 393
  penalty, 390, 391
  scalar, 40, 43
  tensor, 40–43
  test, 380–382
  vector, 40
  weighting, 380

9 780471 823193